Preface

Purpose

This book was written to provide students with a straightforward, readable text that presents trigonometry in an appealing manner. We have tried to "talk" with our readers, without patronizing or lecturing them; and we have avoided sophisticated "mathematical elegance" in favor of an intuitive approach to college trigonometry.

Prerequisites

The material in this book can be understood by the student who has had the equivalent of two years of college-preparatory mathematics in high school, including algebra and some plane geometry, or who has taken a college-level course in introductory algebra. Determined students with less preparation may be able to master the contents of this book, particularly if they use the accompanying *Study Guide* as an aid.

Objectives

This book has three objectives:

First, to provide the reader with a competent working knowledge of college-level trigonometry. We have taken pains to ensure adequate coverage of the traditional topics and techniques, especially those that will be needed in other courses, such as calculus, linear algebra, physics, chemistry, and engineering. Furthermore, we have tried to attune the textbook to the two outstanding mathematical trends of our time—the computer revolution and the burgeoning use of mathematical models in the sciences.

Second, to complement and enhance effective classroom teaching. Our own classroom experience has shown that successful teaching is fostered by a textbook that provides clear and complete explanations, cogent visual presentations, step-by-step procedures, numerous illustrative examples, and a multitude of topical, real-world applications. We have written the textbook with these criteria in mind.

Third, to allay "math anxiety." Many students enrolling in mathematics courses approach the subject with little or no self-confidence. In order to help students build such confidence, we begin with familiar material, explain new concepts in terms of ideas already well understood, offer worked-out examples that show in detail how problems are to be solved, and provide problem sets in which the level of difficulty increases gradually.

Special Features

1. *Use of calculators* In keeping with the recommendations of the National Council of Teachers of Mathematics (NCTM) and the Mathematical Association of America (MAA), we have de-emphasized the use of logarithmic and trigonometric tables in favor of the use of scientific calculators. As classroom teachers, we feel obliged to prepare our students to function in the real world, where virtually everyone who uses mathematics in a practical way—from the actuary to the zoologist—routinely employs a calculator. Today, a scientific calculator can be purchased inexpensively and used throughout the student's tenure in college or university—and beyond.

2. *Logarithmic and Trigonometric Tables* For those teachers who feel that a certain amount of instruction in the use of tables is desirable, we have included in an appendix tables of logarithmic, exponential, and trigonometric functions, along with examples illustrating their use and the technique of linear interpolation.

3. *Mathematical Models* Emphasis on the notion of mathematical models is a unique feature of this book. For instance, the idea of a mathematical model is effectively used in the discussion of oscillatory phenomena in Chapter 2, and in connection with examples of predator–prey relationships in Chapter 3.

4. *Trigonometric Functions* Trigonometric functions are first defined in terms of right triangles. This simple, geometrically appealing approach provides a firm foundation for the ensuing treatment of trigonometric functions of general angles and of circular functions and their graphs.

5. *Examples, Figures, and Problems* The textbook contains 230 examples, worked in detail, with all substitutions shown. The 300 figures in the book are placed next to the text that they illustrate. Problem sets, which appear at the end of each section, contain a total of 1820 problems, most of which correspond to worked-out examples in the text. Problems in each set progress from simple, drill-type exercises to more demanding conceptual questions. Odd-numbered problems require a level of understanding sufficient for most purposes, whereas even-numbered problems, especially those toward the end of each problem set, are more challenging and probe for deeper understanding. Answers to nearly all of the odd-numbered problems are provided in the back of the book.

6. *Homework* You will notice that certain problems at the end of every section are identified with numerals printed in color. This is to indicate a group of problems that could serve as a homework assignment for the section.

7. *Review Problem Sets* Review problem sets at the end of each chapter contain a total of 624 additional problems. These sets of review problems highlight the essential material in the chapter and serve a variety of purposes: Instructors may wish to use them for supplementary or extra-credit assignments or as a source of problems for quizzes and exams; students may wish to scan them to pinpoint areas where further study is needed.

8. *Sections* The book is divided into 34 sections, with an average length of about 5 pages of text (exclusive of problem sets). Each section is designed to be covered in one 50-minute class period. Sections that are considered to be review material can, of course, be covered more rapidly.

9. *Index of Applications* For convenience, an index of applications is provided at the back of the book. The usefulness of trigonometry is amply demonstrated by the abundant applications, not only to engineering, geometry, and the physical sciences, but also to biology, ecology, and navigation.

Pace

The book has been designed for use in either a one-quarter (4-credit) or a one-semester (3-credit) course. The pace of the course, as well as the choice of which topics to cover and which to emphasize, will vary from school to school. There is a more detailed discussion of these matters in the accompanying *Instructor's Test Manual*.

Study Guide

A *Study Guide* is available for students who need more drill or more assistance. It contains study objectives, carefully graded fill-in statements and problems that are broken down into simple units, and self-tests. Answers to all problems in the *Study Guide* are included, in order to provide immediate reinforcement for correct responses.

Acknowledgments

We wish to thank the following people, who reviewed the manuscript and offered many helpful suggestions: William P. Bair, *Pasadena City College;* Ann S. Bumpus, *Langley High School, McLean, Virginia;* Kenneth Chapman, *Michael J. Owens Technical College;* Henry Cohen, *University of Pittsburgh;* Milton Cox, *Miami University;* Albert G. Fadell, *State University of New York at Buffalo;* Anne Fish, *Foothill College;* Leonard E. Fuller, *Kansas State University;* Gerald E. Gannon, *California State University, Fullerton;* August J. Garver, *University of Missouri–Rolla;* Gerald K. Goff, *Oklahoma State University;* Thomas F. Gordon, *North Carolina State University;* B. C. Horne, Jr., *Virginia Polytechnic Institute and State University;* Julia P. Kennedy, *Georgia State University;* George Kosan, *Hillsborough Community College;* Gary Lippman, *California State University, Hayward;* Stanley M. Lukawecki, *Clemson University;* Eldon L. Miller, *University of Mississippi;* Philip R. Montgomery, *University of Kansas;* Gary Oakes, *Butte College;* John Riner, *Ohio State University;* Robert G. Savage, *North Carolina State University;* Mary W. Scott, *University of Florida;* Maury Shurlds, *Mississippi State University;* Howard E. Taylor, *West Georgia College;* and Paul D. Trembeth, *Delaware Valley College of Science and Agriculture*.

We would also like to express our appreciation to Professor Steve Fasbinder of Oakland University for solving all the problems in the book and for assisting in the proofreading; to Paula Ashley and to David B. Foulis for help in the proofreading; and to Hyla Gold Foulis for reviewing each successive stage of the manuscript, and for proofreading, solving problems, and assisting with the preparation of the *Study Guide*. Finally, we would like to thank the staff at Worth Publishers, especially Bob Andrews and Sally Immerman, for their constant help and encouragement.

Mustafa A. Munem
David J. Foulis

January 1982

A Note to the Instructor on the Use of Calculators

Problems and examples for which the use of a calculator is recommended are marked with the symbol Ⓒ. Answers to these problems and solutions for these examples were obtained using an HP-67 calculator—other calculators may give slightly different results because they use different internal routines.

The rule for rounding off numbers presented in Section 3 of Chapter 1 is consistent with the operation of most calculators with round-off capability. Some instructors may wish to mention the popular alternative round-off rule: If the first dropped digit is 5 and there are no nonzero digits to its right, round off so that the digit retained is even.

Because there are so many different calculators on the market, we have made no attempt to give detailed instructions for calculator operation in this textbook. Students should be urged to consult the instruction manuals furnished with their calculators.

Conscientious instructors will wish to encourage their students to learn to use calculators *efficiently;* for instance, to do chain calculations using the memory features of the calculator. In some of our examples we have shown the intermediate results of chain calculations so the students can check their calculator work; however, it should be emphasized that it is not necessary to write down these intermediate results when using the calculator.

As important as it is to encourage students to use their calculators for the examples and problems marked Ⓒ, it is perhaps more important to *restrain* them from attempting to use their calculators when the symbol Ⓒ is not present. For instance, a student who routinely uses a calculator to determine the algebraic sign of a trigonometric function of an angle may fail to learn the connection between the algebraic sign and the quadrant containing the angle.

Finally, we have made no attempt to provide a systematic discussion of the inaccuracies inherent in computations with a calculator. However, in order to make students aware that such inaccuracies exist, we have carried out some computations to the full ten places available on our calculator. An excellent account of calculator inaccuracy can be found in "Calculator Calculus and Round-off Errors," by George Miel in *The American Mathematical Monthly,* 1980, Vol. 87, No. 4, pp. 243–52.

Contents

College Trigonometry

1 Algebraic and Geometric Preliminaries

In this chapter, we review some of the mathematical concepts necessary for the study of trigonometry. These topics are usually dependent upon concepts that have been studied earlier; they include, a review of basic algebra, including inequalities and absolute value, calculator use, approximations, plane geometry and the cartesian coordinate system, and functions and their graphs.

1 Review of Basic Algebra

This section is intended as a concise review of the basic algebra that you will need in the rest of the book. We recommend that you read quickly through this review material, just to make sure it's familiar. To facilitate your review, the material is presented in condensed "outline" form.

Elementary algebra may be thought of as "arithmetic with letters." The letters, which are called **variables** or **unknowns,** stand for numbers. The sum, difference, product, and quotient of x and y are written as

$$x + y, \quad x - y, \quad x \times y = x \cdot y = xy, \quad \text{and} \quad x \div y = x/y = \frac{x}{y}.$$

In a quotient x/y, it is understood that $y \neq 0$.

We call $-x$ the **negative** of x. Be careful; $-x$ isn't necessarily a negative number. For instance, if $x = -2$, then

$$-x = -(-2) = 2,$$

which is positive. Recall the properties of negation:

(i) $-(-x) = x$

(ii) $(-x)y = -xy = x(-y)$

(iii) $(-x)(-y) = xy$

(iv) $\dfrac{-x}{y} = -\dfrac{x}{y} = \dfrac{x}{-y}, \quad y \neq 0.$

The numbers used to measure real-world quantities such as length, area, volume, speed, electrical charge, efficiency, probability of rain, intensity of earthquakes, profit, body temperature, gross national product, growth rate, medicinal dosage, and so forth, are called **real numbers.** They include such numbers as

$$5, \quad -17, \quad \frac{27}{13}, \quad -\frac{2}{3}, \quad 0, \quad 2.71828, \quad \sqrt{2}, \quad -\frac{\sqrt{3}}{2}, \quad 3 \times 10^8, \quad \text{and} \quad \pi.$$

The **natural numbers,** also called the **counting numbers** or the **positive integers,** are the numbers

$$1, 2, 3, 4, 5, \text{ and so on.}$$

The **integers**

$$\ldots, -4, -3, -2, -1, 0, 1, 2, 3, 4, \ldots$$

consist of all the natural numbers, their negatives, and zero. The **rational numbers** are those real numbers that can be written in the form a/b, where a and b are integers and $b \neq 0$. Since b may be equal to 1, every integer is a rational number. Other examples of rational numbers are

$$\tfrac{13}{2}, \quad \tfrac{3}{4}, \quad \text{and} \quad -\tfrac{22}{7}.$$

The **irrational numbers** are the real numbers that are not rational. A real number is an irrational number if and only if its decimal representation is **nonterminating** and **nonrepeating.** Examples of irrational numbers are

$$\sqrt{2} = 1.4142135\ldots \quad \text{and} \quad \pi = 3.1415926\ldots.$$

We denote the set of all real numbers, rational and irrational, by the symbol $\mathbb{R}$.

A positive integer **exponent** $\boldsymbol{n}$ is used to denote repeated multiplication of a quantity by itself:

$$x^n = \overbrace{x \cdot x \cdot x \cdots x}^{n \text{ times}}.$$

If $x \neq 0$, we also define

$$x^0 = 1 \quad \text{and} \quad x^{-n} = \frac{1}{x^n};$$

in particular, $x^{-1} = 1/x$ denotes the **reciprocal of** $\boldsymbol{x}$.

If x is a nonnegative real number and n is a positive integer, then there is exactly one nonnegative real number y such that

$$y^n = x.$$

We call y the **principal** $\boldsymbol{n}$**th root of** $\boldsymbol{x}$ and write

$$y = \sqrt[n]{x} = x^{1/n}.$$

For $n = 2$, we call y the **principal square root of x** and write

$$y = \sqrt{x} = x^{1/2}.$$

The principal square root $\sqrt{x}$ is often called simply the **square root of x**; however, it is always understood that $\sqrt{x}$ is the **nonnegative** number whose square is x. Thus, although there are two numbers, -2 and 2, whose square is 4, it is *incorrect* to write $\sqrt{4} = \pm 2$. Indeed, $\sqrt{4}$ denotes the nonnegative number whose square is 4; hence,

$$\sqrt{4} = 2.$$

If x and y are nonnegative numbers and n is a positive integer, then

$$\sqrt[n]{xy} = \sqrt[n]{x}\sqrt[n]{y}.$$

An algebraic expression such as

$$3x^2y - \frac{4xy^2}{z} + 2\sqrt{w}$$

consisting of various parts connected by plus or minus signs is called an **algebraic sum,** and each part, together with the sign preceding it, is called a **term.** In the algebraic sum above, the terms are $3x^2y$, $-\frac{4xy^2}{z}$, and $2\sqrt{w}$. Numerical multipliers such as 3, -4, and 2 are called the **numerical coefficients,** or sometimes just the *coefficients*, of the terms. Two terms that differ only in their coefficients are called **like terms.** For instance, in the algebraic sum

$$3x^3y - 2x\sqrt{y} - 4x^3y + x\sqrt{y},$$

$3x^3y$ and $-4x^3y$ are like terms, differing only in their coefficients, 3 and -4, as are $-2x\sqrt{y}$ and $x\sqrt{y}$, which differ only in their coefficients, -2 and 1.

To multiply two algebraic sums, you multiply each term of the first sum by each term of the second sum, and then you combine each of the like terms. For instance,

$$\begin{aligned}(3x + 1)(x - 2) &= 3x(x) + 3x(-2) + 1(x) + 1 \cdot (-2) \\ &= 3x^2 - 6x + x - 2 \\ &= 3x^2 - 5x - 2.\end{aligned}$$

Certain **special products** occur so often that it is useful to learn them outright. Among these are the following:

1. $(x + y)(x - y) = x^2 - y^2$
2. $(x + y)^2 = x^2 + 2xy + y^2$
3. $(x - y)^2 = x^2 - 2xy + y^2$
4. $(ax + b)(cx + d) = acx^2 + (ad + bc)x + bd$

Each algebraic expression in a product of two or more such expressions is called a **factor** of the product. For instance, in the product

$$(3x + 1)(x - 2) = 3x^2 - 5x - 2,$$

the factors are $3x + 1$ and $x - 2$. Often we are given a product in its expanded form, for instance $3x^2 - 5x - 2$, and we are required to find the original factors $3x + 1$ and $x - 2$. The process of finding these factors is called **factoring.** Success in factoring depends on your ability to recognize certain patterns in the expression to be factored, an ability that can be cultivated by practice. Special Products 1 through 4 above, read from right to left, suggest especially important patterns. For instance, Special Product 1, read as

$$x^2 - y^2 = (x - y)(x + y),$$

shows that a "difference of two squares" can always be factored.

If p and q are algebraic expressions, then an expression of the form p/q is called a **fraction** with **numerator** p and **denominator** q**.** In dealing with fractions, you must keep in mind that all variables involved must be restricted to numerical values for which the denominators are different from zero.

A fraction is unchanged if both its numerator and denominator are multiplied by the same nonzero quantity; that is,

$$\frac{p}{q} = \frac{pk}{qk}, \qquad q \neq 0, \quad k \neq 0.$$

Reading this "backwards," we obtain the **cancellation property:**

$$\frac{p\not{k}}{q\not{k}} = \frac{p}{q}.$$

Notice how cancellation is indicated by slanted lines drawn through the canceled factors. A fraction in which the numerator and denominator have no common factor (other than 1 or -1), so that cancellation isn't possible, is said to be **reduced to lowest terms.**

It's easy to add or subtract fractions that have the same denominator (called a **common denominator**). Indeed, if $q \neq 0$, then

$$\frac{a}{q} + \frac{b}{q} = \frac{a + b}{q} \quad \text{and} \quad \frac{a}{q} - \frac{b}{q} = \frac{a - b}{q}.$$

Fractions whose denominators aren't the same must first be rewritten with a common denominator before they can be added or subtracted. This is accomplished by multiplying the numerator and denominator of each fraction by a suitable quantity. For instance,

$$\frac{2}{3} + \frac{4}{5} = \frac{2 \cdot 5}{3 \cdot 5} + \frac{3 \cdot 4}{3 \cdot 5} = \frac{10}{15} + \frac{12}{15} = \frac{10 + 12}{15} = \frac{22}{15}.$$

More generally, if $q \neq s$, you can always perform the addition

$$\frac{p}{q} + \frac{r}{s}$$

for $q \neq 0$ and $s \neq 0$ as follows:

$$\frac{p}{q} + \frac{r}{s} = \frac{ps}{qs} + \frac{qr}{qs} = \frac{ps + qr}{qs}.$$

Subtraction is handled similarly.

To multiply two fractions, you just multiply their numerators and multiply their denominators: If $q \neq 0$ and $s \neq 0$, then

$$\frac{p}{q} \cdot \frac{r}{s} = \frac{pr}{qs}.$$

To divide a first fraction by a second fraction, you just invert the second fraction and multiply: If $q \neq 0$, $r \neq 0$, and $s \neq 0$, then

$$\frac{p}{q} \div \frac{r}{s} = \frac{p}{q} \cdot \frac{s}{r} = \frac{ps}{qr}.$$

If a, b, and c are constants and $a \neq 0$, the equation

$$ax^2 + bx + c = 0$$

is called a **second-degree** or **quadratic equation** in standard form. If you can factor the left side of such an equation, you can obtain the roots by setting each factor equal to zero.

Example Solve the equation $15x^2 + 14x = 8$.

Solution We begin by subtracting 8 from both sides of the equation to change it into standard form

$$15x^2 + 14x - 8 = 0.$$

Factoring the polynomial on the left, we obtain

$$(3x + 4)(5x - 2) = 0.$$

Now, we set each factor equal to zero and solve the resulting first-degree equations:

$3x + 4 = 0$	$5x - 2 = 0$
$x = -\frac{4}{3}$	$x = \frac{2}{5}$

Therefore, the solutions are $x = -\frac{4}{3}$ and $x = \frac{2}{5}$.

A second method for solving a quadratic equation is based on the idea of "completing a square." Notice that the square of $x + (k/2)$ is $x^2 + kx + (k/2)^2$. Therefore, by adding $(k/2)^2$ to $x^2 + kx$, we obtain a perfect square:

$$x^2 + kx + \left(\frac{k}{2}\right)^2 = \left(x + \frac{k}{2}\right)^2.$$

The process of obtaining a perfect square $[x + (k/2)]^2$ from an expression of the form $x^2 + kx$ by adding the square of half the coefficient of x is called **completing the square.** Be careful—this only works when the coefficient of x^2 is 1.

Example Complete the square by adding a suitable constant to each expression.

(a) $x^2 + 8x$ (b) $x^2 - 4x$ (c) $x^2 - 3x$

Solution

(a) $x^2 + 8x + (\frac{8}{2})^2 = x^2 + 8x + 16 = (x + 4)^2$
(b) $x^2 - 4x + (-\frac{4}{2})^2 = x^2 - 4x + 4 = (x - 2)^2$
(c) $x^2 - 3x + (-\frac{3}{2})^2 = x^2 - 3x + \frac{9}{4} = (x - \frac{3}{2})^2$

By using the idea of completing the square, we can obtain a formula for the roots of the quadratic equation

$$ax^2 + bx + c = 0, \qquad a \neq 0.$$

First, we subtract c from both sides to obtain

$$ax^2 + bx = -c.$$

Then, to prepare for completing the square, we divide both sides by a

$$x^2 + \frac{b}{a}x = -\frac{c}{a}.$$

To complete the square of the expression on the left, we add $\left[\frac{1}{2}\left(\frac{b}{a}\right)\right]^2 = \frac{b^2}{4a^2}$ to both sides:

$$x^2 + \frac{b}{a}x + \frac{b^2}{4a^2} = \frac{b^2}{4a^2} - \frac{c}{a}$$

$$\left(x + \frac{b}{2a}\right)^2 = \frac{b^2 - 4ac}{4a^2}.$$

Now, provided that $b^2 - 4ac$ is nonnegative, there are exactly two real numbers whose squares are equal to $\frac{b^2 - 4ac}{4a^2}$, namely, the principal square root

$$\sqrt{\frac{b^2 - 4ac}{4a^2}}$$

and its negative. In other words,

$$x + \frac{b}{2a} = \pm\sqrt{\frac{b^2 - 4ac}{4a^2}} \qquad \text{or} \qquad x = \frac{-b}{2a} \pm \frac{\sqrt{b^2 - 4ac}}{2a}.$$

Therefore, provided that $b^2 - 4ac$ is nonnegative, the solutions of the quadratic equation $ax^2 + bx + c = 0$ are given by the **quadratic formula**

$$x = \frac{-b \pm \sqrt{b^2 - 4ac}}{2a}.$$

Example Use the quadratic formula to solve the equation

$$2x^2 - 5x + 1 = 0.$$

Solution The equation is in the standard form

$$ax^2 + bx + c = 0,$$

with $a = 2$, $b = -5$, and $c = 1$. Substituting these values into the quadratic formula, we obtain

$$x = \frac{-b \pm \sqrt{b^2 - 4ac}}{2a} = \frac{-(-5) \pm \sqrt{(-5)^2 - 4(2)(1)}}{2(2)}$$

$$= \frac{5 \pm \sqrt{25 - 8}}{4} = \frac{5 \pm \sqrt{17}}{4}.$$

Hence, there are two solutions,

$$x = \frac{5 - \sqrt{17}}{4} \quad \text{and} \quad x = \frac{5 + \sqrt{17}}{4}.$$

If, in solving an equation, you multiply both sides by an expression containing the unknown, or if you square both sides, you must always check the solutions by substituting them back into the original equation. Fake "solutions" that don't satisfy the original equation are called **extraneous solutions** or **extraneous roots.**

Example Solve the equation $x = \sqrt{x + 2}$.

Solution Squaring both sides of the given equation in order to remove the square root, we obtain

$$x^2 = x + 2 \quad \text{or} \quad x^2 - x - 2 = 0,$$

that is,

$$(x - 2)(x + 1) = 0;$$

hence,

$$x = 2 \quad \text{or} \quad x = -1.$$

Substituting $x = 2$ into the original equation, we obtain $2 = \sqrt{2 + 2}$, which is true. Thus, $x = 2$ is a solution. However, substituting $x = -1$ into the original equation, we obtain $-1 = \sqrt{-1 + 2}$, which is false. Thus, $x = -1$ is an extraneous solution, which must be rejected.

PROBLEM SET 1

In problems 1–20, indicate whether the statement is true or false.

1. For every real number x, the real number $-x$ is negative.
2. For every real number x, $x^0 = 1$.
3. If $x \neq 0$, then $x \cdot x^{-1} = 1$.
4. For any real number x, $x + (-x) = 0$.
5. Every natural number is a rational number.
6. Every irrational number is a real number.
7. $\sqrt{9} = \pm 3$
8. If $x \neq 1$, then $\dfrac{3(x-1)}{(1+x^2)(x-1)} = \dfrac{3}{1+x^2}$.
9. The solutions of the equation $x^2 = 17$ are $x = \pm\sqrt{17}$.
10. If x is positive, then $(\sqrt{x})^2 = x$.
11. If x is any real number, then $\sqrt{x^2} = x$.
12. If x and y are positive, then $\sqrt{x+y} = \sqrt{x} + \sqrt{y}$.
13. $\dfrac{2+3}{5+3} = \dfrac{2}{5}$
14. $\dfrac{2}{3} - \dfrac{3}{3} = \dfrac{2-3}{3}$
15. $\dfrac{3}{5} + \dfrac{3}{4} = \dfrac{3}{9}$
16. $\dfrac{1}{2} - \dfrac{1}{3} = -\dfrac{1}{1}$
17. $\sqrt{\dfrac{2}{3}} = \dfrac{\sqrt{2}}{\sqrt{3}}$
18. If $x \neq 0$, then $\dfrac{x+y+z}{x} = 1 + \dfrac{y}{x} + \dfrac{z}{x}$.
19. If $x \neq 0$, then $\dfrac{3+x}{x} = 3$.
20. $\frac{0}{3}$ is undefined.

In problems 21–26, rewrite each equation as an equivalent quadratic equation in standard form (if it's not already in that form) and then solve it by factoring.

21. $x^2 + 2x - 3 = 0$
22. $z^2 - 2z = 35$
23. $x^2 - 7x = 0$
24. $9 - y^2 = 2y^2$
25. $x^2 - 4x = 21$
26. $(8x + 19)x = 27$

In problems 27–32, complete the square by adding a suitable constant to each expression.

27. $x^2 + 6x$
28. $x^2 - 6x$
29. $x^2 - 5x$
30. $x^2 + \dfrac{b}{a}x$
31. $x^2 + \dfrac{3}{4}x$
32. $x^2 + \sqrt{2}x$

In problems 33–40, use the quadratic formula to solve each equation.

33. $5x^2 - 7x - 6 = 0$
34. $6x^2 - x - 2 = 0$
35. $3x^2 + 5x + 1 = 0$
36. $12y^2 + y - 1 = 0$
37. $2x^2 - x - 2 = 0$
38. $4u^2 - 11u + 3 = 0$
39. $4y^2 - 3y - 3 = 0$
40. $5x^2 + 17x - 3 = 0$

In problems 41–45, solve each equation. Check for extraneous roots.

41. $x = \sqrt{3 - 2x}$
42. $\dfrac{5}{x+4} - \dfrac{3}{x-2} = 4$
43. $x - \sqrt{5x - 6} = 0$
44. $\dfrac{5}{x-5} + 6 = \dfrac{x}{x-5}$
45. $\dfrac{1}{x(x-1)} - \dfrac{1}{x} = \dfrac{1}{x-1}$

46. In order to support a solar collector at the correct angle, each of the roof trusses for a house are designed as a right triangle, with the hypotenuse along the base (Figure 1). If the rafter on the side of the solar collector is 4 feet shorter than the other rafter, and if the base of each truss is 36 feet long, how long are the rafters?

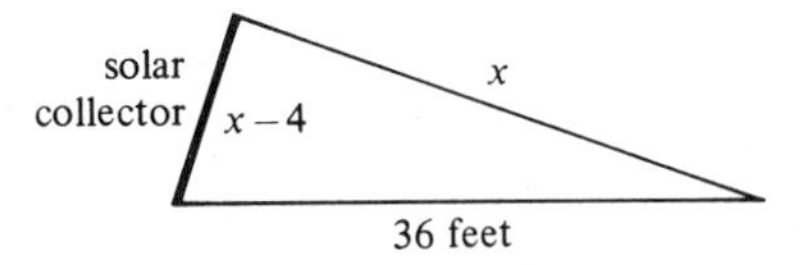

2 Real Numbers, Inequalities, and Absolute Value

Relations among real numbers can often be visualized by using a **number line** or **coordinate axis.** To construct a number line, you begin by fixing a point O called the **origin** and another point U called the **unit point** on a straight line L (Figure 1). The distance between O and U is called the **unit distance,** and may be 1 inch, 1 centimeter, or whatever unit of measure you choose. If the straight line L is horizontal, it is customary to place U to the right of O.

Figure 1

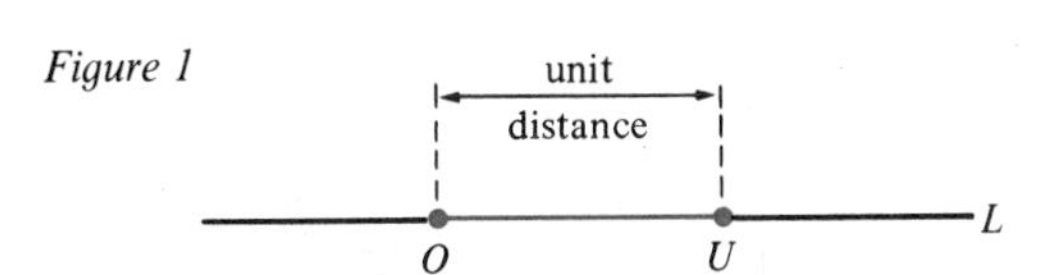

Each point P on the line L is now assigned a "numerical address" or **coordinate** x representing its signed distance from the origin, measured in terms of the given unit. Thus, $x = \pm d$, where d is the distance between O and P, and either the plus or minus sign is used according to whether P is to the right or left of O, respectively (Figure 2). On the resulting number scale (Figure 3), each point P has a corresponding numerical coordinate x. It is convenient to use an arrowhead to indicate the direction (to the right in Figure 3) in which the numerical coordinates are increasing.

Figure 2

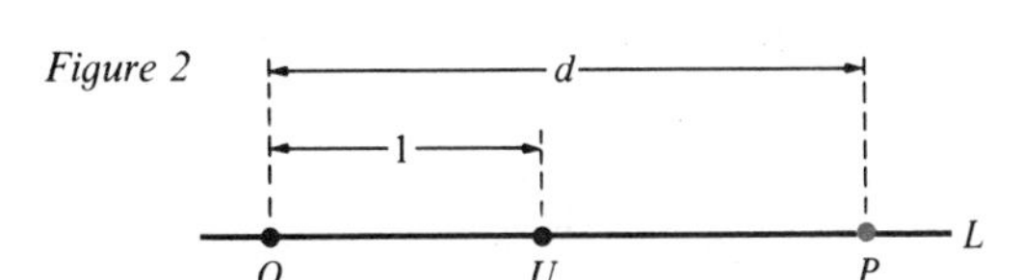

Figure 3

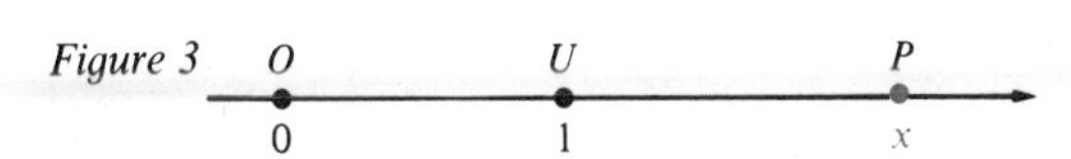

If the point with coordinate x lies to the left of the point with coordinate y (Figure 4), then we say that **y is greater than x** (or equivalently, **x is less than y**) and we write $y > x$ (or $x < y$). In other words, $y > x$ (or $x < y$) means that $y - x$ is positive. Notice that a real number x is positive if and only if $x > 0$, and it is negative if and only if $x < 0$.

Figure 4

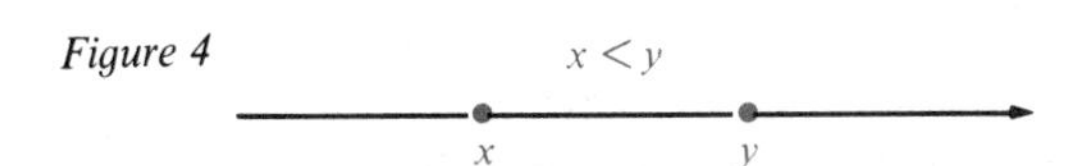

Sometimes we know only that a certain inequality does *not* hold. If $x < y$ does not hold, then either $x > y$ or $x = y$. In this case, we say that **x is greater than or equal to y** and we write $x \geq y$ (or equivalently, $y \leq x$). If $y \leq x$, we say that **y is less than or equal to x.** Statements of the form $y < x$ (or $x > y$) are called **strict** inequalities; those of the form $y \leq x$ (or $x \geq y$) are called **nonstrict** inequalities.

If we write $x < y < z$, we mean that $x < y$ *and* $y < z$. Likewise, $x \geq y > z$ means that $x \geq y$ *and* $y > z$. This notation for combined inequalities is only used when the inequalities run in the *same direction.* We use combined inequalities to define sets of real numbers called **intervals.**

Definition 1 **Bounded Intervals**

Let a and b be fixed real numbers with $a < b$.

(i) The **open interval** (a, b) with *endpoints* a and b is the set of all real numbers x such that $a < x < b$.

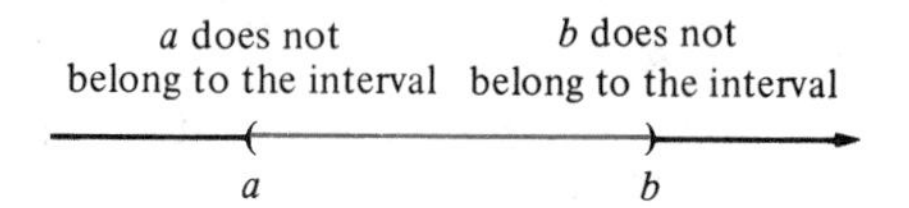

(ii) The **closed interval** $[a, b]$ with *endpoints* a and b is the set of all real numbers x such that $a \leq x \leq b$.

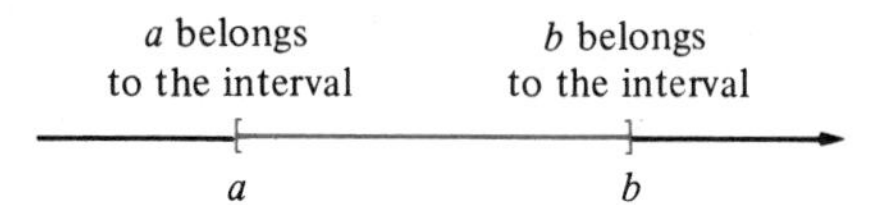

(iii) The **half-open interval** $[a, b)$ with *endpoints* a and b is the set of all real numbers x such that $a \leq x < b$.

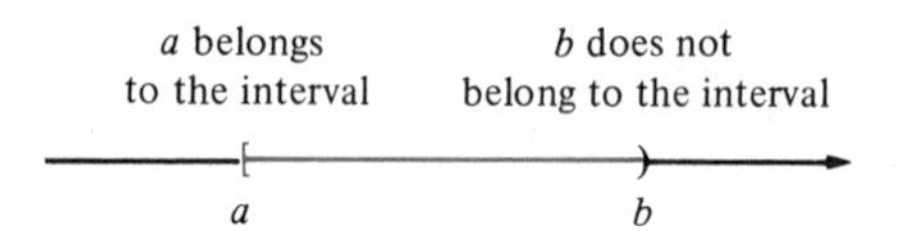

(iv) The **half-open interval** $(a, b]$ with *endpoints* a and b is the set of all real numbers x such that $a < x \leq b$.

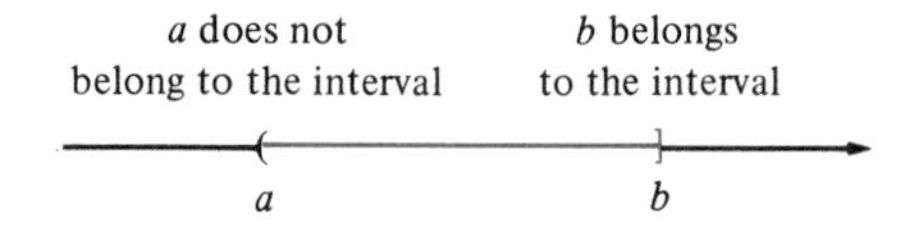

Notice that a closed interval contains its endpoints, but an open interval does not. A half-open interval (also called a **half-closed interval**) contains one of its endpoints, but not the other.

Unbounded intervals, which extend indefinitely to the right or left, are written with the aid of special symbols $+\infty$ and $-\infty$, called **positive infinity** and **negative infinity.**

Definition 2 **Unbounded Intervals**

Let a be a fixed real number.

(i) $(a, +\infty)$ is the set of all real numbers x such that $a < x$.

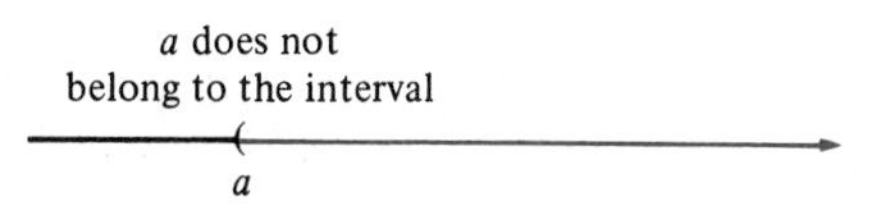

(ii) $(-\infty, a)$ is the set of all real numbers x such that $x < a$.

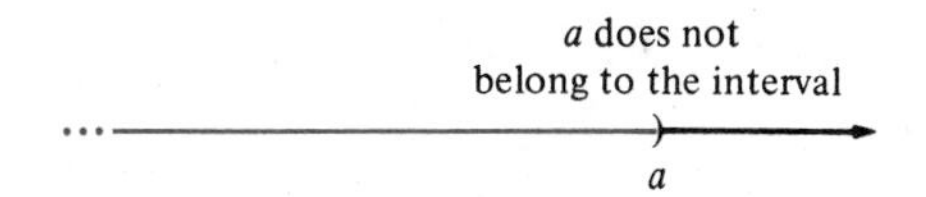

(iii) $[a, +\infty)$ is the set of all real numbers x such that $a \leq x$.

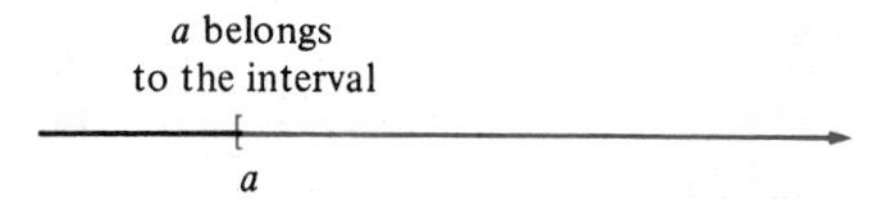

(iv) $(-\infty, a]$ is the set of all real numbers x such that $x \leq a$.

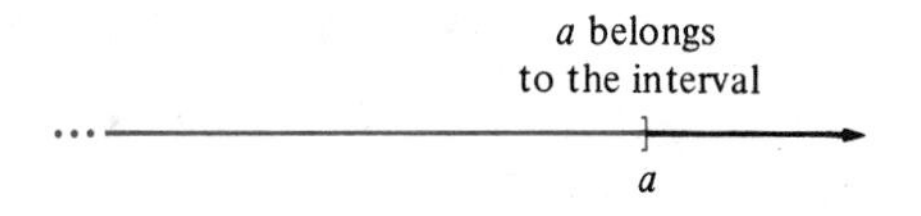

Be careful, $+\infty$ and $-\infty$ are just convenient symbols—they are not real numbers and you must not treat them as if they were. In the notation for unbounded intervals, we usually write ∞ rather than $+\infty$. For instance, $(5, \infty)$ denotes the set of all real numbers that are greater than 5.

Example Illustrate each set on a number line.

(a) $(2, 5]$

(b) $(3, \infty)$

(c) The set A of all real numbers that belong to both intervals $(2, 5]$ and $(3, \infty)$.

(d) The set B of all real numbers that belong to at least one of the intervals $(2, 5]$ and $(3, \infty)$.

Solution (a) The interval $(2, 5]$ consists of all real numbers between 2 and 5, including 5, but excluding 2 (Figure 5a).

Figure 5

(2, 5]

(a) 0 1 2 3 4 5 6 7

(b) The interval $(3, \infty)$ consists of all real numbers that are greater than 3 (Figure 5b).

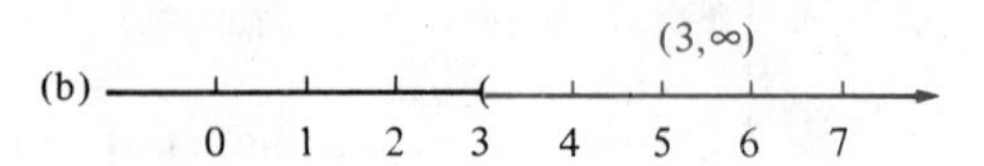

(c) The set A of all real numbers that belong to both intervals $(2, 5]$ and $(3, \infty)$ is the interval $(3, 5]$ (Figure 5c).

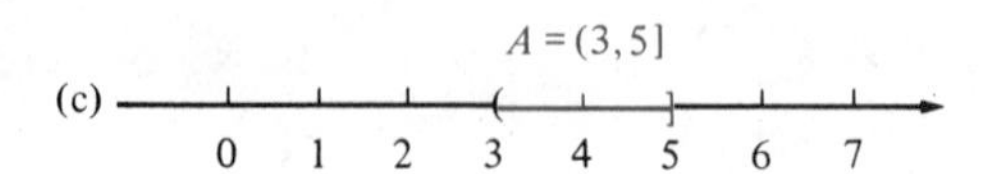

(d) The set B consists of all numbers in the interval $(2, 5]$ together with all numbers in the interval $(3, \infty)$, so B is the interval $(2, \infty)$ (Figure 5d).

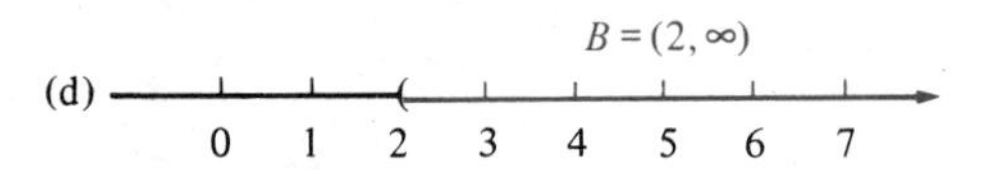

In dealing with inequalities, you will find the following properties useful.

Properties of Inequalities

Let a, b, and c denote real numbers.

1. **Trichotomy** One and only one of the following is true:

$$a < b \quad \text{or} \quad b < a \quad \text{or} \quad a = b.$$

2. **Transitivity**

If $a < b$ and $b < c$, then $a < c$.

3. **Addition**

If $a < b$, then $a + c < b + c$.

4. **Subtraction**

If $a < b$, then $a - c < b - c$.

5. **Multiplication**

If $a < b$ and $c > 0$, then $ac < bc$.

6. **Division**

If $a < b$ and $c > 0$, then $\dfrac{a}{c} < \dfrac{b}{c}$.

According to the addition and subtraction properties, you can add or subtract the same quantity on both sides of an inequality. The multiplication and division properties permit you to multiply or divide both sides of an inequality by a *positive* number. According to the following property, if you multiply or divide both sides of an inequality by a *negative* number, you must *reverse the inequality*.

7. **Order-Reversing**

$$\text{If } a < b \text{ and } c < 0, \text{ then } ac > bc \text{ and } \frac{a}{c} > \frac{b}{c}.$$

For instance, if you multiply both sides of the inequality $2 < 5$ by -3, you must reverse the inequality and write $-6 > -15$, or $-15 < -6$. Indeed, on a number line, -15 lies to the left of -6.

Statements analogous to Properties 2 through 7 can be made for nonstrict inequalities (those involving $\leq$ or $\geq$) and for compound inequalities. For instance, if you know that

$$-1 \leq 3 - \frac{x}{2} < 5,$$

you can multiply all members by -2, provided that you reverse the inequalities:

$$2 \geq x - 6 > -10$$

or

$$-10 < x - 6 \leq 2.$$

Example State one Property of Inequalities to justify each statement:

(a) We know that $\pi \neq \frac{22}{7}$, so it follows that either $\pi < \frac{22}{7}$ or else $\pi > \frac{22}{7}$.

(b) If $-1 < 0$ and $0 < 1$, it follows that $-1 < 1$.

(c) If $5 < 6$, it follows that $5 - 11 < 6 - 11$.

(d) If $4 < 7$ and $2 > 0$, it follows that $2 < \frac{7}{2}$.

Solution (a) Trichotomy (b) Transitivity

(c) Subtraction (d) Division

The idea of absolute value, which we now introduce, allows us to deal with the "magnitude" of a number without regard to its algebraic sign.

Definition 3 **Absolute Value**

If x is a real number, then the **absolute value of *x*,** written $|x|$, is defined by

$$|x| = \begin{cases} x & \text{if } x \geq 0 \\ -x & \text{if } x < 0. \end{cases}$$

For instance,

$$|5| = 5 \quad \text{because} \quad 5 \geq 0,$$
$$|0| = 0 \quad \text{because} \quad 0 \geq 0, \quad \text{and}$$
$$|-5| = -(-5) = 5 \quad \text{because} \quad -5 < 0.$$

Notice that

$$|x| \geq 0 \quad \text{is always true.}$$

If you square a real number and then take the principal square root of the result, you will remove any negative sign that may have been present. Therefore,

$$\sqrt{x^2} = |x|.$$

For instance,

$$\sqrt{(-5)^2} = \sqrt{25} = 5 = |-5|.$$

On a number line $|x|$ gives *the number of units of distance between the point whose coordinate is x and the origin*, regardless of whether the point is to the right or left of the origin (Figure 6a). For instance, both the point with coordinate -3 and the point with coordinate 3 are 3 units from the origin, and we have $|-3| = 3 = |3|$ (Figure 6b).

Figure 6

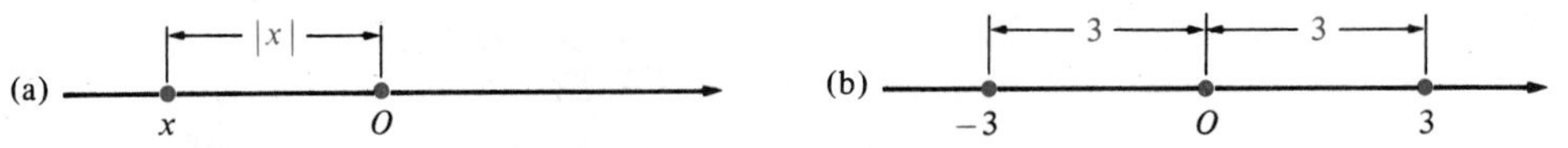

More generally, on a number line $|x - y|$ gives *the number of units of distance between the point whose coordinate is x and the point whose coordinate is y*. This holds regardless of which point is to the left of the other (Figure 7). For simplicity, we refer to the distance $|x - y|$ as *the distance between the numbers x and y*.

Figure 7

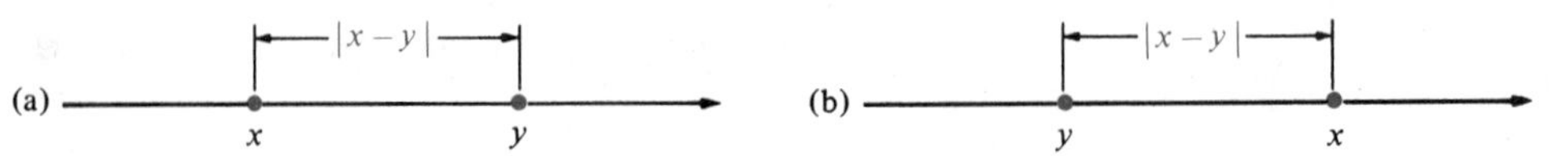

Example Find the distance between the following numbers.

(a) 4 and 7
(b) 7 and 4
(c) -4 and 7
(d) -4 and -7
(e) -4 and 0
(f) -4 and 4

Solution (a) $|4 - 7| = |-3| = 3$ (b) $|7 - 4| = |3| = 3$
(c) $|-4 - 7| = |-11| = 11$ (d) $|-4 - (-7)| = |3| = 3$
(e) $|-4 - 0| = |-4| = 4$ (f) $|-4 - 4| = |-8| = 8$

The absolute value of the product of two real numbers is the same as the product of their absolute values; that is,

$$|xy| = |x|\,|y|.$$

This can be seen as follows:

$$|xy| = \sqrt{(xy)^2} = \sqrt{x^2y^2} = |x|\,|y|.$$

A similar result holds for quotients, so that, if $y \neq 0$,

$$\left|\frac{x}{y}\right| = \frac{|x|}{|y|}.$$

Suppose that d is a positive real number. Because $|x|$ is the distance between the point whose coordinate is x and the origin, the condition

$$|x| < d$$

means that the point whose coordinate is x is less than d units from the origin; that is,

$$-d < x < d$$

(Figure 8). This result is stated more formally in the following theorem.

Figure 8

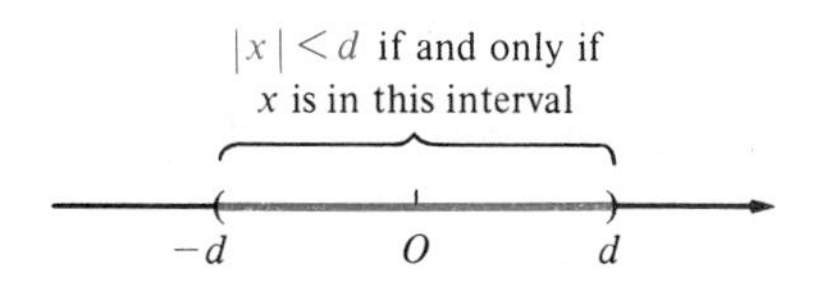

Theorem 1 **The Inequality $|x| < d$**

If $d > 0$, then the inequality

$$|x| < d$$

can be rewritten in the equivalent form

$$-d < x < d.$$

Example Rewrite the inequality $|u - v| < 8$ in a form not involving absolute value.

Solution Using Theorem 1, we can rewrite the given inequality in the equivalent form

$$-8 < u - v < 8.$$

PROBLEM SET 2

In problems 1–8, illustrate each interval on a number line.

1. $(2, 5)$ **2.** $[-1, 3)$ 3. $[-4, 0]$
4. $(-\infty, -3)$ 5. $[1, \infty)$ **6.** $(-\frac{1}{2}, \infty)$
7. $[-\frac{3}{2}, \frac{5}{2}]$ **8.** $(-4, 0]$

In problems 9–14, illustrate each set on a number line.

9. The set A of all real numbers that belong to both of the intervals $(0, 3]$ and $(1, \infty)$.
10. The set B of all real numbers that belong to both of the intervals $(-1, \frac{1}{2}]$ and $[\frac{1}{2}, \frac{3}{4}]$.
11. The set C of all real numbers that belong to at least one of the intervals $[-2, -1]$ and $[1, 2]$.
12. The set D of all real numbers that belong to at least one of the intervals $(-1, 0]$ and $(0, 1]$.
13. The set E of all real numbers that belong to the interval $[-2, 2]$, but not to the interval $(-2, 2)$.
14. The set F of all real numbers x such that x belongs to the interval $[-2\pi, 2\pi]$ and $2x/\pi$ is an integer.

In problems 15–22, state one Property of Inequalities to justify each statement.

15. Since $3 < 5$, it follows that $3 + 2 < 5 + 2$.
16. $-3 < -2$ and $-2 < 2$, it follows that $-3 < 2$.
17. Since $-3 < -2$ and $\frac{1}{3} > 0$, it follows that $-1 < -\frac{2}{3}$.
18. We know that $\sqrt{2} \neq 1.414$, so it follows that either $\sqrt{2} < 1.414$ or else $\sqrt{2} > 1.414$.
19. Since $-5 < -3$ and $5 > 0$, it follows that $-1 < -\frac{3}{5}$.
20. Because $-1 < 0$, the inequality $x < y$ can be rewritten in the equivalent form $-y < -x$.
21. Since $3 < 5$ and $-2 < 0$, it follows that $-6 > -10$.
22. Since $3 < 5$ and $-2 < 0$, it follows that $-\frac{3}{2} > -\frac{5}{2}$.

23. Given that $x > -3$, what equivalent inequality is obtained if (a) 4 is added to both sides? (b) 4 is subtracted from both sides? (c) Both sides are multiplied by 5? (d) Both sides are multiplied by -5?
24. Given that $-2 \leq 3 - 4x < 1$, what equivalent inequality is obtained if (a) 3 is subtracted from all members? (b) 2 is added to all members? (c) All members are divided by 2? (d) All members are divided by -4?

In problems 25–30, rewrite the given condition as an equivalent statement that x belongs to a certain interval.

25. $2x - 5 < 5$ **26.** $2x - 10 \leq 15 - 3x$ 27. $7 - 3x \leq 22$
28. $5 \leq 2x + 3 < 13$ 29. $4 \leq 5x - 1 < 9$ **30.** $\frac{1}{4}(x + 1) \geq -\frac{1}{3}(2 - x)$

31. A student's scores on the first three tests in a sociology course are 65, 78, and 84. What range of scores on a fourth test will give the student an average that is less than 80, but no less than 70, for the four tests?
32. If $c > 0$, show that $1/c > 0$.

In problems 33–38, find the distance between the numbers and illustrate on a number line.

33. 5 and 3 **34.** -3 and 0 35. -3 and -5 **36.** -5 and 3
37. $\frac{5}{2}$ and $\frac{3}{2}$ **38.** $-\frac{2}{3}$ and $\frac{3}{4}$

In problems 39–46, find the value of each expression if $x = -4$ and $y = 6$.

39. $|5x + 2y|$ **40.** $|5x| + |2y|$ 41. $|5x - 2y|$ **42.** $|5x| - |2y|$
43. $||5x| - |2y||$ **44.** $|3xy|$ 45. $|3xy| - 3|x||y|$ **46.** $|5x/(2y)|$

47. If $y \neq 0$, show that $|x/y| = |x|/|y|$.

48. Show that $-|x| \leq x \leq |x|$ holds for all real values of x. [Hint: Consider separately the three cases $x > 0$, $x = 0$, and $x < 0$.]

49. Given an example to show that $|x + y|$ isn't necessarily the same as $|x| + |y|$.

50. Show that if $d > 0$, then the inequality $|u - v| < d$ can be rewritten in the equivalent form $v - d < u < v + d$.

---

3 Calculators, Scientific Notation, and Approximations

Today many students own or have access to an electronic calculator. A scientific calculator with keys for trigonometric functions costs less than many college textbooks and will expedite some of the calculations required in this book. Problems or groups of problems for which the use of a calculator is recommended are marked with the symbol [C]. If you don't have access to a calculator, you can still work most of these problems by using the tables provided in the Appendix at the back of the book.

After acquiring a calculator, learn to use it properly by studying the instruction booklet furnished with it. In particular, practice performing chain calculations so you can do them as efficiently as possible using whatever "memory" features your calculator may possess to store intermediate results. After you learn *how* to use a calculator, it is important to learn *when* to use it and especially when *not* to use it. Attempts to use a calculator for problems that are not marked with the symbol [C] can lead to bad habits, which not only waste time but actually hinder understanding.

3.1 Scientific Notation

In applied mathematics, very large and very small numbers are written in compact form by using integer powers of 10. For instance, the speed of light in vacuum,

$$c = 300{,}000{,}000 \text{ meters per second (approximately)}$$

can be written more compactly as

$$c = 3 \times 10^8 \text{ meters per second.}$$

More generally, a positive real number x is said to be expressed in **scientific notation** if it is written in the form

$$x = p \times 10^n,$$

where n is an integer and $1 \leq p < 10$.

Many calculators automatically switch to scientific notation whenever the number is too large or too small to be displayed in ordinary decimal form. When a number such as

$$2.579 \times 10^{-13}$$

is displayed, the multiplication sign and the base 10 usually do not appear, and the display shows simply

$$2.579 \qquad -13.$$

To change a number from ordinary decimal form to scientific notation, move the decimal point to obtain a number between 1 and 10 and multiply by 10^n or by 10^{-n}, where n is the number of places the decimal point was moved to the left or to the right, respectively. Final zeros after the decimal point can be dropped unless it is necessary to retain them to indicate the accuracy of an approximation.

Example Rewrite each statement so that all numbers are expressed in scientific notation.

(a) The volume of the earth is approximately

1,087,000,000,000,000,000,000 cubic meters.

(b) The earth rotates about its axis with an angular speed of approximately

0.00417 degree per second.

Solution (a) We move the decimal point 21 places to the left

1,087,000,000,000,000,000,000

to obtain a number between 1 and 10, and multiply by 10^{21}, so that

$$1{,}087{,}000{,}000{,}000{,}000{,}000{,}000 = 1.087 \times 10^{21}.$$

Thus, the volume of the earth is approximately 1.087×10^{21} cubic meters.

(b) We move the decimal point 3 places to the right

0.00417

to obtain a number between 1 and 10, and multiply by 10^{-3}, so that

$$0.00417 = 4.17 \times 10^{-3}.$$

Thus, the earth rotates about its axis with an angular speed of approximately 4.17×10^{-3} degree per second.

The procedure we used above can be reversed whenever a number is given in scientific notation and you want to rewrite it in ordinary decimal form.

Example Rewrite the following numbers in ordinary decimal form.

(a) 7.71×10^5 (b) 6.32×10^{-8}

Solution (a) $7.71 \times 10^5 = 7\,7\,1\,0\,0\,0 = 771{,}000$

(b) $6.32 \times 10^{-8} = 0.0\,0\,0\,0\,0\,0\,0\,6\,3\,2 = 0.000{,}000{,}0632$

3.2 Approximations

Numbers produced by a calculator are often inexact because the calculator can work only with a finite number of decimal places. For instance, a 10-digit calculator gives

$$2 \div 3 = 6.666666667 \times 10^{-1} \quad \text{and} \quad \sqrt{2} = 1.414213562,$$

both of which are **approximations** of the true values. Therefore, unless we explicitly ask for numerical approximations or indicate that a calculator is recommended, it's usually best to leave your answers in fractional form or as expressions with radical signs. Don't be too quick to pick up your calculator—answers such as $\frac{2}{3}$, $\sqrt{2}$, $(\sqrt{3} + \sqrt{5})/6$, and $\pi/4$ are often *preferred* to lengthy decimal expressions that are only approximations.

Most numbers obtained from measurements of real-world quantities are subject to error and must also be regarded as approximations. If the result of a measurement (or any calculation involving approximations) is expressed in scientific notation $p \times 10^n$, it is usually understood that p should contain only **significant digits;** that is, digits that—except possibly for the last—are known to be correct or reliable. (The last digit may be off by one unit because the number was rounded off.) For instance, if we read in a physics textbook that

$$1 \text{ electron volt} = 1.60 \times 10^{-19} \text{ joule},$$

we understand that the digits 1, 6, and 0 are significant and we say that, to an accuracy of three significant digits, 1 electron volt is 1.60×10^{-19} joule.

To emphasize that a numerical value is only an approximation, we often use a wave-shaped equal sign $\approx$. For instance,

$$\sqrt{2} \approx 1.414.$$

However, when dealing with inexact quantities, we may use ordinary equal signs simply because it becomes tiresome to indicate repeatedly that approximations are involved.

3.3 Rounding Off

Some scientific calculators can be set to round off all displayed numbers to a particular number of decimal places or significant digits. However, it's easy enough to round off numbers without a calculator: Simply drop all unwanted digits to the right of the digits that are to be retained, and increase the last retained digit by 1 if the first digit dropped is 5 or greater. It may be necessary to replace dropped digits by zeros in order to hold the decimal point; for instance, we round off 5157.3 to the nearest hundred as 5200.

Example Round off the given number as indicated.

(a) 1.327 to the nearest tenth

(b) -19.8735 to the nearest thousandth

(c) 4671 to the nearest hundred

(d) 9.22345×10^7 to 4 significant digits

Solution (a) To the nearest tenth,

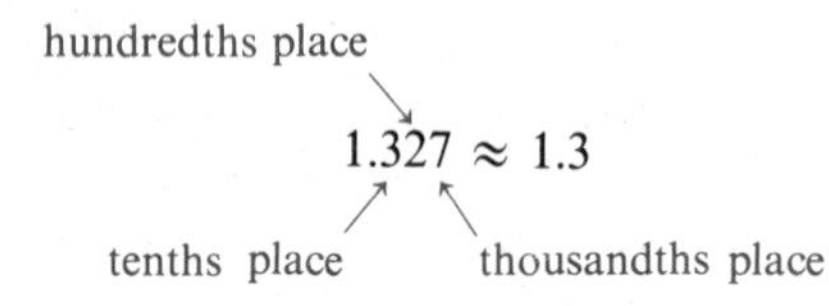

(b) To the nearest thousandth,

$$-19.8735 \approx -19.874$$

thousandths place

(c) To the nearest hundred,

$$4671 \approx 4700$$

hundreds place

(d) To four significant digits, $9.22345 \times 10^7 \approx 9.223 \times 10^7$.

PROBLEM SET 3

In problems 1–8, rewrite each number in scientific notation.

1. 15,500
2. 0.0043
3. 0.000,587
4. 77 million
5. 186,000,000,000
6. 420 trillion
7. 0.000,000,901
8. $(0.025)^{-5}$

In problems 9–14, rewrite each number in ordinary decimal form.

9. 3.33×10^4
10. 1.732×10^{10}
11. 4.102×10^{-5}
12. -8.255×10^{-11}
13. 1.001×10^7
14. -2.00×10^9

In problems 15–20, rewrite each statement so that all numbers are expressed in scientific notation.

15. The image of one frame in a motion-picture film stays on the screen approximately 0.062 second.
16. One liter is defined to be 0.001,000,028 cubic meter.
17. An *astronomical unit* is defined to be the average distance between the earth and the sun, 92,900,000 miles, and a *parsec* is the distance at which one astronomical unit would subtend one second of arc, about 19,200,000,000,000 miles.
18. A *light year* is the distance that light, traveling at approximately 186,200 miles per second, traverses in 1 year. Thus, a light year is approximately 5,872,000,000,000 miles.
19. In physics, the average lifetime of a lambda particle is estimated to be 0.000,000,000,251 second.

20. In thermodynamics, Boltzmann's constant is

0.000,000,000,000,000,000,000,0138 joule per degree Kelvin.

In problems 21–24, convert the given numbers to scientific notation and calculate the indicated quantity. Do not round off your answers.

21. (8000)(2,000,000,000)(0.000,03)

22. $(0.000,006)^3(500,000,000)^{-4}$

23. $\dfrac{(7,000,000,000)^3}{0.0049}$

Ⓒ **24.** $\dfrac{(0.000,000,039)^2(591,000)^3}{(197,000)^2}$

Ⓒ **25.** In electronics, $P = I^2R$ is the formula for the power P in watts dissipated by a resistance of R ohms through which a current of I amperes flows. Calculate P if $I = 1.43 \times 10^{-4}$ ampere and $R = 3.21 \times 10^4$ ohms.

Ⓒ **26.** The mass of the sun is approximately 1.97×10^{29} kilograms and our galaxy (the Milky Way) is estimated to have a total mass of 1.5×10^{11} suns. The mass of the known universe is at least 10^{11} times the mass of our galaxy. Calculate the approximate mass of the known universe.

In problems 27–30, specify the accuracy of the indicated value in terms of significant digits.

27. A drop of water contains 1.7×10^{21} molecules.

28. The binding energy of the earth to the sun is 2.5×10^{33} joules.

29. One mile = 6.3360×10^4 inches.

30. One atmosphere = 1.01×10^5 newtons per square meter.

In problems 31–38, round off the given number as indicated.

31. 5280 to the nearest hundred

32. 9.29×10^7 to the nearest million

33. 0.0145 to the nearest thousandth

34. 999 to the nearest ten

35. 111111.11 to the nearest ten-thousand

36. 5.872×10^{12} to three significant digits

37. 2.1448×10^{-13} to three significant digits

38. π to four significant digits

39. According to the U.S. Bureau of Economic Analysis, the gross national product (GNP) of the United States in 1978 was $\$2.1076 \times 10^{12}$. According to the U.S. Office of Management and Budget, the national debt at the end of fiscal 1978 was $\$7.804 \times 10^{11}$.

(a) How much more was the GNP than the national debt in 1978?

Ⓒ (b) If we estimate the population of the United States in 1978 as 2.2×10^8, find the per capita GNP (that is, GNP ÷ population), to two significant digits, in 1978.

---

4 Plane Geometry and Cartesian Coordinates

This section is intended as a concise review of the basic geometry that you will need in the rest of the book, and as an introduction to the cartesian coordinate system. To facilitate your review of plane geometry, the material is presented in condensed "outline" form.

If the three vertex angles in a first triangle $A_1B_1C_1$ are equal to the corresponding vertex angles in a second triangle $A_2B_2C_2$, we say that the triangles are *similar* (Figure 1). *Corresponding side lengths of two similar triangles are proportional.* For instance, in Figure 1,

$$\frac{a_1}{a_2} = \frac{b_1}{b_2} = \frac{c_1}{c_2}.$$

Figure 1

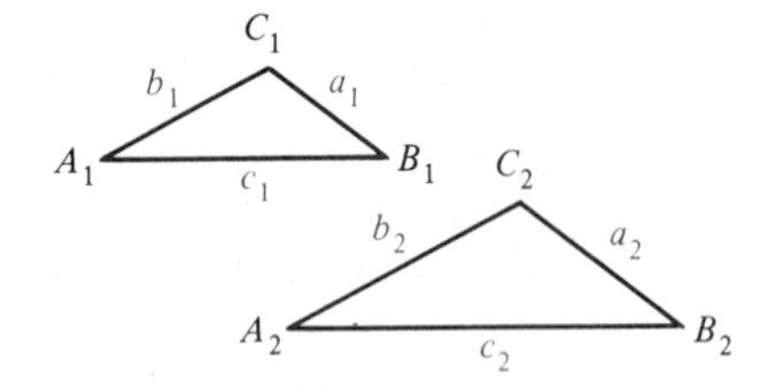

It follows that

$$\frac{a_1}{b_1} = \frac{a_2}{b_2} \quad \text{and} \quad \frac{a_1}{c_1} = \frac{a_2}{c_2} \quad \text{and} \quad \frac{b_1}{c_1} = \frac{b_2}{c_2}.$$

Since *the sum of the three vertex angles of a triangle is always* $180°$, it follows that two vertex angles determine the third one. Therefore, if two of the vertex angles of one triangle are equal to two of the vertex angles of another triangle, then the triangles are similar.

If the lengths of the three sides of a first triangle are the same as the lengths of corresponding sides of a second triangle, then the two triangles are **congruent.** If two triangles are congruent, the three vertex angles in either triangle are equal to the corresponding vertex angles in the other triangle. If two sides and the angle included between them in a first triangle are equal to two sides and the angle included between them in a second triangle, then the two triangles are congruent.

A triangle in which one of the vertex angles is a right angle (that is, a $90°$ angle) is called a **right triangle.** In a right triangle, the side opposite the right angle is called the **hypotenuse** (Figure 2). The word "hypotenuse" is also used to refer to the **length** of the side opposite the right angle. The **Pythagorean theorem** states that the square of the length of the hypotenuse is equal to the sum of the squares of the lengths of the other two sides. Thus, in Figure 2, by the Pythagorean theorem,

$$c^2 = a^2 + b^2.$$

Figure 2

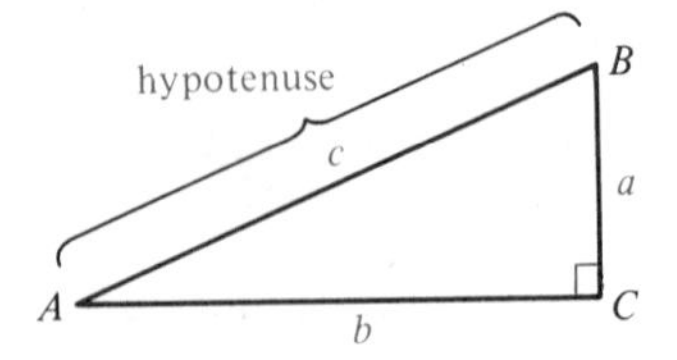

The *converse of the Pythagorean theorem also holds:* If the three sides a, b, and c of a triangle satisfy $a^2 + b^2 = c^2$, then the angle opposite side c is a right angle.

Example 1 In right triangle ACB (Figure 2), suppose $a = 7$ and $c = 25$. Find b.

Solution By the Pythagorean theorem,

$$b^2 = c^2 - a^2 = 25^2 - 7^2 = 625 - 49 = 576,$$

so

$$b = \sqrt{576} = 24.$$

Example 2 The three sides of a triangle have lengths $a = 5$, $b = 12$, and $c = 13$. Is the triangle a right triangle?

Solution Here,

$$a^2 + b^2 = 5^2 + 12^2 = 25 + 144 = 169 = 13^2 = c^2.$$

Hence, by the converse of the Pythagorean theorem, we have a right triangle with hypotenuse c.

In a right triangle ACB (Figure 2), as in any triangle, the sum of the three vertex angles is 180°. Because the angle at vertex C accounts for 90 of these 180 degrees, the sum of the remaining two angles must be 90 degrees. Thus,

$$\text{vertex angle at } A + \text{vertex angle at } B = 90°.$$

Therefore, either of these vertex angles determines the other. It follows that *two right triangles are similar if a vertex angle (other than the right angle) in one triangle is equal to a vertex angle in the other triangle.*

A perpendicular dropped from a vertex of a triangle to the opposite side (extended if necessary) is called an **altitude** of the triangle (Figure 3), its length h is called the **height** of the triangle as measured from that vertex, and the side of the triangle opposite to the vertex is the corresponding **base.** If b is the length of the base, then the **area** of the triangle is given by

$$\text{area} = \tfrac{1}{2}hb.$$

Figure 3

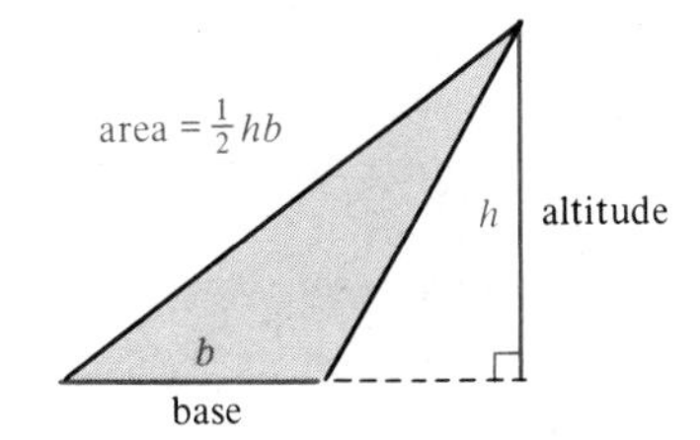

Figure 4

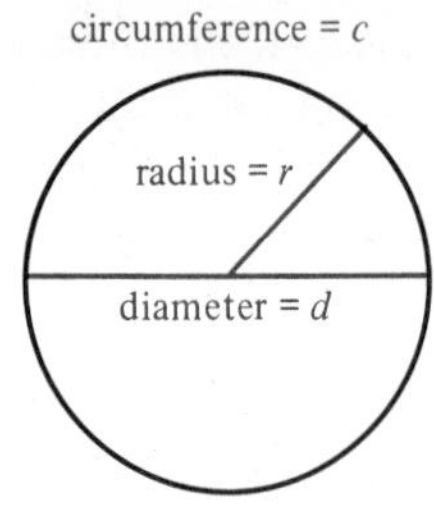

The ratio of the **circumference** c to the **diameter** d of a circle is the constant π (the Greek letter pi) (Figure 4). Because $\pi = c/d$, it follows that

$$c = \pi d \quad \text{or} \quad c = 2\pi r.$$

where $r = d/2$ is the **radius** of the circle. The **area** of the circle is given by the familiar formula

$$\text{area} = \pi r^2$$

4.1 Cartesian Coordinates

In Section 2, we have seen how a point P on a number line can be specified by a real number x called its coordinate. Similarly, by using a **cartesian coordinate system,** named in honor of the French philosopher and mathematician René Descartes (1596–1650), we can specify a point P in the plane by giving *two* real numbers, also called coordinates.

A cartesian coordinate system consists of two perpendicular number lines, called **coordinate axes,** meeting at a common origin O (Figure 5). Ordinarily, one of the number lines, known as the ***x* axis,** is horizontal and the other, known as the ***y* axis,** is vertical. Numerical coordinates increase to the right along the x axis and upward along the y axis. Although we usually use the same scale (that is, the same unit distance) on the two axes, it will be convenient in some of our figures to use different units on the x and y axes.

Figure 5

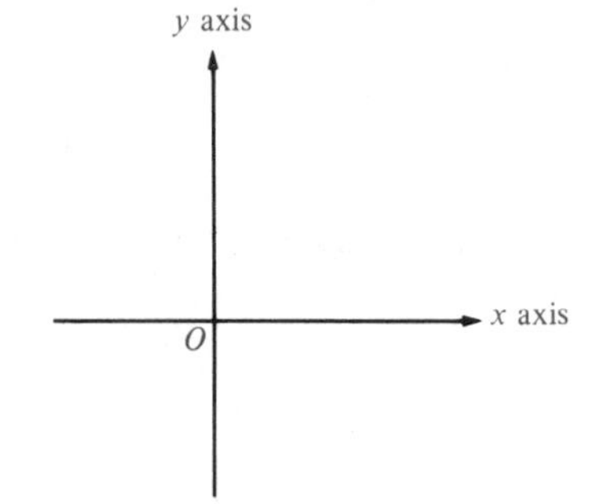

Figure 6

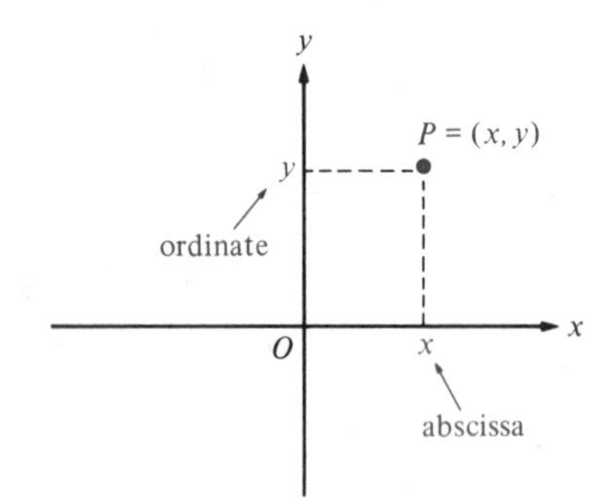

If P is a point in the plane, the **coordinates of *P*** are the coordinates x and y on the horizontal and vertical axes, respectively, of the points where perpendiculars from P meet these axes (Figure 6). The coordinate x is called the **abscissa** or the ***x* coordinate** of P, and the coordinate y is called the **ordinate** or the ***y* coordinate** of P. These coordinates are traditionally written as an ordered pair (x, y), enclosed in parentheses, with the abscissa first and the ordinate second.

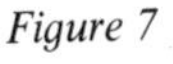

Figure 7

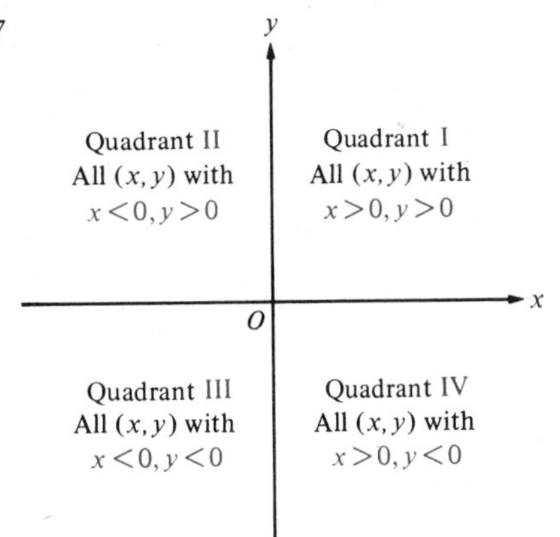

If a geometric point P has abscissa x and ordinate y, we can think of the ordered pair (x, y) as the numerical "address" of P. The correspondence between P and (x, y) seems so natural that we often identify the geometric point P with its "address" (x, y) by writing $P = (x, y)$. With this identification in mind, we call an ordered pair of real numbers (x, y) a *point* and refer to the set of all such ordered pairs as the **cartesian plane** or the ***xy* plane.** To **plot** a point P with coordinates (x, y) means to draw cartesian coordinate axes and to place a dot representing P at the point with abscissa x and ordinate y.

The x and y axes divide the plane into four regions called **quadrants** I, II, III, and IV as indicated in Figure 7. Notice that a point on a coordinate axis belongs to no quadrant.

If P_1 and P_2 are two points in the cartesian plane, we denote the **line segment** from P_1 to P_2 by $\overline{P_1P_2}$. The *length* of the line segment $\overline{P_1P_2}$, that is, the *distance between the points P_1 and P_2*, is denoted by $|\overline{P_1P_2}|$.

Theorem 1 **The Distance Formula**

If $P_1 = (x_1, y_1)$ and $P_2 = (x_2, y_2)$ are two points in the xy plane, then

$$|\overline{P_1P_2}| = \sqrt{(x_2 - x_1)^2 + (y_2 - y_1)^2}.$$

Proof The horizontal distance between P_1 and P_2 is the same as the distance between the points with coordinates x_1 and x_2 on the x axis (Figure 8a). It follows that the *horizontal* distance between P_1 and P_2 is $|x_2 - x_1|$ units. Similarly, the *vertical* distance between P_1 and P_2 is $|y_2 - y_1|$ units (Figure 8b). If the line segment $\overline{P_1P_2}$ is neither horizontal nor vertical, it forms the hypotenuse of a right triangle whose legs have lengths $|x_2 - x_1|$ and $|y_2 - y_1|$ (Figure 9). Therefore, by the Pythagorean theorem,

$$|\overline{P_1P_2}| = \sqrt{|x_2 - x_1|^2 + |y_2 - y_1|^2} = \sqrt{(x_2 - x_1)^2 + (y_2 - y_1)^2}.$$

We leave it to you to verify that the same formula works even if the line segment $\overline{P_1P_2}$ is horizontal or vertical (Problem 32).

Figure 8

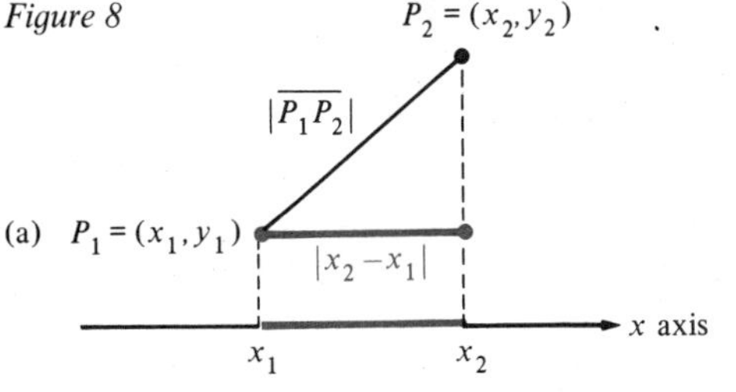

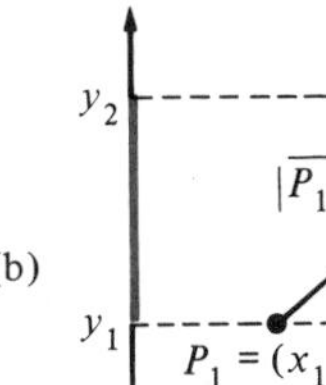

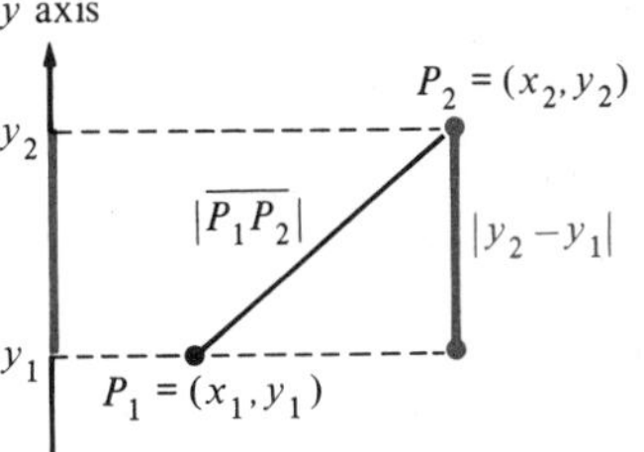

Figure 9

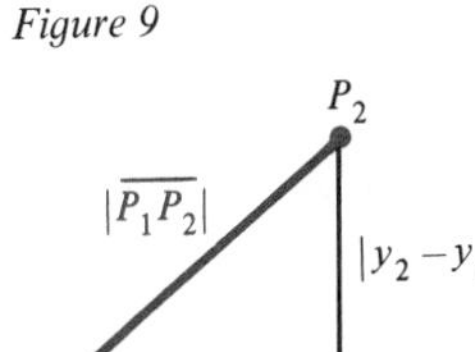

Example Let $A = (-5, 3)$, $B = (6, 0)$, and $C = (5, 5)$.

(a) Plot the points A, B, and C and draw the triangle ABC.

(b) Find the distances $|\overline{AB}|$, $|\overline{BC}|$, and $|\overline{CA}|$.

(c) Show that ABC is a right triangle.

(d) Find the area of triangle ABC.

Solution (a) The points A, B, and C are plotted and triangle ABC is drawn in Figure 10.

Figure 10

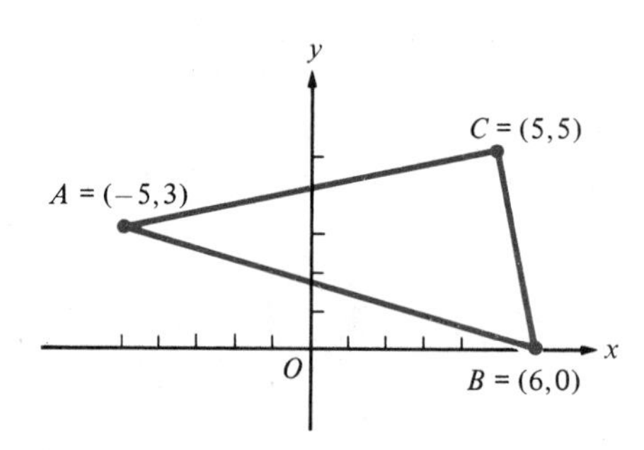

(b) $|\overline{AB}| = \sqrt{[6-(-5)]^2 + (0-3)^2} = \sqrt{11^2 + (-3)^2} = \sqrt{130}$

$|\overline{BC}| = \sqrt{(5-6)^2 + (5-0)^2} = \sqrt{(-1)^2 + 5^2} = \sqrt{26}$

$|\overline{CA}| = \sqrt{(-5-5)^2 + (3-5)^2} = \sqrt{(-10)^2 + (-2)^2} = \sqrt{104} = 2\sqrt{26}$

(c) Figure 10 suggests that the angle at vertex C is a right angle. To confirm this, we use the converse of the Pythagorean theorem. Using the results of (b), we have

$$|\overline{AB}|^2 = 130 = 26 + 104 = |\overline{BC}|^2 + |\overline{CA}|^2,$$

so ACB is indeed a right triangle.

(d) Taking $|\overline{CA}| = 2\sqrt{26}$ as the base and $|\overline{BC}| = \sqrt{26}$ as the height of the triangle, we find that

$$\text{area} = \tfrac{1}{2}\sqrt{26}(2\sqrt{26}) = 26 \text{ square units.}$$

The **graph** of an equation involving x and y is defined to be the set of all points $P = (x, y)$ in the cartesian plane whose coordinates satisfy the equation. Many (but not all) such graphs are smooth curves. For instance, consider the equation

$$x^2 + y^2 = 9.$$

We can rewrite this equation as

$$\sqrt{x^2 + y^2} = \sqrt{9},$$

that is,

$$\sqrt{(x-0)^2 + (y-0)^2} = 3.$$

By the distance formula (Theorem 1), the last equation holds if and only if the point $P = (x, y)$ is 3 units from the origin $O = (0, 0)$. Therefore, the graph of $x^2 + y^2 = 9$ is a circle of radius 3 with center O (Figure 11).

Figure 11

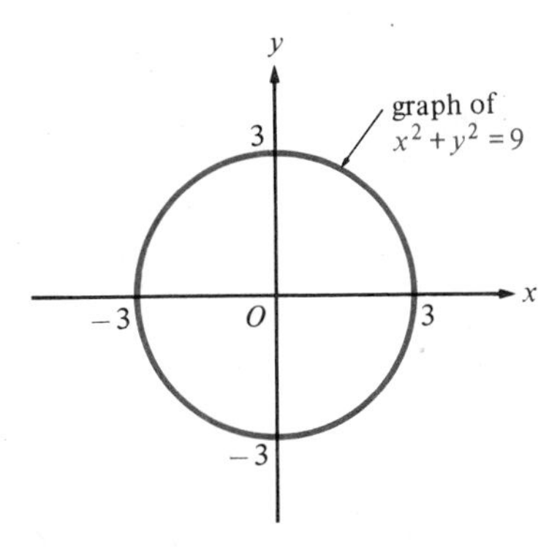

If we are given a curve in the cartesian plane, we can ask whether there is an equation for which it is the graph. Such an equation is called an **equation of the curve** or an **equation for the curve.** For instance, $x^2 + y^2 = 9$ is an equation for the circle in Figure 11. Two equations are said to be **equivalent** if they have the same graph. For example, the equation $\sqrt{x^2 + y^2} = 3$ is equivalent to the equation $x^2 + y^2 = 9$. Sometimes it is convenient to use an equation for a curve to designate the curve; for instance, if we speak of "the circle $x^2 + y^2 = 9$," we mean the circle whose equation is $x^2 + y^2 = 9$.

If $r > 0$, the circle (in the cartesian plane) of radius r with center (h, k) consists of all points $P = (x, y)$ such that the distance between (x, y) and (h, k) is r units (Figure 12). Using the distance formula, we can write an equation for this circle as

$$\sqrt{(x - h)^2 + (y - k)^2} = r,$$

or equivalently as

$$\boxed{(x - h)^2 + (y - k)^2 = r^2.}$$

The last equation is called the **standard form** for the equation of a circle in the xy plane.

Figure 12

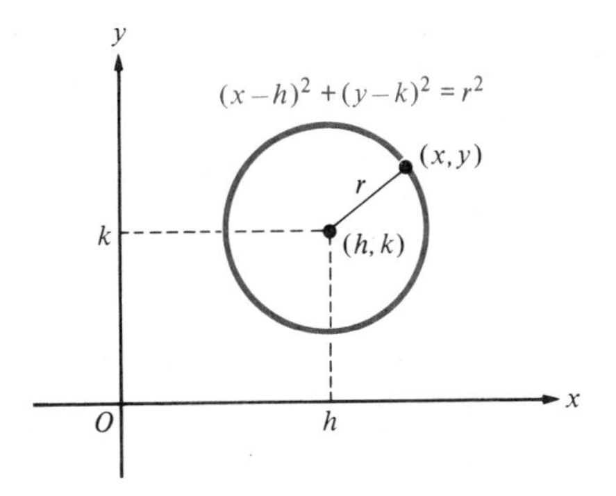

Example 1 Find an equation for the circle of radius 5 with center $(3, -2)$.

Solution Here $r = 5$ and $(h, k) = (3, -2)$, so the equation in standard form is

$$(x - 3)^2 + [y - (-2)]^2 = 5^2 \quad \text{or} \quad (x - 3)^2 + (y + 2)^2 = 25.$$

If desired, we can expand the squares, combine like terms, and rewrite the equation in the equivalent form

$$x^2 + y^2 - 6x + 4y - 12 = 0.$$

Example 2 Sketch the graph of the equation

$$(x - 1)^2 + (y + 3)^2 = 16.$$

Solution The equation has the form

$$(x - h)^2 + (y - k)^2 = r^2$$

with $h = 1$, $k = -3$, and $r = 4$, so its graph is a circle of radius $r = 4$ with center $(h, k) = (1, -3)$ (Figure 13).

Figure 13

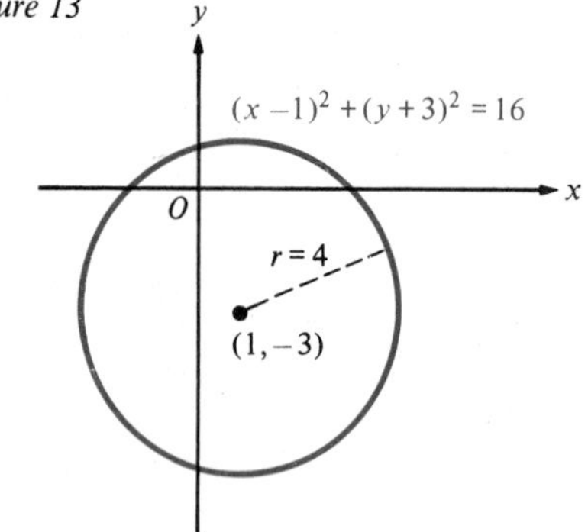

We study circles (and closely related curves called *ellipses*) in more detail in Section 4 of Chapter 5.

If k is a real number, the graph of the equation

$$y = k$$

is the horizontal line consisting of all points $P = (x, k)$ with ordinate k (Figure 14a). Likewise, if h is a real number, the graph of the equation

$$x = h$$

is the vertical line consisting of all points $P = (h, y)$ with abscissa h (Figure 14b). The graph of the equation

$$y = x$$

is the line consisting of all points with the same abscissa and ordinate; it contains the origin $(0, 0)$ and makes a $45°$ angle with the x axis (Figure 14c). We study straight lines and their equations in more detail in Section 3 of Chapter 5.

Figure 14

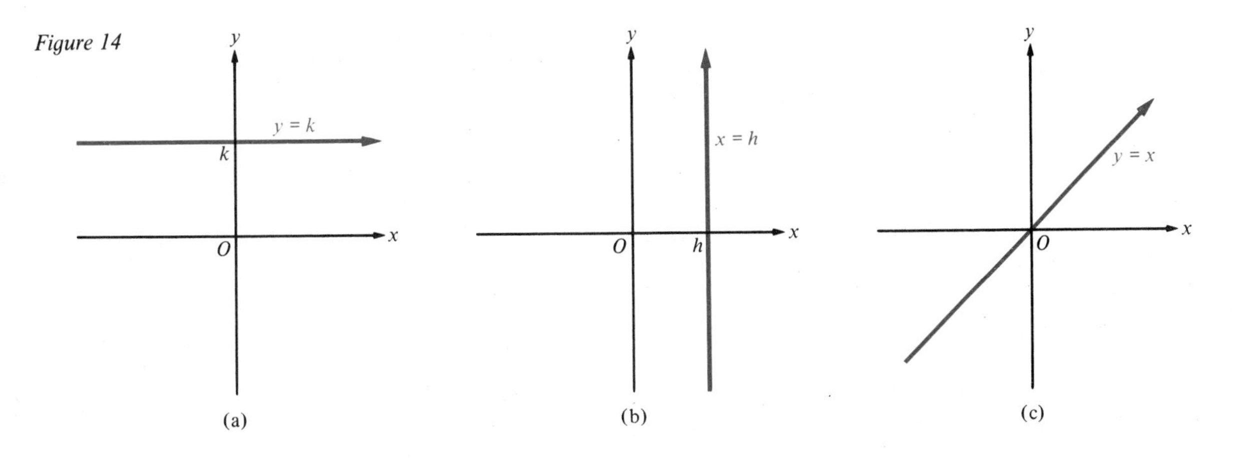

PROBLEM SET 4

1. If one triangle has vertex angles of 20° and 70°, and if a second triangle is a right triangle, one of whose vertex angles is 20°, show that the two triangles are similar.
2. Draw a figure to illustrate the fact that the alternate interior angles formed by a transversal intersecting two parallel lines are equal. (If you have forgotten the meaning of the words, look them up in a plane geometry book or a dictionary.)
3. Draw a figure to illustrate the fact that the vertical angles formed by two lines meeting at a point are equal.
4. If two triangles have corresponding side lengths that are proportional, must they be similar? (No proof is required here; just use your intuition.)
5. If two angles of a first triangle and the length of the side included between them are equal, respectively, to two angles of a second triangle and the length of the side included between them, show that the two triangles are congruent. [Hint: Begin by showing that the two triangles are similar, then use the fact that corresponding side lengths of similar triangles are proportional.]
6. Is it possible to have two triangles ABC and DEF with the angle at vertex B equal to the angle at vertex E, $|\overline{AB}| = |\overline{DE}|$, and $|\overline{AC}| = |\overline{DF}|$ without ABC being congruent to DEF? Explain.

In problems 7–11, c denotes the length of the hypotenuse of a right triangle, and a and b denote the lengths of the other two sides. In each case, use the given information to find the indicated quantity. If use of a calculator is indicated, round off your answer to two decimal places.

7. $a = 3, \quad b = 4, \quad c = ?$
8. $a = \sqrt{7}, \quad c = 3\sqrt{2}, \quad b = ?$
9. [C] $b = 1, \quad c = 1.25, \quad a = ?$
10. [C] $a = 7.07, \quad b = 8.46, \quad c = ?$
11. [C] $a = 41, \quad c = 52, \quad b = ?$

12. [C] René has purchased a secondhand bookcase whose outside dimensions are as follows: height = 84.25 inches, width = 48 inches, depth = 9.5 inches. He intends to lay the bookcase on its side and slide it through his study door. Then he plans to lay it on its back on the floor and swing it into an upright position. The ceiling of René's study is 84.75 inches high. Can he carry out his plans? Explain.

In problems 13–16, suppose that the three sides of a triangle have lengths a, b, and c as indicated. Is the triangle a right triangle?

13. $a = 6, \quad b = 8, \quad c = 10$
14. [C] $a = 2.5, \quad b = 6, \quad c = 6.5$
15. $a = 4, \quad b = 5, \quad c = 6$
16. [C] $a = 1.7, \quad b = 2.3, \quad c = 3.1$

In problems 17–26, find the distance between the two points in the xy plane. If use of a calculator is indicated, round off your answer to four significant digits.

17. $(7, 10)$ and $(1, 2)$
18. $(-1, 7)$ and $(2, 11)$
19. $(7, -1)$ and $(7, 3)$
20. $(-4, 7)$ and $(0, -8)$
21. $(-6, 3)$ and $(3, -5)$
22. $(0, 4)$ and $(-4, 0)$
23. $(-3, -5)$ and $(-7, -8)$
24. $(-\frac{1}{2}, -\frac{3}{2})$ and $(-3, -\frac{5}{2})$
25. [C] $(-2.714, 7.111)$ and $(3.135, 4.982)$
26. [C] $(\pi, \frac{53}{4})$ and $(-\sqrt{17}, \frac{211}{5})$

In problems 27–30: (a) Plot the points A, B, and C and draw the triangle ABC. (b) Find the distances $|\overline{AB}|$, $|\overline{BC}|$, and $|\overline{CA}|$. (c) Show that ABC is a right triangle. (d) Find the area of triangle ABC.

27. $A = (1, 1)$, $B = (5, 1)$, $C = (5, 7)$
28. $A = (-1, -2)$, $B = (3, -2)$, $C = (-1, -7)$
29. $A = (0, 0)$, $B = (-3, 3)$, $C = (2, 2)$
30. $A = (-2, -5)$, $B = (9, \frac{1}{2})$, $C = (4, \frac{21}{2})$

31. If $A = (-5, 1)$, $B = (-6, 5)$, and $C = (-2, 4)$, determine whether or not triangle ABC is isosceles. (An *isosceles* triangle is one with two equal sides.)
32. Verify that the distance formula in Theorem 1 holds even if the line segment $\overline{P_1P_2}$ is horizontal or vertical.
33. On a cartesian coordinate "grid," an aircraft carrier is detected by radar at point $A = (52, 71)$ and a submarine is detected by sonar at point $S = (47, 83)$. If distances are measured in nautical miles, how far is the carrier from a point on the surface of the water directly over the submarine?
34. Show that the points $A = (-2, -3)$, $B = (3, -1)$, $C = (1, 4)$, and $D = (-4, 2)$ are the vertices of a square.
35. Find an equation of (a) the circle of radius 4 with center at the origin and (b) the circle of radius 2 with center at $(-1, 3)$.
36. Find an equation of the circle with center at (h, k) that is tangent to the x axis.

In problems 37–50, sketch the graph of each equation.

37. $x^2 + y^2 = 36$
38. $(x - 3)^2 + (y - 5)^2 = 49$
39. $(x - 2)^2 + (y + 1)^2 = 64$
40. $x^2 - 4x + y^2 - 10y + 4 = 0$ [Hint: Complete the squares for both $x^2 - 4x$ and $y^2 - 10y$ by adding the appropriate quantities to both sides of the equation.]
41. $y = 5$
42. $x^2 + y^2 = 0$
43. $y = -\frac{7}{2}$
44. $y^2 = 25$
45. $x = 7$
46. $(x - 2)^2 = 0$
47. $x = -\frac{3}{4}$
48. $|x| = 2$
49. $x - y = 0$
50. $y = -x$

5 Functions and Their Graphs

If the numerical value of a variable quantity y depends upon the numerical value of another variable quantity x, so that each value of x determines a unique corresponding value of y, we say that y is a **function** of x. For instance, if y denotes the area of a circle of radius x, then the value of y depends on the value of x according to the formula

$$y = \pi x^2.$$

Therefore, the area of a circle is a function of its radius.

Various letters of the alphabet are used to denote functions—the letters f, g, h, F, G, and H are favorites for this purpose. Thus, we have the following definition.

Definition 1 **Function**

A *function* f is a rule or correspondence that assigns one and only one value of a variable y to each value of a variable x. The variable x, which is called the **independent variable,** can take on any value in a set called the **domain** of f. For each value of x in the domain of f, the corresponding value of y is denoted by $f(x)$, which is read "f of x," so that

$$y = f(x).$$

If f is a function and $y = f(x)$, we call y the **dependent variable,** since its value "depends" on the value of x.

This "dependence" is emphasized by the **mapping notation**

$$f: x \longmapsto y \quad \text{or} \quad x \overset{f}{\longmapsto} y \quad \text{or} \quad x \longmapsto f(x),$$

which conveys the idea that each value of x is "mapped onto" the corresponding value of y by the function f.

Scientific calculators have special keys for some of the more important functions. By entering a number x into the display and touching, say, the $\sqrt{x}$ key, you can obtain a vivid demonstration of the function.

$$x \longmapsto \sqrt{x}.$$

For instance,

$$4 \longmapsto 2$$

$$25 \longmapsto 5$$

$$2 \longmapsto \sqrt{2} \approx 1.414$$

and so forth.

If $f: x \longmapsto y$ is a function, then, for each value of x in the domain of f, we have the equation

$$y = f(x).$$

Conversely, an equation of the form

$$y = \text{an expression involving } x$$

may be used to determine a definite function $f: x \longmapsto y$, and we say that the function f is **defined** or **given by** the equation. For instance, the equation

$$y = 3x^2 - 1$$

defines a function $x \longmapsto f(x)$, so that

$$f(x) = 3x^2 - 1.$$

Thus,

$$f(2) = 3(2)^2 - 1 = 11$$

$$f(0) = 3(0)^2 - 1 = -1$$

$$f(-1) = 3(-1)^2 - 1 = 2$$

and so forth.

Example Let g be the function defined by $g(x) = 5x^2 + 3x$. Find the following.

(a) $g(-2)$ (b) $g(s + t)$ (c) $g(-x)$

Solution

(a) $g(-2) = 5(-2)^2 + 3(-2) = 20 - 6 = 14$

(b) $g(s + t) = 5(s + t)^2 + 3(s + t) = 5s^2 + 10st + 5t^2 + 3s + 3t$

(c) $g(-x) = 5(-x)^2 + 3(-x) = 5x^2 - 3x$

Whenever a function $f : x \longmapsto y$ is defined by an equation, it will always be understood (unless we explicitly say otherwise) that its domain consists of all values of x for which the equation makes sense and determines a unique corresponding real number y.

Examples Find the domain of the function defined by each equation.

1 $h(x) = \dfrac{1}{x - 1}$

Solution Since the fraction is undefined only when $x = 1$ (that is, when the denominator $x - 1$ is zero), the domain of h is the set of all real numbers different from 1. In other words, the domain of h consists of the interval $(-\infty, 1)$ together with the interval $(1, \infty)$.

2 $G(x) = \sqrt{4 - x}$

Solution The expression $\sqrt{4 - x}$ is a real number if and only if $4 - x \geq 0$; that is, if and only if $x \leq 4$. Therefore, the domain of G is the interval $(-\infty, 4]$.

3 $F(x) = 3x - 5$

Solution The expression $3x - 5$ is defined for all real values of x, so the domain of F is the set $\mathbb{R}$ of all real numbers.

In dealing with a function f, it is important to distinguish among the **function**

$$f : x \longmapsto y,$$

which is a rule or correspondence; the **value**

$$y \quad \text{or} \quad f(x),$$

which is a number depending on x; and the **equation**

$$y = f(x),$$

which relates the dependent variable y to the independent variable x. Nevertheless, in the interest of brevity, people take shortcuts and speak, incorrectly, of "the function $f(x)$" or "the function $y = f(x)$." Although we avoid these practices when absolute precision is required, we also indulge in them when it seems convenient and harmless.

The particular letters used to denote the dependent and independent variables are of no importance in themselves—the important thing is the **rule** by which a definite value of the dependent variable is made to correspond to each value of the independent variable. For instance, if the radius of a circle is denoted by r and its area by A, then the function f that assigns to each positive value of r the corresponding value of A is $f(r) = \pi r^2$.

5.1 The Graph of a Function

By the *graph* of a function f, we mean the graph of the corresponding equation $y = f(x)$. In other words, the graph of f is the set of all points (x, y) in the cartesian plane such that x is in the domain of f and $y = f(x)$.

The basic technique for sketching the graph of a function f is the following **point-plotting method:** Select several values of x in the domain of f, calculate the corresponding values of $f(x)$, plot the resulting points $(x, y) = (x, f(x))$, and connect the points with a smooth curve. The more points you plot, the more accurate your sketch will be.

The point-plotting method involves a *guess* about the shape of a graph between or beyond known points, and must therefore be used with some caution. If the function is fairly simple, this method usually works pretty well; however, more complicated functions may require more advanced methods studied in calculus.

Examples Use the point-plotting method to sketch the graph of each function.

1 $f(x) = x^2, \quad x > 0$

Solution We begin by selecting several positive values of x and calculating the corresponding values of $f(x) = x^2$ as in the table in Figure 1. In Figure 1, we have plotted the points $(x, f(x))$ from the table and connected them with a smooth curve. Since the domain of f consists only of positive numbers (because of the condition $x > 0$), the point $(0, 0)$ is excluded from the graph. This excluded point is indicated by a small open dot. (Notice that to obtain a figure with reasonable dimensions, we used different scale units on the x and y axes.)

Figure 1

x	$f(x) = x^2$
1	1
2	4
3	9
4	16
5	25
6	36

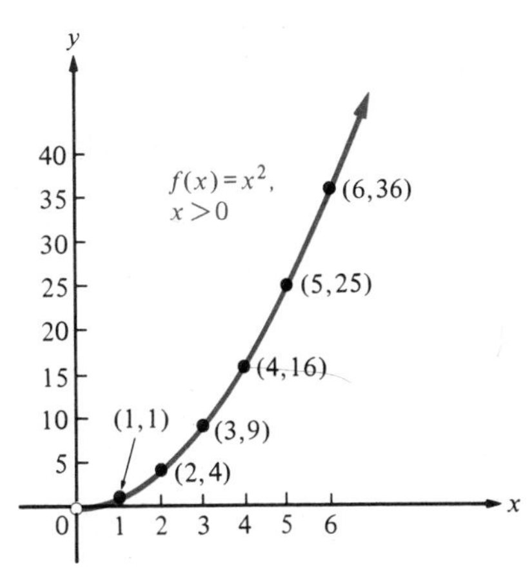

2 $g(x) = 1/x, \quad x > 0$

Solution Again, we select several positive values of x and calculate the corresponding values of $g(x) = 1/x$, as shown in the table in Figure 2. The resulting graph is sketched in Figure 2. Notice that for small positive values of x, the value of $g(x)$ is large, whereas, for large values of x, the value of $g(x)$ is small.

Figure 2

x	$g(x) = \frac{1}{x}$
$\frac{1}{3}$	3
$\frac{1}{2}$	2
1	1
2	$\frac{1}{2}$
3	$\frac{1}{3}$

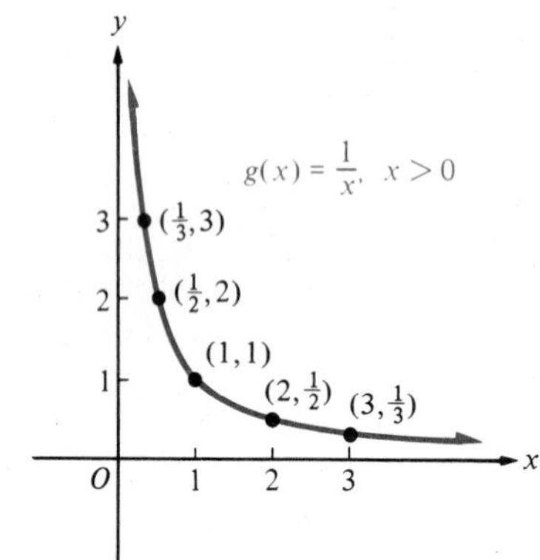

The graph in Figure 1 is always **rising** as we move to the right, a geometric indication that the function f is **increasing** in the sense that, as x increases, so does the value of $f(x)$. On the other hand, the graph in Figure 2 is always **falling** as we move to the right, indicating that the function g is **decreasing** in the sense that, as x increases, the value of $g(x)$ decreases. In Figure 3, the graph neither rises nor falls, indicating that h is a **constant function** whose values $h(x)$ do not change as x increases.

Figure 3

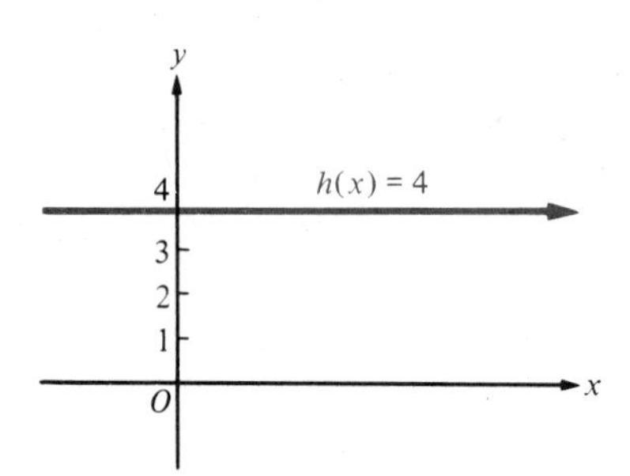

As Figure 4a illustrates, the **domain** of a function f is the set of all abscissas (x coordinates) of points on its graph. The set of all ordinates (y coordinates) of points on the graph of a function f is called the **range** of f (Figure 4b). Notice that the range of f is the set of all values taken on by $y = f(x)$ as x runs through the domain of f.

Figure 4

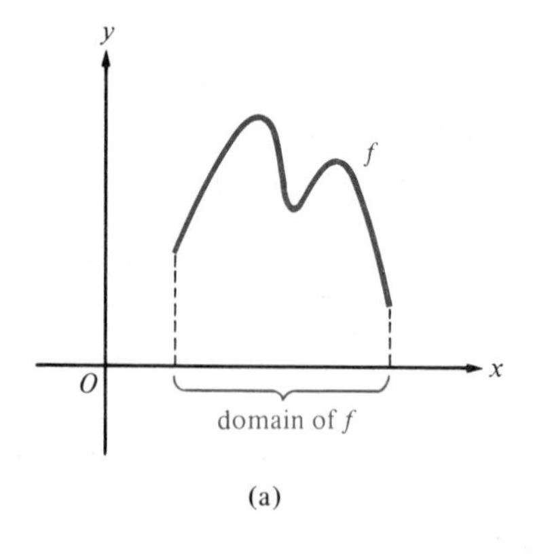

(a)

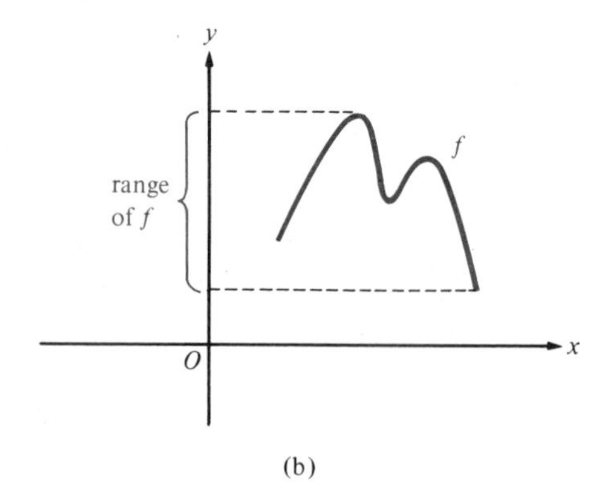

(b)

Example For the function f whose graph is shown in Figure 5, indicate the intervals on which f is increasing or decreasing and the intervals on which it is constant. Also, find the domain and the range of f.

Figure 5

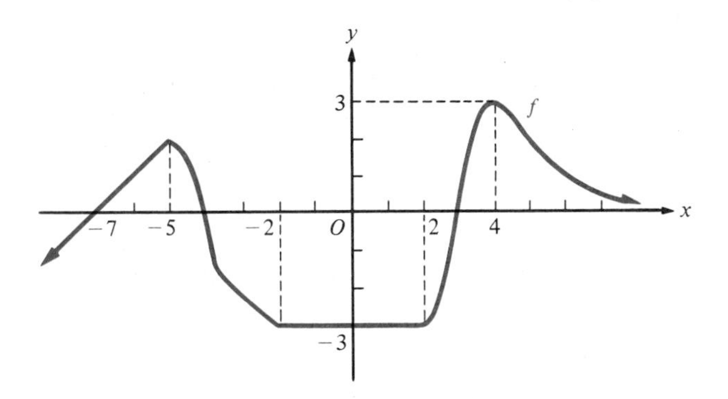

Solution Assuming that the graph continues indefinitely to the left and right as suggested by the arrowheads, the function f is increasing on the intervals $(-\infty, -5]$ and $[2, 4]$, whereas it is decreasing on the intervals $[-5, -2]$ and $[4, \infty)$. On the interval $[-2, 2]$, the function f is constant. Because the graph "covers" the entire x axis, the domain of f is $\mathbb{R}$. The graph (apparently) drops arbitrarily low as we move to the left of -5 on the x axis, but it never climbs any higher than $y = 3$. Therefore, the range of f is the interval $(-\infty, 3]$.

A function f is said to be **even** when for every number x in the domain of f, $-x$ is also in the domain of f and

$$f(-x) = f(x).$$

A function f is even if and only if its graph is **symmetric about the y axis** (Figure 6a).

A function g is said to be **odd** when, for every number x in the domain of g, $-x$ is also in the domain of g and

$$g(-x) = -g(x).$$

A function g is odd if and only if its graph is **symmetric about the origin** (Figure 6b). Of course, there are many functions that are neither even nor odd.

Figure 6

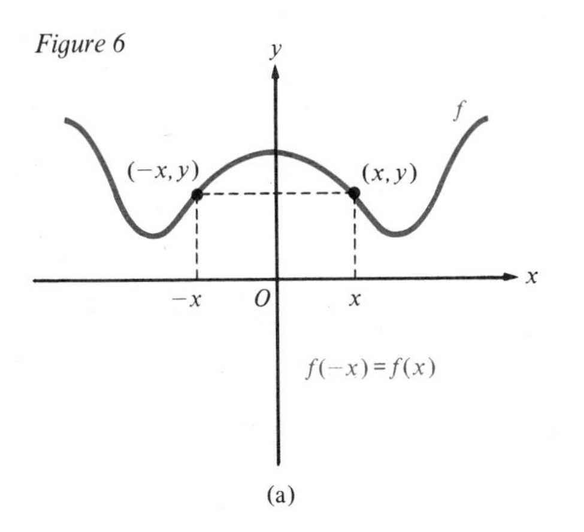

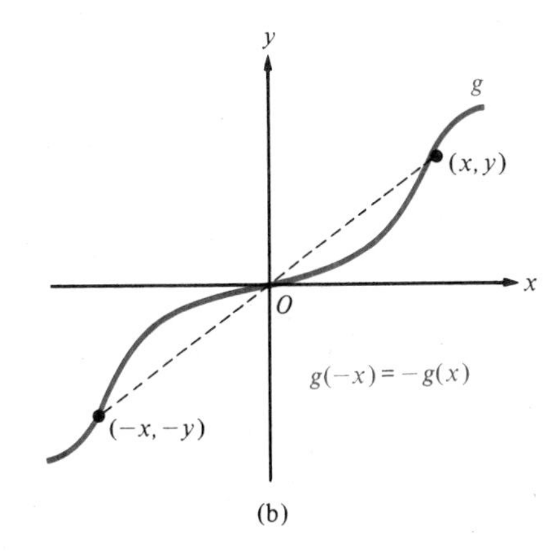

If a function is found to be either even or odd, the job of sketching its graph becomes much easier because of the symmetry involved.

Examples Sketch the graph of each function.

1 $F(x) = x^2$

Solution The domain of F is the set $\mathbb{R}$ of all real numbers. We have

$$F(-x) = (-x)^2 = x^2 = F(x),$$

so F is an even function and its graph is symmetric about the y axis. We have already sketched the portion of this graph for $x > 0$ in Figure 1 on page 33. The full graph includes the mirror image of Figure 1 on the other side of the y axis, as well as the point $(0, 0)$ (Figure 7).

Figure 7

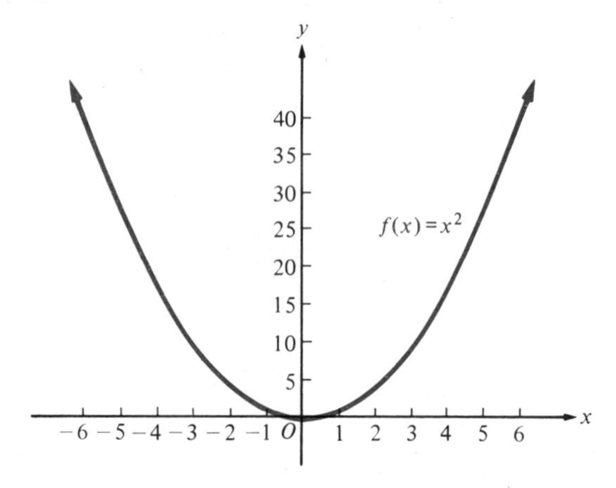

2 $G(x) = \dfrac{1}{x}$

Solution The domain of G is the set of all nonzero real numbers. We have, for $x \neq 0$,

$$G(-x) = \frac{1}{-x} = -\frac{1}{x} = -G(x),$$

so G is an odd function and its graph is symmetric about the origin. Using this fact and the graph of $g(x) = 1/x$ for $x > 0$ already sketched in Figure 2 on page 34, we obtain the graph in Figure 8.

Figure 8

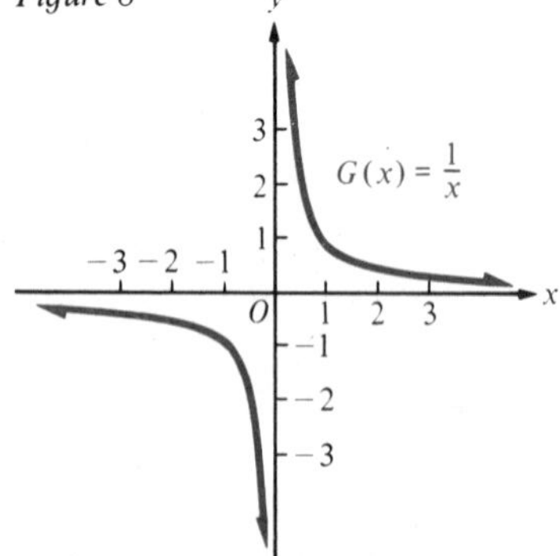

Because the graph of $G(x) = 1/x$ comes closer and closer to the coordinate axes, as shown in Figure 8, the x and y axes are called **asymptotes** of the graph of G. More generally, if the distance between a point P on a curve and a straight line L approaches 0 as P moves farther and farther away from the origin, we say that L is an **asymptote** of the curve (Figure 9).

Figure 9

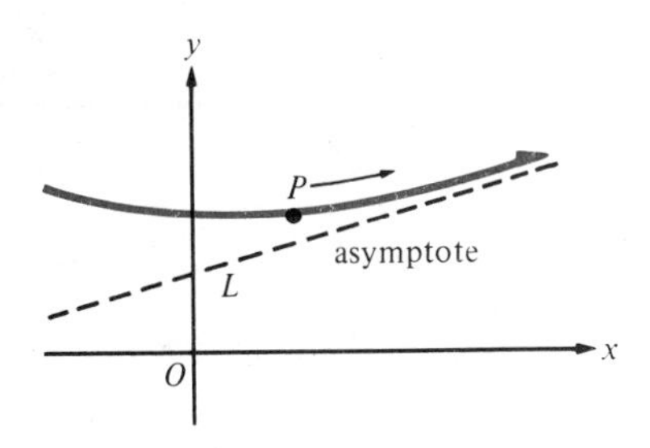

Not every curve in the cartesian plane is the graph of a function. Indeed, the definition of a function (Definition 1) requires that there be one and *only one* value of y corresponding to each value of x in the domain. Therefore, it isn't possible to have two points (x, y_1) and (x, y_2) with the same x coordinate and different y coordinates on the graph of a function. Hence, we have the following test.

Vertical Line Test

> A set of points in the cartesian plane is the graph of a function if and only if no vertical straight line intersects the set more than once.

Example Which of the curves in Figure 10 is the graph of a function?

Solution By the vertical line test, the curve in Figure 10a is the graph of a function, but the curve in Figure 10b is not.

Figure 10

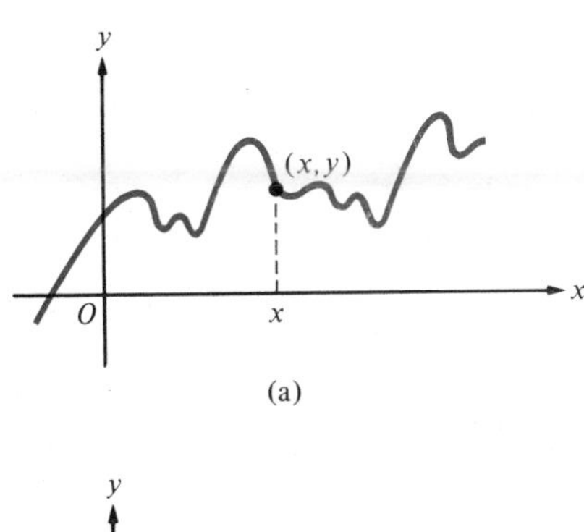

(a)

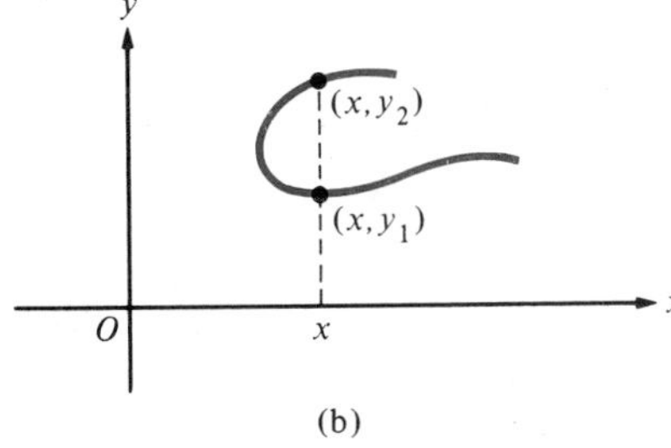

(b)

PROBLEM SET 5

In problems 1–28, let

$f(x) = 2x + 1$ $\qquad g(x) = x^2 - 3x - 4$ $\qquad h(x) = \dfrac{1}{x}$

$F(x) = \dfrac{x - 2}{3x - 7}$ $\qquad G(x) = |2 - 5x|$ $\qquad H(x) = \sqrt{3x + 5}$

Find the indicated values.

1. $f(-3)$
2. $g(0)$
3. $H(-1/3)$
4. Ⓒ $g(4.718)$
5. $F(-1/3)$
6. $[G(-1/2)]^2$
7. $g(4)$
8. $\sqrt{f(4)}$
9. $\sqrt{h(1)}$
10. $f(1/a)$
11. $g(-x)$
12. $F(a/3)$
13. $h(1/a)$
14. $H(2x - 1)$
15. $F(-x)$
16. $G(2/5)$
17. $H(x^4)$
18. $h(-x)$
19. $f(x/2)$
20. $g(x + t)$
21. $f[(x - 1)/2]$
22. $G(1) - G(0)$
23. $g(a) - g(b)$
24. $g(a) - g(-a)$
25. $H(x) - H(0)$
26. $F(x + t) - F(x)$
27. $[f(x) - f(0)]/x$
28. $f[g(x)]$

In problems 29–38, find the domain of each function.

29. $f(x) = 1 - 4x^2$
30. $g(x) = (x + 2)^{-1}$
31. $h(x) = \sqrt{x}$
32. $F(x) = \dfrac{1}{\sqrt{4 - 5x}}$
33. $G(x) = \sqrt{9 - x^2}$
34. $H(x) = \sqrt{9 + x^2}$
35. $f(x) = \dfrac{1}{x + |x|}$
36. $g(x) = \dfrac{x^3 - 8}{x^2 - 4}$
37. $h(x) = \dfrac{-3}{x^3}$
38. The function $r \mapsto A$ that gives the area A of a circle of radius r.

In problems 39–46, (a) determine the domain of the function; (b) determine whether the function is even, odd, or neither even nor odd; (c) sketch the graph of the function; (d) indicate the intervals on which the function is increasing, decreasing, or constant; and (e) find the range of the function.

39. $f(x) = x$
40. $g(x) = 1/x^2$
41. $h(x) = x^3$
42. $F(x) = \sqrt{9 - x^2}$
43. $G(x) = 2$
44. $H(x) = \sqrt{|x|}$
45. $f(x) = \sqrt{x}$
46. $g(x) = \dfrac{x - 1}{x + 1}$

47. For each function in Figure 11, find the domain and range; indicate the intervals on which the function is increasing, decreasing, or constant; and determine whether the function is even, odd, or neither even nor odd.

Figure 11

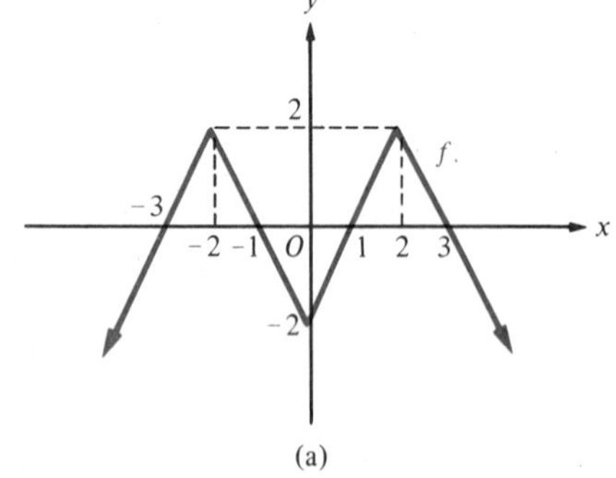

(a)

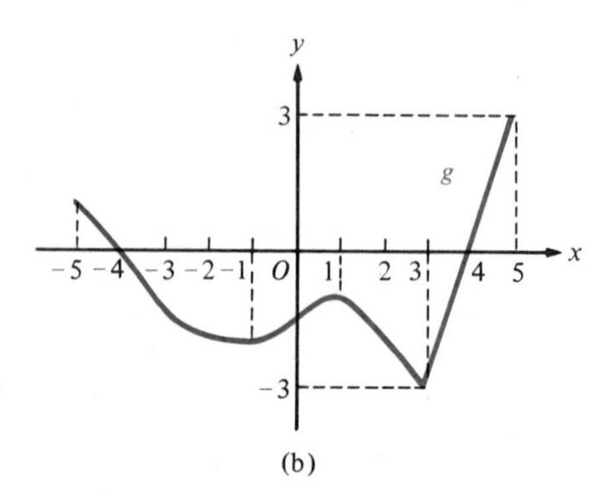

(b)

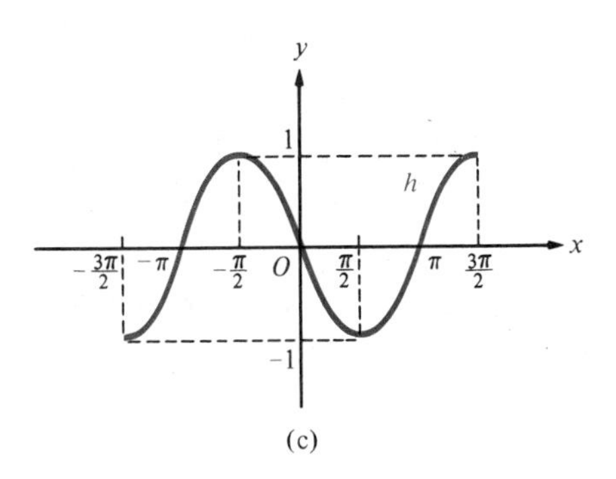

(c)

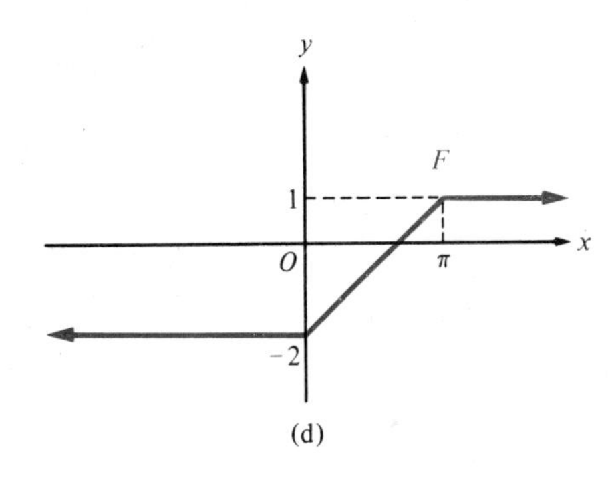

(d)

48. Is the function $f(x) = \sqrt{x}$ even, odd, or neither? Explain.

In problems 49–54, determine whether each function is even, odd, or neither, and discuss the symmetry of its graph.

49. $f(x) = x^4 + 3x^2$

50. $g(x) = x^2 + |x|$

51. $h(x) = 5x^2 - x^3$

52. $F(x) = \dfrac{x^3}{x^4 + 1}$

53. $G(x) = x|x|$

54. $H(x) = \sqrt{8x^4 + 1}$

55. Use the vertical line test to determine which of the curves in Figure 12 are graphs of functions.

Figure 12

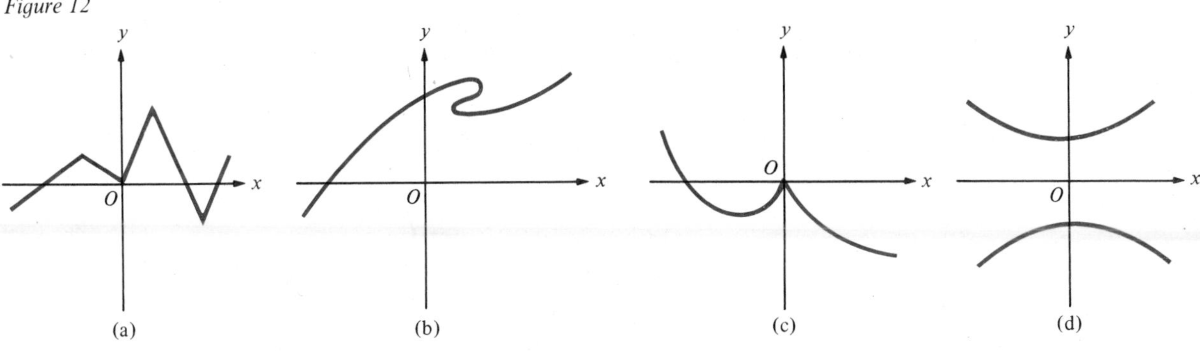

6 Techniques of Graphing

In Section 5.1, we introduced the basic point-plotting method for graph sketching. In this section, we introduce some additional techniques of graphing, such as shifting, stretching, reflecting, and adding or subtracting ordinates.

6.1 Shifting, Stretching, and Reflecting

If f is a function and k is a constant, then the graph of the function F defined by

$$F(x) = f(x) + k$$

is obtained by **shifting** the graph of f **vertically** by $|k|$ units, **upward** if $k > 0$ and **downward** if $k < 0$ (Figure 1). Do you see why?

Figure 1

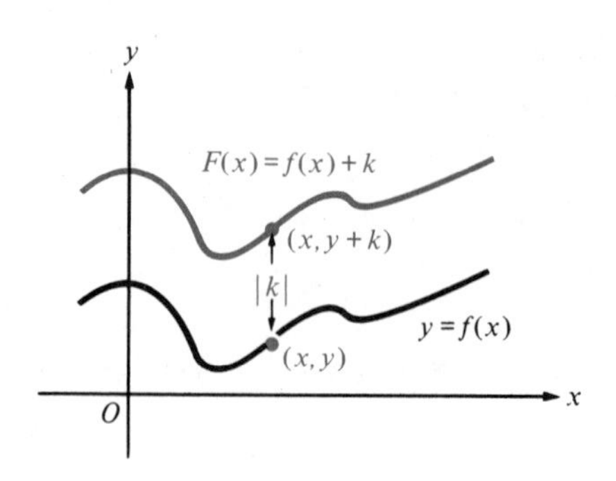

Example Sketch the graphs of

$$F(x) = x^2 + 20 \quad \text{and} \quad G(x) = x^2 - 30.$$

Solution The graph of $y = x^2$ was sketched in Figure 7 of Section 5 (page 36). By shifting this graph upward by 20 units and downward by 30 units, we obtain the graphs of $F(x) = x^2 + 20$ and $G(x) = x^2 - 30$, respectively (Figure 2).

Figure 2

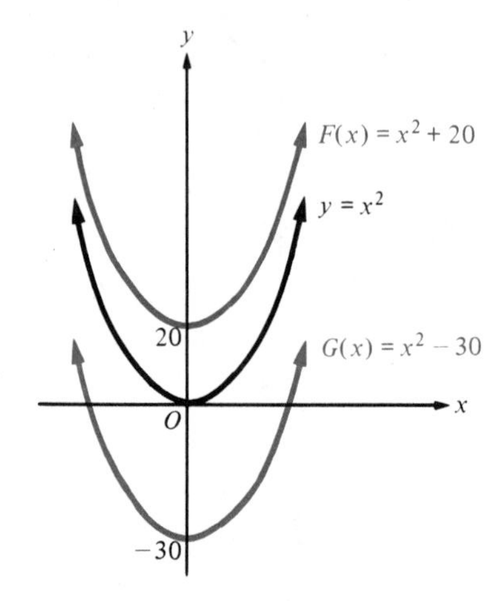

Now, let f be a function, let h be a constant, and define the function F by

$$F(x) = f(x - h).$$

Thus, a point (x, y) belongs to the graph of F if and only if $(x - h, y)$ belongs to the graph of f (Figure 3). Therefore, the graph of $F(x) = f(x - h)$ is obtained by **shifting** the graph of f **horizontally** by $|h|$ units, to the **right** if $h > 0$ and to the **left** if $h < 0$.

Figure 3

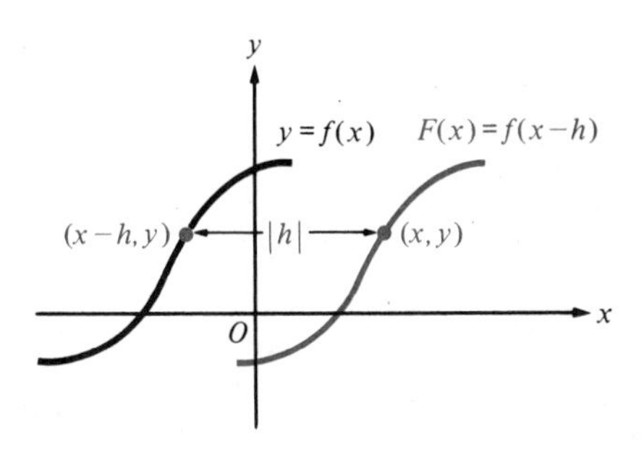

Example Sketch the graph of

$$F(x) = (x - 3)^2.$$

Solution By shifting the graph of $f(x) = x^2$ three units to the right, we obtain the graph of $F(x) = (x - 3)^2$ (Figure 4).

Figure 4

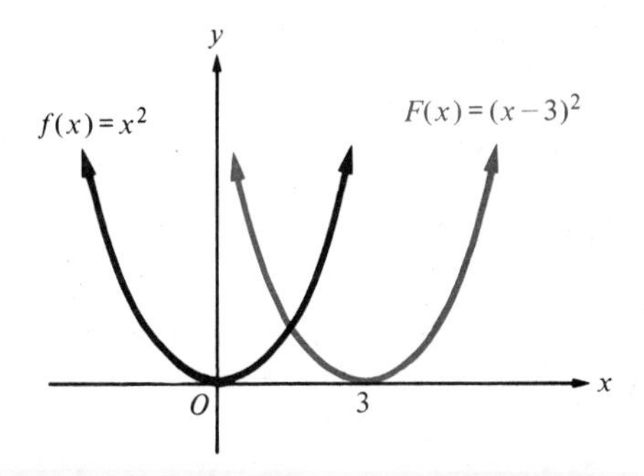

Of course, we can combine horizontal and vertical shifts.

Theorem 1 **Graph Shifting**

> If h and k are constants and f is a function, then the graph of
>
> $$F(x) = f(x - h) + k$$
>
> is obtained by shifting the graph of f horizontally by $|h|$ units and vertically by $|k|$ units. The horizontal shift is to the right if $h > 0$ and to the left if $h < 0$. The vertical shift is upward if $k > 0$ and downward if $k < 0$.

Example Sketch the graph of

$$F(x) = (x - 3)^2 - 30.$$

Solution By Theorem 1, the graph of $F(x) = (x - 3)^2 - 30$ is obtained by shifting the graph of $f(x) = x^2$ three units to the right and thirty units downward (Figure 5). Notice, for instance, that the "lowest point" (0,0) on the graph of $f(x) = x^2$ shifts 3 units to the right and 30 units downward to become the "lowest point" on the graph of F.

Figure 5

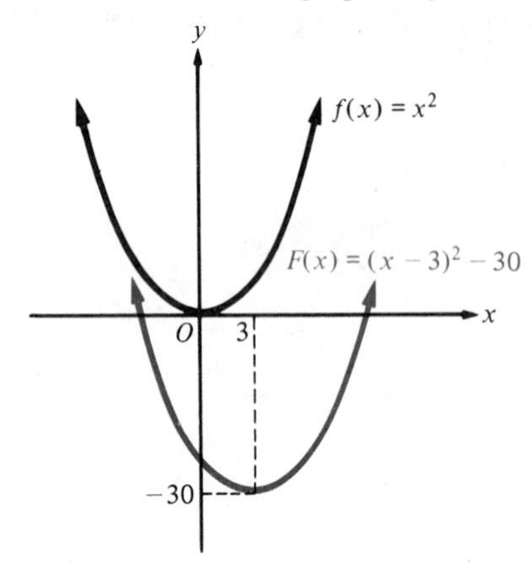

If f is a function and a is a positive constant, the graph of the function F defined by

$$F(x) = af(x)$$

is obtained from the graph of f by multiplying each ordinate by a. If $a > 1$, the result is to "stretch" the graph vertically by a factor of a. If $0 < a < 1$, the result is to "flatten" the graph.

Example 1 Sketch the graph of

$$F(x) = 3x^2.$$

Solution We begin with the graph of $f(x) = x^2$, and we triple the ordinates of points on this graph to obtain points on the graph of $F(x) = 3x^2$ (Figure 6).

Figure 6

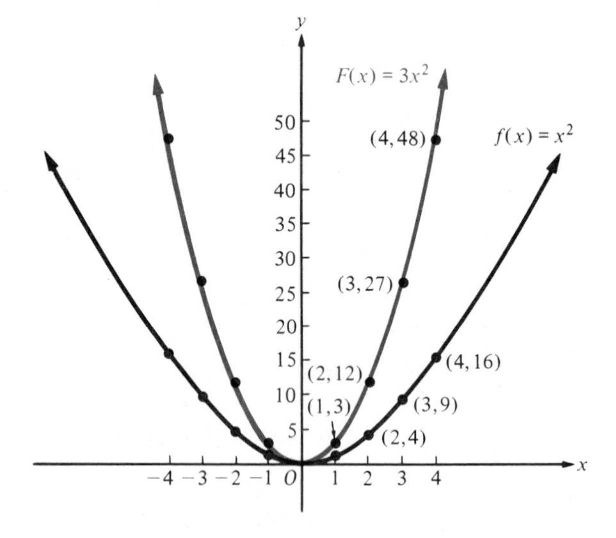

Example 2 Given the graph of f in Figure 7a, sketch the graph of $F(x) = \frac{1}{2}f(x)$.

Solution We select a number of points on the graph of f, take $\frac{1}{2}$ of the ordinate of each point, and thus obtain a number of points on the graph of F. Connecting these points with a smooth curve, we obtain the graph of $F(x) = \frac{1}{2}f(x)$ (Figure 7b).

Figure 7

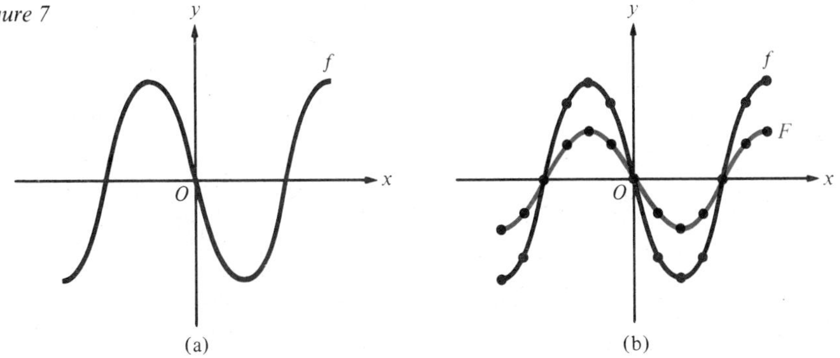

(a) (b)

If f is a function, then the graph of the function F defined by

$$F(x) = -f(x)$$

is obtained by reflecting the graph of f across the x axis. If you imagine that the graph of f is drawn with wet ink, then the graph of F would be the imprint obtained by folding the paper along the x axis.

Example Sketch the graph of

$$F(x) = -3x^2.$$

Solution The graph of $y = 3x^2$ was sketched in Figure 6. By reflecting this graph across the x axis, we obtain the graph of $F(x) = -3x^2$ (Figure 8). Notice that each point on the graph of F is obtained by multiplying the ordinate of the corresponding point on the graph of $y = 3x^2$ by -1.

Figure 8

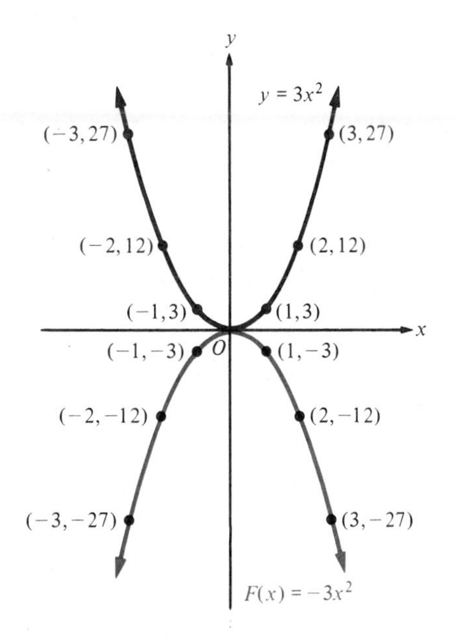

If h, k, and a are constants, $a \neq 0$, and f is a function, then the graph of the function F defined by

$$F(x) = af(x - h) + k$$

is obtained from the graph of f by multiplying each ordinate by a, shifting horizontally by $|h|$ units, and shifting vertically by $|k|$ units. Of course, if $a < 0$, a reflection across the x axis is involved.

A particularly important example is obtained when $f(x) = x^2$. Then,

$$F(x) = a(x - h)^2 + k,$$

and the graph of F has the same general shape as the graph of $y = x^2$, but (possibly) stretched, shifted, and, if $a < 0$, turned upside down (Figure 9). These curves are called **parabolas.** If $a > 0$, the parabola opens upward and has a lowest point, called its **vertex,** at (h, k) (Figure 9a). If $a < 0$, the parabola opens downward and has a highest point, also called its **vertex,** at (h, k) (Figure 9b). In either case, the parabola is symmetric about the vertical line $x = h$ through its vertex. This vertical line is called the **axis** of the parabola.

Figure 9

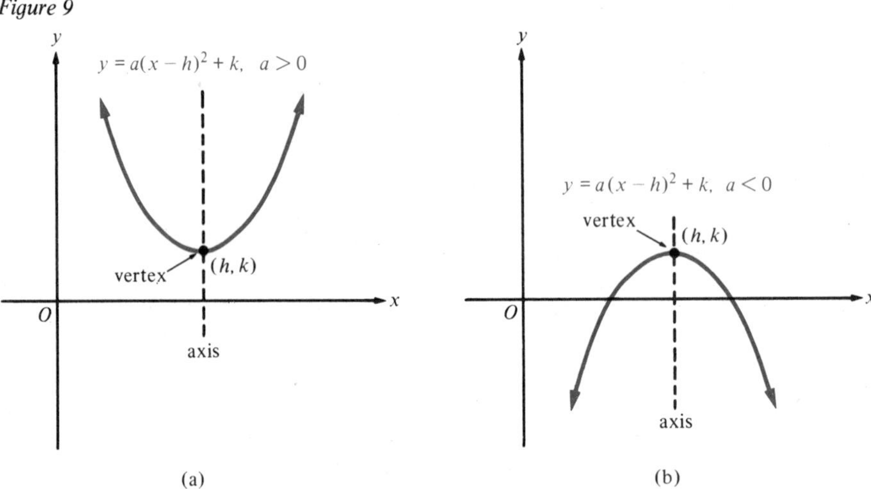

Example Sketch the graph of

$$g(x) = 2(x - 3)^2 - 5.$$

Solution The function g has the form

$$g(x) = a(x - h)^2 + k$$

with $a = 2$, $h = 3$, and $k = -5$, so its graph is a parabola, opening upward, with vertex $(h, k) = (3, -5)$. This information tells us the general shape of the graph.

By plotting a few points, we obtain a reasonably accurate sketch of the graph (Figure 10).

Figure 10

x	$g(x)$
0	13
1	3
2	−3
3	−5
4	−3
5	3
6	13

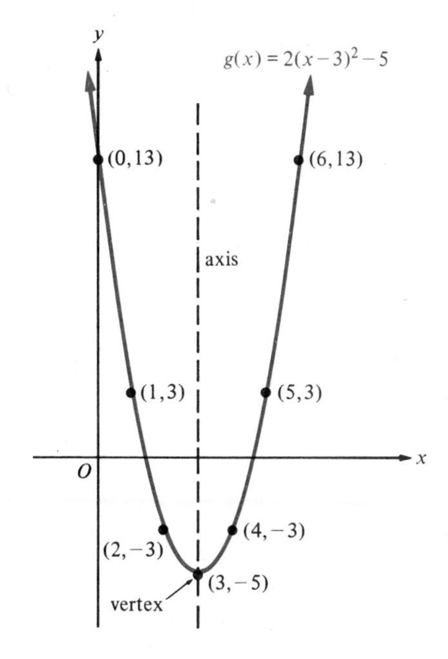

6.2 Adding and Subtracting Ordinates

If $F(x) = f(x) + g(x)$, we naturally refer to F as the **sum** of the functions f and g. Likewise, if $G(x) = f(x) - g(x)$, we call G the **difference** of the functions f and g. The graph of the sum F of two functions f and g is obtained by **adding the ordinates** of the points $(x, f(x))$ and $(x, g(x))$ to obtain the ordinate of the point $(x, F(x))$ on the graph of F (Figure 11a). Likewise, by **subtracting ordinates,** we obtain the graph of the difference G of f and g (Figure 11b).

Figure 11

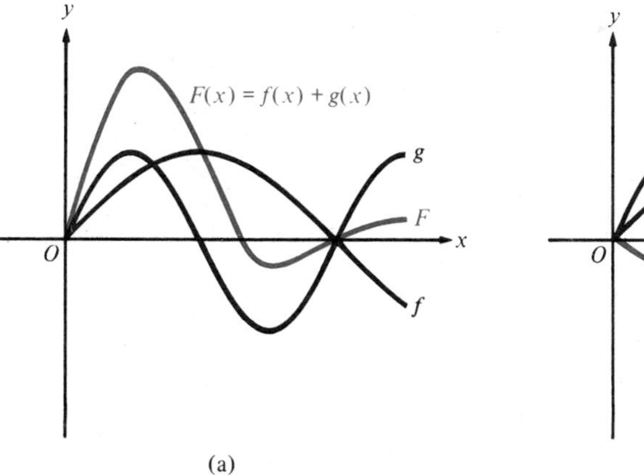

(a)

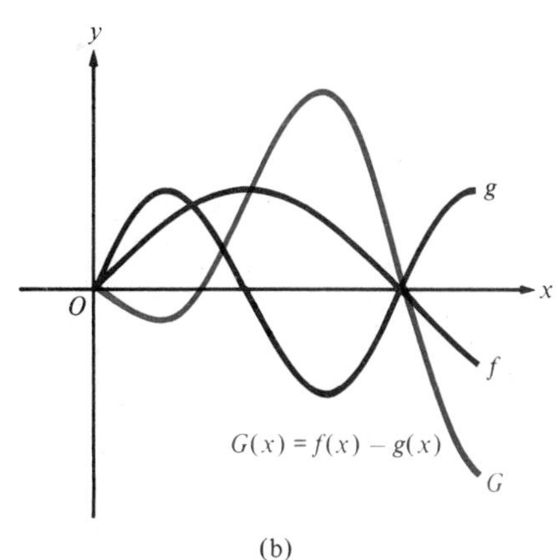

(b)

Example Sketch the graphs of

$$f(x) = x, \quad g(x) = \frac{1}{x}, \quad \text{and} \quad F(x) = x + \frac{1}{x}$$

for $x > 0$ on the same coordinate system. Use the method of adding ordinates to obtain the graph of F.

Solution Let $y_1 = f(x) = x$ and let $y_2 = g(x) = \frac{1}{x}$, so that

$$y_1 + y_2 = f(x) + g(x) = x + \frac{1}{x} = F(x).$$

We calculate values of y_1, y_2, and $y_1 + y_2$ corresponding to selected values of x in the table accompanying Figure 12. Points (x, y_1) belong to the graph of f, points (x, y_2) belong to the graph of g, and points $(x, y_1 + y_2)$ belong to the graph of F. By plotting these points and connecting them with smooth curves, we obtain the graphs of f, g, and F (Figure 12).

Figure 12

x	y_1	y_2	$y_1 + y_2$
$\frac{1}{2}$	$\frac{1}{2}$	2	$\frac{5}{2}$
$\frac{3}{4}$	$\frac{3}{4}$	$\frac{4}{3}$	$\frac{25}{12}$
1	1	1	2
$\frac{3}{2}$	$\frac{3}{2}$	$\frac{2}{3}$	$\frac{13}{6}$
2	2	$\frac{1}{2}$	$\frac{5}{2}$
$\frac{5}{2}$	$\frac{5}{2}$	$\frac{2}{5}$	$\frac{29}{10}$

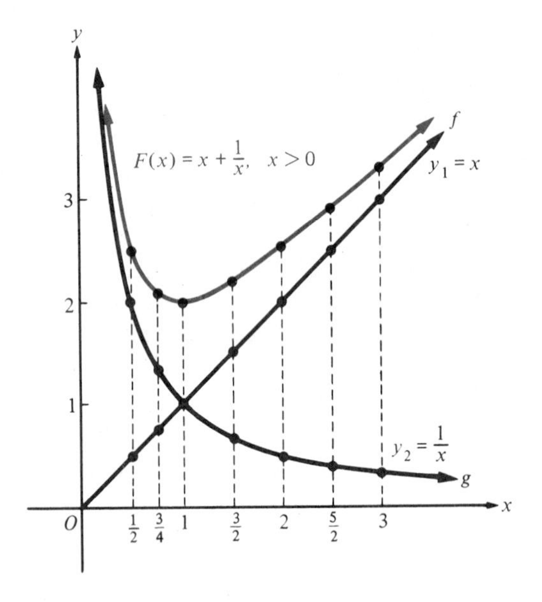

6.3 Reflecting a Graph Across the Line $y = x$

As Figure 13 shows, the points (a, b) and (b, a) are symmetrically located on either side of the line $y = x$. In other words, the "mirror image" of a point across the line $y = x$ is obtained by interchanging the abscissa and the ordinate of the point.

Figure 13

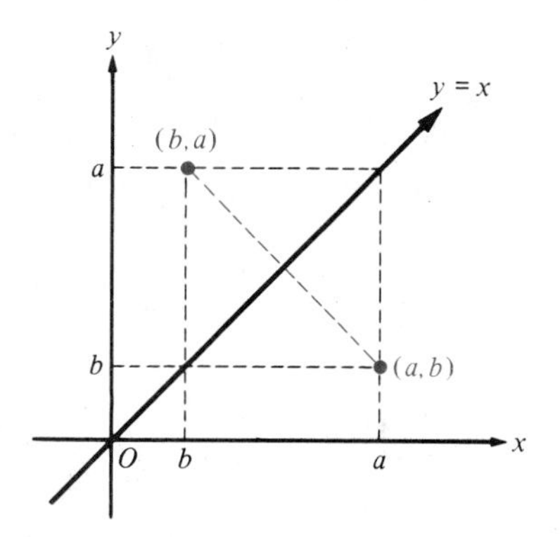

If the curve obtained by reflecting the graph of a function f across the line $y = x$ is the graph of a function g (Figure 14), then we say that f is an **invertible function** and that g is the **inverse** of f. If g is the inverse of f, then (b, a) belongs to the graph of g if and only if (a, b) belongs to the graph of f; that is,

$$g(b) = a \quad \text{if and only if} \quad b = f(a).$$

The inverse g of an invertible function f is often written as f^{-1} (read, "f inverse"), so that

$$f^{-1}(b) = a \quad \text{if and only if} \quad b = f(a).$$

Figure 14

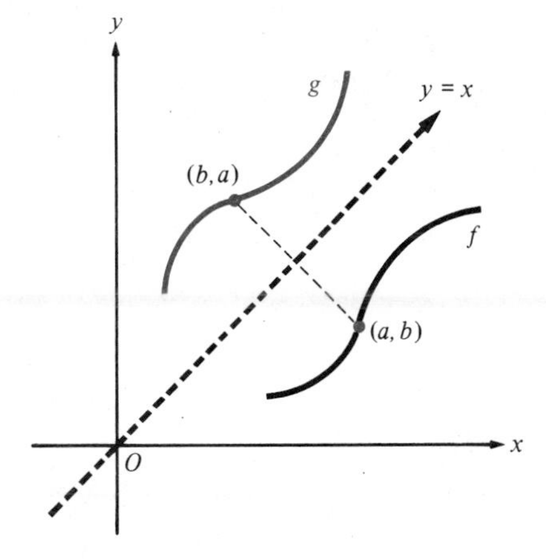

Using the conventional letters x and y for variables, we can rewrite this as

$$f^{-1}(x) = y \quad \text{if and only if} \quad x = f(y).$$

In other words,

$$x \overset{f^{-1}}{\longmapsto} y \quad \text{if and only if} \quad y \overset{f}{\longmapsto} x,$$

so that f^{-1} "undoes" whatever f "does."

Not every function is invertible. Indeed, consider the function f whose graph appears in Figure 15. The mirror image of the graph of f across the line $y = x$ isn't the graph of a function because there is a vertical line ℓ that intersects it more than once. (Recall the vertical line test, page 37).

Figure 15

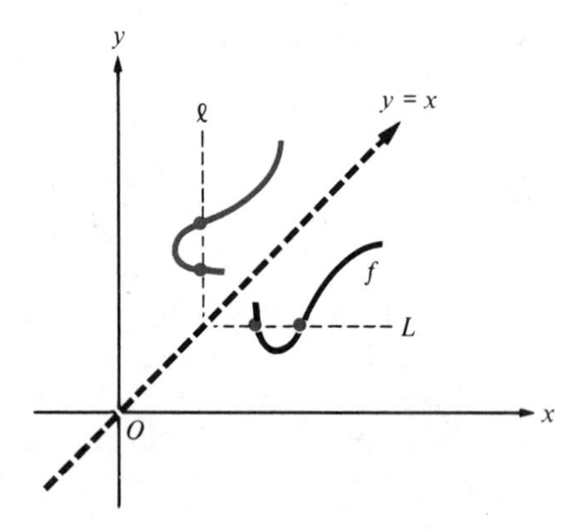

Notice that the horizontal line L obtained by reflecting ℓ across $y = x$ intersects the graph of f more than once. These considerations provide the basis of the following test.

Horizontal Line Test

> A function f is invertible if and only if no horizontal straight line intersects its graph more than once.

Example In Figure 16, use the horizontal line test to determine whether or not the functions are invertible. If an inverse function exists, sketch its graph.

Figure 16

(a)

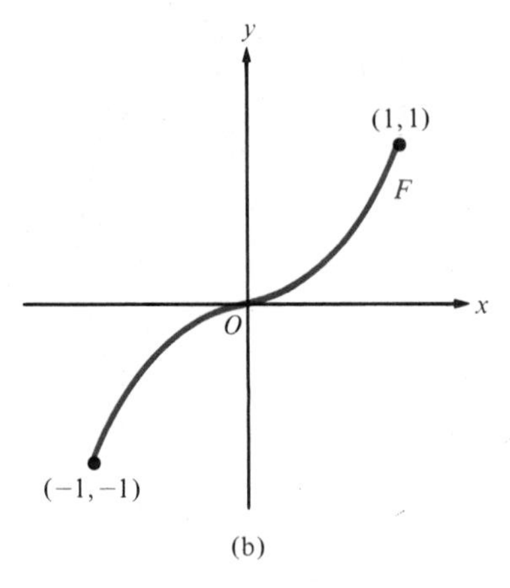

(b)

Solution (a) Any horizontal line drawn above the origin intersects the graph of f twice, so f is not invertible.

(b) No horizontal line intersects the graph of F more than once, so F is invertible. The graph of F^{-1} is obtained by reflecting the graph of F across the line $y = x$ (Figure 17). If we imagine that the graph of F is drawn with wet ink, then the graph of F^{-1} would be the imprint obtained by folding the paper along the line $y = x$.

Figure 17

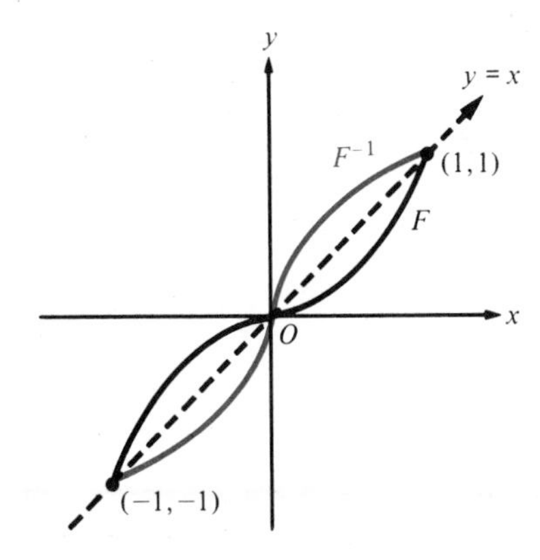

PROBLEM SET 6

1. Sketch graphs of $f(x) = x^2$, $F(x) = x^2 + 10$, and $G(x) = x^2 - 20$ on the same coordinate system.
2. Sketch graphs of $g(x) = 2x^2$, $G(x) = 2x^2 + 15$, and $H(x) = 2x^2 - 10$ on the same coordinate system.

In problems 3–10, sketch the graph of each function.

3. $f(x) = (x - 2)^2$
4. $g(x) = (x + 2)^2 - 2$
5. $h(x) = 2(x + 1)^2$
6. $F(x) = 3(x - 4)^2 + 12$
7. $G(x) = -3(x - 4)^2$
8. $H(x) = -2(x + 2)^2 - 3$
9. $p(x) = -\frac{1}{2}(x + 1)^2 - 4$
10. $q(x) = \frac{1}{2}(x + 1)^2 - \frac{1}{2}$

In problems 11–14, use the given graph of the function f to obtain the graph of:

(a) $F(x) = f(x) + 2$ (b) $G(x) = f(x - 2)$ (c) $H(x) = 2f(x)$ (d) $Q(x) = -f(x)$

11.

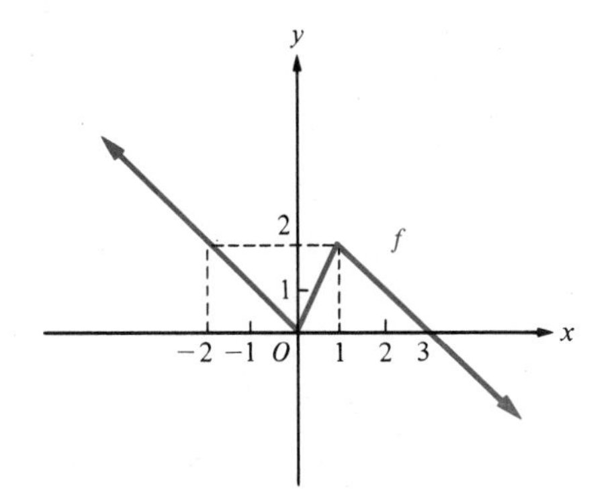

12.

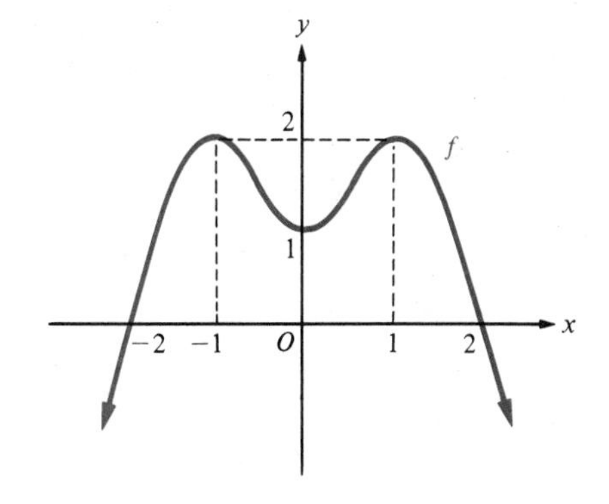

13.

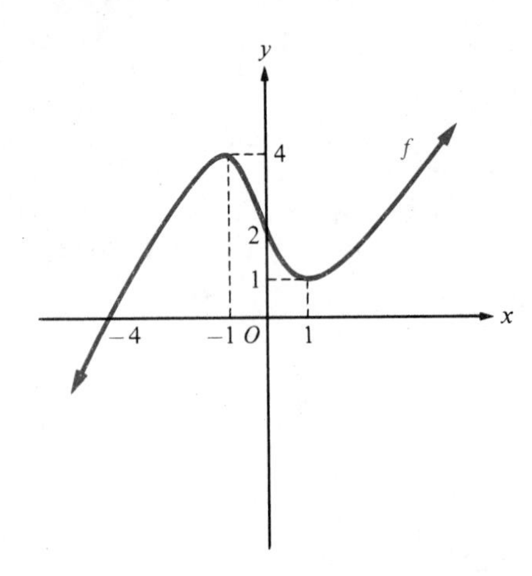

14.

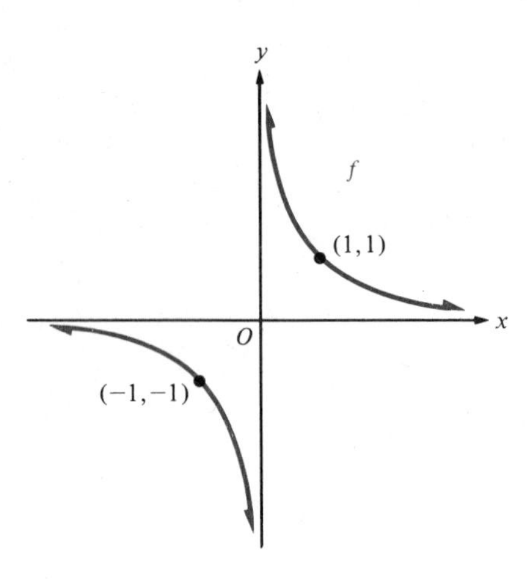

In problems 15–18, sketch graphs of f, g, F, and G, where $F(x) = f(x) + g(x)$ and $G(x) = f(x) - g(x)$, on the same coordinate system. Use the method of adding or subtracting ordinates to obtain the graphs of F and G.

15. $f(x) = x^2$, $g(x) = \frac{1}{x}$, $x > 0$

16. $f(x) = x$, $g(x) = (x - 1)^2$

17. $f(x) = -x^2$, $g(x) = x$

18. $f(x) = x$, $g(x) = \frac{1}{x^2}$, $x > 0$

In problems 19–22, the function whose graph is shown is invertible. In each case, obtain the graph of the inverse function by reflecting the given graph across the line $y = x$.

19.

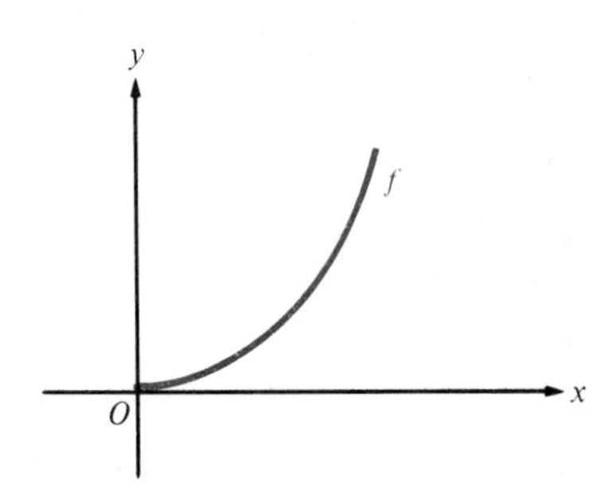

20.

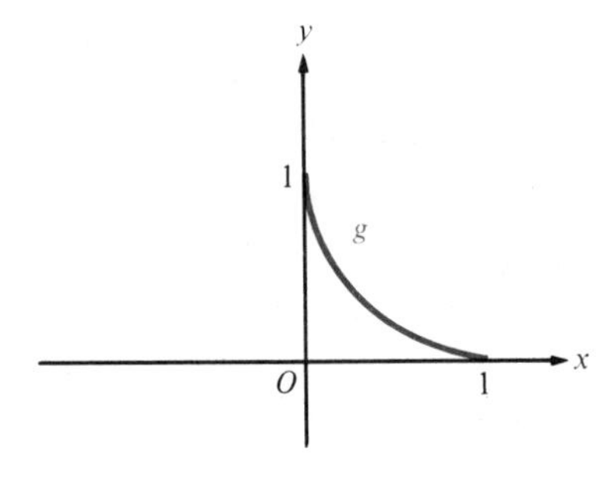

21.

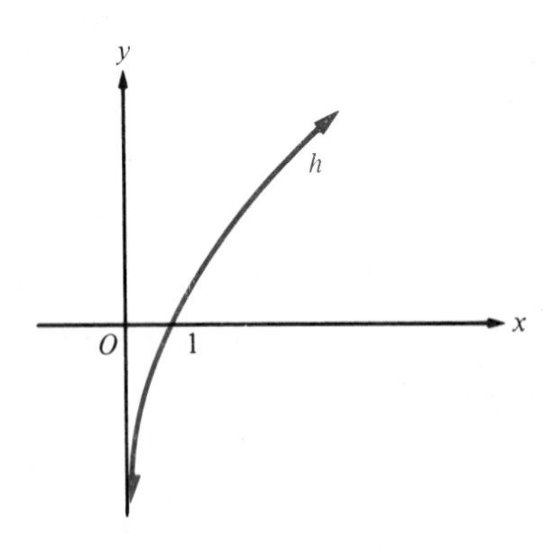

22.

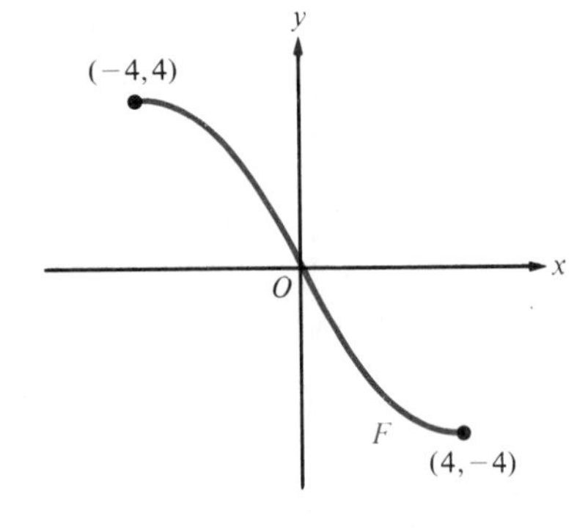

In problems 23–26, use the horizontal line test to determine whether or not the function with the indicated graph is invertible. If it is invertible, sketch the graph of its inverse.

23.

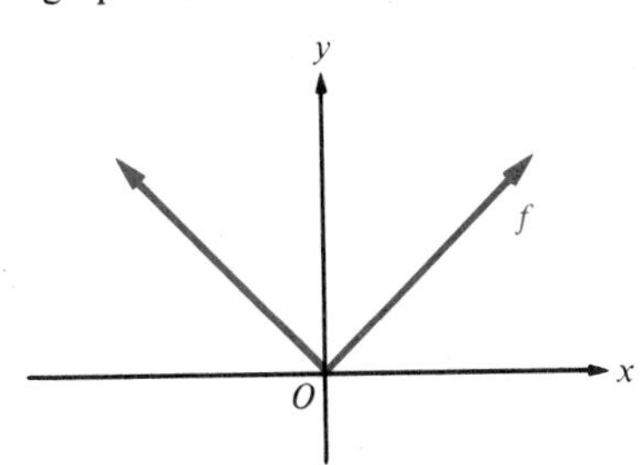

24.

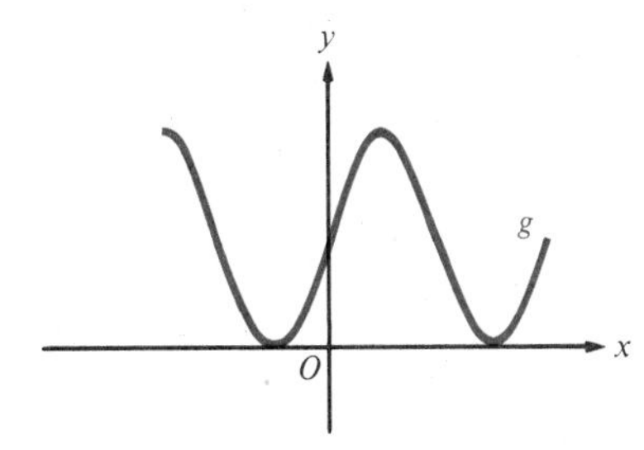

25.

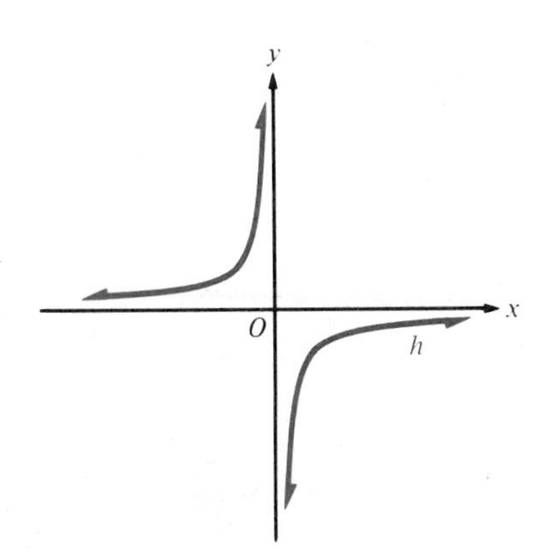

26.

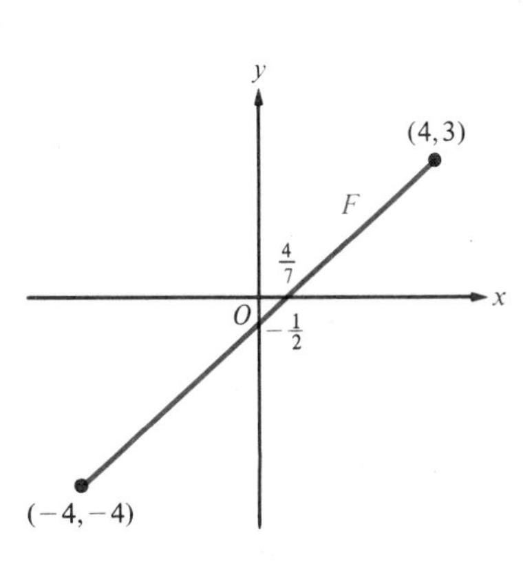

27. Let f be the function defined by $f(x) = x^2$ for $x \geq 0$. Show that $f^{-1}(x) = \sqrt{x}$.

28. Let $f(x) = \dfrac{2x+1}{x-3}$. Find $f^{-1}(x)$.

REVIEW PROBLEM SET

In problems 1–14, indicate whether the statement is true or false.

1. $(3+5)^2 = 3^2 + 5^2$

2. $\dfrac{2}{3} + \dfrac{3}{2} = \dfrac{4+9}{6}$

3. $\sqrt{9+16} = 3+4$

4. $\sqrt{9 \cdot 16} = 3 \cdot 4$

5. $\sqrt{4} = \pm 2$

6. $\dfrac{1}{3+5} = \dfrac{1}{3} + \dfrac{1}{5}$

7. The solutions of the equation $x^2 = 4$ are $x = \pm 2$.

8. If $|x| = a$, where $a \geq 0$, then $x = \pm a$.

9. $\dfrac{2+3}{7+3} = \dfrac{2}{7}$

10. $-(x+y) = -x-y$

11. Every rational number is an integer.

12. ∞ belongs to the interval $(1, \infty)$.

13. 2 belongs to the interval $(-\frac{1}{2}, 2]$.

14. $\frac{2}{0}$ is undefined.

In problems 15–20, rewrite each equation as an equivalent quadratic equation in standard form (if it's not already in that form) and then solve it by factoring. Check for extraneous solutions if necessary.

15. $x^2 - 7x + 12 = 0$ **16.** $7x^2 = 2x$ **17.** $x - 16 = 105/x$

18. $10 + \frac{19}{y} = \frac{15}{y^2}$ **19.** $\frac{2y - 5}{2y + 1} + \frac{6}{2y - 3} = \frac{7}{4}$ **20.** $\frac{t - 4}{t + 1} - \frac{15}{4} = \frac{t + 1}{t - 4}$

In problems 21–24, complete the square by adding a suitable constant to each expression.

21. $x^2 + 9x$ **22.** $u^2 + u$
23. $x^2 - \frac{1}{2}x$ **24.** $t^2 + \sqrt{3}t$

In problems 25–30, use the quadratic formula to solve each quadratic equation.

25. $2x^2 - 5x - 7 = 0$ **26.** $2t^2 - 7t + 5 = 0$ **27.** $5x^2 - 7x = 2$

28. $x^2 + 2x = 5$ **29.** $2r = 1 + \frac{2}{r}$ **30.** $x^2 + \sqrt{30}x = -1$

In problems 31–34, solve each equation. Be sure to check for extraneous roots.

31. $\frac{3 - x}{3x} + \frac{1}{4} = \frac{1}{2x}$ **32.** $\frac{2y}{y + 7} - 1 = \frac{y}{y + 3} + \frac{1}{y^2 + 10y + 21}$

33. $\frac{2}{x - 2} + \frac{1}{x + 1} = \frac{1}{x^2 - x - 2}$ **34.** $\frac{9}{z - 3} - \frac{4}{z - 6} = \frac{18}{z^2 - 9z + 18}$

In problems 35–38, illustrate each set on a number line.

35. (a) $(-1, 3)$ (b) $[-2, 5]$ (c) $[-7, \infty)$ (d) $(-\infty, 4]$
(e) $[-\frac{1}{3}, 5]$ (f) $(-\frac{5}{2}, \frac{3}{2})$ (g) $[-\frac{2}{3}, \frac{1}{3})$ (h) $(-\infty, \sqrt{2})$

36. The set of all real numbers x such that $-3 \leq x \leq 0$ and $3x$ is an integer.

37. The set A of all real numbers that belong to both of the intervals $(-\infty, 5]$ and $(-\frac{2}{3}, 10]$.

38. The set B of all real numbers that belong to at least one of the intervals $(-3, -1]$ and $[1, 3)$.

In problems 39–42, state one Property of Inequalities to justify each statement.

39. Since $-7 < -4$ and $-4 < 4$, it follows that $-7 < 4$.

40. Since $-12 < -5$ and $-6 < 0$, it follows that $2 > \frac{5}{6}$.

41. We know that $\sqrt{3} \neq 1.732$, so it follows that either $\sqrt{3} < 1.732$ or else $\sqrt{3} > 1.732$.

42. Since $2x - 5 < 5$, it follows that $2x < 10$.

In problems 43–46, find the distance between the numbers and illustrate on a number line.

43. -3 and 4 **44.** -3.2 and 4.1 **45.** $\frac{2}{3}$ and $\frac{5}{2}$ **46.** $-\frac{4}{5}$ and $-\frac{5}{4}$

In problems 47–50, determine which equations or inequalities are true for all values of the variables involved.

47. $|x^2 - 4| = |x - 2||x + 2|$ **48.** $|-x| = -|x|$
49. $|x + 1| = |x| + 1$ **50.** $|x + y| \leq |x| + |y|$

In problems 51–54, rewrite each number in scientific notation.

51. 57,120,000,000 **52.** 731 billion **53.** 0.000,000,714 **54.** 33 millionths

In problems 55–58, rewrite each number in ordinary decimal form.

55. 1.732×10^7 **56.** -1.066×10^4 **57.** 3.12×10^{-8} **58.** -3.05×10^{-11}

In problems 59–62, specify the accuracy of the indicated value in terms of significant digits.

59. A tobacco mosaic virus weighs 6.6×10^{-17} gram.

60. The diameter of the star Betelguese is 3.584×10^8 kilometers.

61. In a game of bridge, there is one chance in 1.587534×10^{11} that a player will be dealt a hand in which all cards are of the same suit.

62. In a game of bridge, there is one chance in 2.24×10^{27} that all four players will be dealt hands in which the cards are all of the same suit.

In problems 63–66, round off the given number as indicated.

63. 17,450 to two significant digits.

64. 0.00251 to the nearest thousandth.

65. 7.2283×10^5 to three significant digits.

66. 2.71828 to four significant digits.

67. If one triangle has vertex angles of 30° and 60°, and if a second triangle is a right triangle, one of whose vertex angles is 30°, show that the two triangles are similar.

68. Explain why the sum of the four vertex angles in a parallelogram is always 360°.

Ⓒ **69.** The hypotenuse of a right triangle is 3.57 meters long and one of the sides is 1.23 meters long. Using a calculator, and rounding off your answer to two decimal places, find the length of the remaining side.

70. The three sides of a triangle have lengths 1 meter, 0.8 meter, and 0.6 meter. Is it a right triangle?

In problems 71–74, find the distance between the two points in the cartesian plane.

71. $(1, 1)$ and $(4, 5)$

72. $(-1, 2)$ and $(5, -7)$

73. $(-3, 2)$ and $(2, 14)$

74. $(-\frac{2}{5}, -1)$ and $(-2, 3)$

75. If $A = (-1, 5)$, $B = (8, -7)$, and $C = (4, 1)$, sketch the triangle ABC and find its perimeter.

Ⓒ **76.** If $A = \left(\dfrac{3\pi}{4}, \sqrt{5}\right)$ and $B = \left(-\sqrt{3}, -\dfrac{4}{3}\right)$, find the distance between A and B rounded off to four decimal places.

In problems 77 and 78, find the equation of the circle that satisfies the given conditions.

77. Center at $(-5, 2)$ and radius of 6

78. Tangent to the x axis at $(3, 0)$ and tangent to the y axis at $(0, 3)$

In problems 79–84, sketch the graph of each equation.

79. $(x - 2)^2 + (y + 3)^2 = 25$

80. $(x^2 + 2x + 1) + (y^2 - 2y + 1) = 9$

81. $x^2 + (y - 2)^2 = 36$

82. $x^2 - 2xy + y^2 = 0$

83. $y = -4$

84. $x^2 = y^2$

In problems 85–92, let $f(x) = 3x^2 - 4$, $g(x) = 6 - 5x$, and $h(x) = 1/x$. Find the indicated values.

85. $f(-3)$

86. $h(\frac{1}{2})$

87. $g(\frac{6}{5})$

88. $f(x) - f(2)$

89. $f(\sqrt{3}/3)$

90. $h(x + k) - h(x)$

91. $g\left(\dfrac{1}{4 + k}\right)$

92. $\sqrt{f(-|x|)}$

In problems 93–96, find the domain of each function.

93. $f(x) = \dfrac{1}{x + 1}$

94. $g(x) = \dfrac{1}{\sqrt{4 - x^2}}$

95. $h(x) = \sqrt{1 + x}$

96. $F(x) = \dfrac{1}{|x| - x}$

In problems 97–102, (a) determine the domain of the function; (b) determine whether the function is even, odd, or neither even nor odd; (c) sketch the graph of the function; (d) indicate the intervals on which the function is increasing, decreasing, or constant; and (e) find the range of the function.

97. $f(x) = \dfrac{x^4}{16}$ **98.** $g(x) = \dfrac{1 + x}{1 + x^2}$ **99.** $h(x) = \dfrac{1}{x^3}$

100. $F(x) = x + 1$ **101.** $G(x) = -16$ **102.** $H(x) = 1 - \sqrt{x}$

103. For each function whose graph is shown in Figure 1, find the domain and range; indicate the intervals on which the function is increasing, decreasing, or constant; and determine whether the function is even, odd, or neither.

Figure 1

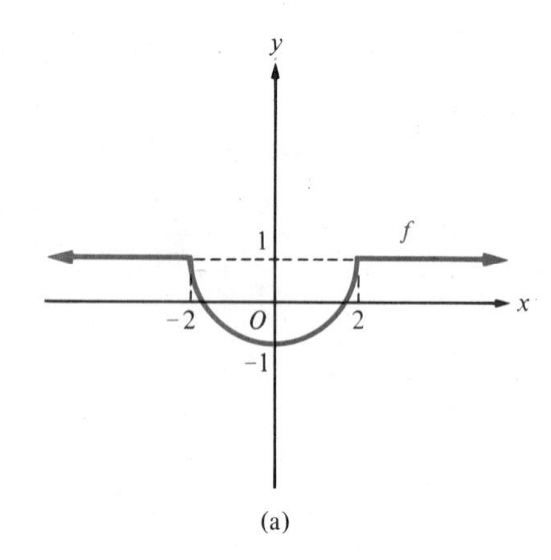

(a)

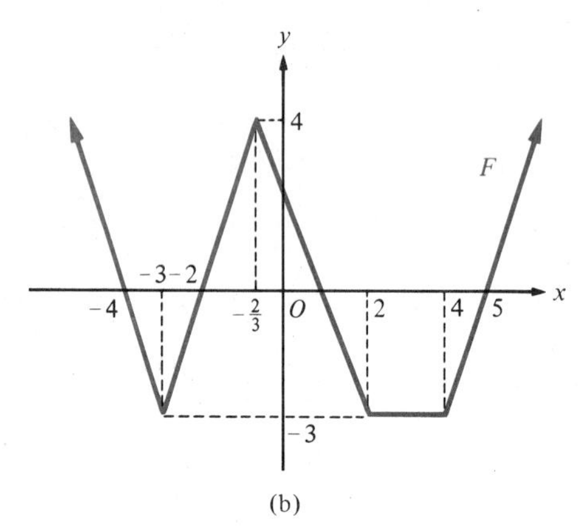

(b)

104. If n is an integer and $f(x) = x^n$, determine the conditions under which f is (a) an even function and (b) an odd function.

105. Determine whether each function is even or odd and discuss the symmetry (if any) of the graph.

(a) $f(x) = 5x^5 + 3x^3 + x$ (b) $g(x) = (x^4 + 1)^{-1}$ (c) $h(x) = (x + 1)/x$

(d) $F(x) = -x^3|x|$ (e) $G(x) = x^{80} - 5x^6 + 9$ (f) $H(x) = \sqrt{x}/(x + 1)$

106. If f is a function with domain $\mathbb{R}$, show that the function $h(x) = \frac{1}{2}[f(x) + f(-x)]$ is even.

107. Which of the curves in Figure 2 are graphs of functions?

Figure 2

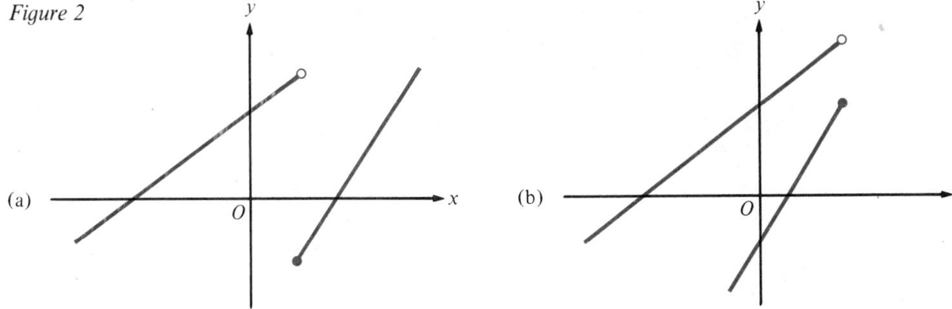

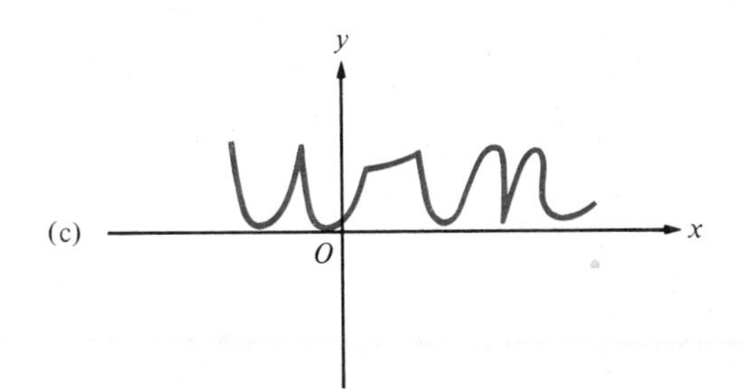

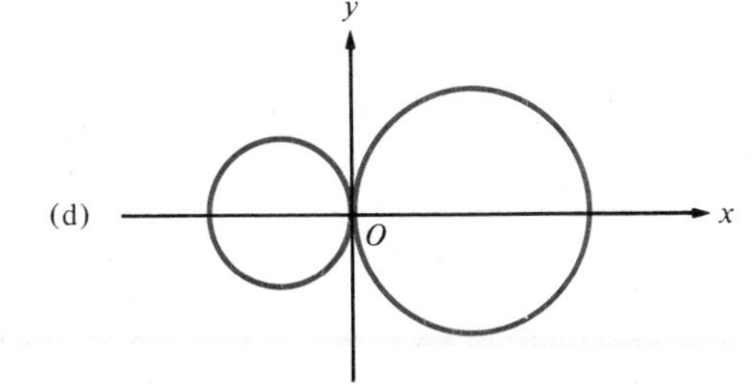

108. The graph of a function f is shown in Figure 3.

Figure 3

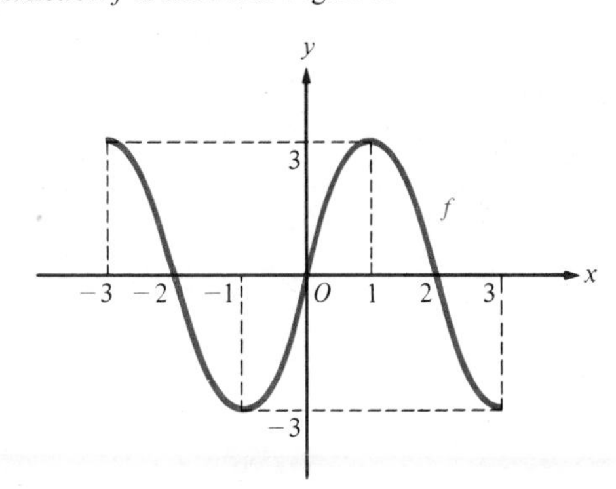

Sketch the graph of the function F defined by:

(a) $F(x) = f(x) + 1$ (b) $F(x) = f(x) - 2$ (c) $F(x) = f(x - 1)$
(d) $F(x) = f(x + 2)$ (e) $F(x) = -f(x)$.

In problems 109–112, compare the graph of the first function with the graph of the second function by sketching them both on the same coordinate system and explaining in words how they are related.

109. $f(x) = x, \quad F(x) = x + 1$

110. $g(x) = x^2, \quad G(x) = (x + 1)^2$

111. $h(x) = x^2, \quad H(x) = \frac{1}{2}(x - 2)^2 + 1$

112. $p(x) = \sqrt{x}, \quad P(x) = 2\sqrt{x}$

In problems 113 and 114, sketch graphs of f, g, F, and G, where $F(x) = f(x) + g(x)$ and $G(x) = f(x) - g(x)$, on the same coordinate system by using the method of adding or subtracting ordinates.

113. $f(x) = (x + 1)^2, \quad g(x) = (x - 1)^2$

114. $f(x) = x^2, \quad g(x) = \dfrac{-1}{x + 1}$

115. Each function whose graph appears in Figure 4 is invertible. In each case, sketch the graph of the inverse function.

Figure 4

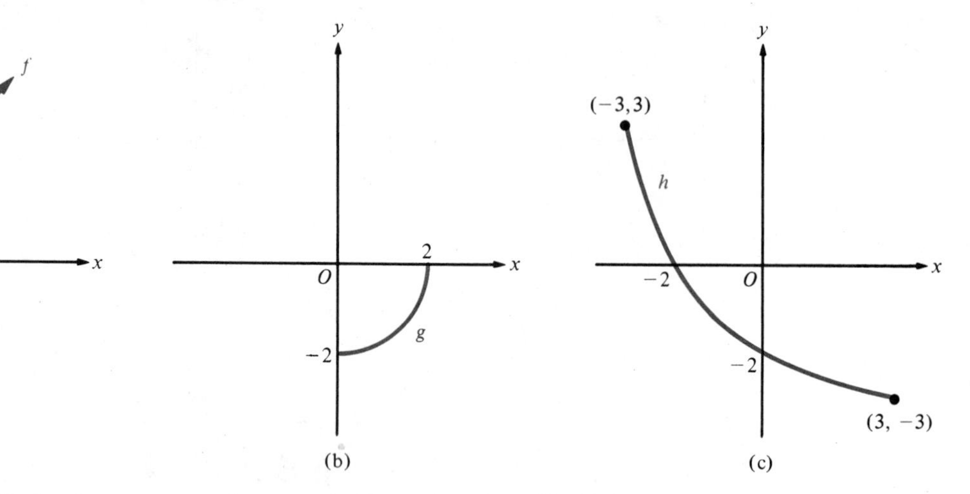

116. Determine which of the functions whose graphs appear in Figure 5 are invertible. If the function is invertible, sketch the graph of its inverse.

Figure 5

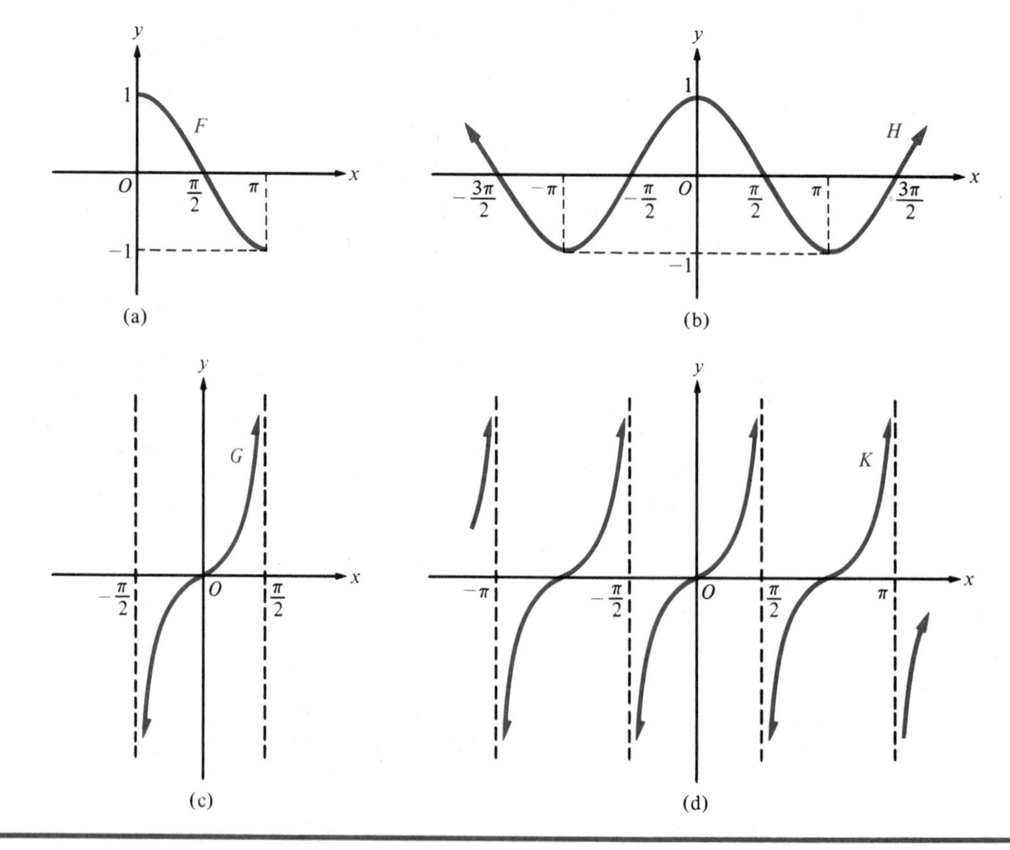

2 Trigonometric Functions

The word "trigonometry" is derived from the Greek words "trigon" for triangle and "metra" for measurement. Trigonometry in the sense of "triangle measurement" is based on the relationship between the vertex angles of a right triangle and ratios of sides of the triangle. The six *trigonometric functions*, which are defined in terms of these ratios, are used routinely in calculations made by surveyors, engineers, and navigators. Trigonometric functions also have applications in the physical and life sciences, where they provide mathematical models for periodic phenomena —for instance, the ebb and flow of the tides, electromagnetic waves, seasonal variations in an animal's food supply, or the alpha rhythms of the human brain. We begin this chapter by considering angles and their measure. We define the trigonometric functions, first for acute angles in a right triangle and then for general angles. We study the evaluation of trigonometric functions, circular functions, graphs of trigonometric functions, and the simple harmonic model.

1 Angles and Their Measure

In order to define the trigonometric functions so that they can be used not only for triangle measurement but also for modeling periodic phenomena, we must give a definition of an angle that is somewhat more general than a vertex angle of a triangle.

If A and B are distinct points, the portion of the straight line that starts at A and continues indefinitely through B is called a **ray** with **endpoint** A (Figure 1a). An **angle** is determined by rotating a ray about its endpoint. This rotation can be indicated by a curved arrow as in Figure 1b. The endpoint of the rotated ray is called the **vertex** of the angle. The position of the ray before the rotation is called the **initial side** of the angle, and the position of the ray after the rotation is called the **terminal side.** Angles will often be denoted by small Greek letters, such as angle α in Figure 1b (α is the Greek letter *alpha*).

Figure 1

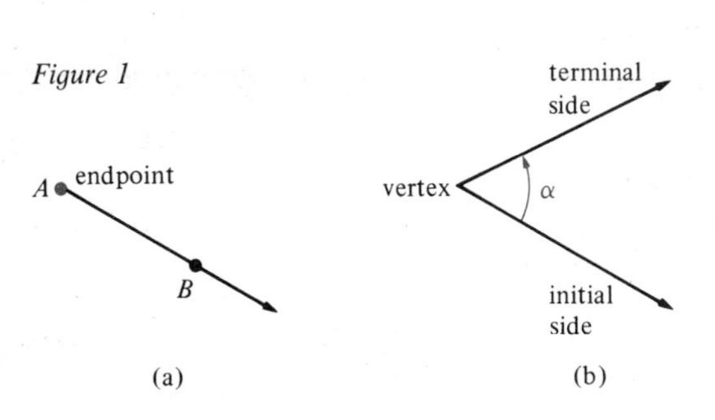

Angles determined by a *counterclockwise* rotation are said to be **positive** and angles determined by a *clockwise* rotation are said to be **negative.** In Figure 2a, angle β is positive (β is the Greek letter *beta*); in Figure 2b, angle γ is negative (γ is the Greek letter *gamma*).

Figure 2

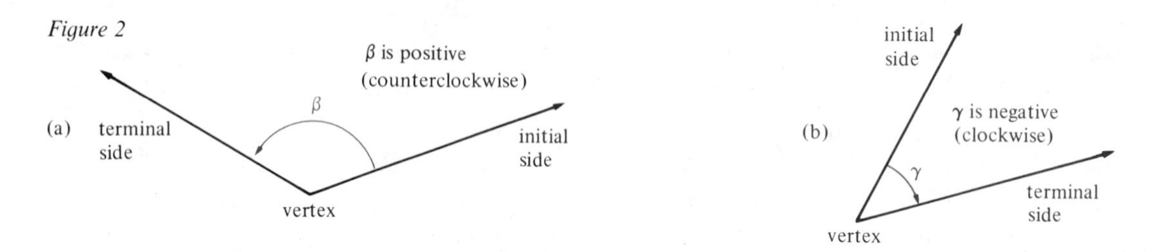

One full turn of a ray about its endpoint is called **one revolution** (Figure 3a); one-half of a revolution is called a **straight angle** (Figure 3b); and one-quarter of a revolution is called a **right angle** (Figure 3c). The initial and terminal sides of a right angle are perpendicular; hence, in drawing a right angle, we often replace the usual curved arrow by two perpendicular line segments (Figure 3c).

Figure 3

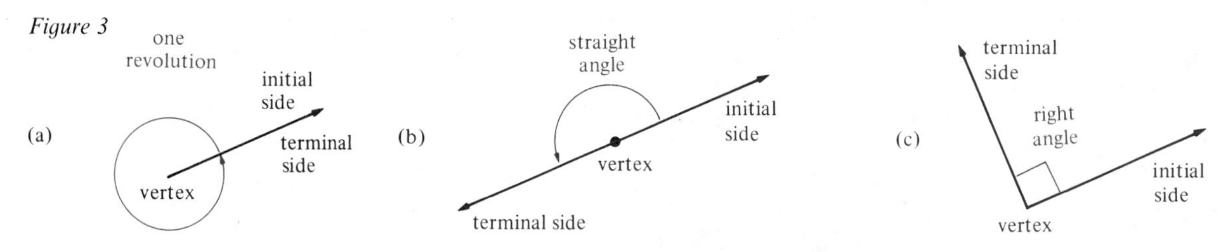

Angles that have the same initial sides and the same terminal sides are called **coterminal** angles. For instance, Figure 4 shows three coterminal angles α, β, and γ. Because different angles can be coterminal, an angle is not completely determined merely by specifying its initial and terminal sides. Nevertheless, angles are often named by using three letters such as ABC to denote a point A on the initial side, the vertex B, and a point C on the terminal side (Figure 5). Usually, the intended angle is the *smallest* rotation about the vertex that carries the initial side around to the terminal side. Sometimes, when it is perfectly clear which angle is intended, we refer to the angle by simply naming its vertex. For instance, the angle in Figure 5 could be called angle B.

Figure 4 *Figure 5*

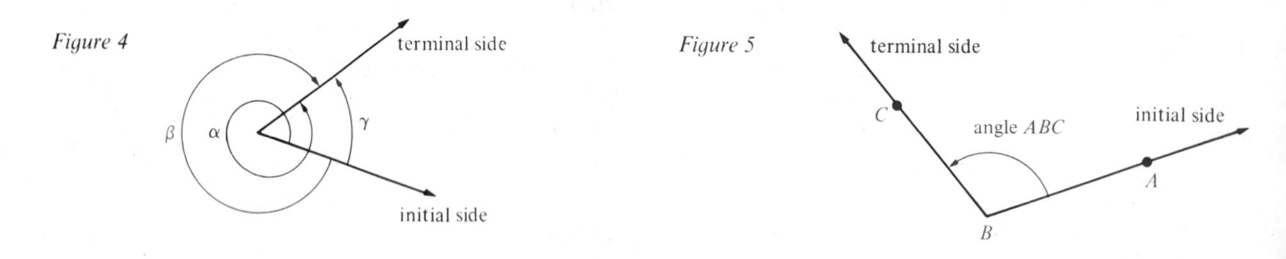

There are various ways to assign numerical measures to angles. Of these, the most familiar is the *degree* measure: **One degree** (1°) is the measure of an angle formed by $\frac{1}{360}$ of a counterclockwise revolution. Negative angles are measured by a negative number of degrees; for instance, -360° is the measure of one clockwise revolution. A positive right angle is $\frac{1}{4}$ of a counterclockwise revolution, and it therefore has a measure of $\frac{1}{4}(360^\circ) = 90^\circ$. Similarly, 180° represents $180^\circ/360^\circ$ or $1/2$ of a counterclockwise revolution.

Example 1 What is the degree measure of one-eighth of a clockwise revolution?

Solution $\frac{1}{8}(-360^\circ) = -45^\circ$.

Example 2 What fraction of a revolution is an angle of 30°?

Solution $30^\circ/360^\circ = 1/12$ of a counterclockwise revolution.

Although fractions of a degree can be expressed as decimals, such fractions are sometimes given in "minutes" and "seconds." **One minute** ($1'$) is defined to be $1/60$ of a degree and **one second** ($1''$) is defined to be $1/60$ of a minute. Hence, one second is $1/3600$ of a degree. Using the relationships

$$1' = \left(\frac{1}{60}\right)^\circ, \quad 1'' = \left(\frac{1}{3600}\right)^\circ, \quad 1^\circ = 60', \quad \text{and} \quad 1' = 60'',$$

you can convert from degrees, minutes, and seconds to decimals and vice versa.

Example 1 Ⓒ Express the angle measure $72^\circ 13' 59''$ as a decimal. Round off your answer to four decimal places.

Solution
$$72^\circ 13' 59'' = 72^\circ + \left(\frac{13}{60}\right)^\circ + \left(\frac{59}{3600}\right)^\circ \approx 72.2331^\circ$$

Example 2 Express the angle measure 173.372° in degrees, minutes, and seconds. Round off to the nearest second.

Solution
$$0.372^\circ = 0.372(1^\circ) = 0.372(60') = 22.32'$$

and
$$0.32' = 0.32(1') = 0.32(60'') = 19.2'' \approx 19''.$$

Hence,
$$173.372^\circ = 173^\circ + 0.372^\circ = 173^\circ + 22.32' = 173^\circ + 22' + 0.32'$$
$$\approx 173^\circ + 22' + 19'' = 173^\circ 22' 19''.$$

Because calculators work with angles expressed in decimal form, minutes and seconds are not as popular as they once were.

Figure 6

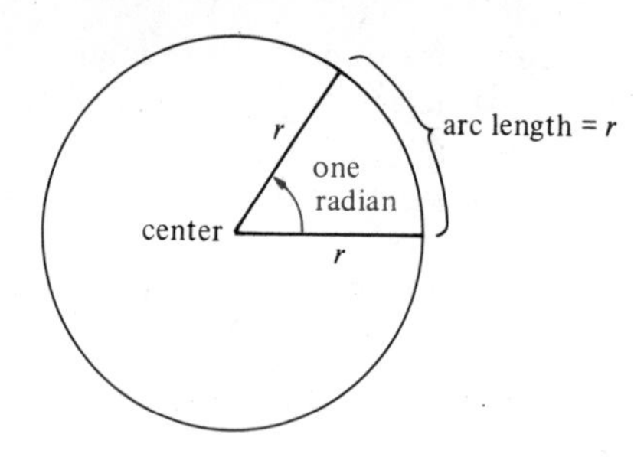

Although the degree measure of angles is used in most elementary applications of trigonometry, more advanced applications (especially those that involve calculus) require radian measure. **One radian** is the measure of an angle that has its vertex at the center of a circle (that is, a **central angle**) and intercepts an arc on the circle equal in length to the radius r (Figure 6).

A central angle of 2 radians in a circle of radius r intercepts an arc of length $2r$ on the circle, a central angle of $\frac{3}{4}$ radian intercepts an arc of length $\frac{3}{4}r$, and so forth. More generally, if a central angle AOB of θ radians (θ is the Greek letter *theta*) intercepts an arc $\widehat{AB}$ of length s on a circle of radius r (Figure 7), then we have

Figure 7

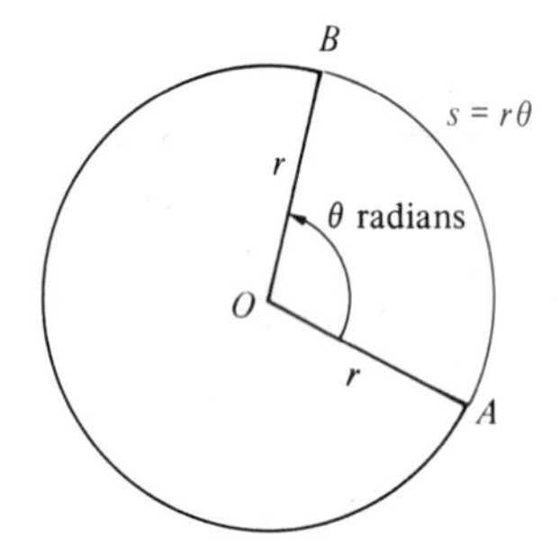

$$s = r\theta.$$

It follows that angle AOB has radian measure θ given by the formula

$$\theta = \frac{s}{r}.$$

Example 1 Find the length s of the arc intercepted on a circle of radius $r = 3$ meters by a central angle whose measure is $\theta = 4.75$ radians.

Solution $s = r\theta = 3(4.75) = 14.25$ meters.

Example 2 A central angle in a circle of radius 27 inches intercepts an arc of length 9 inches. Find the measure θ of the angle in radians.

Solution Here $s = 9$ inches, $r = 27$ inches, and

$$\theta = \frac{s}{r} = \frac{9}{27} = \frac{1}{3} \text{ radian.}$$

Figure 8

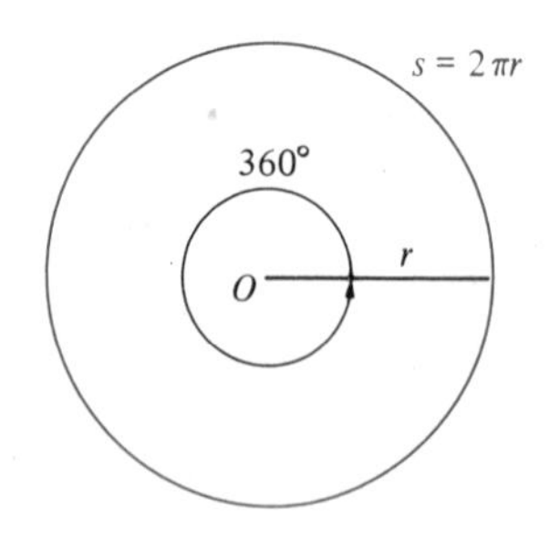

A central angle of 360° corresponds to one revolution; hence it intercepts an arc $s = 2\pi r$ equal to the entire circumference of the circle (Figure 8). Therefore, if θ is the radian measure of the 360° angle,

$$\theta = \frac{s}{r} = \frac{2\pi r}{r} = 2\pi \text{ radians;}$$

that is, $360^\circ = 2\pi$ radians, or

$$180^\circ = \pi \text{ radians.}$$

You can use this relationship to convert degrees into radians and vice versa. In particular,

$$1^\circ = \frac{\pi}{180} \text{ radian} \quad \text{and} \quad 1 \text{ radian} = \left(\frac{180}{\pi}\right)^\circ.$$

Thus,

> multiply degrees by $\frac{\pi}{180^\circ}$ to convert to radians and multiply radians by $\frac{180^\circ}{\pi}$ to convert to degrees.

Example 1 Convert each degree measure to radian measure.

(a) 60° (b) -22.5°

Solution

(a) $60^\circ = \frac{\pi}{180^\circ}(60^\circ) \text{ radians} = \frac{\pi}{3} \text{ radians}$

(b) $-22.5^\circ = \frac{\pi}{180^\circ}(-22.5^\circ) \text{ radian} = -\frac{\pi}{8} \text{ radian}$

Example 2 Convert each radian measure to degree measure.

(a) $\frac{13\pi}{10}$ radians (b) $-\frac{\pi}{4}$ radian

Solution

(a) $\frac{13\pi}{10} \text{ radians} = \frac{180^\circ}{\pi}\left(\frac{13\pi}{10}\right) = 234^\circ$ (b) $-\frac{\pi}{4} \text{ radian} = \frac{180^\circ}{\pi}\left(-\frac{\pi}{4}\right) = -45^\circ$

As the previous examples illustrate, when radian measures are expressed as rational multiples of π, we may leave them in that form rather than writing them as decimals. Also, when angles are measured in radians, the word "radian" is often omitted. For instance, rather than writing $60^\circ = \pi/3$ radians, we write $60^\circ = \pi/3$. Therefore, *when no unit of angular measure is indicated, it is always understood that radian measure is intended.*

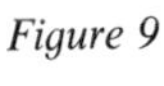

Figure 9

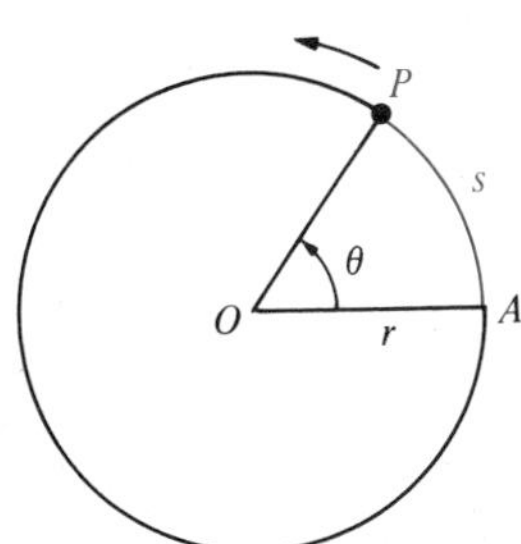

Angles measured in radians are especially useful in studying the motion of a particle moving at constant speed around a circle of radius r with center O (Figure 9). Suppose that the moving particle starts at the point A and that t units of time later it is at the point P. If the arc $\widehat{AP}$ has length s, then the particle has moved s units of distance in t units of time, and we say that its *speed* v is given by

$$v = \frac{s}{t} \text{ units of distance per unit of time.}$$

Here we assume that v is a constant.

Even though the particle in Figure 9 is moving around the circle, we speak of $v = s/t$ as its **linear speed** (since we could imagine the circular path straightened into a line). Now suppose that the radial line segment $\overline{OP}$ sweeps through an angle of θ radians in t units of time. Then we say that the **angular speed** of the particle is θ/t radians per unit of time. Angular speed is usually denoted by the Greek letter ω (*omega*), so we have

$$\omega = \frac{\theta}{t} \text{ radians per unit of time.}$$

To find the relationship between linear speed v and angular speed ω, we begin by recalling the formula

$$s = r\theta.$$

Dividing both sides of this equation by t, we obtain

$$\frac{s}{t} = r\frac{\theta}{t}$$

or

$$v = r\omega.$$

In words, *linear speed is the product of the radius and the angular speed.*

Example A stone is whirled around in a circle at the end of a string 70 centimeters long. If it makes 5 revolutions in 2 seconds, find (a) its angular speed ω and (b) its linear speed v.

Solution (a) Since the stone travels five times around the circle in 2 seconds, the radial line segment (the string) sweeps through $5(2\pi) = 10\pi$ radians in 2 seconds. Therefore, the angular speed of the stone is

$$\omega = \frac{\theta}{t} = \frac{10\pi}{2} = 5\pi \text{ radians per second.}$$

(b) The linear speed of the stone is

$$v = r\omega = 70(5\pi) = 350\pi$$
$$\approx 1100 \text{ centimeters per second.}$$

Until now, we have been careful to distinguish between an *angle* (a rotation) and the *measure* of the angle in degrees or radians (a real number). However, continual use of such phrases as "the measure of the angle α is 30°" can become tedious, so people often write "$\alpha = 30°$" as an abbreviation for the correct statement. We shall follow this practice whenever it seems convenient and harmless.

PROBLEM SET 1

In problems 1–4, find the degree measure of each angle.

1. One-sixth of a counterclockwise revolution.
2. Two-thirds of a clockwise revolution.
3. Eleven-fifths of a clockwise revolution.
4. The angle through which the small hand on a clock turns in 24 hours.

In problems 5–12, indicate the number of revolutions or the fraction of a revolution represented by the angle whose degree measure is given. Indicate whether the rotation is clockwise or counterclockwise.

5. 45° **6.** -1080° **7.** 120° **8.** 800°
9. -330° **10.** 15° **11.** -12° **12.** -700°

Ⓒ In problems 13–18, express each angle measure as a decimal.

13. $2'$ **14.** $5^\circ 6' 0''$ **15.** $3''$
16. $20^\circ 45''$ **17.** $100^\circ 30' 20''$ **18.** $-28^\circ 70' 3''$

Ⓒ In problems 19–24, express each angle measure in degrees, minutes, and seconds.

19. 62.25° **20.** -371.12° **21.** -0.125° **22.** 880.23°
23. 21.16° **24.** 1/17 of a counterclockwise revolution

In problems 25–30, s denotes the length of the arc intercepted on a circle of radius r by a central angle of θ radians. Find the missing quantity.

25. $r = 2$ meters, $\theta = 1.65$ radians, $s = ?$
26. $r = 1.8$ centimeters, $\theta = 8$ radians, $s = ?$
27. $r = 9$ feet, $s = 12$ feet, $\theta = ?$
28. $s = 4\pi$ kilometers, $\theta = \frac{\pi}{2}$ radians, $r = ?$
29. $r = 12$ inches, $\theta = \frac{5\pi}{18}$ radian, $s = ?$
30. $r = 5$ meters, $s = 13\pi$ meters, $\theta = ?$

31. Convert each degree measure to radians. Do not use a calculator. Write your answers as rational multiples of π.
(a) 30° (b) 45° (c) 90° (d) 120° (e) -150° (f) 520°
(g) 72° (h) 67.5° (i) -330° (j) 450° (k) 21° (l) -360°

Ⓒ **32.** Use a calculator to convert each degree measure to an approximate radian measure expressed as a decimal. Round off all answers to five significant digits.
(a) 7° (b) 33.333° (c) -11.227° (d) 571° (e) 1229° (f) 0.017320°

33. Convert each radian measure to degrees. Do not use a calculator.
(a) $\frac{\pi}{2}$ (b) $\frac{\pi}{3}$ (c) $\frac{\pi}{4}$ (d) $\frac{\pi}{6}$ (e) $\frac{2\pi}{3}$ (f) $-\pi$
(g) $\frac{3\pi}{5}$ (h) $-\frac{5\pi}{2}$ (i) $\frac{9\pi}{4}$ (j) $-\frac{3\pi}{8}$ (k) 7π (l) $-\frac{\pi}{14}$

Ⓒ **34.** Use a calculator to convert each radian measure to an approximate degree measure. Round off all answers to five significant digits.
(a) $\frac{2}{3}$ (b) -2 (c) 200 (d) $\frac{7\pi}{12}$ (e) 2.7333 (f) 1.5708

35. Indicate both the degree measure and the radian measure of the angle formed by (a) three-eighths of a clockwise revolution, (b) four and one-sixth counterclockwise revolutions, and (c) the hands of a clock at 4:00.

36. (a) Find a formula for the length s of the arc intercepted by a central angle of α *degrees*. (b) Use part (a) to find s if $r = 10$ meters and $\alpha = 35°$.
37. As a monkey swings on a 10-meter rope from one platform to another, the rope turns through an angle of $\pi/3$. How far does the monkey travel along the arc of the swing?
[C] 38. What radius should be used for a circular monorail track if the track is to change its direction by 7° in a distance of 24 meters (measured along the arc of the track)?
[C] 39. A motorcycle is moving at a speed of 88 kilometers per hour. If the radius of its wheels is 0.38 meter, find the angle in *degrees* through which a spoke of a wheel turns in 3 seconds.
40. A wheel of radius r is turning at a constant speed of n revolutions per second. Find a formula for (a) the angular speed ω and (b) the linear speed v of a point on the rim of the wheel.
[C] 41. A satellite is orbiting a certain planet in a perfectly circular orbit of radius 7680 kilometers. If it makes two-thirds of a revolution every hour, find (a) its angular speed ω and (b) its linear speed v.
[C] 42. A **nautical mile** may be defined as the arc length intercepted on the surface of the earth by a central angle of measure 1 minute. The radius of the earth is 2.09×10^7 feet. How many feet are there in a nautical mile?
[C] 43. A girl rides her bicycle to school at a speed of 12 miles per hour. If the wheel diameter is 26 inches, what is the angular speed of the wheels?
[C] 44. To determine the approximate speed of the water in a river, a water wheel of radius 1.5 meters is used. If the wheel makes 13 revolutions per minute, what is the speed of the river in kilometers per hour?
[C] 45. A $33\frac{1}{3}$-rpm (revolutions per minute) phonograph record has a radius of 14.6 centimeters. What is the linear speed (in centimeters per second) of a point on the rim of the record?
46. A first pulley of radius r_1 and a second pulley of radius r_2 are connected by a belt (Figure 10). The first pulley makes n_1 revolutions per minute and the second makes n_2 revolutions per minute. A point on the rim of either pulley will be moving at the same linear speed v as the belt. (a) Find formulas for n_1 and n_2 in terms of v, r_1, and r_2. (b) Show that $n_1/n_2 = r_2/r_1$.

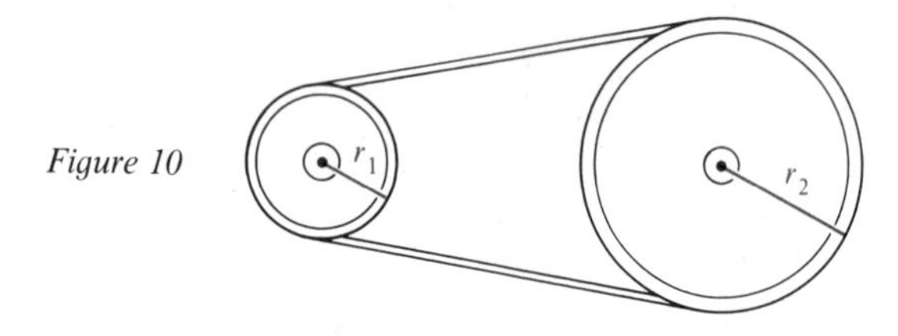

Figure 10

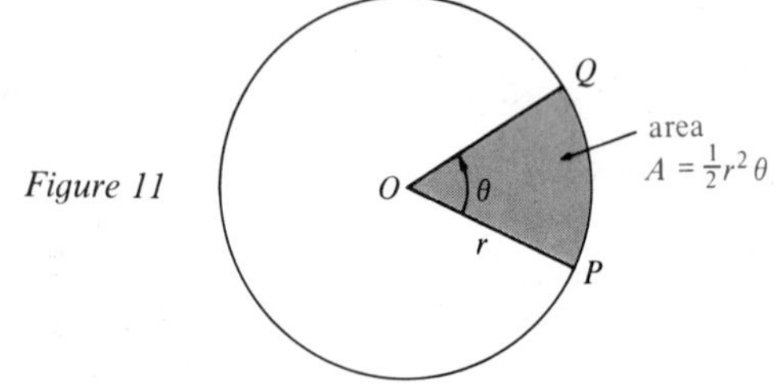

Figure 11

[C] 47. The earth, which is approximately 93,000,000 miles from the sun, revolves about the sun in a nearly circular orbit in approximately 365 days. Find the approximate linear speed (in miles per hour) of the earth in its orbit.
48. A **sector** OPQ of a circle is the region inside the circle bounded by the arc $\widehat{PQ}$ and the radial segments $\overline{OP}$ and $\overline{OQ}$ (Figure 11). If the central angle POQ has a measure of θ radians and r is the radius of the circle, show that the area A of the sector is given by $A = \frac{1}{2}r^2\theta$. [Hint: The area of the circle is πr^2. What fraction of this is the area of the sector?]
49. Use the formula $A = \frac{1}{2}r^2\theta$ (problem 48) to find the area A of a circular sector of radius r with a central angle of θ radians if (a) $r = 7$ centimeters and $\theta = 3\pi/14$ and (b) $r = 9$ inches and $\theta = 13\pi/9$.
50. A radar beam has an effective range of 70 miles and sweeps through an angle of $3\pi/4$ radians. What is the effective area in square miles swept by the radar beam?

2 Trigonometric Functions of Acute Angles in Right Triangles

In this section, we introduce the six functions—sine, cosine, tangent, cotangent, secant, and cosecant—upon which trigonometry is based. Here we define the values of these functions for acute angles in right triangles; in Section 3, we extend the definitions to general angles.

An angle with a measure greater than 0° but less than 90° is called an **acute angle.** For instance, in a right triangle ACB (Figure 1a) the vertex angle C is a right angle (90°), and the other two vertex angles A and B are acute angles. Recall that the side $\overline{AB}$ opposite the right angle C is called the *hypotenuse.* If we denote one of the two acute angles by θ, we can abbreviate the lengths of the side **opposite** θ, the **hypotenuse,** and the side **adjacent** to θ as **opp, hyp,** and **adj,** respectively (Figure 1b). The six trigonometric functions of the acute angle θ are defined as follows:

Figure 1

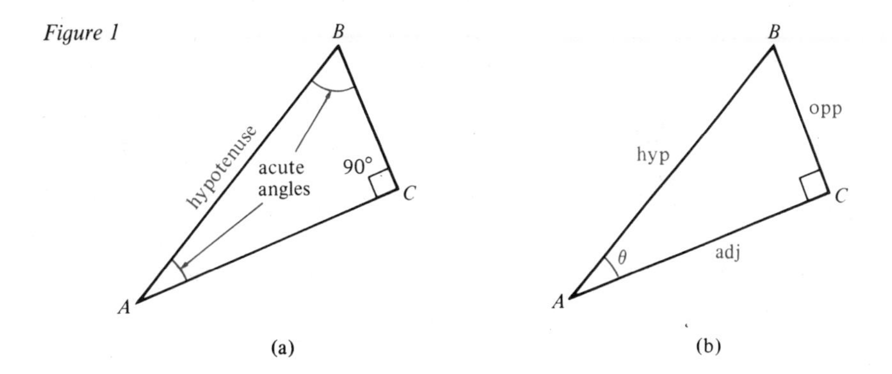

Name of Function	Abbreviation	Value at θ
sine	sin	$\sin\theta = \dfrac{\text{opp}}{\text{hyp}}$
cosine	cos	$\cos\theta = \dfrac{\text{adj}}{\text{hyp}}$
tangent	tan	$\tan\theta = \dfrac{\text{opp}}{\text{adj}}$
cotangent	cot	$\cot\theta = \dfrac{\text{adj}}{\text{opp}}$
secant	sec	$\sec\theta = \dfrac{\text{hyp}}{\text{adj}}$
cosecant	csc	$\csc\theta = \dfrac{\text{hyp}}{\text{opp}}$

The values of the six trigonometric functions depend only on the angle θ and not on the size of the right triangle ACB. Indeed, two right triangles with the same acute vertex angle θ are similar (Figure 2); hence, ratios of corresponding sides are equal.

Figure 2

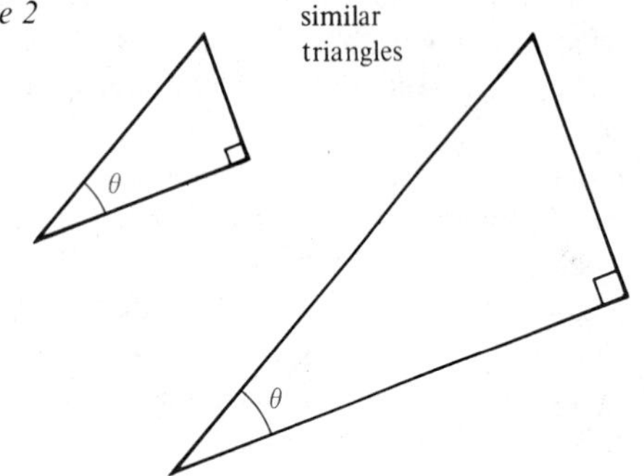

Figure 3

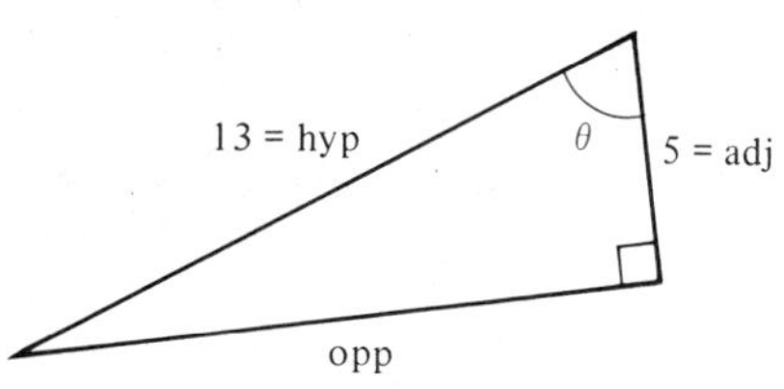

Example Find the values of the six trigonometric functions of the acute angle θ in Figure 3.

Solution For angle θ, we have

$$\text{adj} = 5 \quad \text{and} \quad \text{hyp} = 13,$$

but opp isn't given. However, by the Pythagorean theorem,

$$(\text{adj})^2 + (\text{opp})^2 = (\text{hyp})^2,$$

so that

$$(\text{opp})^2 = (\text{hyp})^2 - (\text{adj})^2 = 13^2 - 5^2 = 144.$$

Therefore,

$$\text{opp} = \sqrt{144} = 12.$$

It follows that

$$\sin\theta = \frac{\text{opp}}{\text{hyp}} = \frac{12}{13} \qquad \csc\theta = \frac{\text{hyp}}{\text{opp}} = \frac{13}{12}$$

$$\cos\theta = \frac{\text{adj}}{\text{hyp}} = \frac{5}{13} \qquad \sec\theta = \frac{\text{hyp}}{\text{adj}} = \frac{13}{5}$$

$$\tan\theta = \frac{\text{opp}}{\text{adj}} = \frac{12}{5} \qquad \cot\theta = \frac{\text{adj}}{\text{opp}} = \frac{5}{12}.$$

If you look at the ratios that define the six trigonometric functions, you will see that

$$\tan\theta = \frac{\sin\theta}{\cos\theta} \qquad \cot\theta = \frac{\cos\theta}{\sin\theta}$$

$$\sec\theta = \frac{1}{\cos\theta} \qquad \csc\theta = \frac{1}{\sin\theta}.$$

For instance,

$$\tan\theta = \frac{\text{opp}}{\text{adj}} = \frac{\text{opp/hyp}}{\text{adj/hyp}} = \frac{\sin\theta}{\cos\theta},$$

and the remaining equations are derived similarly (Problem 42). Perhaps the easiest way for you to become familiar with all six of the trigonometric functions is to learn the two basic definitions

$$\sin\theta = \frac{\text{opp}}{\text{hyp}} \quad \text{and} \quad \cos\theta = \frac{\text{adj}}{\text{hyp}}$$

and use the equations obtained above to write the values of the other four functions in terms of sines and cosines. Thus, in the example above, once you know that

$$\sin\theta = \frac{12}{13} \quad \text{and} \quad \cos\theta = \frac{5}{13},$$

you can immediately write

$$\tan\theta = \frac{\sin\theta}{\cos\theta} = \frac{12/13}{5/13} = \frac{12}{5} \qquad \cot\theta = \frac{\cos\theta}{\sin\theta} = \frac{5/13}{12/13} = \frac{5}{12}$$

$$\sec\theta = \frac{1}{\cos\theta} = \frac{1}{5/13} = \frac{13}{5} \qquad \csc\theta = \frac{1}{\sin\theta} = \frac{1}{12/13} = \frac{13}{12}.$$

Two angles are called **complementary** if their sum is a right angle. For instance, *the two acute angles of a right triangle are* **complementary angles**. (Why?) If θ is the degree measure of an angle, then the degree measure of the complementary angle is $90^\circ - \theta$. For radian measure, the angle $(\pi/2) - \theta$ is complementary to θ.

Example Find the angle complementary to θ if (a) $\theta = 37^\circ$ and (b) $\theta = \pi/5$.

Solution (a) $90^\circ - \theta = 90^\circ - 37^\circ = 53^\circ$ (b) $\frac{\pi}{2} - \frac{\pi}{5} = \frac{5\pi}{10} - \frac{2\pi}{10} = \frac{3\pi}{10}$

In Figure 4, the two acute angles α and β of right triangle ABC are complementary, a is the length of the side opposite α, b is the length of the side opposite β, and c is the length of the hypotenuse. Notice that the side opposite either one of the acute angles is adjacent to the other. Therefore,

Figure 4

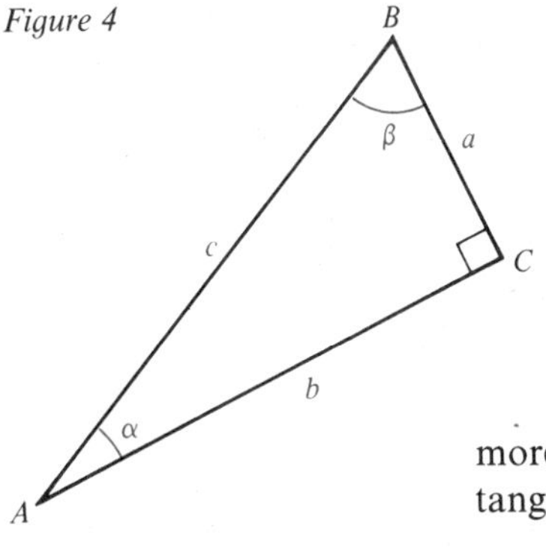

$$\sin\alpha = \frac{a}{c} = \cos\beta \qquad \csc\alpha = \frac{c}{a} = \sec\beta$$

$$\cos\alpha = \frac{b}{c} = \sin\beta \qquad \sec\alpha = \frac{c}{b} = \csc\beta$$

$$\tan\alpha = \frac{a}{b} = \cot\beta \qquad \cot\alpha = \frac{b}{a} = \tan\beta.$$

By using the idea of *cofunctions*, you can remember the equations above more easily. We say that sine and cosine are **cofunctions** of one another. Likewise, tangent and cotangent are said to be cofunctions, and so are secant and cosecant.

Thus, *the value of any trigonometric function of an acute angle is the same as the value of the cofunction of the complementary angle:*

> "Cosine" is "sine of the complementary angle."
> "Cotangent" is "tangent of the complementary angle."
> "Cosecant" is "secant of the complementary angle."

Example Express the value of each trigonometric function as the value of the cofunction of the complementary angle.

(a) $\cos 70^\circ$ (b) $\cot \frac{\pi}{12}$ (c) $\sin 10^\circ$ (d) $\sec \frac{\pi}{4}$

Solution (a) $\cos 70^\circ = \sin(90^\circ - 70^\circ) = \sin 20^\circ$ (b) $\cot \frac{\pi}{12} = \tan\left(\frac{\pi}{2} - \frac{\pi}{12}\right) = \tan \frac{5\pi}{12}$

(c) $\sin 10^\circ = \cos(90^\circ - 10^\circ) = \cos 80^\circ$ (d) $\sec \frac{\pi}{4} = \csc\left(\frac{\pi}{2} - \frac{\pi}{4}\right) = \csc \frac{\pi}{4}$

The special angles with measures of 30° ($\pi/6$ radian), 45° ($\pi/4$ radian), and 60° ($\pi/3$ radians) have a way of showing up over and over again in various applications of trigonometry. Notice that 30° and 60° angles are complementary, whereas a 45° angle is complementary to itself.

Using some elementary geometry, we can find the values of the six trigonometric functions of the special angles. To begin with, consider an equilateral triangle ABC having sides of length 2 units (Figure 5a). Recall from plane geometry that the vertex angles of an equilateral triangle all measure 60°, and that the bisector $\overline{AD}$ of vertex angle A is the perpendicular bisector of the opposite side $\overline{BC}$ (Figure 5b). Therefore,

$$|\overline{BD}| = \tfrac{1}{2}|\overline{BC}| = \tfrac{1}{2}(2) = 1 \quad \text{and} \quad \text{angle } DAB = \tfrac{1}{2}(60^\circ) = 30^\circ.$$

Applying the Pythagorean theorem to right triangle ADB, we have

$$|\overline{AD}|^2 + |\overline{BD}|^2 = |\overline{AB}|^2 \quad \text{or} \quad |\overline{AD}|^2 + 1^2 = 2^2,$$

so that

$$|\overline{AD}|^2 = 2^2 - 1^2 = 3 \quad \text{or} \quad |\overline{AD}| = \sqrt{3}.$$

Figure 5

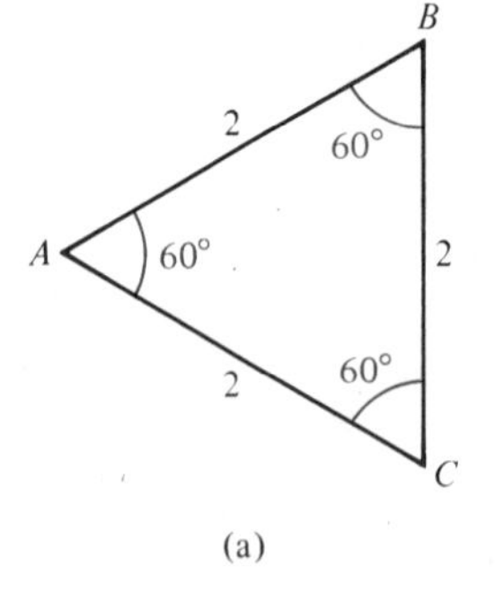

(a)

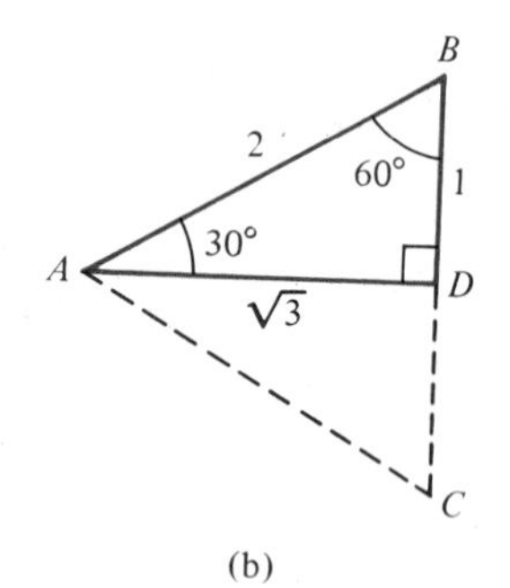

(b)

Thus, from triangle ADB in Figure 5b, we have:

$$\sin 30° = \cos 60° = 1/2 \qquad \csc 30° = \sec 60° = 2$$
$$\cos 30° = \sin 60° = \sqrt{3}/2 \qquad \sec 30° = \csc 60° = 2/\sqrt{3} = 2\sqrt{3}/3$$
$$\tan 30° = \cot 60° = 1/\sqrt{3} = \sqrt{3}/3 \qquad \cot 30° = \tan 60° = \sqrt{3}.$$

To find the values of the trigonometric functions of a 45° angle, we construct an isosceles right triangle with two equal sides of length 1 unit (Figure 6). We know from plane geometry that both of the acute angles in this right triangle measure 45°. Also, by the Pythagorean theorem,

$$1^2 + 1^2 = (\text{hyp})^2 \quad \text{or} \quad 2 = (\text{hyp})^2$$

so that

$$\text{hyp} = \sqrt{2}.$$

Figure 6

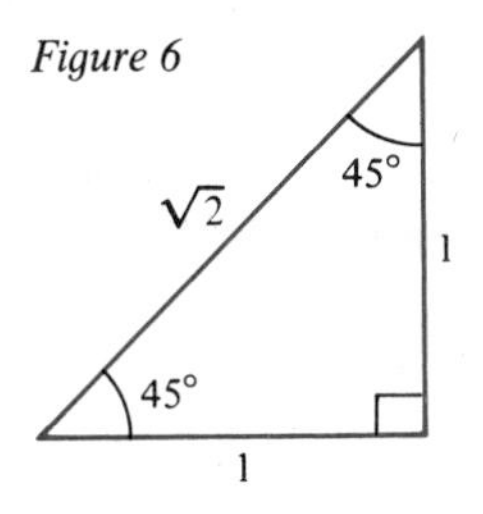

It follows that

$$\sin 45° = 1/\sqrt{2} = \sqrt{2}/2 \qquad \csc 45° = \sqrt{2}$$
$$\cos 45° = 1/\sqrt{2} = \sqrt{2}/2 \qquad \sec 45° = \sqrt{2}$$
$$\tan 45° = 1 \qquad \cot 45° = 1.$$

For convenience, the values of the trigonometric functions for the special angles are gathered in Table 1.

Table 1

θ degrees	θ radians	$\sin\theta$	$\cos\theta$	$\tan\theta$	$\cot\theta$	$\sec\theta$	$\csc\theta$
30°	$\frac{\pi}{6}$	$\frac{1}{2}$	$\frac{\sqrt{3}}{2}$	$\frac{\sqrt{3}}{3}$	$\sqrt{3}$	$\frac{2\sqrt{3}}{3}$	2
45°	$\frac{\pi}{4}$	$\frac{\sqrt{2}}{2}$	$\frac{\sqrt{2}}{2}$	1	1	$\sqrt{2}$	$\sqrt{2}$
60°	$\frac{\pi}{3}$	$\frac{\sqrt{3}}{2}$	$\frac{1}{2}$	$\sqrt{3}$	$\frac{\sqrt{3}}{3}$	2	$\frac{2\sqrt{3}}{3}$

You can find values of the trigonometric functions for angles other than the special angles in Table 1 by using a scientific calculator. If a calculator isn't available, you can use the tables provided in Section 3 of the Appendix found at the back of the book. Instructions for using these tables are included in the Appendix. Before using a calculator, you must decide whether you are going to measure angles in *degrees* or *radians* and set the calculator accordingly. (Refer to the instruction booklet furnished with your calculator.) Most calculators can deal with fractions of a degree only in decimal form. If angles are given in degrees, minutes, and seconds, you must first convert to decimal form before pressing a trigonometric key.

Examples [C] Use a calculator to find the approximate value of each trigonometric function.

1 $\sin 41.3^\circ$

Solution We make sure the calculator is set in degree mode, enter 41.3, and press the sine (SIN) key. On a 10-digit calculator we get

$$\sin 41.3^\circ = 0.660001668.$$

An 8-digit calculator gives this result rounded off as

$$\sin 41.3^\circ = 0.6600017.$$

If a calculator isn't available, Table 1 in Section 3 of the Appendix can be used to obtain the result rounded off to four decimal places:

$$\sin 41.3^\circ = 0.6600.$$

2 $\cos 19^\circ 21' 17''$

Solution We begin by using a calculator to convert to decimal form:

$$19^\circ 21' 17'' = 19^\circ + \left(\frac{21}{60}\right)^\circ + \left(\frac{17}{3600}\right)^\circ = 19.35472222^\circ.$$

Now, we make sure the calculator is set in degree mode, and press the cosine (COS) key to get

$$\cos 19^\circ 21' 17'' = \cos 19.35472222^\circ = 0.943484853.$$

3 $\tan \frac{5\pi}{36}$

Solution We set the calculator in *radian mode*, press the π key (or enter 3.141592654), multiply by 5, divide by 36, and press the tangent (TAN) key to obtain

$$\tan \frac{5\pi}{36} = 0.466307658.$$

4 $\sin 0.8432$

Solution Because degrees are not indicated, we know that the angle is measured in radians. We set the calculator in radian mode, enter 0.8432, and press the sine (SIN) key to obtain

$$\sin 0.8432 = 0.746775185.$$

Most scientific calculators have keys only for sine, cosine, and tangent. To evaluate the cotangent function on such a calculator, you can use the fact that

$$\cot \theta = \frac{1}{\tan \theta}$$

(Problem 43). Similarly, to evaluate the secant or cosecant functions, you can use the relations

$$\sec\theta = \frac{1}{\cos\theta} \quad \text{and} \quad \csc\theta = \frac{1}{\sin\theta}.$$

Examples Ⓒ Use a calculator to find the approximate values of each trigonometric function.

1 cot 50°

Solution With the calculator in degree mode, we enter 50, press the tangent (TAN) key, then we take the reciprocal of the result by pressing the 1/x key. (We're using the fact that $\cot 50° = 1/\tan 50°$.) The result is

$$\cot 50° = 0.839099631.$$

2 sec 1.098

Solution With the calculator in radian mode, we enter 1.098, press the cosine (COS) key, and then press the 1/x key. The result is

$$\sec 1.098 = 2.195979642.$$

PROBLEM SET 2

In problems 1–8, find the values of the six trigonometric functions of the acute angle θ in the indicated figure.

1.

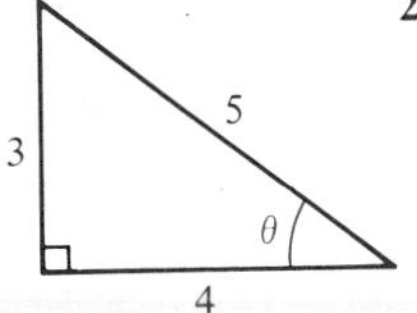

2.

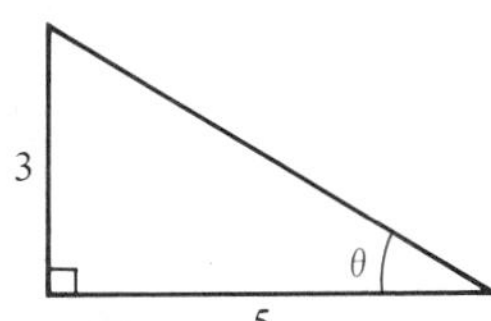

3.

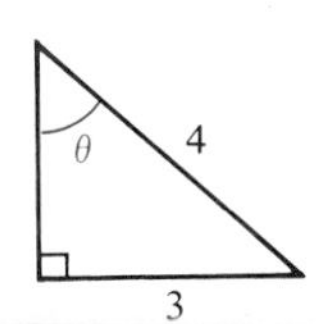

4.

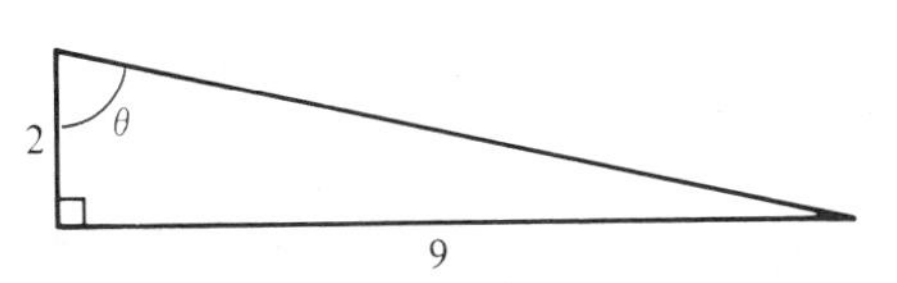

5.

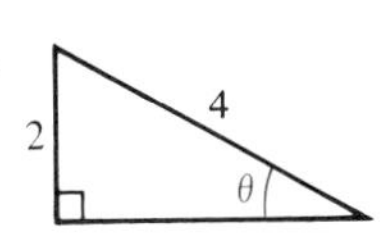

6.

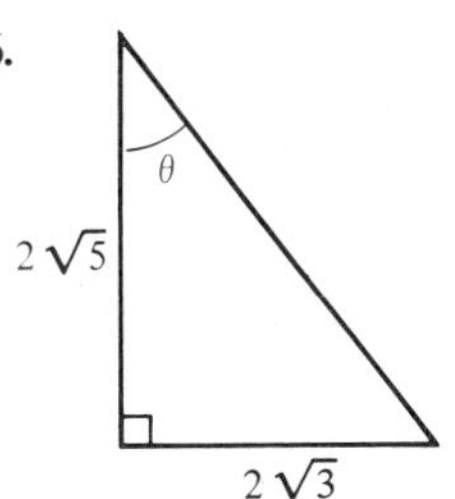

7.

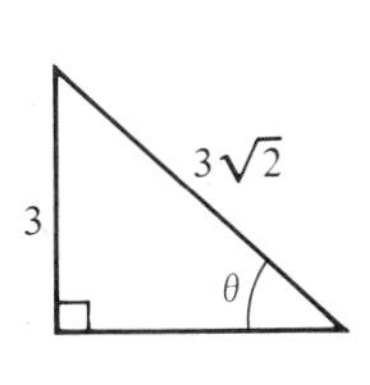

8.

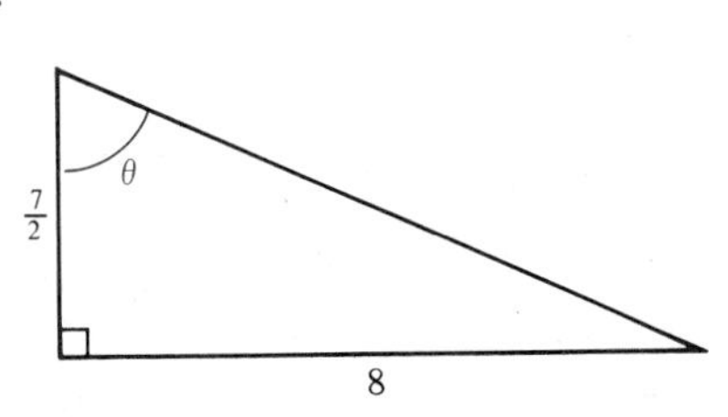

In problems 9–16, find the angle complementary to each of the given angles.

9. $35°$ **10.** $89°$ 11. $\frac{\pi}{5}$ **12.** $\frac{\pi}{7}$ 13. $\frac{5\pi}{12}$

14. $\frac{7\pi}{33}$ 15. $22.34°$ **16.** one-sixth of a counterclockwise revolution

In problems 17–24, express the value of each trigonometric function as the value of the cofunction of the complementary angle.

17. $\sin 31°$ **18.** $\cos 13°$ 19. $\tan \frac{\pi}{9}$ **20.** $\csc \frac{\pi}{11}$

21. $\cot \frac{2\pi}{5}$ **22.** $\sin 9.03°$ 23. $\csc 77.03°$ **24.** $\tan 0.774$

C In problems 25–34, use a calculator to find the approximate values of the six trigonometric functions of each angle.

25. $48°$ **26.** $7°$ 27. $23°12'33''$ **28.** $82°55'55''$ 29. $16.19°$

30. $\frac{9\pi}{10}$ 31. $\frac{2\pi}{7}$ **32.** $\frac{5\pi}{21}$ 33. 0.7764 **34.** 1.33

C In problems 35–38, use a calculator to verify that the given equation is true for the indicated value of the angle.

35. $\cos\theta = \sin(90° - \theta)$ for $\theta = 33°$

36. $\sin\theta = \cos(90° - \theta)$ for $\theta = 77.77°$

37. $\sin\theta = \cos\left(\frac{\pi}{2} - \theta\right)$ for $\theta = \frac{2\pi}{9}$

38. $\cos\theta = \sin\left(\frac{\pi}{2} - \theta\right)$ for $\theta = 1.0785$

39. How could you check to make sure that a calculator is set in degree mode? [Hint: We know that $\sin 30° = \frac{1}{2}$.]

40. Justify the following procedure: To convert from degrees, minutes, and seconds to degrees in decimal form, divide the number of seconds by 60, add the quotient to the number of minutes, divide this sum by 60, and add the result to the number of degrees.

C 41. Using a calculator, verify the entries in Table 1 on page 69.

42. Show that

(a) $\cot\theta = \frac{\cos\theta}{\sin\theta}$ (b) $\sec\theta = \frac{1}{\cos\theta}$ (c) $\csc\theta = \frac{1}{\sin\theta}$

43. Show that $\cot\theta = \frac{1}{\tan\theta}$.

C **44.** Use a calculator to verify that $\tan\theta = \frac{\sin\theta}{\cos\theta}$ for

(a) $\theta = 10°$ (b) $\theta = \frac{3\pi}{13}$ (c) $\theta = 0.9474$

45. Using the Pythagorean theorem, show that, if θ is an acute angle in a right triangle, then

$$(\cos\theta)^2 + (\sin\theta)^2 = 1.$$

C **46.** Use a calculator to verify that $(\cos\theta)^2 + (\sin\theta)^2 = 1$ for

(a) $\theta = 20°$ (b) $\theta = \frac{4\pi}{17}$ (c) $\theta = 0.5678$

3 Trigonometric Functions of General Angles

In order to extend the definitions of the six trigonometric functions to general angles, we shall make use of the following ideas: In a cartesian coordinate system, an angle α is said to be in **standard position** if its vertex is at the origin O and its initial side coincides with the positive x axis (Figure 1). An angle is said to be *in* a certain quadrant if, when the angle is in standard position, the terminal side lies in that quadrant. For instance, a 65° angle is in quadrant I (Figure 2a). Slight variations of this terminology are often used; for instance, we might also say that a 65° angle **lies in quadrant I** or simply that it is a **quadrant I angle.** As Figure 2b shows, an angle of $-187°$ is a quadrant II angle.

Figure 1

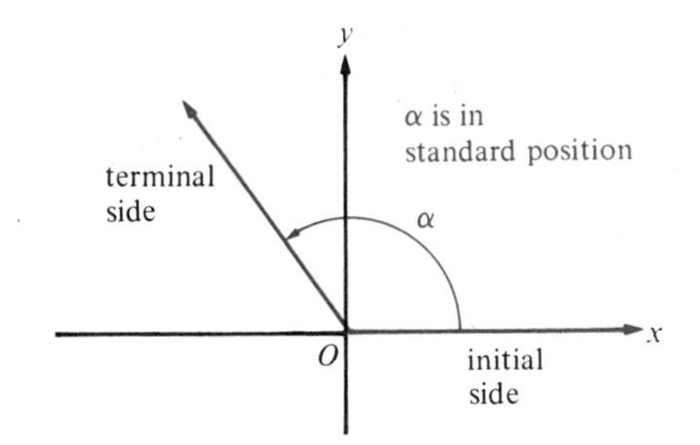

If the terminal side of an angle in standard position lies along either the x axis or the y axis, then the angle is called **quadrantal.** For example, $-360°$, $-270°$, $-180°$, $-90°$, $0°$, $90°$, $180°$, $270°$, and $360°$ are quadrantal angles. Evidently, an angle is quadrantal if and only if its measure is an integer multiple of 90° ($\pi/2$ radians).

Figure 2

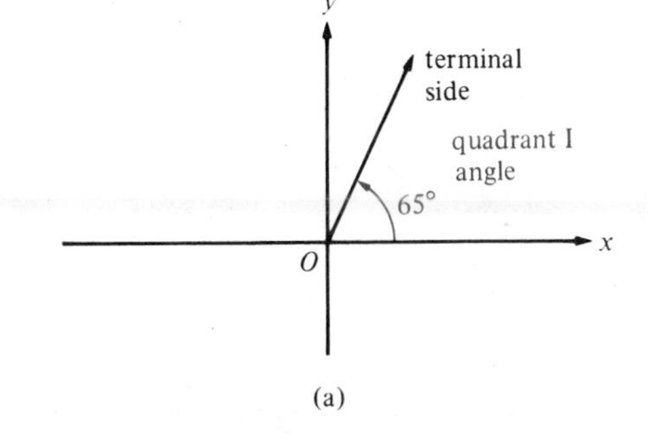

(a)

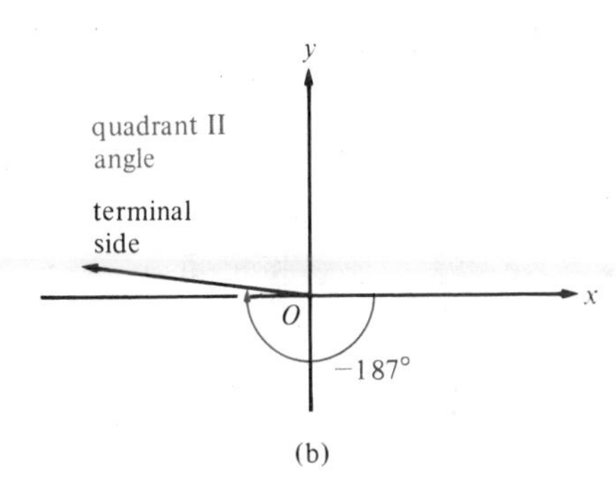

(b)

Example 1 For each point given, sketch two coterminal angles α and β in standard position whose terminal side contains the point. Arrange it so that α is positive, β is negative, and neither angle exceeds one revolution. In each case, name the quadrant in which the angle lies, or indicate that the angle is quadrantal.

(a) $(4, 3)$ (b) $(-4, 4)$ (c) $(-3, 0)$ (d) $(3, -4)$

Figure 3

(a)

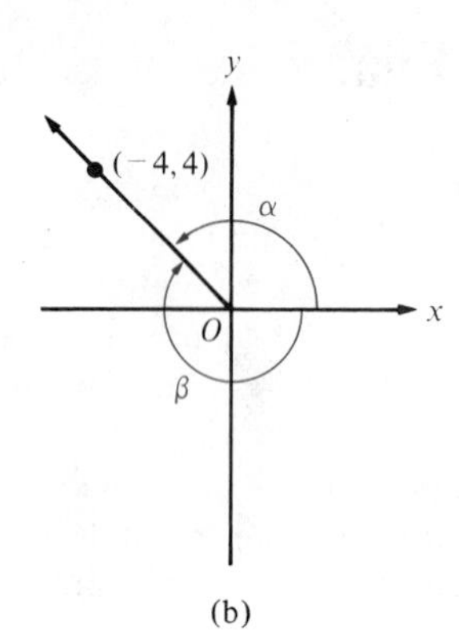

(b)

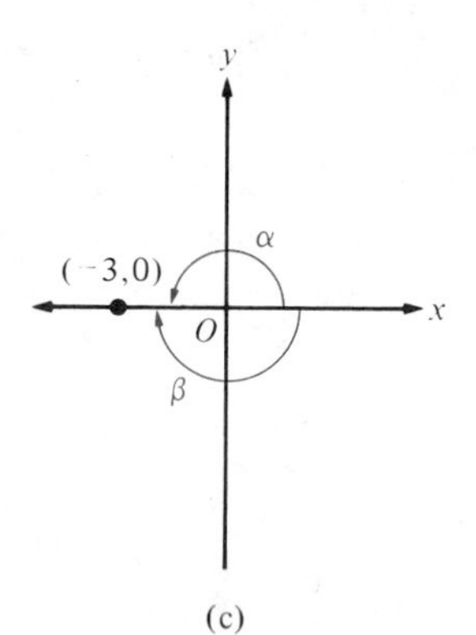

(c)

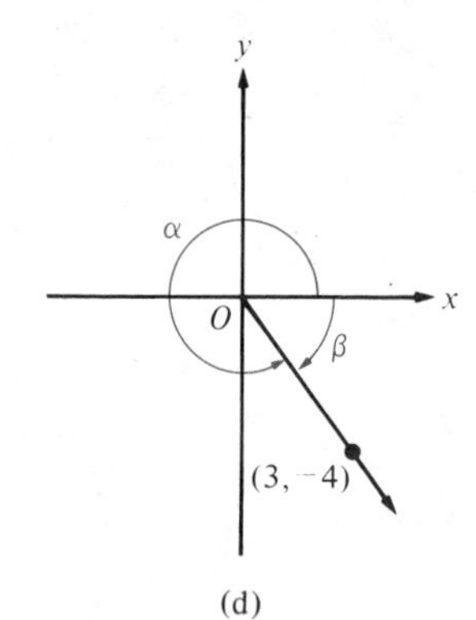

(d)

Solution (a) α and β are quadrant I angles (Figure 3a).

(b) α and β are quadrant II angles (Figure 3b).

(c) α and β are quadrantal angles (Figure 3c).

(d) α and β are quadrant IV angles (Figure 3d).

Example 2 Specify and sketch three angles that are coterminal with a 30° angle in standard position.

Solution The three angles shown in Figure 4 are obtained as follows:

$$30^\circ + 360^\circ = 390^\circ,$$
$$30^\circ + 2(360^\circ) = 750^\circ,$$
$$30^\circ - 360^\circ = -330^\circ.$$

Of course, there are other possibilities, such as

$$30^\circ + 5(360^\circ) = 1830^\circ, \quad 30^\circ - 13(360^\circ) = -4650^\circ,$$

and so on.

Figure 4

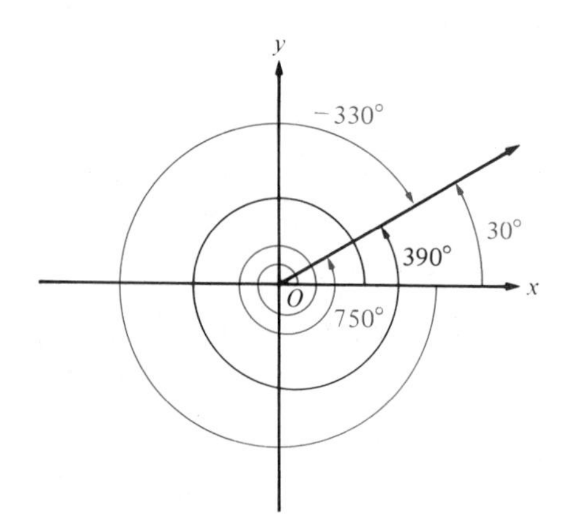

Figure 5

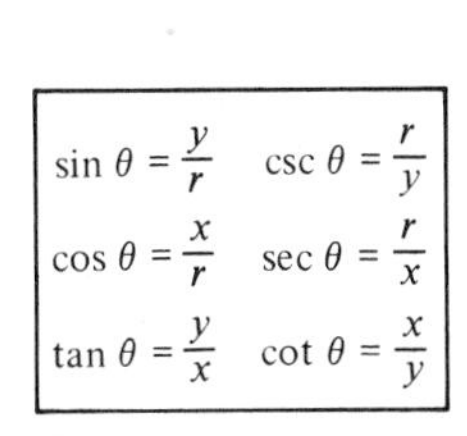

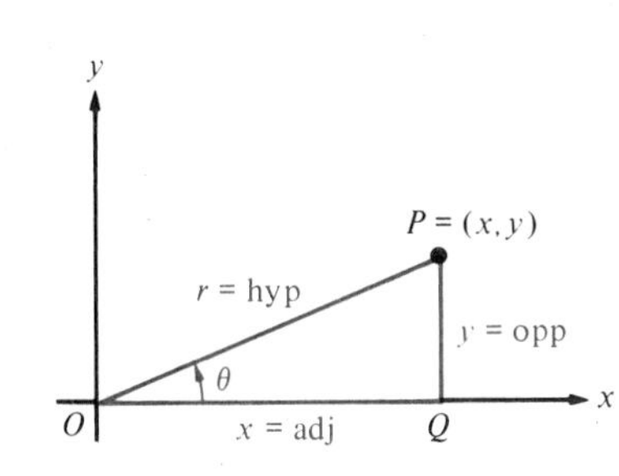

Now, let θ be one of the acute angles of a right triangle OQP placed on a cartesian coordinate system with θ in standard position and the hypotenuse $\overline{OP}$ in quadrant I (Figure 5). Then $P = (x, y)$ is on the terminal side of θ, and we have

$$x = \text{adj} \quad \text{and} \quad y = \text{opp}.$$

If we let

$$r = \text{hyp} = \sqrt{x^2 + y^2},$$

then the six trigonometric functions of θ are given by the ratios shown in Figure 5.

Figure 5 suggests the following definition.

Definition 1 **Trigonometric Functions of a General Angle**

Let θ be an angle in standard position and suppose that (x, y) is any point other than $(0, 0)$ on the terminal side of θ (Figure 6). If $r = \sqrt{x^2 + y^2}$ is the distance between (x, y) and $(0, 0)$, then the six trigonometric functions of θ are defined by

$$\sin\theta = \frac{y}{r} \qquad \csc\theta = \frac{r}{y}$$

$$\cos\theta = \frac{x}{r} \qquad \sec\theta = \frac{r}{x}$$

$$\tan\theta = \frac{y}{x} \qquad \cot\theta = \frac{x}{y},$$

Figure 6

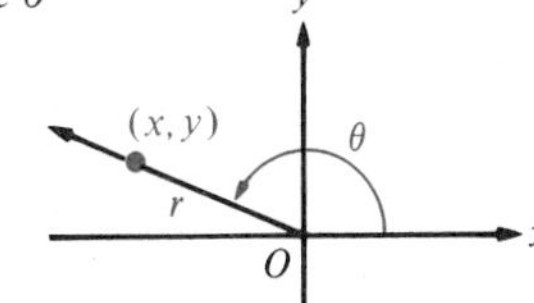

provided that the denominators are not zero.

By using similar triangles, you can see that the values of the six trigonometric functions in Definition 1 depend only on the angle θ and not on the choice of the point (x, y) on the terminal side of θ (Problem 30).

Example Evaluate the six trigonometric functions of the angle θ in standard position if the terminal side of θ contains the point $(x, y) = (2, -1)$ (Figure 7).

Figure 7

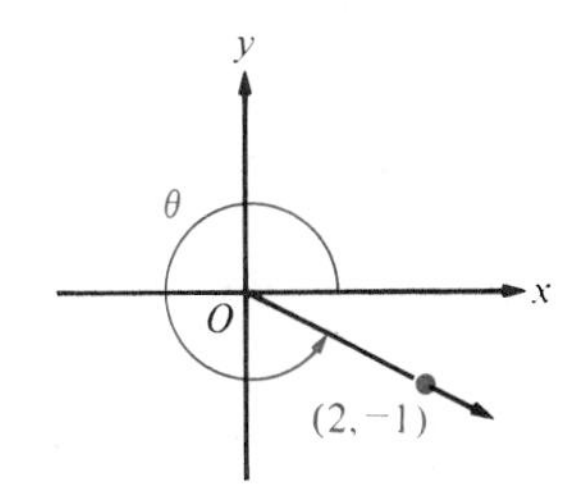

Solution Here $x = 2$, $y = -1$, and

$$r = \sqrt{x^2 + y^2} = \sqrt{2^2 + (-1)^2} = \sqrt{5}.$$

Thus,

$$\sin\theta = \frac{y}{r} = \frac{-1}{\sqrt{5}} = -\frac{\sqrt{5}}{5} \qquad \csc\theta = \frac{r}{y} = \frac{\sqrt{5}}{-1} = -\sqrt{5}$$

$$\cos\theta = \frac{x}{r} = \frac{2}{\sqrt{5}} = \frac{2\sqrt{5}}{5} \qquad \sec\theta = \frac{r}{x} = \frac{\sqrt{5}}{2}$$

$$\tan\theta = \frac{y}{x} = \frac{-1}{2} = -\frac{1}{2} \qquad \cot\theta = \frac{x}{y} = \frac{2}{-1} = -2.$$

Figure 8

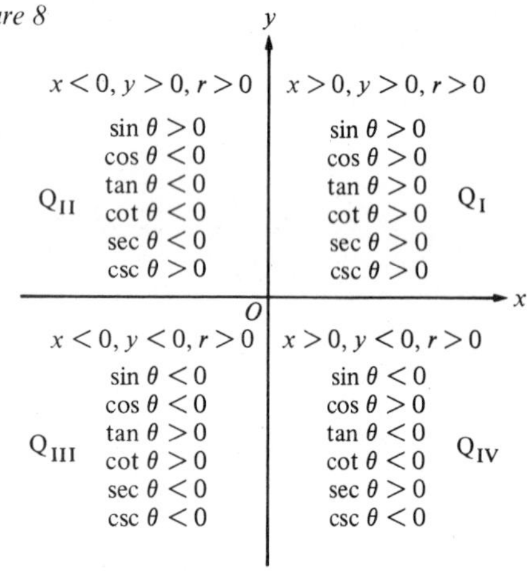

Q_{II}: $x<0, y>0, r>0$	Q_I: $x>0, y>0, r>0$
$\sin\theta>0$	$\sin\theta>0$
$\cos\theta<0$	$\cos\theta>0$
$\tan\theta<0$	$\tan\theta>0$
$\cot\theta<0$	$\cot\theta>0$
$\sec\theta<0$	$\sec\theta>0$
$\csc\theta>0$	$\csc\theta>0$
Q_{III}: $x<0, y<0, r>0$	**Q_{IV}: $x>0, y<0, r>0$**
$\sin\theta<0$	$\sin\theta<0$
$\cos\theta<0$	$\cos\theta>0$
$\tan\theta>0$	$\tan\theta<0$
$\cot\theta>0$	$\cot\theta<0$
$\sec\theta<0$	$\sec\theta>0$
$\csc\theta<0$	$\csc\theta<0$

If (x, y) is a point in quadrant I, then x, y, and $r = \sqrt{x^2 + y^2}$ are positive, so all six trigonometric functions have positive values for quadrant I angles. In quadrant II, x is negative while y and r are positive; so the values of the sine and cosecant are positive, whereas the values of the cosine, tangent, cotangent, and secant are negative. The algebraic signs of the trigonometric functions for angles in quadrants III and IV are easily found by observing the algebraic signs of x and y in these quadrants and keeping in mind that r is always positive (Figure 8).

Example Find the quadrant in which θ lies if $\tan\theta > 0$ and $\sin\theta < 0$.

Solution Let (x, y) be a point on the terminal side of θ (in standard position). Since $\tan\theta = \dfrac{y}{x} > 0$, it follows that x and y have the same algebraic sign. In addition, since $\sin\theta = \dfrac{y}{r} < 0$, it follows that $y < 0$; therefore, $x < 0$. Because $x < 0$ and $y < 0$, the point (x, y) is in quadrant III; hence, θ is a quadrant III angle.

The same basic relationships that we obtained in Section 2 for the trigonometric functions of acute angles continue to hold for general angles.

Theorem 1 **Reciprocal Identities**

If θ is an angle for which the functions are defined, then:

(i) $\csc\theta = \dfrac{1}{\sin\theta}$ (ii) $\sec\theta = \dfrac{1}{\cos\theta}$ (iii) $\cot\theta = \dfrac{1}{\tan\theta}$

Proof

Let (x, y) be a point on the terminal side of θ in standard position and let $r = \sqrt{x^2 + y^2}$ (Figure 9). Then, provided that the denominators are not zero:

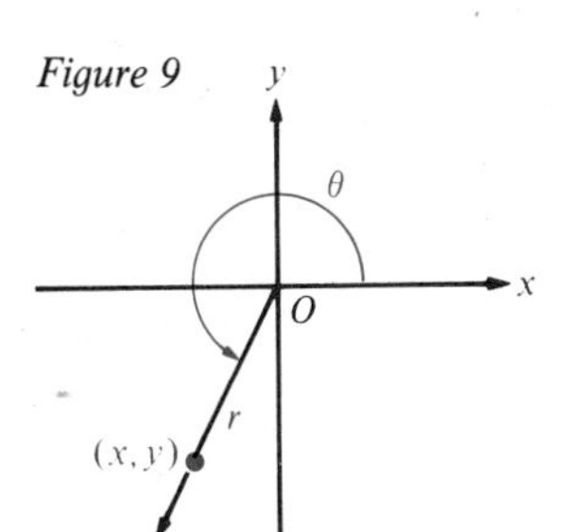

Figure 9

$$\text{(i)} \quad \csc\theta = \frac{r}{y} = \frac{1}{y/r} = \frac{1}{\sin\theta}$$

$$\text{(ii)} \quad \sec\theta = \frac{r}{x} = \frac{1}{x/r} = \frac{1}{\cos\theta}$$

$$\text{(iii)} \quad \cot\theta = \frac{x}{y} = \frac{1}{y/x} = \frac{1}{\tan\theta}$$

In a similar way, you can prove the following theorem (Problem 35).

Theorem 2 **Quotient Identities**

If θ is an angle for which the functions are defined, then:

$$\tan\theta = \frac{\sin\theta}{\cos\theta} \quad \text{and} \quad \cot\theta = \frac{\cos\theta}{\sin\theta}.$$

Example

If $\sin\theta = 1/2$ and $\cos\theta = -\sqrt{3}/2$, find the values of the other four trigonometric functions of θ.

Solution

Using Theorems 1 and 2, we have

$$\tan\theta = \frac{\sin\theta}{\cos\theta} = \frac{1/2}{(-\sqrt{3}/2)} = -\frac{\sqrt{3}}{3} \qquad \cot\theta = \frac{1}{\tan\theta} = \frac{1}{(-\sqrt{3}/3)} = -\sqrt{3}$$

$$\sec\theta = \frac{1}{\cos\theta} = \frac{1}{(-\sqrt{3}/2)} = -\frac{2\sqrt{3}}{3} \qquad \csc\theta = \frac{1}{\sin\theta} = \frac{1}{1/2} = 2$$

Another important relationship is derived as follows: Again suppose that θ is an angle in standard position and that (x, y) is a point on the terminal side of θ (Figure 9). Because $r = \sqrt{x^2 + y^2}$, we have $x^2 + y^2 = r^2$, so

$$(\cos\theta)^2 + (\sin\theta)^2 = \left(\frac{x}{r}\right)^2 + \left(\frac{y}{r}\right)^2 = \frac{x^2}{r^2} + \frac{y^2}{r^2} = \frac{x^2 + y^2}{r^2} = \frac{r^2}{r^2} = 1.$$

The relationship

$$(\cos\theta)^2 + (\sin\theta)^2 = 1$$

is called the **fundamental Pythagorean identity** because its derivation involves the fact that $x^2 + y^2 = r^2$, which is a consequence of the Pythagorean theorem.

The fundamental Pythagorean identity is used quite often, and it would be bothersome to write the parentheses each time for $(\cos\theta)^2$ and $(\sin\theta)^2$; yet, if the parentheses were simply omitted, the resulting expressions would be misunderstood. (For instance, $\cos\theta^2$ is usually understood to mean the cosine of the square of θ.) Therefore, it is customary to write $\cos^2\theta$ and $\sin^2\theta$ to mean $(\cos\theta)^2$ and $(\sin\theta)^2$. Similar notation is used for the remaining trigonometric functions and for powers other than 2. Thus, $\cot^4\theta$ means $(\cot\theta)^4$, $\sec^n\theta$ means $(\sec\theta)^n$, and so forth. With this notation, the fundamental Pythagorean identity becomes

$$\cos^2\theta + \sin^2\theta = 1.$$

Actually, there are three Pythagorean identities—the fundamental identity and two others derived from it.

Theorem 3 **Pythagorean Identities**

If θ is an angle for which the functions are defined, then:

(i) $\cos^2\theta + \sin^2\theta = 1$

(ii) $1 + \tan^2\theta = \sec^2\theta$

(iii) $\cot^2\theta + 1 = \csc^2\theta$

Proof We have already proved (i). To prove (ii), we divide both sides of (i) by $\cos^2\theta$ to obtain

$$1 + \frac{\sin^2\theta}{\cos^2\theta} = \frac{1}{\cos^2\theta}$$

or

$$1 + \left(\frac{\sin\theta}{\cos\theta}\right)^2 = \left(\frac{1}{\cos\theta}\right)^2,$$

provided that $\cos\theta \neq 0$. Since

$$\frac{\sin\theta}{\cos\theta} = \tan\theta \quad \text{and} \quad \frac{1}{\cos\theta} = \sec\theta,$$

we have

$$1 + \tan^2\theta = \sec^2\theta.$$

Identity (iii) is proved by dividing both sides of (i) by $\sin^2\theta$ (Problem 34).

Examples In the following examples, the value of one of the trigonometric functions of an angle θ is given along with information about the quadrant in which θ lies. Find the values of the other five trigonometric functions of θ.

1 $\sin\theta = \dfrac{5}{13}$, θ in quadrant II.

Solution By the fundamental Pythagorean identity, $\cos^2\theta + \sin^2\theta = 1$, so

$$\cos^2\theta = 1 - \sin^2\theta = 1 - \left(\frac{5}{13}\right)^2 = 1 - \frac{25}{169} = \frac{169 - 25}{169} = \frac{144}{169}.$$

Therefore,

$$\cos\theta = \pm\sqrt{\frac{144}{169}} = \pm\frac{12}{13}.$$

Because θ is in quadrant II, we know that $\cos\theta$ is negative; hence,

$$\cos\theta = -\frac{12}{13}.$$

It follows that

$$\tan\theta = \frac{\sin\theta}{\cos\theta} = \frac{5/13}{(-12/13)} = -\frac{5}{12} \qquad \cot\theta = \frac{1}{\tan\theta} = \frac{1}{(-5/12)} = -\frac{12}{5}$$

$$\sec\theta = \frac{1}{\cos\theta} = \frac{1}{(-12/13)} = -\frac{13}{12} \qquad \csc\theta = \frac{1}{\sin\theta} = \frac{1}{5/13} = \frac{13}{5}.$$

2 $\tan\theta = -8/7$ and $\sin\theta < 0$.

Solution Because $\tan\theta < 0$ only in quadrants II and IV, and $\sin\theta < 0$ only in quadrants III and IV, it follows that θ must be in quadrant IV. By part (ii) of Theorem 3, $\sec^2\theta = 1 + \tan^2\theta$, so

$$\sec\theta = \pm\sqrt{1+\tan^2\theta} = \pm\sqrt{1+\left(-\frac{8}{7}\right)^2} = \pm\sqrt{1+\frac{64}{49}} = \pm\sqrt{\frac{113}{49}} = \pm\frac{\sqrt{113}}{7}.$$

Since θ is in quadrant IV, $\sec\theta > 0$; hence,

$$\sec\theta = \frac{\sqrt{113}}{7}.$$

Because

$$\sec\theta = \frac{1}{\cos\theta},$$

it follows that

$$\cos\theta = \frac{1}{\sec\theta} = \frac{1}{\sqrt{113}/7} = \frac{7}{\sqrt{113}} = \frac{7\sqrt{113}}{113}.$$

Now,

$$\tan\theta = \frac{\sin\theta}{\cos\theta}$$

so

$$\sin\theta = (\tan\theta)(\cos\theta) = \left(-\frac{8}{7}\right)\left(\frac{7\sqrt{113}}{113}\right) = -\frac{8\sqrt{113}}{113}.$$

Finally,

$$\csc\theta = \frac{1}{\sin\theta} = \frac{1}{(-8\sqrt{113}/113)} = -\frac{\sqrt{113}}{8}$$

and

$$\cot\theta = \frac{1}{\tan\theta} = \frac{1}{(-8/7)} = -\frac{7}{8}.$$

PROBLEM SET 3

In problems 1–10, sketch two coterminal angles α and β in standard position whose terminal side contains the given point. Arrange it so that α is positive, β is negative, and neither angle exceeds one revolution. In each case, name the quadrant in which the angle lies, or indicate that the angle is quadrantal.

1. $(1, 2)$ **2.** $(2, -\frac{4}{3})$ **3.** $(-5, 0)$ **4.** $(-3, -4)$
5. $(5, -3)$ **6.** $(0, 4)$ **7.** $(-1, 1)$ **8.** $(\sqrt{2}/2, 0)$
9. $(-\frac{3}{4}, -\frac{3}{4})$ **10.** $(0, -3)$

In problems 11–18, specify and sketch three angles that are coterminal with the given angle in standard position.

11. 60° **12.** -15° **13.** $-\dfrac{\pi}{4}$ **14.** $\dfrac{4\pi}{3}$
15. -612° **16.** $-\dfrac{7\pi}{4}$ **17.** $\dfrac{5\pi}{6}$ **18.** 1440°

In problems 19–28, evaluate the six trigonometric functions of the angle θ in standard position if the terminal side of θ contains the given point (x, y). [Do not use a calculator—leave all answers in the form of a fraction or an integer.] In each case, sketch one of the coterminal angles θ.

19. $(4, 3)$ **20.** $(2, 7)$ **21.** $(-5, 12)$ **22.** $(-2, 4)$
23. $(-3, -4)$ **24.** $(1, 1)$ **25.** $(7, 3)$ **26.** $(1, -3)$
27. $(-\sqrt{3}, -\sqrt{2})$ **28.** $(20, 21)$

29. Is there any angle θ for which $\sin\theta = \frac{5}{4}$? Explain.

30. Using similar triangles, show that the values of the six trigonometric functions in Definition 1 depend only on the angle θ and not on the choice of the point (x, y) on the terminal side of θ.

31. In each case, assume that θ is an angle in standard position and find the quadrant in which it lies.
(a) $\tan\theta > 0$ and $\sec\theta > 0$ (b) $\sin\theta > 0$ and $\sec\theta < 0$
(c) $\sin\theta > 0$ and $\cos\theta < 0$ (d) $\sec\theta > 0$ and $\tan\theta < 0$
(e) $\tan\theta > 0$ and $\csc\theta < 0$ (f) $\cos\theta < 0$ and $\csc\theta < 0$
(g) $\sec\theta > 0$ and $\cot\theta < 0$ (h) $\cot\theta > 0$ and $\sin\theta > 0$

32. Is there any angle θ for which $\sin\theta > 0$ and $\csc\theta < 0$? Explain.

33. Give the algebraic sign of each of the following.
(a) $\cos 163^\circ$ (b) $\sin 211^\circ$ (c) $\sec(-355^\circ)$ (d) $\tan\dfrac{7\pi}{4}$
(e) $\cot\left(-\dfrac{28\pi}{45}\right)$ (f) $\csc\left(-\dfrac{5\pi}{12}\right)$ (g) $\sec\dfrac{38\pi}{15}$

34. Prove part (iii) of Theorem 3.

35. Prove Theorem 2.

36. If θ is an angle for which the functions are defined, show that $\sec\theta - (\sin\theta)(\tan\theta) = \cos\theta$.

37. If $\sin\theta = -\frac{5}{13}$ and $\cos\theta = \frac{12}{13}$, use the reciprocal and quotient identities to find (a) $\sec\theta$, (b) $\csc\theta$, (c) $\tan\theta$, and (d) $\cot\theta$.

38. If $\sec\theta = -\frac{5}{4}$ and $\csc\theta = -\frac{5}{3}$, use the reciprocal and quotient identities to find (a) $\sin\theta$, (b) $\cos\theta$, (c) $\tan\theta$, and (d) $\cot\theta$.

In problems 39–56, the value of one of the trigonometric functions of an angle θ is given along with information about the quadrant (Q) in which θ lies. Find the values of the other five trigonometric functions of θ.

39. $\sin\theta = \frac{4}{5}$, θ in Q_{I}
40. $\cos\theta = \frac{12}{13}$, θ in Q_{IV}
41. $\sin\theta = -\frac{3}{4}$, θ in Q_{III}
42. $\sin\theta = \frac{3}{4}$, θ not in Q_{I}
43. $\cos\theta = \frac{4}{7}$, $\sin\theta < 0$
44. $\cos\theta = \frac{1}{3}$, θ not in Q_{I}
45. $\csc\theta = \frac{3}{2}$, θ in Q_{I}
46. $\sec\theta = -\frac{5}{3}$, θ in Q_{III}
47. $\tan\theta = \frac{4}{3}$, θ in Q_{I}
48. $\tan\theta = 2$, $\sin\theta < 0$
49. $\cot\theta = -\frac{12}{5}$, $\csc\theta > 0$
50. $\csc\theta = -\frac{25}{7}$, $\sec\theta < 0$
51. $\sec\theta = -3$, $\csc\theta > 0$
52. $\cot\theta = -\sqrt{15/7}$, $\sin\theta > 0$
Ⓒ 53. $\sin\theta = 0.4695$, θ in Q_{I}
Ⓒ 54. $\tan\theta = -0.5095$, $\sin\theta < 0$
Ⓒ 55. $\csc\theta = 2.613$, θ in Q_{II}
Ⓒ 56. $\sec\theta = 2.924$, $\tan\theta > 0$

57. According to Theorem 1, the six trigonometric functions can be grouped in three reciprocal pairs. They can also be grouped in three pairs of cofunctions. Is any reciprocal pair also a pair of cofunctions?

4 Evaluation of Trigonometric Functions

In this section, we obtain values of the trigonometric functions for quadrantal angles, we introduce the idea of reference angles, and we discuss the use of a calculator to evaluate trigonometric functions of general angles.

In Definition 1, Section 3, the domain of each trigonometric function consists of all angles θ for which the denominator in the corresponding ratio is not zero. Because $r > 0$, it follows that $\sin\theta = y/r$ and $\cos\theta = x/r$ are defined for all angles θ. However, $\tan\theta = y/x$ and $\sec\theta = r/x$ are not defined when the terminal side of θ lies along the y axis (so that $x = 0$). Likewise, $\cot\theta = x/y$ and $\csc\theta = r/y$ are not defined when the terminal side of θ lies along the x axis (so that $y = 0$). Therefore, when you deal with a trigonometric function of a quadrantal angle, you must check to be sure that the function is actually defined for that angle.

Example Find the values (if they are defined) of the six trigonometric functions for the quadrantal angle $\theta = 90°$ (or $\theta = \pi/2$).

Solution In order to use Definition 1 in Section 3, we begin by choosing any point $(0, y)$, with $y > 0$, on the terminal side of the 90° angle (Figure 1). Because $x = 0$, it follows that $\tan 90°$ and $\sec 90°$ are undefined. Since $y > 0$, we have

Figure 1

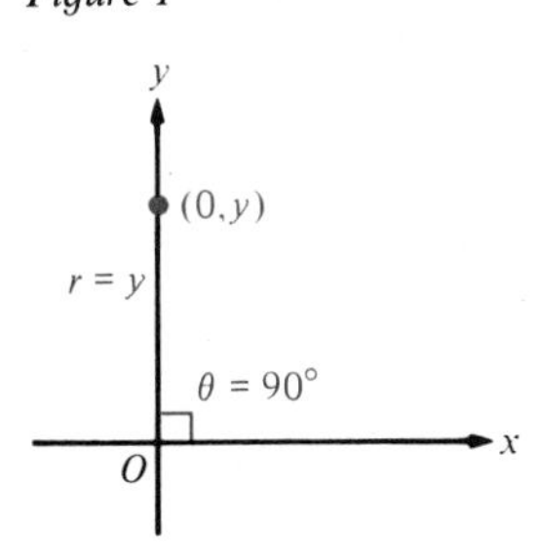

$$r = \sqrt{x^2 + y^2} = \sqrt{0^2 + y^2} = \sqrt{y^2} = |y| = y.$$

Therefore,

$$\sin 90° = \frac{y}{r} = \frac{y}{y} = 1 \qquad \cos 90° = \frac{x}{r} = \frac{0}{y} = 0$$

$$\csc 90° = \frac{r}{y} = \frac{y}{y} = 1 \qquad \cot 90° = \frac{x}{y} = \frac{0}{y} = 0.$$

The values of the trigonometric functions for other quadrantal angles are found in a similar manner (Problem 1). The results appear in Table 1. Dashes in the table indicate that the function is undefined for that angle.

Table 1

θ degrees	θ radians	$\sin\theta$	$\cos\theta$	$\tan\theta$	$\cot\theta$	$\sec\theta$	$\csc\theta$
$0°$	0	0	1	0	—	1	—
$90°$	$\frac{\pi}{2}$	1	0	—	0	—	1
$180°$	π	0	-1	0	—	-1	—
$270°$	$\frac{3\pi}{2}$	-1	0	—	0	—	-1
$360°$	2π	0	1	0	—	1	—

It follows from Definition 1 in Section 3 that the values of each of the six trigonometric functions remain unchanged if the angle is replaced by a coterminal angle. If an angle exceeds one revolution or is negative, you can change it to a nonnegative coterminal angle that is less than one revolution by adding or subtracting an integer multiple of $360°$ (or 2π radians). For instance,

$$\sin 450° = \sin(450° - 360°) = \sin 90° = 1$$

$$\sec 7\pi = \sec[7\pi - (3 \times 2\pi)] = \sec\pi = -1$$

$$\cos(-660°) = \cos[-660° + (2 \times 360°)] = \cos 60° = \tfrac{1}{2}.$$

Examples Replace each angle by a nonnegative coterminal angle that is less than one revolution and then find the values of the six trigonometric functions (if they are defined).

1 $1110°$

Solution By trial and error (or by dividing 1110 by 360), we find that the largest integer multiple of $360°$ that is less than $1110°$ is $3 \times 360° = 1080°$. Thus,

$$1110° - (3 \times 360°) = 1110° - 1080° = 30°.$$

(Or, we could have started with $1110°$ and repeatedly subtracted $360°$ until we obtained $30°$.) It follows that

$$\sin 1110° = \sin 30° = 1/2 \qquad \csc 1110° = \csc 30° = 2$$

$$\cos 1110° = \cos 30° = \sqrt{3}/2 \qquad \sec 1110° = \sec 30° = 2\sqrt{3}/3$$

$$\tan 1110° = \tan 30° = \sqrt{3}/3 \qquad \cot 1110° = \cot 30° = \sqrt{3}.$$

2 $-5\pi/2$

Solution We repeatedly add 2π to $-5\pi/2$ until we obtain a nonnegative coterminal angle:

$$(-5\pi/2) + 2\pi = -\pi/2 \qquad \text{(still negative)}$$

$$(-\pi/2) + 2\pi = 3\pi/2.$$

Therefore, by Table 1 for quadrantal angles,

$$\sin(-5\pi/2) = \sin 3\pi/2 = -1 \qquad \cot(-5\pi/2) = \cot 3\pi/2 = 0$$

$$\cos(-5\pi/2) = \cos 3\pi/2 = 0 \qquad \csc(-5\pi/2) = \csc 3\pi/2 = -1,$$

and both $\tan(-5\pi/2)$ and $\sec(-5\pi/2)$ are undefined.

In Section 2 we found exact values of the six trigonometric functions for the special acute angle 30°, 45°, and 60° (see Table 1 on page 69). This information can be applied to certain angles in quadrants II, III, and IV by using the idea of a "reference angle." If θ is an angle in standard position and if θ is not a quadrantal angle, then the **reference angle θ_R** for θ is defined to be the positive acute angle formed by the terminal side of θ and the x axis. Figure 2 shows the reference angle θ_R for an angle θ in each of the four quadrants.

Figure 2

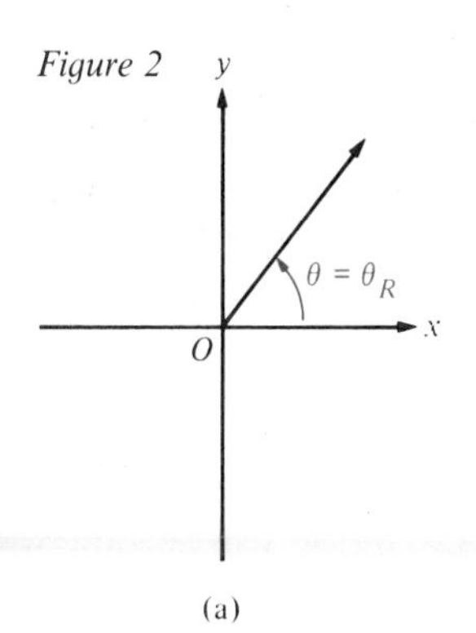

(a)

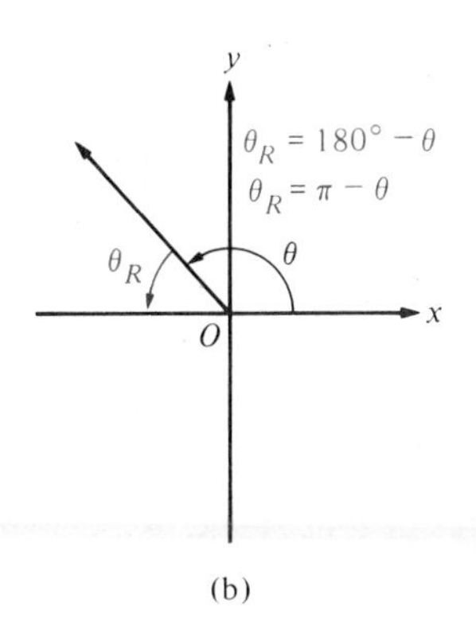

(b)

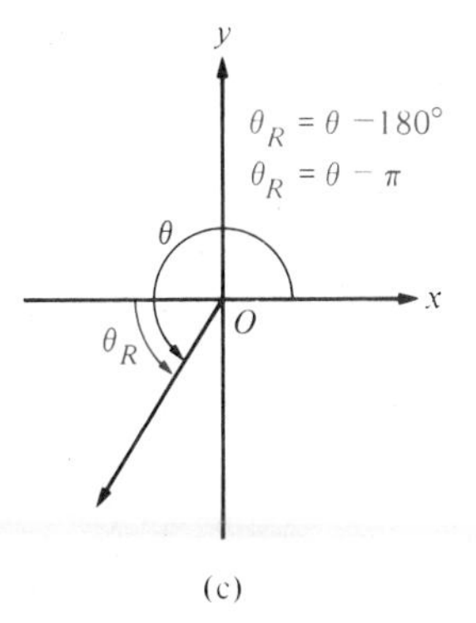

(c)

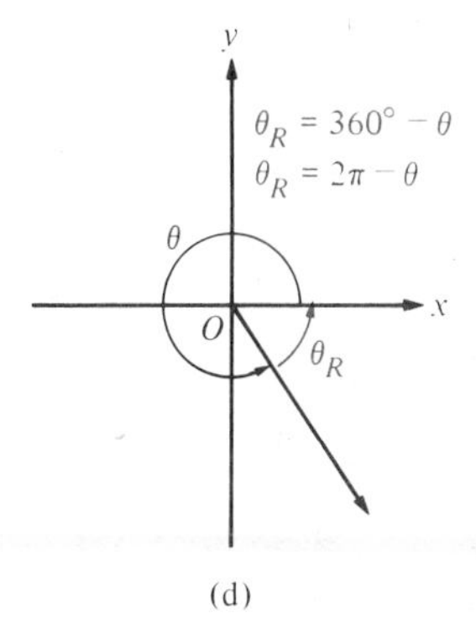

(d)

Example Find the reference angle θ_R for each angle θ.

(a) $\theta = 60°$ (b) $\theta = \dfrac{3\pi}{4}$ (c) $\theta = 210°$ (d) $\theta = \dfrac{5\pi}{3}$

Solution (a) By Figure 2a, $\theta_R = \theta = 60°$.

(b) By Figure 2b, $\theta_R = \pi - \theta = \pi - \dfrac{3\pi}{4} = \dfrac{\pi}{4}$.

(c) By Figure 2c, $\theta_R = \theta - 180° = 210° - 180° = 30°$.

(d) By Figure 2d, $\theta_R = 2\pi - \theta = 2\pi - \dfrac{5\pi}{3} = \dfrac{\pi}{3}$.

The value of any trigonometric function of an angle θ is the same as the value of the function for the reference angle θ_R, except possibly for a change of algebraic sign. Thus,

$$\sin\theta = \pm\sin\theta_R, \qquad \cos\theta = \pm\cos\theta_R,$$

and so forth. You can always determine the correct algebraic sign by considering which quadrant θ lies in.

Figure 3

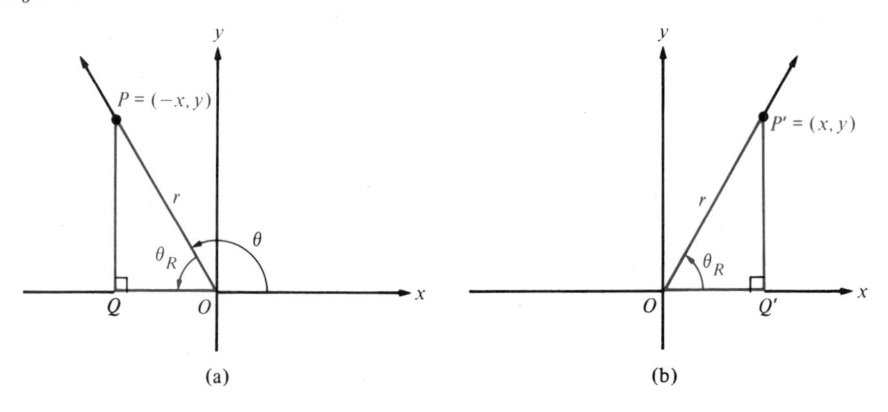

We shall verify the statement above for θ in quadrant II, and leave the remaining cases as an exercise (Problem 16). Figure 3a shows a quadrant II angle θ in standard position; the corresponding reference angle θ_R is shown in standard position in Figure 3b. In the two figures, we have chosen points P and P' as shown, both r units from the origin on the terminal sides of θ and θ_R. Dropping perpendiculars $\overline{PQ}$ and $\overline{P'Q'}$ from P and P' to the x axis, we find that the right triangles PQO and $P'Q'O$ are congruent. (Why?) Therefore, if the coordinates of P' are (x, y), it follows that the coordinates of P are $(-x, y)$. Applying Definition 1 in Section 3 to Figures 3a and 3b, we have:

$$\sin\theta = \frac{y}{r} = \sin\theta_R \qquad \csc\theta = \frac{r}{y} = \csc\theta_R$$

$$\cos\theta = \frac{-x}{r} = -\frac{x}{r} = -\cos\theta_R \qquad \sec\theta = \frac{r}{-x} = -\frac{r}{x} = -\sec\theta_R$$

$$\tan\theta = \frac{y}{-x} = -\frac{y}{x} = -\tan\theta_R \qquad \cot\theta = \frac{-x}{y} = -\frac{x}{y} = -\cot\theta_R.$$

Examples Find the reference angle θ_R for the given angle θ, and then use the information in Table 1 on page 69 to find the values of the six trigonometric functions of θ.

1 $\theta = 135^\circ$

Solution The reference angle is $\theta_R = 180° - 135° = 45°$ (Figure 4). Since 135° is a quadrant II angle, we have

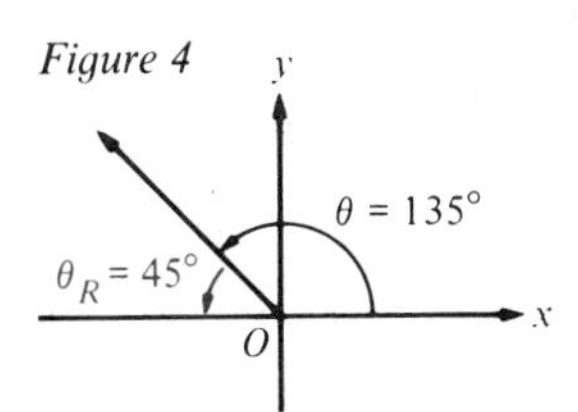

Figure 4

$$\sin 135° = \sin 45° = \frac{\sqrt{2}}{2} \qquad \csc 135° = \csc 45° = \sqrt{2}$$

$$\cos 135° = -\cos 45° = -\frac{\sqrt{2}}{2} \qquad \sec 135° = -\sec 45° = -\sqrt{2}$$

$$\tan 135° = -\tan 45° = -1 \qquad \cot 135° = -\cot 45° = -1.$$

2 $\theta = 210°$

Solution The reference angle is $\theta_R = 210° - 180° = 30°$ (Figure 5). Since 210° is a quadrant III angle, we have

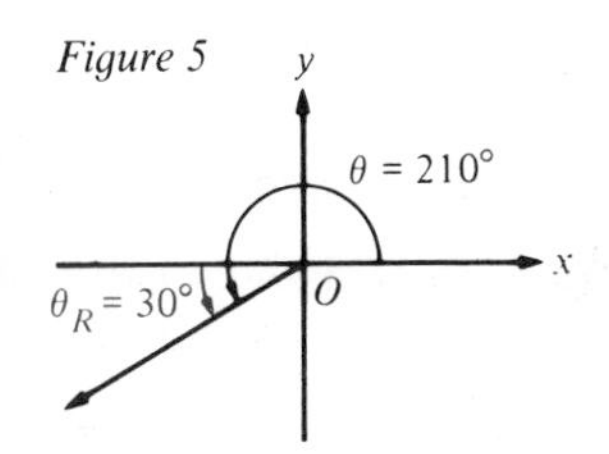

Figure 5

$$\sin 210° = -\sin 30° = -\frac{1}{2} \qquad \csc 210° = -\csc 30° = -2$$

$$\cos 210° = -\cos 30° = -\frac{\sqrt{3}}{2} \qquad \sec 210° = -\sec 30° = -\frac{2\sqrt{3}}{3}$$

$$\tan 210° = \tan 30° = \frac{\sqrt{3}}{3} \qquad \cot 210° = \cot 30° = \sqrt{3}.$$

3 $\theta = \frac{5\pi}{3}$

Solution The reference angle is $\theta_R = 2\pi - \frac{5\pi}{3} = \frac{\pi}{3}$ (Figure 6). Since $\frac{5\pi}{3}$ is a quadrant IV angle, we have

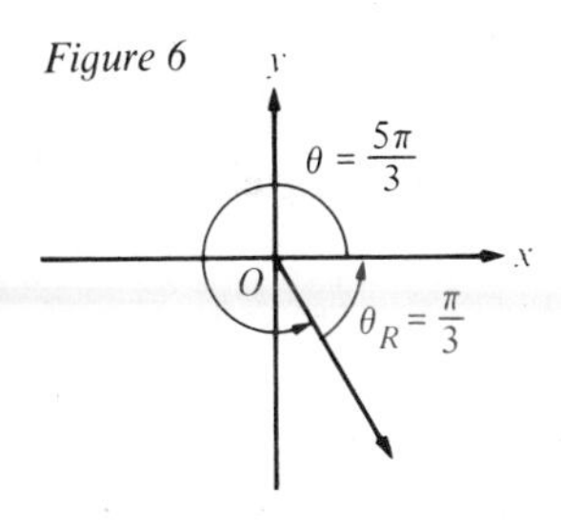

Figure 6

$$\sin \frac{5\pi}{3} = -\sin \frac{\pi}{3} = -\frac{\sqrt{3}}{2} \qquad \csc \frac{5\pi}{3} = -\csc \frac{\pi}{3} = -\frac{2\sqrt{3}}{3}$$

$$\cos \frac{5\pi}{3} = \cos \frac{\pi}{3} = \frac{1}{2} \qquad \sec \frac{5\pi}{3} = \sec \frac{\pi}{3} = 2$$

$$\tan \frac{5\pi}{3} = -\tan \frac{\pi}{3} = -\sqrt{3} \qquad \cot \frac{5\pi}{3} = -\cot \frac{\pi}{3} = -\frac{\sqrt{3}}{3}.$$

To find the reference angle θ_R for an angle θ in standard position when θ is negative or greater than one revolution, you can replace θ by a nonnegative coterminal angle that is less than one revolution, and then proceed as in the examples above.

Example Find the reference angle θ_R for the given angle θ, and then use the information in Table 1 on page 281 to find the values of the six trigonometric functions of θ.

(a) $\theta = \frac{21\pi}{4}$ (b) $\theta = -300°$

Solution (a) Here $\theta = \frac{21\pi}{4}$ is coterminal with $\frac{21\pi}{4} - 2(2\pi) = \frac{5\pi}{4}$, a quadrant III angle (Figure 7a). The corresponding reference angle is $\theta_R = \frac{5\pi}{4} - \pi = \frac{\pi}{4}$, so

$$\sin\frac{21\pi}{4} = -\sin\frac{\pi}{4} = -\frac{\sqrt{2}}{2} \qquad \csc\frac{21\pi}{4} = -\csc\frac{\pi}{4} = -\sqrt{2}$$

$$\cos\frac{21\pi}{4} = -\cos\frac{\pi}{4} = -\frac{\sqrt{2}}{2} \qquad \sec\frac{21\pi}{4} = -\sec\frac{\pi}{4} = -\sqrt{2}$$

$$\tan\frac{21\pi}{4} = \tan\frac{\pi}{4} = 1 \qquad \cot\frac{21\pi}{4} = \cot\frac{\pi}{4} = 1.$$

Figure 7

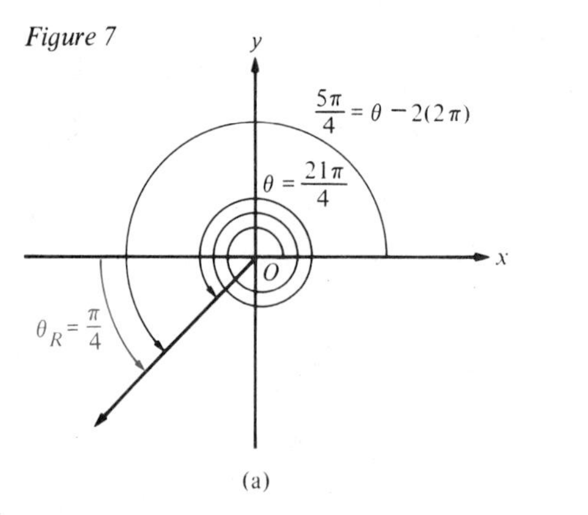

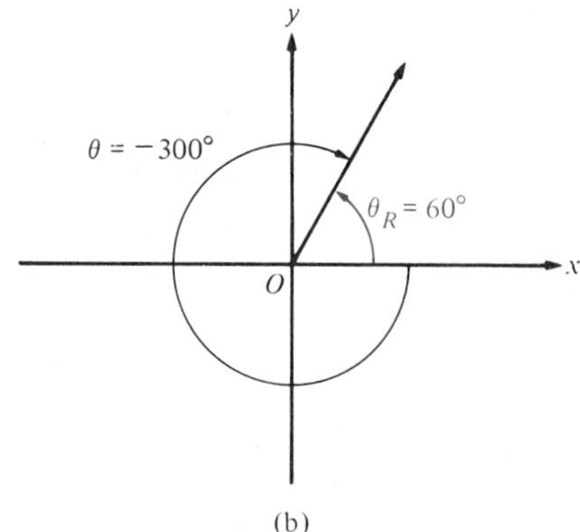

(b) The angle $\theta = -300^\circ$ is coterminal with $-300^\circ + 360^\circ = 60^\circ$, a quadrant I angle (Figure 7b). Here the reference angle is also $\theta_R = 60^\circ$, and we have

$$\sin(-300^\circ) = \sin 60^\circ = \frac{\sqrt{3}}{2} \qquad \csc(-300^\circ) = \csc 60^\circ = \frac{2\sqrt{3}}{3}$$

$$\cos(-300^\circ) = \cos 60^\circ = \frac{1}{2} \qquad \sec(-300^\circ) = \sec 60^\circ = 2$$

$$\tan(-300^\circ) = \tan 60^\circ = \sqrt{3} \qquad \cot(-300^\circ) = \cot 60^\circ = \frac{\sqrt{3}}{3}.$$

In the examples above, we obtained *exact* values of the trigonometric functions for certain angles. Approximate values of these functions for all other angles can be found by using a scientific calculator. No special techniques are required beyond those already presented in Section 2 for evaluating the trigonometric functions of acute angles—correct algebraic signs are automatically provided. (If a calculator isn't available, use the tables in Section 3 of the Appendix.)

Examples Ⓒ Use a calculator to find the approximate value of each trigonometric function.

1 $\cot 300.42^\circ$

Solution With the calculator in degree mode, we enter 300.42, press the tangent (TAN) key, then press the 1/x key. (We're using the fact that $\cot 300.42^\circ = \dfrac{1}{\tan 300.42^\circ}$.) On a 10-digit calculator, the result is

$$\cot 300.42^\circ = -0.587165830.$$

2 $\sec \dfrac{40\pi}{7}$

Solution With the calculator in radian mode, we enter π, multiply by 40, divide by 7, press the cosine (COS) key, and then press the 1/x key. On a 10-digit calculator, we get

$$\sec \frac{40\pi}{7} = 1.603875452.$$

3 $\sin(-2.007)$

Solution Because degrees are not indicated, we know that the angle is measured in radians. We set the calculator in radian mode, enter -2.007, and press the sine (SIN) key to obtain

$$\sin(-2.007) = -0.906362145.$$

PROBLEM SET 4

In problems 1 and 2, find the values (if they are defined) of the six trigonometric functions of the given quadrantal angles.

1. (a) 0° (b) 180° (c) 270° (d) 360°
 [When you have finished, compare your answers with the results in Table 1 on page 82.]
2. (a) 5π (b) 6π (c) -7π (d) $\dfrac{5\pi}{2}$ (e) $\dfrac{7\pi}{2}$

In problems 3–14, replace each angle by a nonnegative coterminal angle that is less than one revolution, and then find the exact values of the six trigonometric functions (if they are defined) for the angle.

3. 1440°
4. 810°
5. 900°
6. -720°
7. 750°
8. 1845°
9. -675°
10. $\dfrac{19\pi}{2}$
11. 5π
12. $\dfrac{25\pi}{6}$
13. $-\dfrac{17\pi}{3}$
14. $-\dfrac{31\pi}{4}$

Ⓒ **15.** What happens when you try to evaluate $\tan 90^\circ$ on a calculator? [Try it.]

16. Let θ be a quadrant III angle in standard position and let θ_R be its reference angle. Show that the value of any trigonometric function of θ is the same as the value of the function for θ_R, except possibly for a change of algebraic sign. Repeat for θ in quadrant IV.

In problems 17–36, find the reference angle θ_R for each angle θ, and then find the exact values of the six trigonometric functions of θ.

17. $\theta = 150°$ **18.** $\theta = 120°$ **19.** $\theta = 240°$ **20.** $\theta = 225°$ **21.** $\theta = 315°$

22. $\theta = \frac{9\pi}{4}$ **23.** $\theta = -150°$ **24.** $\theta = 675°$ **25.** $\theta = -60°$ **26.** $\theta = -\frac{7\pi}{6}$

27. $\theta = -\frac{\pi}{4}$ **28.** $\theta = -\frac{13\pi}{6}$ **29.** $\theta = -\frac{2\pi}{3}$ **30.** $\theta = \frac{53\pi}{6}$ **31.** $\theta = \frac{7\pi}{4}$

32. $\theta = -\frac{50\pi}{3}$ **33.** $\theta = \frac{11\pi}{3}$ **34.** $\theta = -\frac{147\pi}{4}$ **35.** $\theta = -420°$ **36.** $\theta = -5370°$

37. Complete the following table. (Do not use a calculator.)

θ degrees	θ radians	$\sin\theta$	$\cos\theta$	$\tan\theta$	$\cot\theta$	$\sec\theta$	$\csc\theta$
210°	$7\pi/6$						
225°	$5\pi/4$						
240°	$4\pi/3$						
300°	$5\pi/3$						
315°	$7\pi/4$						
330°	$11\pi/6$						

38. A calculator is set in radian mode, π is entered, and the sine (SIN) key is pressed. The display shows -4.1×10^{-10}. But we know that $\sin\pi = 0$. Explain.

Ⓒ In problems 39–58, use a calculator to find the approximate values of the six trigonometric functions of each angle.

39. 46° **40.** 162° **41.** 143° **42.** −225.31° **43.** −61.37°

44. 852.48° **45.** 61°35′ **46.** −31°8′ **47.** −97°9′8″ **48.** $\frac{12\pi}{5}$

49. $-\frac{2\pi}{7}$ **50.** $\frac{\sqrt{2}\pi}{3}$ **51.** 1.67 **52.** 3.725 **53.** −2.436

54. 16.57 **55.** −10.79 **56.** −6.203 **57.** 9.673 **58.** −3985.2

Ⓒ In problems 59–62, use a calculator to verify that the equation is true for the indicated value of the angle θ.

59. $\tan\theta = \sin\theta/\cos\theta$ for $\theta = 35°$

60. $(\cos\theta)(\tan\theta) = \sin\theta$ for $\theta = 5\pi/7$

61. $\sin^2\theta + \cos^2\theta = 1$ for $\theta = 3\pi/5$

62. $1 + \tan^2\theta = \sec^2\theta$ for $\theta = 17.75°$

63. Verify that for $\theta = 0°, 30°, 45°, 60°$, and $90°$, we have $\sin\theta = \sqrt{0}/2, \sqrt{1}/2, \sqrt{2}/2, \sqrt{3}/2$, and $\sqrt{4}/2$, respectively. [Although there is no theoretical significance to this pattern, people often use it as a memory aid to help recall these values of $\sin\theta$.]

5 Graphs of the Sine and Cosine Functions

Until now, we have considered trigonometric functions of *angles*, measured either in degrees or radians. In advanced work, particularly in calculus and its applications, it is necessary to modify the trigonometric functions so that their domains consist of **real numbers.** For reasons that we shall soon explain, the resulting functions are sometimes called the **circular functions.**

If t is a real number, then the value of $\sin t$ is defined by

$$\sin t = \sin \theta,$$

where θ is an angle whose measure is t radians. Similarly, the values of $\cos t$, $\tan t$, $\cot t$, $\sec t$, $\csc t$ are defined to be the values of the corresponding trigonometric functions of an angle θ whose measure is t radians. We shall now explain just what these so-called circular functions have to do with circles.

Consider the **unit circle;** that is, the circle with center at the origin and radius $r = 1$ (Figure 1a). Let t be any real number. Starting at the point $(1, 0)$ on the unit circle, we move $|t|$ units around the circle, counterclockwise if $t > 0$ and clockwise if $t < 0$, to the point (x, y) (Figure 1a). Now let θ be the angle in standard position whose terminal side contains the point (x, y). Recall from Section 1 that $t = r\theta$. Since $r = 1$, it follows that angle θ has radian measure t. Hence,

$$\sin t = \sin \theta = \frac{y}{r} = \frac{y}{1} = y \quad \text{and} \quad \cos t = \cos \theta = \frac{x}{r} = \frac{x}{1} = x.$$

Therefore,

$$(x, y) = (\cos t, \sin t)$$

(Figure 1b).

In Figure 1b, the coordinates of $(\cos t, \sin t)$ repeat themselves each time the point goes completely around the circle; that is, each time t increases by 2π (the circumference of the unit circle). Thus, the sine and cosine functions are **periodic** in the sense that their values repeat themselves whenever the independent variable increases by 2π, that is,

$$\sin(t + 2\pi) = \sin t \quad \text{and} \quad \cos(t + 2\pi) = \cos t.$$

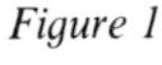
Figure 1

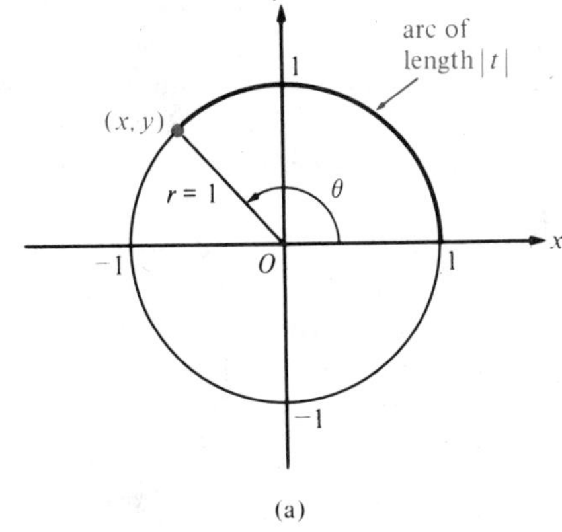

(a)

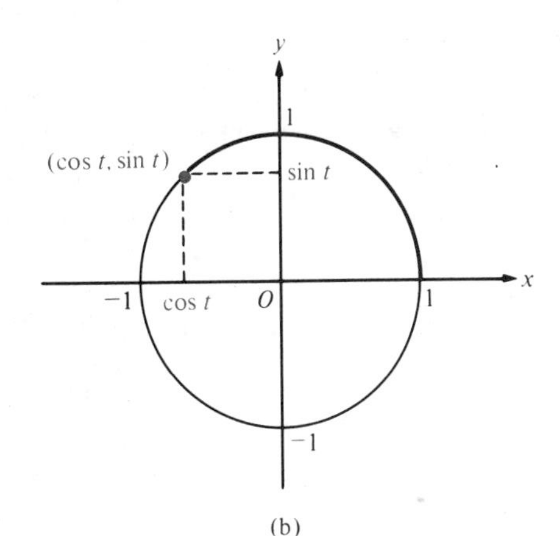

(b)

As t increases uniformly from 0 to 2π, the point $(\cos t, \sin t)$ (Figure 1b) moves once around the unit circle in a counterclockwise direction, and the coordinates $\cos t$ and $\sin t$ change as shown in Table 1.

Table 1

As t Increases from	$\sin t$	$\cos t$
0 to $\frac{\pi}{2}$	is positive and increases from 0 to 1	is positive and decreases from 1 to 0
$\frac{\pi}{2}$ to π	is positive and decreases from 1 to 0	is negative and decreases from 0 to -1
π to $\frac{3\pi}{2}$	is negative and decreases from 0 to -1	is negative and increases from -1 to 0
$\frac{3\pi}{2}$ to 2π	is negative and increases from -1 to 0	is positive and increases from 0 to 1

The idea of a circular function (that is, a trigonometric function of a real number rather than of an angle) is more a point of view than a new mathematical concept. For instance, put a calculator in radian mode, enter $\pi/3$, and press the sine key. The display will show the approximate value of $\sin \pi/3$, and it makes no difference whether you have in mind that $\pi/3$ is the measure of an angle in radians, the length of an arc on the unit circle, or simply a real number. In any case,

$$\sin\frac{\pi}{3} = \frac{\sqrt{3}}{2} \approx 0.866025404.$$

For this reason, people speak of "trigonometric functions" and "circular functions" more or less interchangeably.

On most scientific calculators, the number in the display register is denoted by x. Pressing the 1/x key gives you the reciprocal of x, pressing the x^2 key gives you the square of x, and, likewise, pressing the sine or cosine key gives you $\sin x$ or $\cos x$. This notation is consistent with the custom of denoting the independent variable of a function by x (unless there are good reasons for doing otherwise). Therefore, in what follows we shall sketch graphs of $y = \sin x$ and of $y = \cos x$ on the usual xy coordinate system. Notice that x, used in this way as the independent variable for a circular function, must not be confused with the abscissa of the point (x, y) in Figure 1.

We can make a rough sketch of the graph of $y = \sin x$ for $0 \leq x \leq 2\pi$ by using only the information in Table 1, if we replace t by x. However, we can make a more accurate sketch by using the values of $\sin x$ shown in Table 2. If even more accuracy is desired, additional points can be found with the aid of a calculator. The graph of $y = \sin x$ for $0 \leq x \leq 2\pi$ is sketched in Figure 2. This curve is called **one cycle** of the graph of $y = \sin x$.

Table 2

x	$y = \sin x$
0	0
$\pi/6$	$1/2$
$\pi/4$	$\sqrt{2}/2$
$\pi/3$	$\sqrt{3}/2$
$\pi/2$	1
$2\pi/3$	$\sqrt{3}/2$
$3\pi/4$	$\sqrt{2}/2$
$5\pi/6$	$1/2$
π	0
$7\pi/6$	$-1/2$
$5\pi/4$	$-\sqrt{2}/2$
$4\pi/3$	$-\sqrt{3}/2$
$3\pi/2$	-1
$5\pi/3$	$-\sqrt{3}/2$
$7\pi/4$	$-\sqrt{2}/2$
$11\pi/6$	$-1/2$
2π	0

Figure 2

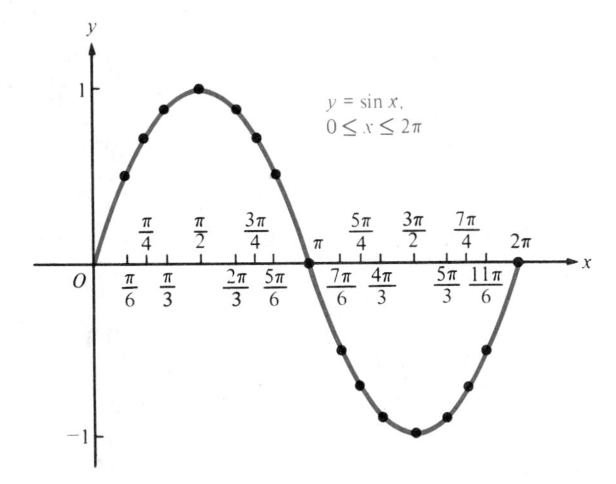

Because the sine is a periodic function, the complete graph of $y = \sin x$ consists of an endless sequence of copies of the cycle in Figure 2, repeated to the right and left over successive intervals of length 2π (Figure 3). The graph shows that the range of the sine function is the closed interval $[-1, 1]$. In particular,

$$-1 \leq \sin x \leq 1$$

holds for all real numbers x. Other features of the sine function are indicated by its graph. For instance, the graph appears to be symmetric about the origin, which suggests that the sine is an odd function,

$$\sin(-x) = -\sin x.$$

Figure 3

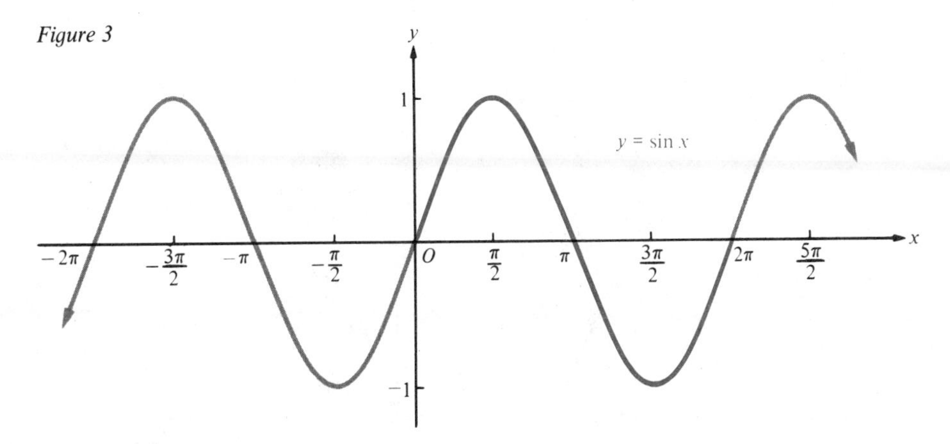

(We prove this in Section 1 of Chapter 3.) Another interesting feature suggested by the graph is that

$$\sin(x + \pi) = -\sin x$$

(Figure 4, page 92). This property is proved in Section 3 of Chapter 3.

Figure 4

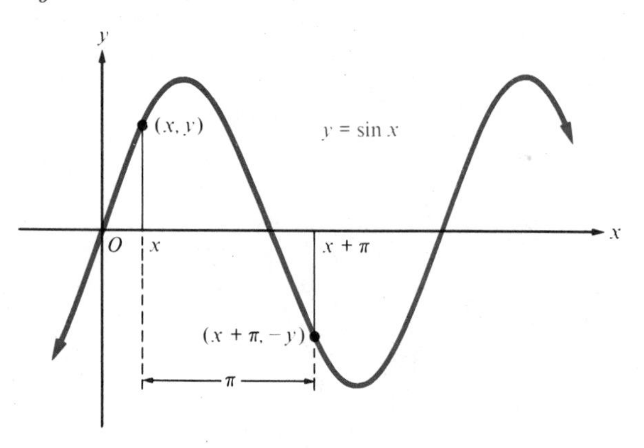

By using the information in Table 1 and the values of the cosine function for angles such as 0, $\pi/6$, $\pi/4$, $\pi/3$, $\pi/2$, and so on, you can sketch the graph of $y = \cos x$ for $0 \leq x \leq 2\pi$ (Problem 1). This curve (Figure 5) is called one *cycle* of the graph of $y = \cos x$. By extending the cycle to the right and to the left in a periodic fashion, we obtain the complete graph of $y = \cos x$ (Figure 6). Figure 6 also includes the graph of $y = \sin x$, which, as you can see, is the graph of $y = \cos x$ shifted $\pi/2$ units to the right. This indicates that

$$\sin x = \cos\left(x - \frac{\pi}{2}\right),$$

a fact which is proved in Section 3 of Chapter 3.

Like the sine function, the range of the cosine function is the interval $[-1, 1]$; hence,

$$-1 \leq \cos x \leq 1$$

holds for all real numbers x. The graph in Figure 6 appears to be symmetric about the y axis, indicating that the cosine is an even function,

$$\cos(-x) = \cos x.$$

(Again, this is proved in Section 1 of Chapter 3.) By drawing a figure for $y = \cos x$ similar to Figure 4, you can also see graphically that

$$\cos(x + \pi) = -\cos x$$

(Problem 2).

Figure 5

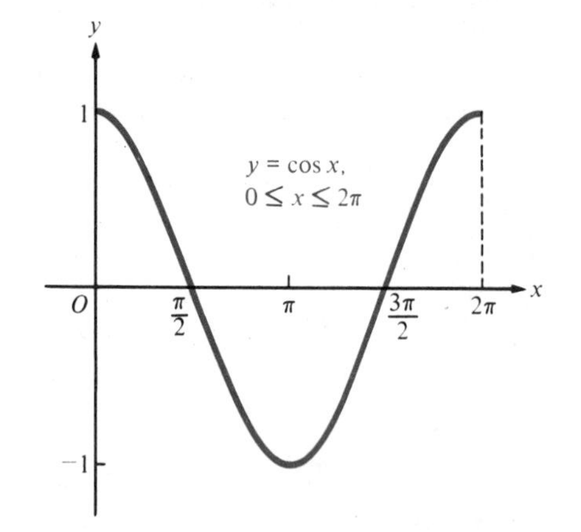

Figure 6

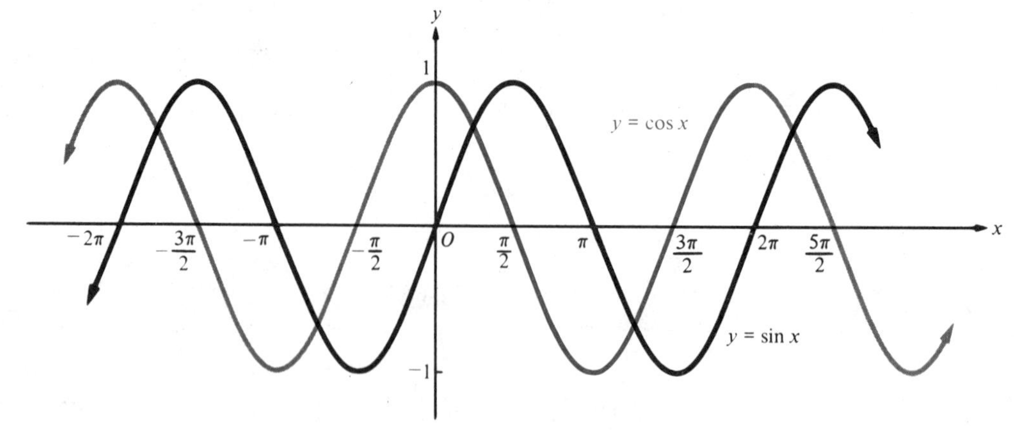

If the techniques of graphing discussed in Section 6 of Chapter 1 are used to stretch or shift the graphs of sine or cosine functions, the resulting curves always have the wavelike form shown in Figure 7. Such a curve is called a **simple harmonic curve,** a **sine wave,** or a **sinusoidal curve.** The highest points on a simple harmonic curve are called **crests** and the lowest points are called **troughs.** The crests lie along one straight line, the troughs lie along another, and the two lines are parallel. The straight line midway between the crest line and the trough line is called the **axis** of the simple harmonic curve, and the points where the curve crosses the axis are called **nodes.** The distance between the axis and the crest line (or trough line) is called the **amplitude** of the simple harmonic curve.

Figure 7

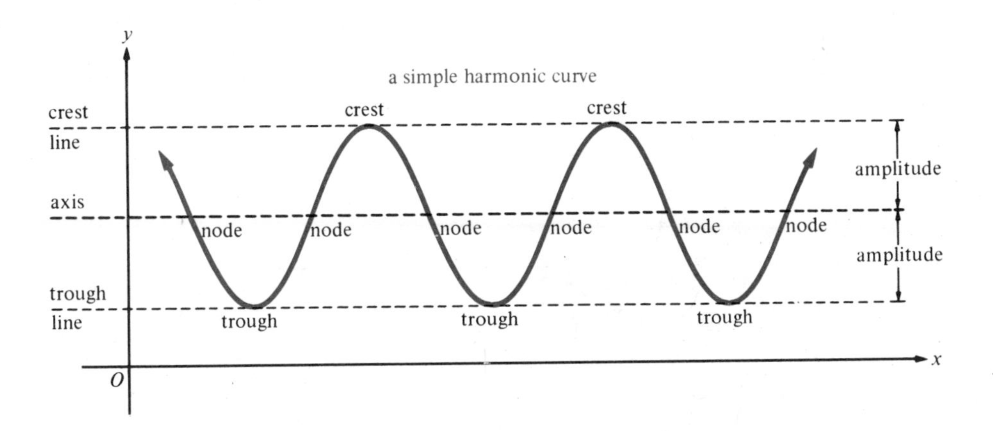

The portion of a simple harmonic curve from any one point P to the next point Q at which the curve starts to repeat itself is called a **cycle,** and the distance between P and Q is called the **period** of the simple harmonic curve (Figure 8). In sketching a cycle, we usually begin at a crest, node, or trough. If you are given (or have already sketched) one cycle, you can easily sketch as much of the rest of the curve as you please by repeating the cycle to the right and to the left over successive intervals, each of which has length equal to the period. It is often convenient to use different units of length along the x and y axes.

Figure 8

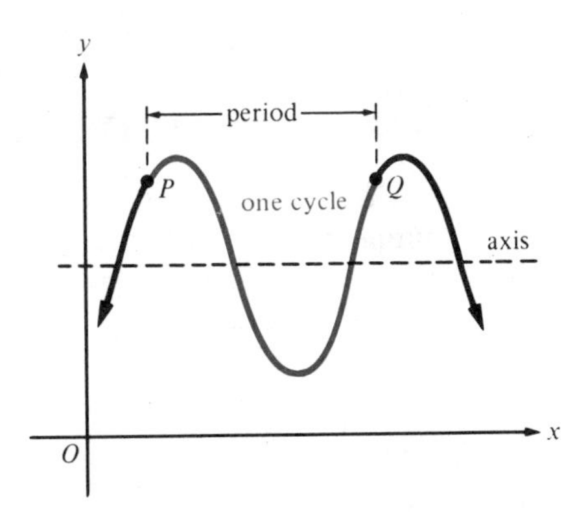

Examples Sketch the graph of the function defined by each equation, find the amplitude and the period, and indicate one cycle (starting at a node) on your graph.

1 (a) $y = 3 \sin x$

(b) $y = -3 \sin x$

Solution (a) We begin by sketching the graph of $y = \sin x$ (Figure 3); then we multiply each ordinate by 3 (Figure 9a). The x axis is the axis of the simple harmonic curve. The crests are 3 units above the axis, so the amplitude is 3. The period is 2π. One cycle, starting at $(0, 0)$ and ending at $(2\pi, 0)$, is indicated on the graph.

Figure 9

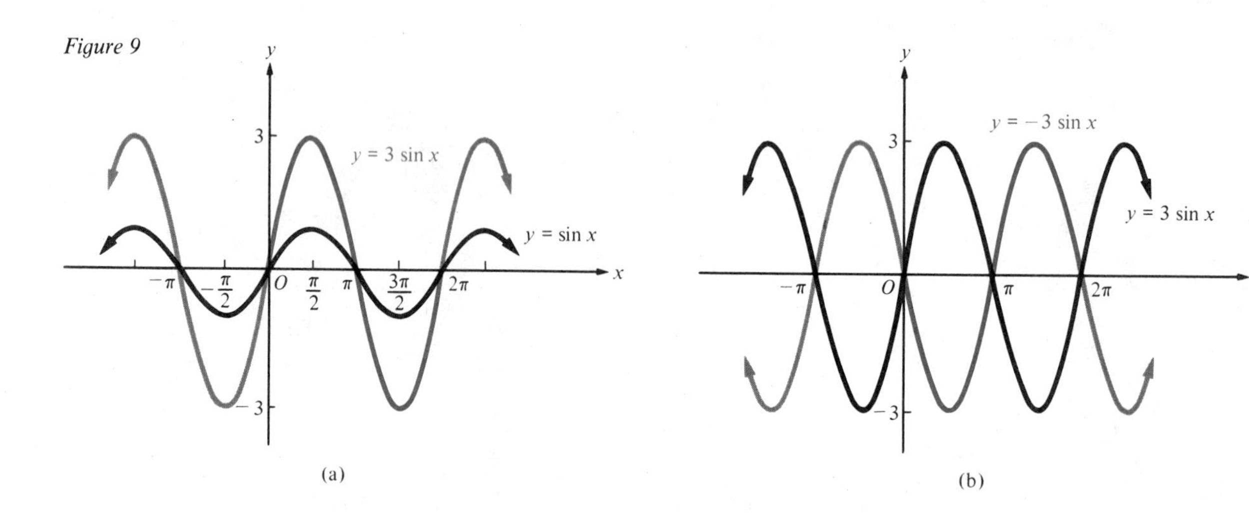

(b) We obtain the graph of $y = -3 \sin x$ by reflecting the graph of $y = 3 \sin x$ across the x axis (Figure 9b). The amplitude of the reflected graph is still 3, and its period is still 2π. One cycle, starting at $(0, 0)$ and ending at $(2\pi, 0)$, is indicated on the graph.

2 (a) $y = \frac{1}{3} \cos x$

(b) $y = 1 + \frac{1}{3} \cos x$

Solution (a) The graph of $y = \frac{1}{3} \cos x$ is obtained from the graph of $y = \cos x$ (Figure 6) by multiplying each ordinate by $\frac{1}{3}$ (Figure 10a). The amplitude is $\frac{1}{3}$, the period is 2π, and one cycle [starting at $x = \pi/2$ and ending at $x = (\pi/2) + 2\pi = 5\pi/2$] is indicated on the graph.

(b) The graph of $y = 1 + \frac{1}{3} \cos x$ is obtained by shifting the graph of $y = \frac{1}{3} \cos x$ one unit upward. This shift does not affect the amplitude, which is still $\frac{1}{3}$, or the period, which is still 2π. One cycle is indicated on the graph (Figure 10b).

Figure 10

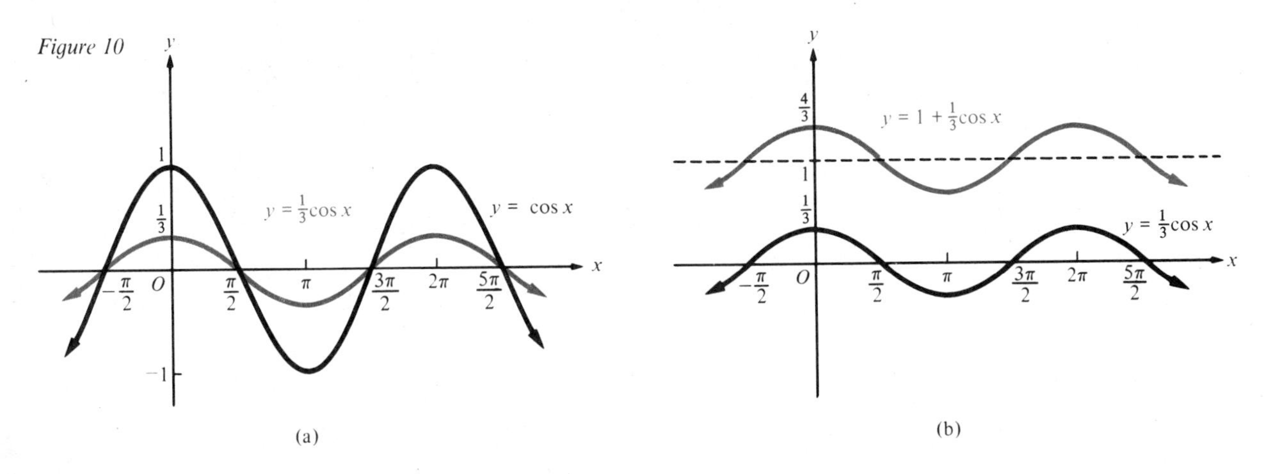

3 $y = 3\sin 2x$

Solution The graph of $y = 3\sin 2x$ will cover one cycle as $2x$ varies from 0 to 2π. When $2x = 0$, we have $x = 0$; when $2x = 2\pi$, we have $x = \pi$. Therefore, the graph covers one cycle as x varies from 0 to π, and the period is π. As x varies from 0 to π, $\sin 2x$ will take on maximum and minimum values of $+1$ and -1; hence, $3\sin 2x$ will take on maximum and minimum values of $+3$ and -3. Thus, the amplitude is 3. We use this information to sketch one cycle, and then repeat the cycle on either side (Figure 11). Figure 11 includes the graph of $y = 3\sin x$ for comparison. Notice that the graph of $y = 3\sin 2x$ oscillates up and down twice as fast as the graph of $y = 3\sin x$.

Figure 11

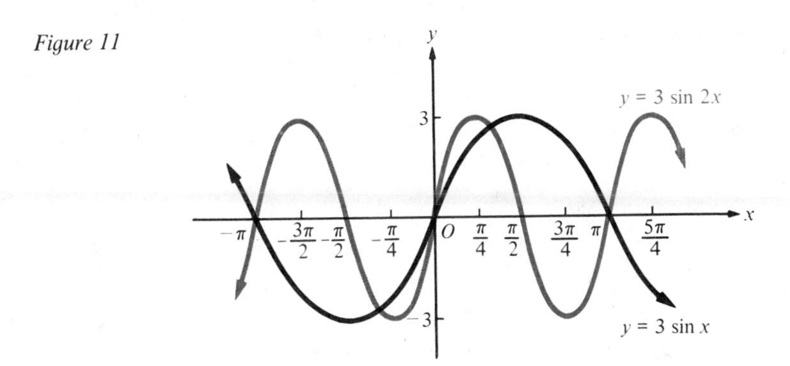

4 $y = 3\sin\left(2x - \frac{\pi}{2}\right)$

Solution The graph will cover one cycle as $2x - \frac{\pi}{2}$ varies from 0 to 2π. When $2x - \frac{\pi}{2} = 0$, we have $x = \frac{\pi}{4}$; when $2x - \frac{\pi}{2} = 2\pi$, we have $2x = 2\pi + \frac{\pi}{2} = \frac{5\pi}{2}$, so $x = \frac{5\pi}{4}$.

Figure 12

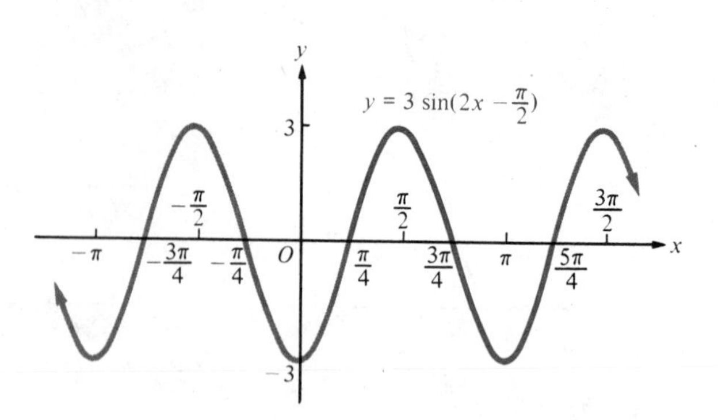

Therefore, the graph covers one cycle as x varies from $\frac{\pi}{4}$ to $\frac{5\pi}{4}$. It follows that the period is $\frac{5\pi}{4} - \frac{\pi}{4} = \pi$. The amplitude is 3 (Figure 12). Notice that the graph of $y = 3\sin\left(2x - \frac{\pi}{2}\right)$ is the graph of $y = 3\sin 2x$ (Figure 11) shifted $\frac{\pi}{4}$ unit to the right.

PROBLEM SET 5

1. Use the information in Table 1 (page 69) and the values of the cosine function for the angles $0, \frac{\pi}{6}, \frac{\pi}{4}, \frac{\pi}{3}, \frac{\pi}{2}, \frac{2\pi}{3}, \frac{3\pi}{4}, \frac{5\pi}{6}, \pi, \frac{7\pi}{6}, \frac{5\pi}{4}, \frac{4\pi}{3}, \frac{3\pi}{2}, \frac{5\pi}{3}, \frac{7\pi}{4}, \frac{11\pi}{6}$, and 2π to sketch an accurate graph of $y = \cos x$ for $0 \le x \le 2\pi$. When you have finished your sketch compare it with Figure 5. Ⓒ (For enhanced accuracy, you may wish to use a calculator to obtain additional points on the graph.)
2. By sketching a figure for $y = \cos x$ similar to Figure 4, show graphical evidence for the fact that $\cos(x + \pi) = -\cos x$.

Ⓒ In problems 3–6, use a calculator to verify that the equation is true for the indicated value of the variable.

3. $\sin(-x) = -\sin x, \quad x = 4.203$
4. $\cos(-x) = \cos x, \quad x = \frac{5\pi}{9}$
5. $\sin(x + \pi) = -\sin x, \quad x = 2.771$
6. $\cos(x + \pi) = -\cos x, \quad x = \frac{11\pi}{7}$

In problems 7–12, use the graphs of $y = \sin x$ (Figure 3) and $y = \cos x$ (Figure 6) to help answer each question.

7. If $-2\pi \le x \le 2\pi$, for what values of x does $\sin x$ reach its maximum value and what is this maximum value?
8. If $-2\pi \le x \le 2\pi$, for what values of x is $\cos x = \frac{1}{2}$?
9. If $-2\pi \le x \le 2\pi$, for what values of x is $\cos x = 0$?
10. If h is a constant with $0 < h < 1$ and if $-2\pi \le x \le 2\pi$, how many different values of x are there for which $\sin x = h$?
11. If $-2\pi \le x \le 2\pi$, for what values of x does $\sin x$ reach its minimum value and what is this minimum value?
12. Complete the following sentence: $\sin x$ reaches its maximum and minimum values at the same values of x for which the graph of $y = \cos x$ ____________.

13. Suppose a friend who is just beginning to study trigonometry says, "I understand how to take the sine of an *angle*, but how do you take the sine of a *number*?" Answer your friend's question in your own words.
14. Suppose you are going to measure an acute angle with a protractor and then use a calculator to find the sine of the angle. Of course, if there is an error in your measurement, there will be an error in the calculated value of the sine. Will the error in the sine value be more pronounced when the acute angle is small or when it is large? Why?

In problems 15–34, sketch the graph of the function defined by each equation, find the amplitude and the period, and indicate one cycle on your graph. Start the cycle at a node for the sine functions and at a crest for the cosine functions.

15. $y = 2 \sin x$ **16.** $y = -2 \sin x$ **17.** $y = \frac{1}{2} \sin x$

18. $y = 1 + \cos x$ **19.** $y = 1 - \frac{2}{3} \cos x$ **20.** $y = 3 \cos x - \frac{1}{2}$

21. $y = \sin 2x$ **22.** $y = 2 \sin \frac{x}{3}$ **23.** $y = \cos 6x$

24. $y = 1 + \cos \pi x$ **25.** $y = 1 + \cos \frac{x}{2}$ **26.** $y = \cos \frac{\pi x}{2} + 1$

27. $y = 2 - \pi \cos \pi x$ **28.** $y = \frac{1}{3} - \frac{1}{3} \cos \frac{3\pi x}{2}$ **29.** $y = \cos\left(x - \frac{\pi}{6}\right)$

30. $y = 1 + \sin\left(x + \frac{\pi}{4}\right)$ **31.** $y = 2 \cos\left(x - \frac{\pi}{3}\right)$ **32.** $y = -\cos\left(x - \frac{7\pi}{8}\right)$

33. $y = 1 - \cos\left(x - \frac{\pi}{2}\right)$ **34.** $y = -2 + 2 \cos\left(2x - \frac{\pi}{2}\right)$

35. Suppose that a, b, and c are constants, $a > 0$, and $b > 0$. Show that the simple harmonic curve $y = a \sin(bx - c)$ has amplitude a and period $2\pi/b$.

---

6 Graphs of the Tangent, Cotangent, Secant, and Cosecant Functions

In the previous section, we studied the graphs of the sine and cosine functions. In this section, we consider the graphs of the remaining four trigonometric functions.

Example Ⓒ Sketch the graph of

$$y = \tan x \quad \text{for} \quad -\frac{\pi}{2} < x < \frac{\pi}{2}.$$

Solution Table 1 shows values of tan x rounded off to two decimal places and obtained by using a calculator. Of course, $\tan x = \dfrac{\sin x}{\cos x}$ is undefined when $\cos x$ is zero; for instance, when $x = -\pi/2$ or when $x = \pi/2$. For values of x slightly smaller than $\pi/2$, the numerator sin x is close to 1, the denominator cos x is small, and thus,

$\tan x = \sin x/\cos x$ is very large. (Why?) Similarly, for values of x slightly larger than $-\pi/2$, $\tan x$ is negative with a very large absolute value. Thus the vertical lines

$$x = -\frac{\pi}{2} \quad \text{and} \quad x = \frac{\pi}{2}$$

are asymptotes of the graph. Plotting the points in Table 1, and using the information about the asymptotes, we obtain a sketch of the graph (Figure 1).

Table 1

x	$\tan x$
$-\pi/2$	undefined
$-5\pi/12$	-3.73
$-\pi/3$	-1.73
$-\pi/4$	-1
$-\pi/6$	-0.58
$-\pi/12$	-0.27
0	0
$\pi/12$	0.27
$\pi/6$	0.58
$\pi/4$	1
$\pi/3$	1.73
$5\pi/12$	3.73
$\pi/2$	undefined

Figure 1

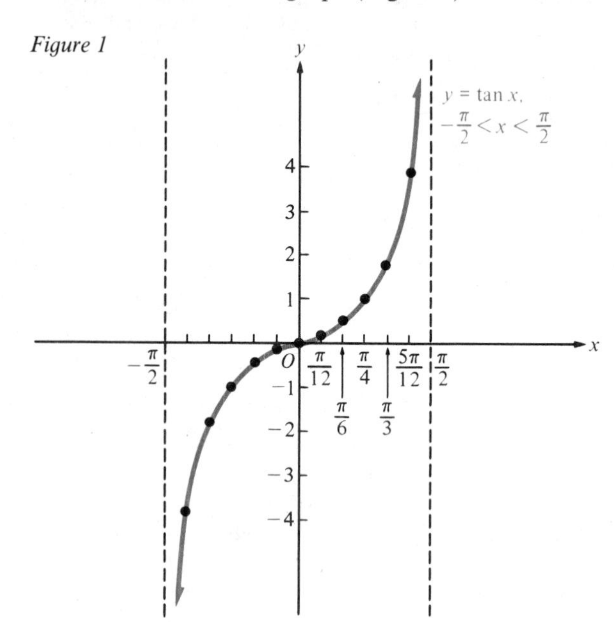

Like the sine and cosine functions, the tangent function is periodic; however, its values repeat themselves whenever the independent variable increases by π. In other words,

$$\tan(x + \pi) = \tan x$$

holds for every value of x in the domain of the tangent function. This can be seen by combining the identities

$$\sin(x + \pi) = -\sin x \quad \text{and} \quad \cos(x + \pi) = -\cos x,$$

which were mentioned in Section 5. Thus,

$$\tan(x + \pi) = \frac{\sin(x + \pi)}{\cos(x + \pi)} = \frac{-\sin x}{-\cos x} = \frac{\sin x}{\cos x} = \tan x.$$

Notice that the interval from $-\pi/2$ to $\pi/2$ in Figure 1 has length π units. Therefore, the complete graph of $y = \tan x$ consists of an endless sequence of copies of the curve in Figure 1, repeated to the right and left over successive intervals of length π (Figure 2).

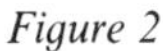

Figure 2

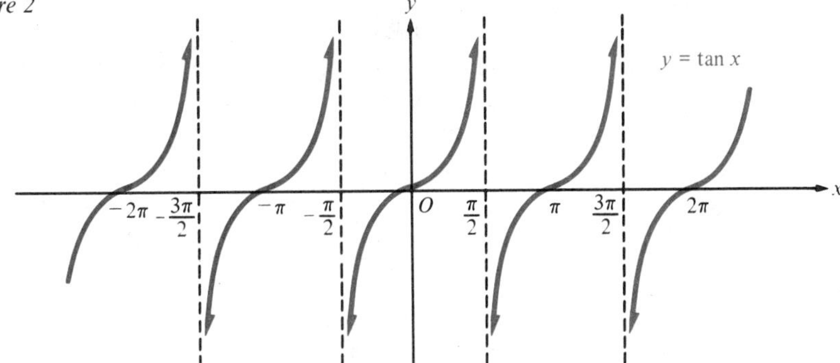

As you can see, the range of the tangent function is ℝ. Also, the graph appears to be symmetric about the origin, suggesting that the tangent is an odd function,

$$\tan(-x) = -\tan x$$

for all values of x in the domain.

Like the tangent function, the values of the cotangent function repeat themselves periodically whenever the independent variable increases by π; that is,

$$\cot(x + \pi) = \cot x$$

holds for every value of x in the domain of the cotangent function (Problem 4). Because $\cot x = 1/\tan x$, the graph of the cotangent function has vertical asymptotes where the graph of the tangent function has x intercepts and vice versa. To sketch the graph of $y = \cot x$, start by sketching the graph for $0 < x < \pi$ (Problem 1), then draw copies of this curve to the left and right over successive intervals of length π (Problem 2). The resulting graph is shown in Figure 3.

Figure 3

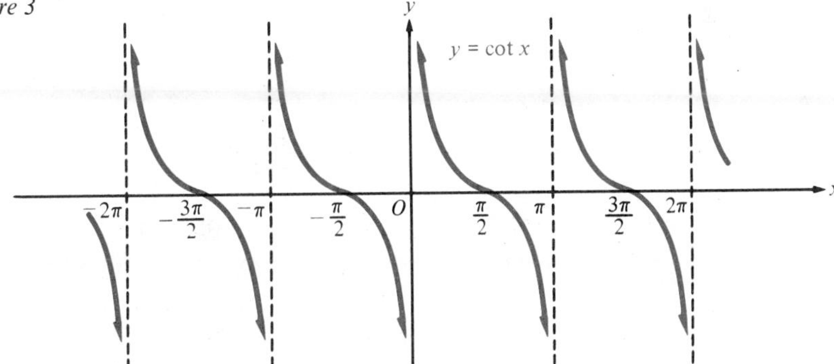

As you can see, the range of the cotangent function is ℝ. Also, the graph appears to be symmetric about the origin, suggesting that the cotangent is an odd function,

$$\cot(-x) = -\cot x$$

for all values of x in the domain.

Example Ⓒ Sketch the graph of $y = \sec x$.

Solution Since $\sec x = 1/\cos x$, we obtain the graph of the secant function by using a calculator to find the reciprocals of nonzero ordinates of points on the graph of the cosine function (Figure 4). The graph of the secant function has vertical asymptotes where the graph of the cosine function has x intercepts.

Figure 4

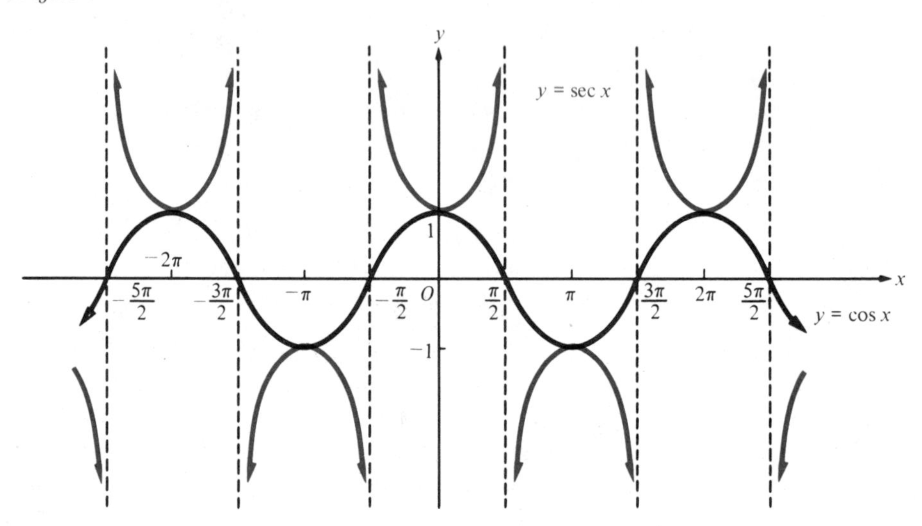

As you can see, the range of the secant function consists of the intervals $(-\infty, -1]$ and $[1, \infty)$. In particular,

$$|\sec x| \geq 1$$

holds for all values of x in the domain of the secant function.

You can sketch the graph of $y = \csc x$ (Figure 5) by using a calculator to find the reciprocals of nonzero ordinates of points on the graph of $y = \sin x$ (Problem 3). Like the secant function, the range of the cosecant function consists of the two intervals $(-\infty, -1]$ and $[1, \infty)$, and

$$|\csc x| \geq 1$$

holds for all values of x in the domain of the function.

You can use the techniques of graphing discussed in Section 6 of Chapter 1 to stretch, shift, or reflect the graphs of the tangent, cotangent, secant, and cosecant functions.

Figure 5

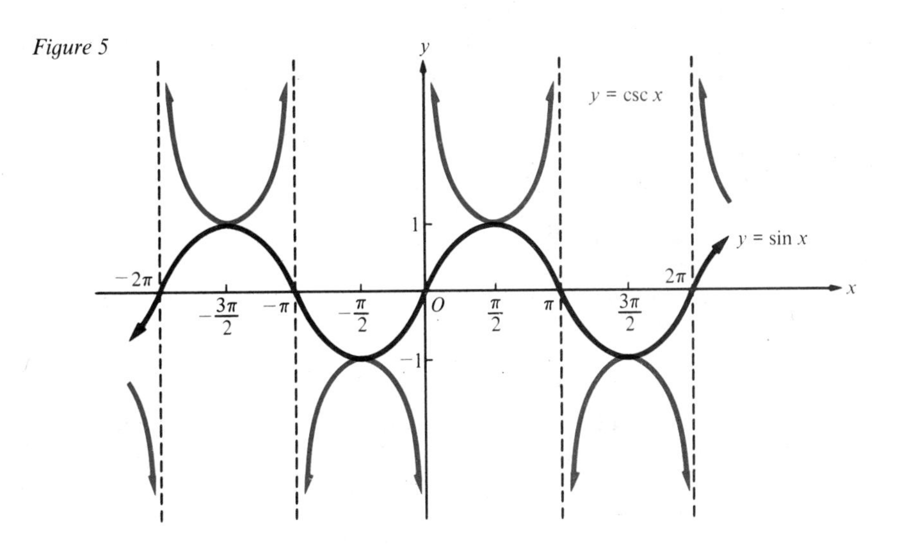

Example Sketch the graph of $y = 3\sec(x - \pi)$.

Solution The graph is obtained by multiplying ordinates of points on the graph of $y = \sec x$ (Figure 4) by 3 and then shifting the resulting graph π units to the right (Figure 6).

Figure 6

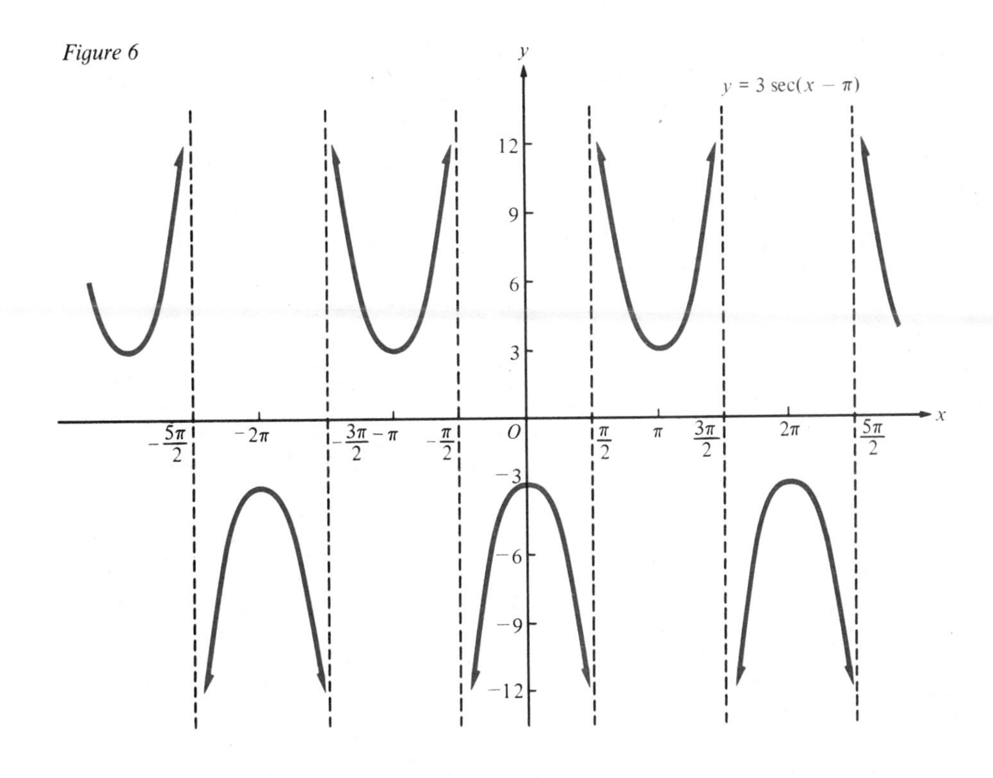

PROBLEM SET 6

Ⓒ 1. (a) Make a table showing the values of $\cot x$, rounded off to two decimal places, for the following values of x: $\frac{\pi}{12}, \frac{\pi}{6}, \frac{\pi}{4}, \frac{\pi}{3}, \frac{5\pi}{12}, \frac{\pi}{2}, \frac{7\pi}{12}, \frac{2\pi}{3}, \frac{3\pi}{4}, \frac{5\pi}{6}$, and $\frac{11\pi}{12}$. (b) Use the table in part (a) to sketch an accurate graph of $y = \cot x$ for $0 < x < \pi$.

2. Using the result of problem 1, sketch the graph of $y = \cot x$. When you have finished your sketch, compare it with Figure 3.

Ⓒ 3. With the aid of a calculator, sketch an accurate graph of $y = \csc x$. When you have finished, compare your graph with Figure 5.

4. Assuming that $\sin(x + \pi) = -\sin x$ and $\cos(x + \pi) = -\cos x$, show that $\cot(x + \pi) = \cot x$ holds for all values of x in the domain of the cotangent function.

Ⓒ In problems 5–10, use a calculator to verify that the equation is true for the indicated value of the variable.

5. $\tan(x + \pi) = \tan x, \quad x = \frac{2\pi}{3}$

6. $\cot(x + \pi) = \cot x, \quad x = -\frac{5\pi}{12}$

7. $\tan(-x) = -\tan x, \quad x = 1.334$

8. $\cot(-x) = -\cot x, \quad x = 0.7075$

9. $\sec(-x) = \sec x, \quad x = \frac{5\pi}{12}$

10. $\csc(-x) = -\csc x, \quad x = 0.3544$

In problems 11–16, use the graphs of $y = \tan x$, $y = \cot x$, $y = \sec x$, and $y = \csc x$ to help answer each question.

11. If $-2\pi \le x \le 2\pi$, for what values of x is $\tan x = 0$?

12. If h is a positive constant and if $-2\pi \le x \le 2\pi$, how many different values of x are there for which $\tan x = h$?

13. If $-2\pi \le x \le 2\pi$, for what values of x does $\sec x$ reach its smallest positive value and what is this value?

14. How is the graph of $y = -\tan x$ related to the graph of $y = \cot x$?

15. For the graph of $y = \sec x$, what is the distance from one vertical asymptote to the next?

16. How is the graph of $y = \csc x$ related to the graph of $y = \sec x$?

17. Assuming that the cosine is an even function, show that the secant is also an even function.

18. The portion of the graph of

$$y = \sec x \qquad \text{for } (-\pi/2) < x < (\pi/2)$$

looks vaguely like a parabola. Could it possibly be a parabola? Why?

In problems 19–32, use the graphs of $y = \tan x$, $y = \cot x$, $y = \sec x$, and $y = \csc x$ and the techniques of shifting, stretching, and reflecting to sketch the graph of each equation.

19. $y = 2\tan x$

20. $y = -2\tan x$

21. $y = 1 - \frac{1}{2}\cot x$

22. $y = 3 - \sec x$

23. $y = 1 + \frac{1}{2}\sec x$

24. $y = \frac{2}{3}\csc\left(x - \frac{\pi}{3}\right)$

25. $y = \tan\left(x - \frac{\pi}{4}\right)$

26. $y = \tan\left(x + \frac{\pi}{4}\right)$

27. $y = \frac{2}{3}\cot\left(x - \frac{\pi}{2}\right)$

28. $y = 5\sec\left(x - \frac{\pi}{2}\right)$

29. $y = \sec\left(x + \frac{\pi}{6}\right)$

30. $y = 1 - \csc\left(x + \frac{5\pi}{6}\right)$

31. $y = 2\csc\left(x - \frac{\pi}{3}\right)$

32. $y = 2 + \frac{1}{2}\csc\left(x + \frac{\pi}{3}\right)$

7 The Simple Harmonic Model

In the real world there are a number of quantities that oscillate or vibrate in a uniform manner, repeating themselves periodically in definite intervals of time. Examples include alternating electrical currents; sound waves; light waves, radio waves, and other electromagnetic waves; pendulums, mass-spring systems, and other mechanical oscillators; tides, geysers, seismic waves, the seasons, climatic cycles, and other periodic phenomena of interest in the earth sciences; and biological phenomena ranging from a human heartbeat to periodic variation of the population of a plant or animal species. Simple harmonic curves often provide useful mathematical models for such oscillations. It is only necessary to replace the variable x in the equation of a simple harmonic curve by the variable t representing units of time.

The mathematical model for a quantity y that is oscillating in a **simple harmonic** manner is the equation

$$y = a\cos(\omega t - \phi) + k$$

where a and ω are *positive* constants and ϕ and k are constants. Here a is the **amplitude,** ω (the Greek letter omega) is called the **angular frequency** (see Problem 36), ϕ (the Greek letter phi) is called the **phase angle,** k is the **vertical shift,** and the independent variable t is the **time.**

To sketch a graph of

$$y = a\cos(\omega t - \phi) + k,$$

we begin by noticing that y will go through one cycle as $\omega t - \phi$ goes from 0 to 2π. When $\omega t - \phi = 0$, we have $t = \phi/\omega$; when $\omega t - \phi = 2\pi$, we have $t = (\phi/\omega) + (2\pi/\omega)$. Therefore, one cycle of the graph of $y = a\cos(\omega t - \phi) + k$ covers the interval from $t = \phi/\omega$ to $t = (\phi/\omega) + (2\pi/\omega)$ (Figure 1). The vertical shift k moves the axis of the simple harmonic curve from $y = 0$ to $y = k$ (Figure 1). The **period** of the oscillation is the amount of time T required for y to proceed through one cycle. From Figure 1, you can see that

Figure 1

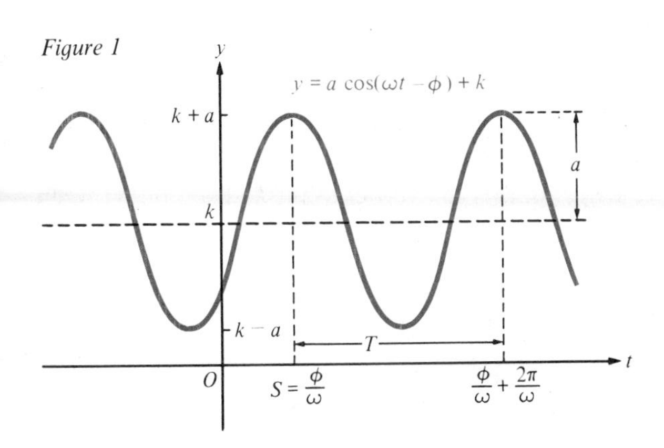

$$T = \left(\frac{\phi}{\omega} + \frac{2\pi}{\omega}\right) - \frac{\phi}{\omega} = \frac{2\pi}{\omega}.$$

The equation

$$T = \frac{2\pi}{\omega}$$

is often written in the alternative form

$$\omega = \frac{2\pi}{T}.$$

Notice that one cycle (from crest to crest) of the graph of $y = a\cos(\omega t - \phi)$ has the same general shape as one cycle (from crest to crest) of the cosine function (Figure 5 in Section 5), but shifted horizontally by an amount ϕ/ω. For this reason, the quantity ϕ/ω is called the **phase shift.** We denote the phase shift by S, so that

$$S = \frac{\phi}{\omega}.$$

The last equation is often written in the alternative form

$$\phi = \omega S.$$

Example 1 If y is oscillating according to the simple harmonic model

$$y = 2\cos\left(\frac{\pi}{4}t - \frac{\pi}{2}\right) - 1,$$

find (a) the amplitude a, (b) the angular frequency ω, (c) the period T, (d) the phase angle ϕ, (e) the phase shift S, and (f) the vertical shift k. Then (g) sketch the graph of one cycle.

Solution

(a) $a = 2$ (b) $\omega = \dfrac{\pi}{4}$

(c) $T = \dfrac{2\pi}{\omega} = \dfrac{2\pi}{(\pi/4)} = 8$ (d) $\phi = \dfrac{\pi}{2}$

(e) $S = \dfrac{\phi}{\omega} = \dfrac{\pi/2}{\pi/4} = 2$ (f) $k = -1$

(g) See Figure 2.

Example 2 The curve in Figure 3 is one cycle of the graph of a simple harmonic model. Determine (a) the amplitude a, (b) the period T, (c) the phase shift S, (d) the angular frequency ω, (e) the phase angle ϕ, and (f) the vertical shift k. Then (g) write an equation for y in terms of t.

Figure 2

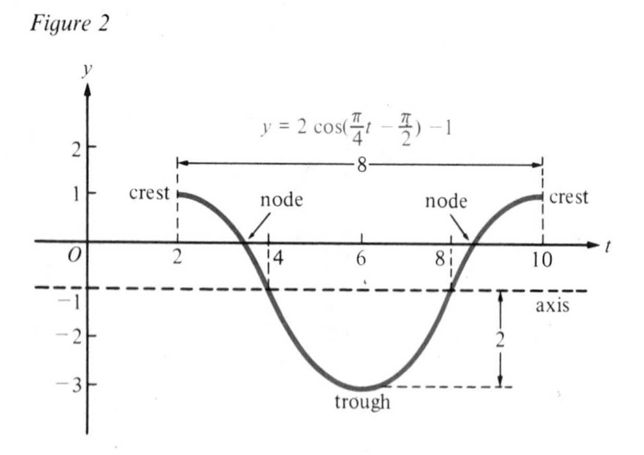

Figure 3

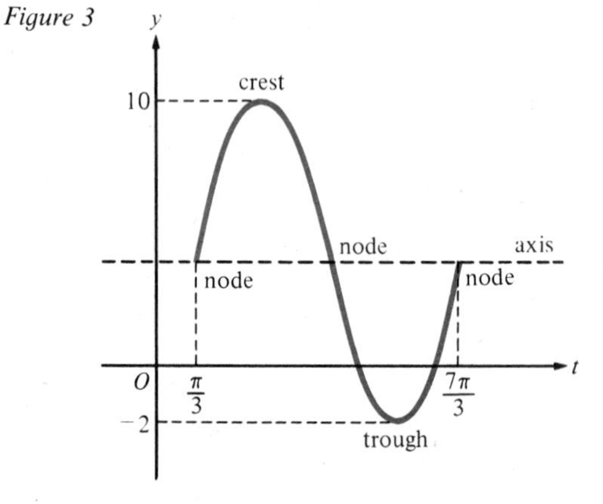

Solution (a) The vertical distance between the crest and the trough is $10 - (-2) = 12$ units, so $a = \frac{12}{2} = 6$.

(b) $T = \dfrac{7\pi}{3} - \dfrac{\pi}{3} = 2\pi$

(c) We need to calculate the distance S from the y axis to the crest. The distance from the y axis to the first node is $\pi/3$ units. The horizontal distance from this node to the crest is one-fourth of the period T, or $T/4 = 2\pi/4 = \pi/2$. Hence,

$$S = \frac{\pi}{3} + \frac{\pi}{2} = \frac{5\pi}{6}.$$

(d) $\omega = \dfrac{2\pi}{T} = \dfrac{2\pi}{2\pi} = 1$ (e) $\phi = \omega S = 1\left(\dfrac{5\pi}{6}\right) = \dfrac{5\pi}{6}$

(f) The axis is a units below the crest, so $k = 10 - a = 10 - 6 = 4$. (The axis is the horizontal line $y = 4$.)

(g) $y = a\cos(\omega t - \phi) + k = 6\cos\left(t - \dfrac{5\pi}{6}\right) + 4$

In the simple harmonic model

$$y = a\cos(\omega t - \phi) + k,$$

one cycle requires $T = 2\pi/\omega$ units of time, so y **oscillates** *through* $1/T = \omega/(2\pi)$ *cycles in one unit of time.* We call $1/T$ the **frequency of the oscillation.** The Greek letter ν (nu) is often used to denote frequency:

$$\nu = \frac{1}{T} = \frac{\omega}{2\pi}.$$

For instance, a quantity oscillating with a period of $T = 1/60$ second will oscillate through $\nu = 1/T = 60$ cycles per second.

One cycle per second is called a **hertz** (abbreviated Hz) in honor of the German physicist Heinrich Hertz (1857–94), who discovered radio waves in the late 1880s. Broadcast AM radio waves have frequencies of thousands of hertz (kilohertz), whereas FM and TV waves have frequencies of millions of hertz (megahertz). Alternating current generated by electrical utilities in the United States has a standard frequency of 60 Hz (that is, 60 cycles per second).

Figure 4

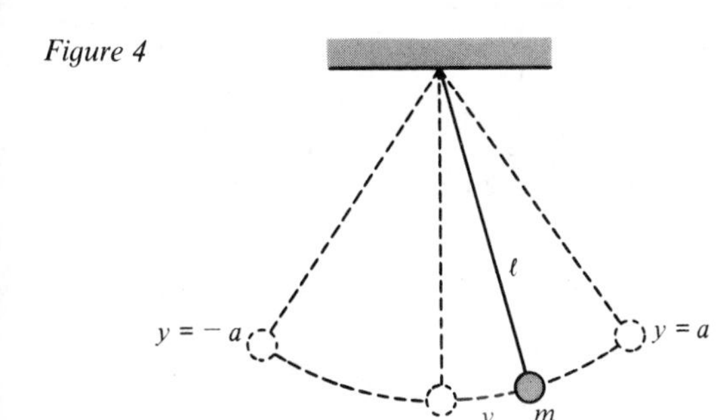

The idea of simple harmonic oscillation is nicely illustrated by a **simple pendulum,** which is an idealized object consisting of a point mass m suspended by a weightless string of length ℓ (Figure 4). When pulled to one side of its vertical position and released, it moves periodically to and fro. Let y denote the displacement of the mass from its vertical position, measured along the arc of the swing (positive to the right and negative to the left) at time t. Suppose that $y = a$

when $t = 0$, the instant of release. Then, if a is not too large, the quantity y will oscillate (approximately) according to a simple harmonic model

$$y = a \cos \omega t,$$

with amplitude a, phase angle $\phi = 0$, vertical shift $k = 0$, and period

$$T = 2\pi \sqrt{\frac{\ell}{g}},$$

where g is the acceleration of gravity. (At the surface of the earth $g \approx 32$ feet/sec^2 or $g \approx 9.8$ meters/sec^2.)

Example Ⓒ A pendulum of length $\ell = 1.2$ meters is pulled to the right through an arc of $a = 0.05$ meter and released at $t = 0$ seconds. Find (a) the period T, (b) the frequency ν, (c) the angular frequency ω, and (d) the equation for y as a function of t. Round off your answers in parts (a), (b), and (c) to 2 significant digits.

Solution

(a) $T = 2\pi \sqrt{\frac{\ell}{g}} = 2\pi \sqrt{\frac{1.2}{9.8}} \approx 2.2$ seconds

(b) $\nu = \frac{1}{T} \approx \frac{1}{2.2} \approx 0.45$ Hz

(c) $\omega = \frac{2\pi}{T} \approx \frac{2\pi}{2.2} \approx 2.9$

(d) $y = a \cos \omega t = 0.05 \cos(2.9t)$

Figure 5

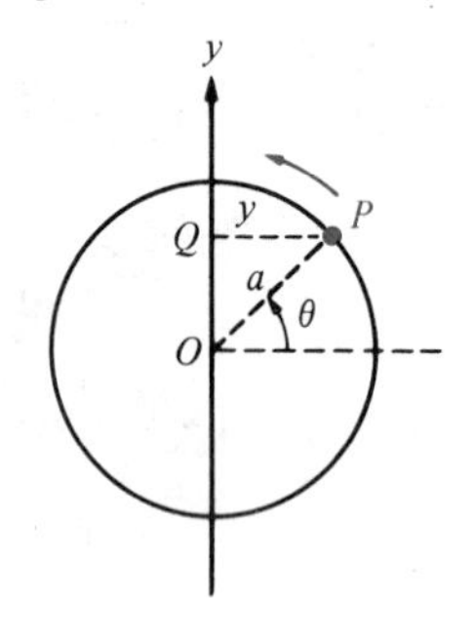

The simple harmonic model does not accurately represent the motion of a pendulum swinging with a large amplitude a. But if a is no larger than $\ell/4$, the formula $T = 2\pi\sqrt{\ell/g}$ gives the true period with an error of less than 1%.

There is an important relationship between simple harmonic motion and circular motion with constant speed (called **uniform circular motion**) which was discussed in Section 1. *Simple harmonic motion is the perpendicular projection along a diameter of uniform circular motion.* Indeed, in Figure 5, suppose that the point P is moving counterclockwise with constant angular speed ω around a circle of radius $r = a$ with center at the origin O of the y axis. Let Q be the perpendicular projection of P on the y axis, and let y denote the y coordinate of Q. Then y oscillates according to the simple harmonic model

$$y = a \cos(\omega t - \phi)$$

with amplitude a, angular frequency ω, and vertical shift $k = 0$ (Problem 36). If the central angle θ (Figure 5) has the value $\theta = \theta_0$ when $t = 0$, then the phase angle ϕ is given by

$$\phi = \frac{\pi}{2} - \theta_0$$

(Problem 36).

PROBLEM SET 7

In problems 1–10, suppose that y is oscillating according to the given simple harmonic model. Find (a) the amplitude a, (b) the angular frequency ω, (c) the period T, (d) the phase angle ϕ, (e) the phase shift S, and (f) the vertical shift k. Then (g) sketch the graph of one cycle.

1. $y = 2\cos\left(t - \frac{\pi}{3}\right) + 1$

2. $y = 3\cos\left(t + \frac{\pi}{6}\right) - 5$

3. $y = 4\cos\left(t + \frac{\pi}{4}\right) + 2$

4. $y = 2\cos(2t - \pi) + 2$

5. $y = 3\cos\left(3t + \frac{5\pi}{2}\right)$

6. $y = \frac{1}{2}\cos\left(2t - \frac{\pi}{2}\right) + 1$

7. $y = \frac{3}{4}\cos(4t + 12) - \frac{3}{4}$

8. $y = \frac{1}{3}\cos\left(\frac{\pi t}{4} - \frac{7\pi}{8}\right) + \frac{2}{3}$

9. $y = 110\cos\left(120\pi t + \frac{3\pi}{2}\right)$

10. $y = \sqrt{3}\cos(t - \sqrt{10}) - \frac{\sqrt{3}}{3}$

11. In the odd problems from 1 to 9, find the frequency v.

12. For the simple harmonic model $y = a\cos(\omega t - \phi) + k$, show that $\phi = 2\pi S/T$, where S is the phase shift and T is the period.

In problems 13–18, the figure shows one cycle of the graph of a simple harmonic model. Determine (a) the amplitude a, (b) the period T, (c) the phase shift S, (d) the angular frequency ω, (e) the phase angle ϕ, and (f) the vertical shift k. Then (g) write an equation for y in terms of t.

13.

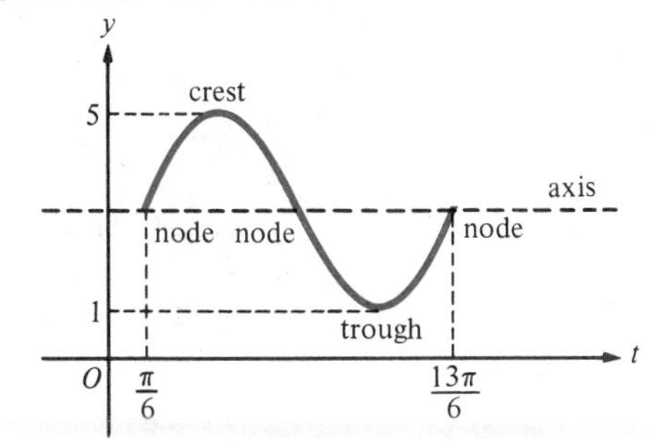

14.

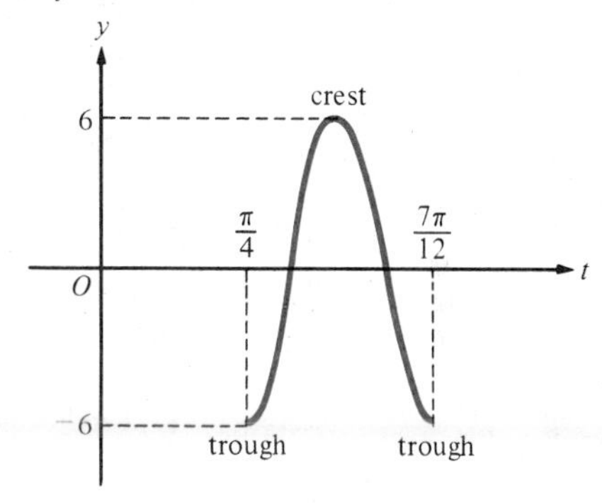

15.

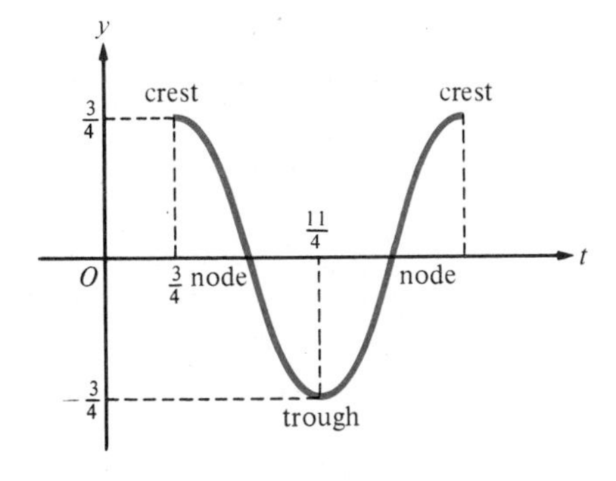

16.

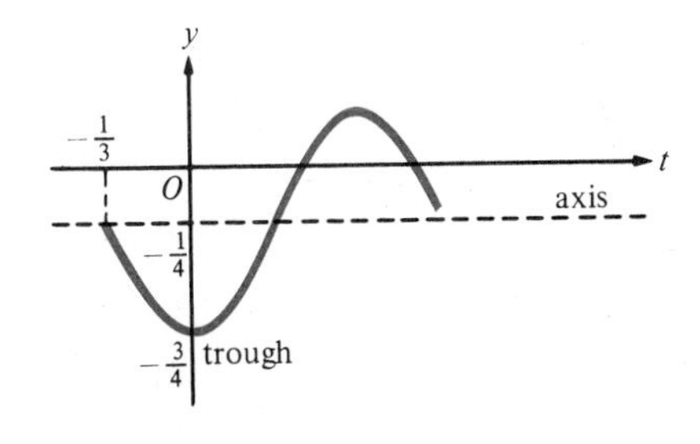

17.

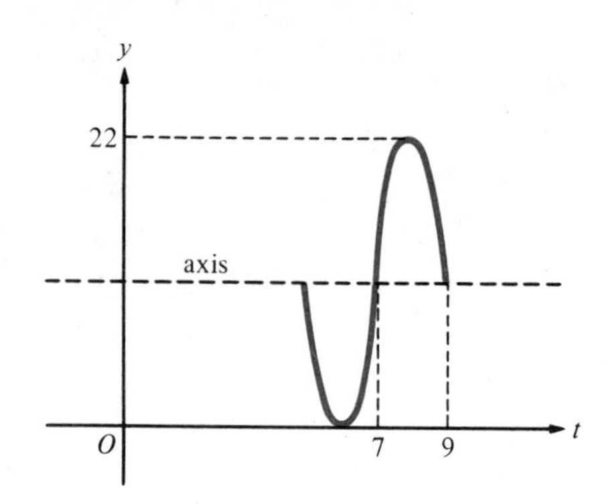

18.

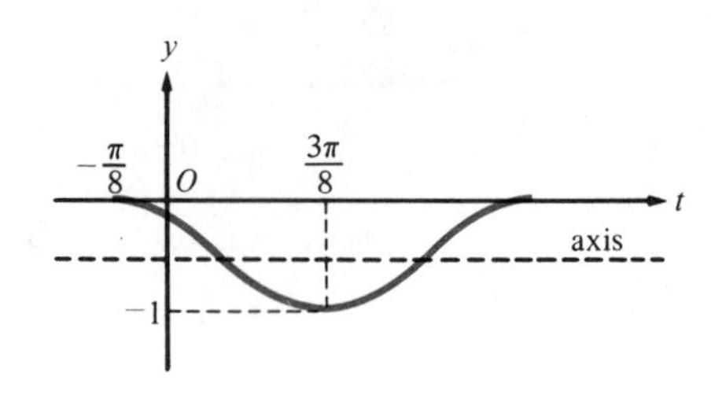

19. In problems 13, 15, and 17, find the frequency v.

20. For the graph of a simple harmonic model with period T, complete the following sentences: (a) The distance between two successive nodes is ____ units. (b) The horizontal distance between a node and the next crest or trough is ____ units.

21. Figure 6 shows one cycle, from crest to crest, of the graph of a simple harmonic model

$$y = a\cos(\omega t - \phi) + k.$$

Show that

(a) $a = \dfrac{d - c}{2}$

(b) $\omega = \dfrac{2\pi}{\beta - \alpha}$

(c) $\phi = \dfrac{2\pi\alpha}{\beta - \alpha}$

(d) $k = \dfrac{d + c}{2}$

Figure 6

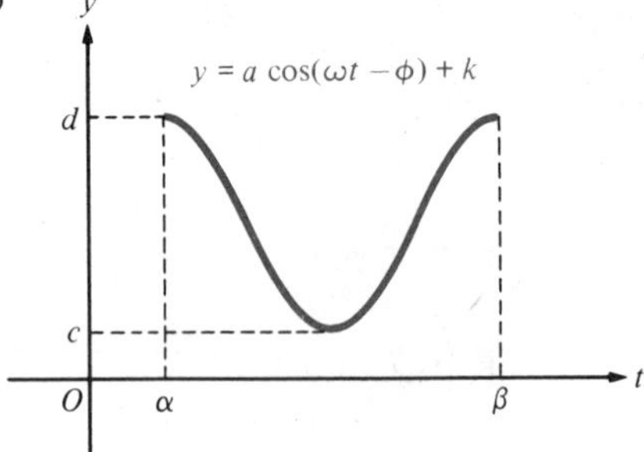

22. Assuming that $\sin x = \cos[x - (\pi/2)]$, show that the graph of $y = a\sin(\omega t - \mu) + k$ is the same as that of the simple harmonic model $y = a\cos(\omega t - \phi) + k$, where $\phi = \mu + (\pi/2)$. [μ is the Greek letter mu.]

23. Figure 7 shows one cycle, starting at a node, of the graph of a simple harmonic model

$$y = a\cos(\omega t - \phi) + k.$$

Show that

(a) $a = \dfrac{d - c}{2}$

(b) $\omega = \dfrac{2\pi}{\beta - \alpha}$

(c) $\phi = \dfrac{(3\alpha + \beta)\pi}{2(\beta - \alpha)}$

(d) $k = \dfrac{d + c}{2}$

Figure 7

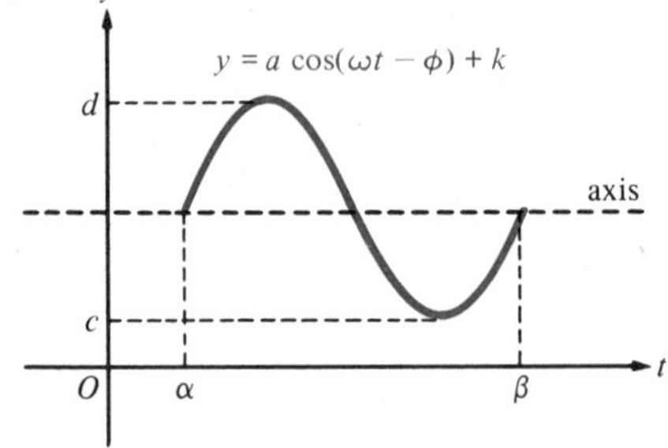

24. What does *amplitude modulation* (AM) mean? What does *frequency modulation* (FM) mean? [If you don't know, use a dictionary.]

In problems 25–30, sketch one cycle of the graph of a quantity y oscillating according to a simple harmonic model $y = a\cos(\omega t - \phi) + k$, with amplitude a, frequency ν, phase angle ϕ, and vertical shift k.

25. $a = 1$, $\nu = 50$ Hz, $\phi = \frac{\pi}{2}$, $k = 0$

26. $a = 220$, $\nu = 60$ Hz, $\phi = 0$, $k = 0$

27. $a = \frac{2}{3}$, $\nu = \frac{1}{4\pi}$ Hz, $\phi = \frac{\pi}{4}$, $k = \frac{2}{3}$

28. $a = 5$, $\nu = \frac{1}{28}$ cycle per day, $\phi = \frac{-3\pi}{14}$, $k = 0$

29. $a = 8.4$, $\nu = 0.18$ cycle per week, $\phi = 0$, $k = 0$

30. $a = 5 \times 10^{-7}$, $\nu = 88.5$ megahertz, $\phi = 0.8$, $k = 0$

31. A pendulum of length $\ell = 0.5$ foot is pulled to the right through an arc of $a = 0.1$ foot and released at $t = 0$ seconds. Find the period T, the frequency ν, the angular frequency ω, and the equation giving the displacement y at time t.

Ⓒ **32.** An astronaut on the surface of the moon finds that a pendulum has a period 2.45 times as long as it does on the surface of the earth. From this information, she is able to calculate the acceleration of gravity on the moon. How does she do this, and what is her answer?

33. You are lost on a desert island and your watch is broken. You have managed to salvage a toolbox containing, among other things, a tape measure and a ball of string. Explain how you go about constructing a crude "clock" that will measure time in seconds.

Ⓒ **34.** In the northern sky, the constellation Cepheus includes a number of stars that pulsate in brightness or magnitude. One of these stars, Beta Cephei, has a magnitude y that varies from a minimum of 3.141 to a maximum of 3.159 with a period of 4.5 hours. Using a simple harmonic model with a phase angle of $\phi = 6.33$, (a) sketch a graph showing the variation of y over the time interval $t = 0$ to $t = 8$ hours, and (b) determine the magnitude of Beta Cephei at $t = 4$ hours.

Ⓒ 35. A typical mass-spring system consists of a mass m suspended on a spring (Figure 8). A vertical y axis is set up, with the equilibrium level of the mass at $y = k$. Suppose that the mass is raised to a level $y = k + a$ and released at time $t = 0$. It is shown in physics that, neglecting friction and air resistance, the mass will bob up and down according to a simple harmonic model $y = a\cos(\omega t - \phi) + k$ with phase angle $\phi = 0$ and frequency

$$\nu = \frac{1}{2\pi}\sqrt{\frac{K}{m}}.$$

Here K, a constant that measures the stiffness of the spring, is the amount of force required to stretch it by one unit of distance. If $m = 0.05$ kilogram, $a = 0.04$ meter, and $K = 1.25$ newtons per meter, sketch two cycles of the graph of y.

Figure 8

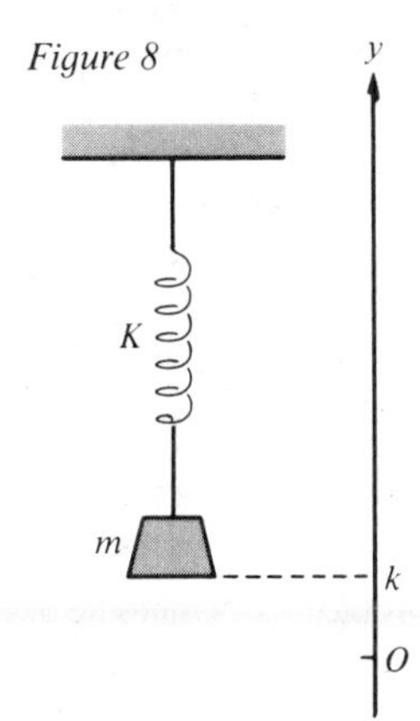

36. In Figure 5, suppose that the point P is moving counterclockwise around the circle with angular speed ω and that $\theta = \theta_0$ when $t = 0$. Show that $y = a\cos(\omega t - \phi)$, where $\phi = (\pi/2) - \theta_0$. Hint: Show that $\theta = \omega t + \theta_0$, then use $\sin\theta = \cos[\theta - (\pi/2)]$.

37. An electrical circuit consisting of a coil with inductance L henries and a capacitor with a capacitance of C farads is called an LC-circuit (Figure 9). If the capacitor is charged and the switch is closed at time $t = 0$ seconds, then the potential difference E volts across the plates of the capacitor will oscillate in a simple harmonic manner with amplitude E_0, angular frequency $\omega = 1/\sqrt{LC}$, and phase angle $\phi = 0$. If $E_0 = 100$ volts, $L = 10$ henries, and $C = 0.0005$ farad, find the period T, the frequency ν, and the equation giving the voltage E at time t seconds.

Figure 9

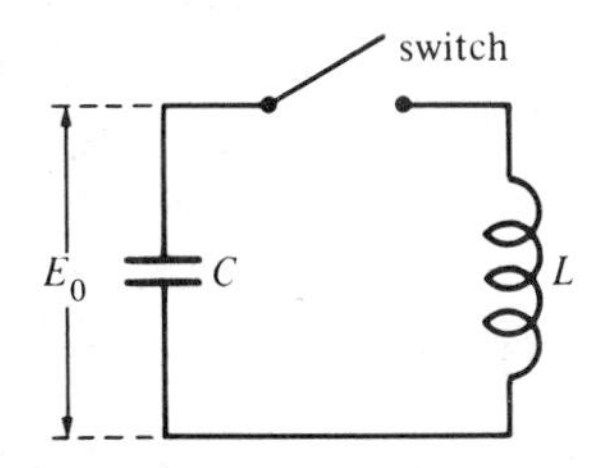

Ⓒ **38.** A **sound wave** in air consists of a continuous train of alternate compressions and rarefactions traveling at a speed of approximately 331 meters per second. The distance between any two successive compressions is called the **wavelength.** If the air pressure at any one fixed point varies in a simple harmonic manner, the sound heard by a human ear at that point is perceived as a **pure tone.** The amplitude of this variation determines the loudness of the tone, and its frequency determines the pitch. The pure tone middle C has a frequency of 256 Hz. (a) Find the wavelength of a sound wave that produces the pure tone middle C. (b) Suppose the sound wave in (a) causes an air pressure variation of amplitude 0.05 newton/meter^2 at a certain point. Using a phase angle $\phi = \pi/2$, write the equation for the air pressure y at this point at time t, if normal undisturbed atmospheric pressure is 10^5 newtons/meter^2.

Ⓒ **39.** Some people believe that we are all subject to periodic variations, called **biorhythms,** in our physical stamina, emotional well-being, and intellectual ability. The physical cycle is supposed to have a period of 23 days, the emotional cycle a period of 28 days, and the intellectual cycle a period of 33 days. Three graphs plotted on the same coordinate system (using, say, three different colors) showing the variations of these three factors for a particular individual over a period of a month or more make up a **biorhythm chart.** Most practitioners use simple harmonic curves all of the same (arbitrary) amplitude a. Taking $a = 1$, plot a biorhythm chart over a 62-day period for an individual with the following phase angles: physical phase angle = 5.804, emotional phase angle = 3.030, and intellectual phase angle = 6.045.

Ⓒ **40.** Believers in biorhythms (see problem 39) refer to the nodes of any one of the three biorhythm curves as **critical points,** and to the corresponding days as **critical days.** Contrary to what one might at first think, it is believed that the critical points are more significant than the crests or the troughs of the biorhythm curves. For the biorhythm chart plotted in problem 39, locate the physical, emotional, and intellectual critical days. Are there any doubly critical days—that is, days on which two curves pass through a node simultaneously? What about triply critical days?

REVIEW PROBLEM SET

In problems 1–4, find the degree measure of the given angle.

1. One-tenth of a counterclockwise revolution.
2. Five-sixths of a counterclockwise revolution.
3. Four-fifths of a clockwise revolution.
4. The angle through which the minute hand on a clock turns in 5 hours.

In problems 5–10, indicate the number of revolutions or the fraction of a revolution represented by the given angle. Indicate whether the rotation is clockwise or counterclockwise.

5. 60° **6.** 140° **7.** -1440° **8.** -310°
9. 930° **10.** -610°

Ⓒ In problems 11–14, express each angle measure as a decimal. Round off to four decimal places.

11. $2^\circ 3'$ **12.** $23^\circ 19' 13''$ **13.** $55^\circ 45' 35''$ **14.** $5''$

In problems 15–18, express each angle measure in degrees, minutes, and seconds.

15. 87.35° **16.** -62.45° Ⓒ **17.** -24.53° Ⓒ **18.** 165.37°

In problems 19–24, s denotes the length of the arc intercepted on a circle of radius r by a central angle of θ radians. Find the missing quantity.

19. $r = 5$ meters, $\theta = 0.57$ radian, $s = ?$

20. $r = 40$ centimeters, $s = 4$ centimeters, $\theta = ?$

21. $s = 3\pi$ feet, $\theta = \pi$ radians, $r = ?$

22. $r = 13$ kilometers, $\theta = \dfrac{3\pi}{7}$ radians, $s = ?$

23. $r = 2$ meters, $s = \pi$ meters, $\theta = ?$

24. $s = 17\pi$ microns, $\theta = \dfrac{5\pi}{6}$ radians, $r = ?$

In problems 25–30, convert each degree measure to radian measure. Do not use a calculator. Write your answer as a rational multiple of π.

25. 80° **26.** 570° **27.** -355° **28.** -810° **29.** -310° **30.** 765°

In problems 31–36, convert each radian measure to degree measure. Do not use a calculator.

31. $\dfrac{2\pi}{5}$ **32.** $-\dfrac{13\pi}{4}$ **33.** $-\dfrac{7\pi}{8}$ **34.** $\dfrac{35\pi}{3}$ **35.** $\dfrac{51\pi}{4}$ **36.** $\dfrac{18\pi}{5}$

[C] **37.** Use a calculator to convert each degree measure to an approximate radian measure rounded off to four decimal places.
(a) 5° (b) 27.7533° (c) -17.173° (d) $35^\circ 16' 55''$

[C] **38.** Use a calculator to convert each radian measure to an approximate degree measure rounded off to four decimal places.
(a) 5 (b) 3.9 (c) -7.63 (d) -21.403

[C] In problems 39–42, s denotes the length of the arc intercepted on a circle of radius r by a central angle θ measured in *degrees*. Find the missing quantity. Round off your answers to three significant digits.

39. $r = 10$ feet, $\theta = 36^\circ$, $s = ?$

40. $s = 111$ centimeters, $\theta = 135^\circ$, $r = ?$

41. $r = 12$ meters, $s = 3\pi$ meters, $\theta = ?$

42. $r = 5$ kilometers, $\theta = 65^\circ$, $s = ?$

[C] In problems 43 and 44, find the area $A = \frac{1}{2}r^2\theta$ of a sector of a circle of radius r with central angle θ. (See problem 48 in Problem Set 1.)

43. $r = 25$ centimeters, $\theta = \dfrac{\pi}{6}$

44. $r = 3.5$ meters, $\theta = 60^\circ$

[C] **45.** The minute hand on a tower clock is 0.6 meter long. How far does the tip of the hand travel in 4 minutes?

[C] **46.** In problem 45, what is the area of the sector swept out by the minute hand in 4 minutes?

[C] **47.** A wheel with radius 2 feet makes 30 revolutions per minute. Find the linear speed of a point on the rim of the wheel.

[C] **48.** Find the approximate diameter of the moon if its disk subtends an angle of $30'$ at a point on the earth 240,000 miles away (Figure 1). [Hint: Approximate the diameter $|\overline{DE}|$ by the length of the arc $\widehat{BC}$.]

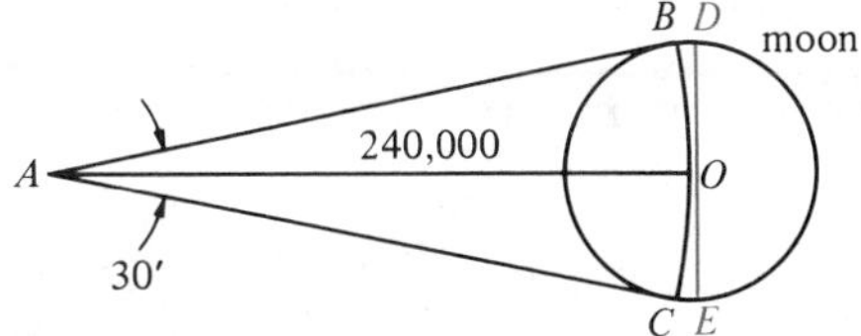

Figure 1

[C] **49.** A satellite in a circular orbit above the earth is known to have a linear speed of 9.92 kilometers per second. In 10 seconds it moves along an arc that subtends an angle of 0.75° at the center of the earth. If the radius of the earth is 6371 kilometers, how high is the satellite above the surface of the earth?

[C] **50.** A laser beam is projected from a point on the surface of the earth to a reflector on the surface of the moon 384,000 kilometers away. If the laser beam diverges at an angle of $1''$, find the approximate diameter of the beam when it strikes the surface of the moon.

[C] **51.** A phonograph record is rotating at 33.33 revolutions per minute. When the needle is in a groove 10 centimeters from the center, what is the linear speed of the groove with respect to the needle?

[C] **52.** A small sprocket (toothed wheel) of radius 7 centimeters is connected by a chain to a large sprocket of radius 11 centimeters. The small sprocket is turning at 60 revolutions per minute. (a) Find the linear speed of the chain connecting the sprockets. (b) Find the rate of rotation of the large sprocket.

Figure 2

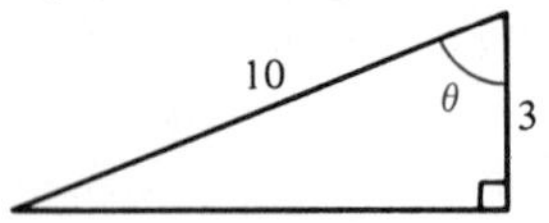

53. Find the values of the six trigonometric functions of the acute angle θ in Figure 2.

54. Given that $\sin 67^\circ = 0.9205$, $\tan 67^\circ = 2.356$, and $\sec 67^\circ = 2.559$, find the values of
(a) $\cos 23^\circ$ (b) $\cot 23^\circ$ (c) $\csc 23^\circ$

55. Express the value of each trigonometric function as the value of the cofunction of the complementary angle.

(a) $\sin 50^\circ$ (b) $\cos \dfrac{\pi}{7}$ (c) $\sec 89^\circ$ (d) $\cot \dfrac{3\pi}{28}$

[C] **56.** Suppose that α and β are complementary acute angles and that $\sin \alpha = 0.9063$. Find:
(a) $\cos \beta$ (b) $\csc \alpha$ (c) $\sec \beta$.

57. Sketch two coterminal angles α and β in standard position whose terminal side contains each point. Arrange it so that α is positive, β is negative, and neither angle exceeds one revolution. In each case, name the quadrant in which the angle lies or indicate that it is quadrantal.
(a) $(5, 12)$ (b) $(-3, 5)$ (c) $(-7, -6)$ (d) $(0, -4)$
(e) $(\sqrt{2}, -\sqrt{3})$ (f) $(\sqrt{5}, 0)$

58. Sketch each angle in standard position and name the quadrant in which it lies or specify that it is quadrantal.
(a) 60° (b) -210° (c) 110° (d) -2160° (e) -340° (f) -750°

59. In each case, specify and sketch three different angles that are coterminal with the given angle in standard position.

(a) -15° (b) 460° (c) 170° (d) -980° (e) $\dfrac{5\pi}{3}$

60. Indicate which of the six trigonometric functions are *not* defined for each angle.

(a) 1260° (b) 37π (c) 38π (d) $\dfrac{19\pi}{2}$

61. For each point (x, y), evaluate the six trigonometric functions of an angle θ in standard position if the terminal side of θ contains (x, y). In each case, sketch one of the coterminal angles θ thus formed. Do not use a calculator.
(a) $(-3, 5)$ (b) $(2, -3)$ (c) $(-6, -8)$ (d) $(\sqrt{3}, -1)$

62. Without using a calculator, give the algebraic sign of:

(a) $\sin 183^\circ$ (b) $\tan \dfrac{37\pi}{39}$ (c) $\cos(-269^\circ)$ (d) $\cot\left(-\dfrac{27\pi}{17}\right)$.

63. Indicate the quadrant in which θ lies if
(a) $\sin \theta > 0$ and $\cos \theta < 0$ (b) $\tan \theta > 0$ and $\sec \theta < 0$
(c) $\csc \theta < 0$ and $\tan \theta < 0$ (d) $\sin \theta < 0$ and $\sec \theta > 0$

64. If n is an integer, explain why $\cos(n\pi) = (-1)^n$.

65. If $\sin \theta = -\frac{4}{5}$ and $\cos \theta = \frac{3}{5}$, use the reciprocal and quotient identities to find the values of the other four trigonometric functions.

Ⓒ **66.** If $\sin\theta = 0.3145$ and $\cos\theta = -0.9493$, use the reciprocal and quotient identities to find the values of the other four trigonometric functions.

In problems 67–74, the value of one of the trigonometric functions is given along with information about the quadrant (Q) in which θ lies. Find the values of the other five trigonometric functions of θ.

67. $\sin\theta = -\frac{5}{13}$, θ in Q_{IV}
68. $\tan\theta = \frac{3}{2}$, θ not in Q_I
69. $\csc\theta = \frac{13}{12}$, θ in Q_{II}
70. $\sec\theta = -\frac{5}{4}$, θ in Q_{III}
71. $\sin\theta = \frac{3}{5}$, $\cos\theta < 0$
72. $\sec\theta = -5$, $\csc\theta < 0$
Ⓒ **73.** $\cot\theta = 0.6249$, $\sin\theta < 0$
Ⓒ **74.** $\tan\theta = 0.5543$, $\sec\theta < 0$

75. Without using a calculator, find the values (if they are defined) of the six trigonometric functions of

(a) $-180°$ (b) $540°$ (c) $990°$ (d) $-360°$ (e) $\frac{7\pi}{2}$ (f) -5π

(g) 17π (h) 18π

76. If n is an integer, explain why $\sin\left[\frac{(2n-1)\pi}{2}\right] = (-1)^{n+1}$.

77. Using reference angles and the values of the six trigonometric functions for 30°, 45°, and 60° (or $\pi/6$, $\pi/4$, and $\pi/3$), find the values of the six trigonometric functions of:

(a) $-150°$ (b) $-315°$ (c) $780°$ (d) $\frac{13\pi}{3}$ (e) $-\frac{15\pi}{4}$

78. Trigonometric tables ordinarily give approximate values of the trigonometric functions for acute angles only. Explain in your own words how to use such a table to find values of these functions for angles that are not acute.

Ⓒ In problems 79–96, use a calculator to find the approximate values of the indicated trigonometric functions.

79. $\sin 27°20'$
80. $\sin(421°15')$
81. $\cos 53.47°$
82. $\cos(-113.81°)$
83. $\tan(-117°15'30'')$
84. $\tan(-281°31'25'')$
85. $\sec 16.43°$
86. $\sec(-248.2°)$
87. $\csc\frac{4\pi}{5}$
88. $\csc 5.132$
89. $\cot(-3.18)$
90. $\cot(-7.167)$
91. $\cos(-19.213)$
92. $\csc 18.113$
93. $\sin 5.015$
94. $\cot\sqrt{3}$
95. $\cos\frac{\sqrt{2}\pi}{4}$
96. $\sin[\tan(-71.32)]$

In problems 97–106, use the graphs of the six trigonometric functions (Sections 5 and 6) and the techniques of graphing (Section 6 of Chapter 1) to sketch the graph of each equation.

97. $y = 1 + \frac{1}{3}\sin x$
98. $y = 1 - \frac{2}{5}\sin(x + \pi)$
99. $y = 2 + \frac{1}{2}\cos\left(x - \frac{\pi}{2}\right)$
100. $y = -0.1\cos\left(x - \frac{\pi}{4}\right)$
101. $y = -\tan\left(x - \frac{\pi}{6}\right)$
102. $y = \sec\left(x - \frac{\pi}{4}\right) + 1$
103. $y = \sec(x - \pi)$
104. $y = 1 - 4\csc\left(x - \frac{\pi}{3}\right)$
105. $y = \frac{2}{3}\cot\left(x - \frac{\pi}{2}\right)$
106. $y = 2 - \tan(x + \pi)$

107. If $-2\pi \le x \le 2\pi$, for what values of x does $\cos x$ reach its maximum and minimum values and what are these values?

108. If $-2\pi \le x \le 2\pi$, for what values of x is $\sin x = \sqrt{3}/2$?

109. Sketch the graphs of

$$y = \sin\left(\frac{3x}{2} - \frac{\pi}{4}\right) \quad \text{and} \quad y = \cos\left(\frac{3x}{2} - \frac{\pi}{4}\right)$$

on the same coordinate system.

110. Sketch the graph of $y = |\sin x|$.

In problems 111–116, suppose that y is a quantity that is oscillating according to the given simple harmonic model. Find (a) the amplitude a, (b) the angular frequency ω, (c) the period T, (d) the phase angle ϕ, (e) the phase shift S, (f) the vertical shift k, and (g) the frequency v. Then (h) sketch the graph of one cycle.

111. $y = 2\cos(\pi t + \pi) + 1$

112. $y = 3\cos(0.4t - 1) + 3$

113. $y = 3\cos\left(\frac{t}{2} - \frac{\pi}{2}\right) - 3$

114. $y = 5\cos\left(\frac{t}{3} + \frac{5\pi}{6}\right) - 2$

115. $y = 0.2\cos(0.25t - \pi)$

116. $y = 0.1\cos\left[3\left(t - \frac{\pi}{3}\right)\right]$

Ⓒ **117.** A pendulum of length $\ell = 5.2$ meters is pulled to the right through an arc of $a = 0.2$ meter and released at $t = 0$ seconds. Find the period T, the frequency v, the angular frequency ω, and the equation giving the displacement y at time t. Sketch a graph showing two cycles of the motion starting from $t = 0$.

Ⓒ **118.** The long-period variable star T Cephei has a magnitude y that varies from a minimum of 5.13 to a maximum of 10.87 with a period of 390 days. Using a simple harmonic model with a phase angle of $\phi = 0.16$, (a) sketch a graph showing the variation of y over the time interval $t = 0$ to $t = 600$ days, and (b) determine the magnitude of T Cephei at $t = 24$ days.

119. A sound wave in air with a frequency of 440 Hz is perceived as the pure tone "concert A." Suppose that this sound wave causes a sinusoidal air-pressure variation of amplitude 0.03 newton/meter^2 at a certain point. Using a phase angle of $\phi = \pi/2$, write the equation for the air pressure y at this point at time t if normal undisturbed atmospheric pressure is 10^5 newtons/meter.

Ⓒ **120.** An LC-circuit (problem 37, Section 7) contains an inductance of $L = 5 \times 10^{-4}$ henry and a variable capacitor that allows the circuit to be tuned from a frequency of 5.5×10^5 Hz to a frequency of 1.6×10^6 Hz. Sketch a graph showing the capacitance C as a function of the frequency v over this range of values.

121. People who believe in biorhythms (see problem 39, Section 7) claim that the physical, emotional, and intellectual curves start with a common node at birth, $t = 0$ days. Assuming this, determine the number of years that will elapse before the three biorhythms again reach a common node.

3 Trigonometric Identities and Equations

In Section 3 of Chapter 2, we derived the fundamental *reciprocal*, *quotient*, and *Pythagorean* identities. In this chapter, we establish the remaining standard trigonometric identities, including the important formulas for *sums*, *differences*, and *multiples* of angles. (For convenience, all of the standard trigonometric identities are listed inside the back cover of this book.) In many applications, ranging from calculus, engineering, and physics to economics and the life sciences, these identities and formulas are routinely used to simplify complicated expressions and to help solve equations involving trigonometric functions.

1 The Fundamental Trigonometric Identities

A **trigonometric equation** is, by definition, an equation that involves at least one trigonometric function of a variable. Such an equation is called a **trigonometric identity** if it is true for all values of the variable for which both sides of the equation are defined. An equation that is not an identity is called a **conditional equation.**

For instance, the trigonometric equation

$$\csc t = \frac{1}{\sin t}$$

is an identity, since it is true for all values of t (except, of course, for those values for which $\csc t$ or $1/\sin t$ is undefined). On the other hand, the trigonometric equation

$$\sin t = \cos t$$

is a conditional equation, since there are values of t (for instance, $t = 0$) for which it isn't true.

Now we are going to derive the trigonometric identities

$$\sin(-\theta) = -\sin\theta \quad \text{and} \quad \cos(-\theta) = \cos\theta$$

suggested by the graphs of the sine and cosine functions in Section 5 of Chapter 2. Figure 1 shows an angle θ and the corresponding angle $-\theta$, both in standard position. Evidently, the points P and Q,

Figure 1

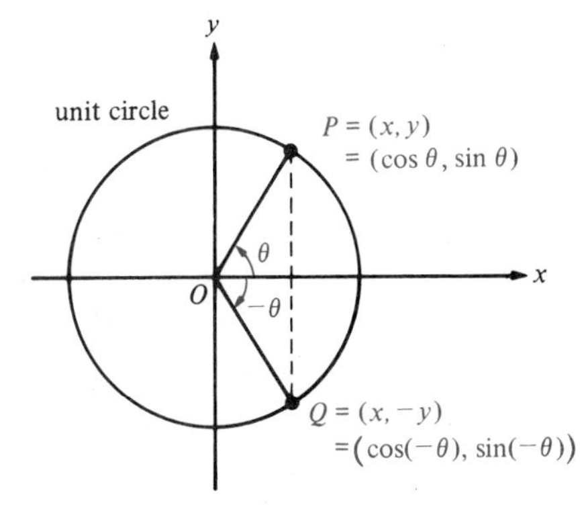

where the terminal sides of these angles intersect the unit circle, are mirror images of each other across the x axis. Therefore, if $P = (x, y)$ it follows that $Q = (x, -y)$. In Section 5 of Chapter 2, we showed that

$$P = (x, y) = (\cos\theta, \sin\theta).$$

Likewise,

$$Q = (x, -y) = (\cos(-\theta), \sin(-\theta)).$$

Therefore,

$$\sin(-\theta) = -y = -\sin\theta$$

and

$$\cos(-\theta) = x = \cos\theta.$$

If we now combine the identities obtained above with the quotient identity, $\tan\theta = \sin\theta/\cos\theta$, we find that

$$\tan(-\theta) = \frac{\sin(-\theta)}{\cos(-\theta)} = \frac{-\sin\theta}{\cos\theta} = -\tan\theta.$$

Similar arguments apply to $\cot(-\theta)$, $\sec(-\theta)$, and $\csc(-\theta)$ (Problem 51). The results are summarized in the following theorem.

Theorem 1 **Even–Odd Identities**

For all values of θ in the domains of the functions:

(i) $\sin(-\theta) = -\sin\theta$ (ii) $\cos(-\theta) = \cos\theta$ (iii) $\tan(-\theta) = -\tan\theta$

(iv) $\cot(-\theta) = -\cot\theta$ (v) $\sec(-\theta) = \sec\theta$ (vi) $\csc(-\theta) = -\csc\theta$

Notice that only the cosine and its reciprocal the secant are even functions—the remaining four trigonometric functions are odd. The even–odd identities are often used to simplify expressions as the following example illustrates.

Example Use the even–odd identities to simplify each expression.

(a) $\dfrac{\sin(-\theta) + \cos(-\theta)}{\sin(-\theta) - \cos(-\theta)}$ (b) $1 + \tan^2(-t)$

Solution (a)
$$\frac{\sin(-\theta) + \cos(-\theta)}{\sin(-\theta) - \cos(-\theta)} = \frac{-\sin\theta + \cos\theta}{-\sin\theta - \cos\theta} = \frac{-(\sin\theta - \cos\theta)}{-(\sin\theta + \cos\theta)} = \frac{\sin\theta - \cos\theta}{\sin\theta + \cos\theta}$$

(b) $1 + \tan^2(-t) = 1 + [\tan(-t)]^2 = 1 + (-\tan t)^2 = 1 + \tan^2 t = \sec^2 t$

The reciprocal, quotient, and Pythagorean identities obtained in Chapter 2, Section 3, together with the even–odd identities, are often called the *fundamental trigonometric identities*. For convenience, we summarize these identities here.

Fundamental Trigonometric Identities

1. $\csc\theta = \dfrac{1}{\sin\theta}$
2. $\sec\theta = \dfrac{1}{\cos\theta}$
3. $\cot\theta = \dfrac{1}{\tan\theta}$
4. $\tan\theta = \dfrac{\sin\theta}{\cos\theta}$
5. $\cot\theta = \dfrac{\cos\theta}{\sin\theta}$
6. $\cos^2\theta + \sin^2\theta = 1$
7. $1 + \tan^2\theta = \sec^2\theta$
8. $\cot^2\theta + 1 = \csc^2\theta$
9. $\sin(-\theta) = -\sin\theta$
10. $\cos(-\theta) = \cos\theta$
11. $\tan(-\theta) = -\tan\theta$
12. $\cot(-\theta) = -\cot\theta$
13. $\sec(-\theta) = \sec\theta$
14. $\csc(-\theta) = -\csc\theta$

Not only should you memorize these fourteen fundamental identities, but they should become so familiar to you that you can recognize them quickly even when they are written in equivalent forms. For instance, $\csc\theta = 1/\sin\theta$ can also be written as

$$(\sin\theta)(\csc\theta) = 1 \quad \text{or} \quad \sin\theta = \frac{1}{\csc\theta}.$$

Incidentally, a product of values of trigonometric functions such as $(\sin\theta)(\csc\theta)$ is usually written simply as $\sin\theta\csc\theta$, unless the parentheses are necessary to prevent confusion.

Examples Simplify each trigonometric expression by using the fundamental identities.

1 $\csc\theta\cos\theta$

Solution

$$\csc\theta\cos\theta = \frac{1}{\sin\theta}\cos\theta = \frac{\cos\theta}{\sin\theta} = \cot\theta$$

2 $\tan^2 t - \sec^2 t$

Solution Because $1 + \tan^2 t = \sec^2 t$, it follows that

$$\tan^2 t - \sec^2 t = -1.$$

3 $\csc^4 x - 2\csc^2 x\cot^2 x + \cot^4 x$

Solution The given expression is the square of $\csc^2 x - \cot^2 x$. Because $\cot^2 x + 1 = \csc^2 x$, we have $\csc^2 x - \cot^2 x = 1$. Therefore,

$$\csc^4 x - 2\csc^2 x\cot^2 x + \cot^4 x = (\csc^2 x - \cot^2 x)^2 = 1^2 = 1.$$

The reciprocal and quotient identities enable us to write $\csc\theta$, $\sec\theta$, $\tan\theta$, and $\cot\theta$ in terms of $\sin\theta$ and $\cos\theta$. Therefore, *any trigonometric expression can be rewritten in terms of sines and cosines only*. This fact and the Pythagorean identity $\cos^2\theta + \sin^2\theta = 1$ can often be used to simplify trigonometric expressions.

Examples Rewrite each trigonometric expression in terms of sines and cosines, and then simplify the result.

1 $\csc t - \dfrac{\cot t}{\sec t}$

Solution

$$\csc t - \frac{\cot t}{\sec t} = \frac{1}{\sin t} - \frac{\dfrac{\cos t}{\sin t}}{\dfrac{1}{\cos t}} = \frac{1}{\sin t} - \frac{\cos t}{\sin t}\cos t$$

$$= \frac{1}{\sin t} - \frac{\cos^2 t}{\sin t} = \frac{1 - \cos^2 t}{\sin t} = \frac{\sin^2 t}{\sin t} = \sin t$$

2 $\dfrac{\csc^2 x \sec^2 x}{\csc^2 x + \sec^2 x}$

Solution

$$\frac{\csc^2 x \sec^2 x}{\csc^2 x + \sec^2 x} = \frac{\left(\dfrac{1}{\sin^2 x}\right)\left(\dfrac{1}{\cos^2 x}\right)}{\dfrac{1}{\sin^2 x} + \dfrac{1}{\cos^2 x}} = \frac{\sin^2 x \cos^2 x\left(\dfrac{1}{\sin^2 x}\right)\left(\dfrac{1}{\cos^2 x}\right)}{\sin^2 x \cos^2 x\left(\dfrac{1}{\sin^2 x} + \dfrac{1}{\cos^2 x}\right)}$$

$$= \frac{1}{\cos^2 x + \sin^2 x} = \frac{1}{1} = 1$$

The Pythagorean identity $\cos^2 \theta + \sin^2 \theta = 1$ can be rewritten as

$$\sin^2 \theta = 1 - \cos^2 \theta \quad \text{or} \quad \cos^2 \theta = 1 - \sin^2 \theta.$$

Therefore, we have

$$\sin \theta = \pm\sqrt{1 - \cos^2 \theta}$$

and

$$\cos \theta = \pm\sqrt{1 - \sin^2 \theta}.$$

In either case, the correct algebraic sign is determined by the quadrant or coordinate axis containing the terminal side of the angle θ in standard position. After you have rewritten a trigonometric expression in terms of sines and cosines, you can use these equations to bring the expression into a form involving *only the sine* or *only the cosine.*

Example Rewrite the expression $\cot \theta \csc^2 \theta$ in terms of $\sin \theta$ only.

Solution

$$\cot \theta \csc^2 \theta = \frac{\cos \theta}{\sin \theta} \cdot \frac{1}{\sin^2 \theta} = \frac{\cos \theta}{\sin^3 \theta}$$

$$= \frac{\pm\sqrt{1 - \sin^2 \theta}}{\sin^3 \theta}$$

Algebraic expressions not originally containing trigonometric functions can often be simplified by substituting trigonometric expressions for the variable. This technique, called **trigonometric substitution,** is routinely used in calculus to rewrite radical expressions as trigonometric expressions containing no radicals.

Example If a is a positive constant, rewrite the radical expression $\sqrt{a^2 - u^2}$ as a trigonometric expression containing no radical by using the trigonometric substitution $u = a\sin\theta$. Assume that $-\frac{\pi}{2} \le \theta \le \frac{\pi}{2}$ so that $\cos\theta \ge 0$.

Solution

$$\begin{aligned}\sqrt{a^2 - u^2} &= \sqrt{a^2 - (a\sin\theta)^2} = \sqrt{a^2 - a^2\sin^2\theta}\\ &= \sqrt{a^2(1 - \sin^2\theta)} = \sqrt{a^2\cos^2\theta}\\ &= a\cos\theta\end{aligned}$$

PROBLEM SET 1

In problems 1–6, use the even–odd identities to simplify each expression.

1. $\sin(-\theta)\cos(-\theta)$
2. $\cot^2(-u) + 1$
3. $\tan t + \tan(-t)$
4. $\cos(-x)\sec x$
5. $\dfrac{1 + \csc(-\alpha)}{1 - \cot(-\beta)}$
6. $[1 + \sin\gamma][1 + \sin(-\gamma)]$

In problems 7–28, use the fundamental identities to simplify each expression.

7. $\sec\theta\sin\theta$
8. $1 + \dfrac{\tan\alpha}{\cot\alpha}$
9. $\cot v\sec v$
10. $\dfrac{\csc^2 u}{1 + \tan^2 u}$
11. $\dfrac{\csc\beta}{\sec\beta}$
12. $\dfrac{\sin^2\theta - 1}{\sec\theta}$
13. $\cot^2\alpha - \csc^2\alpha$
14. $\dfrac{\sec^2 t - 1}{\sec^2 t}$
15. $(\csc u - 1)(\csc u + 1)$
16. $\dfrac{1 + \cot^2 y}{1 + \tan^2 y}$
17. $\dfrac{1}{\sec^2 x} + \dfrac{1}{\csc^2 x}$
18. $\dfrac{(\sec\gamma - 1)(\sec\gamma + 1)}{\tan\gamma}$
19. $\sin^4 t + 2\cos^2 t\sin^2 t + \cos^4 t$
20. $\sin^4 u + 2\cos^2 u - \cos^4 u$
21. $\tan^4\alpha - 2\tan^2\alpha\sec^2\alpha + \sec^4\alpha$
22. $(1 + \tan^2\theta)(1 - \sin^2\theta)$
23. $\cos x\sin^3 x + \sin x\cos^3 x$
24. $(1 - \cos^2\beta)(1 + \cot^2\beta)$
25. $\dfrac{1}{\sin t\cos t} - \dfrac{\cos t}{\sin t}$
26. $\dfrac{\cos\gamma}{1 - \sin\gamma} + \dfrac{\cos\gamma}{1 + \sin\gamma}$
27. $\dfrac{\sin t}{1 + \cos t} + \dfrac{1 + \cos t}{\sin t}$
28. $\dfrac{\sin\alpha + \sin\beta}{\cos\alpha + \cos\beta} + \dfrac{\cos\alpha - \cos\beta}{\sin\alpha - \sin\beta}$

In problems 29–38, rewrite each trigonometric expression in terms of sines and cosines, and then simplify the result.

29. $\dfrac{\tan x}{\sec x}$
30. $(\cos\theta + \tan\theta\sin\theta)\cot\theta$
31. $\dfrac{\csc(-t)}{\sec(-t)\cot t}$
32. $\dfrac{\csc^2 x + \sec^2 x}{\csc^2 x\sec x}$
33. $\dfrac{\sec\alpha}{\csc\alpha(\tan\alpha + \cot\alpha)}$
34. $\dfrac{\sin y + \tan y}{1 + \sec y}$

35. $\dfrac{1 + \tan\theta}{\sec\theta}$

36. $\dfrac{\cot(-t) - 1}{1 - \tan(-t)}$

37. $\dfrac{\tan u + \sin u}{\cot u + \csc u}$

38. $\dfrac{\csc\beta}{\csc\beta + \tan\beta} + \dfrac{\csc\beta}{\csc\beta - \tan\beta}$

In problems 39–44, rewrite each expression in terms of the indicated function only.

39. $\sec^2\theta \tan\theta$ in terms of $\cos\theta$

40. $\dfrac{\sin t + \cot t \cos t}{\cot t}$ in terms of $\sec t$

41. $\dfrac{1 + \cot^2 x}{\cot^2 x}$ in terms of $\cos x$

42. $\dfrac{\csc^2 y + \sec^2 y}{\csc y \sec y}$ in terms of $\tan y$

43. $\dfrac{\sin(-\alpha) + \tan(-\alpha)}{1 + \sec(-\alpha)}$ in terms of $\sin\alpha$

44. $(\cot u + \csc u)(\tan u - \sin u)$ in terms of $\sec u$

In problems 45–48, rewrite each radical expression as a trigonometric expression containing no radical, by making the indicated trigonometric substitution. Assume that a is a positive constant.

45. $\sqrt{a^2 + u^2}$, $u = a\tan\theta$, $-\pi/2 < \theta < \pi/2$

46. $\sqrt{u^2 - a^2}$, $u = a\sec\theta$, $0 \le \theta < \pi/2$

47. $\sqrt{(4 - x^2)^3}$, $x = 2\cos t$, $0 \le t \le \pi$

48. $\sqrt{9 - 25x^2}$, $x = (3/5)\sin u$, $-\pi/2 \le u \le \pi/2$

49. Show that every trigonometric expression can be rewritten in terms of each of the six trigonometric functions alone, provided that ambiguous $\pm$ signs are allowed.

50. Let (x, y) be a point 1 unit from the origin (that is, on the unit circle) on the terminal side of an angle θ in standard position. Sketch figures to show that $(x, -y)$ is the point 1 unit from the origin on the terminal side of $-\theta$ for the following cases: (a) θ lies in quadrants II, III, or IV; (b) θ is a quadrantal angle; (c) θ is a negative angle.

51. Prove parts (iv), (v), and (vi) of Theorem 1.

2 Verifying Trigonometric Identities

By combining the fundamental trigonometric identities, it is possible to derive a large number of related identities. It isn't necessary to memorize these derived identities—when you need one of them, you can obtain it by manipulating the fundamental identities.

Although there is no universal step-by-step procedure for proving that a trigonometric equation is an identity, there are some useful guidelines. Perhaps the most important one concerns what *not* to do: *Do not treat a trigonometric equation as if it were an identity until after you have proved that it really is one.* The easiest way to avoid this pitfall is to *treat both sides of the equation separately.* To see the necessity for being careful, consider the following simple example. It's clear that

$$\sin x = -\sin x$$

is *not* an identity. Yet, if we square both sides, we get

$$\sin^2 x = \sin^2 x,$$

which *is* an identity.

Here are some additional suggestions.

Suggestion 1. Take one side of the equation and try to reduce it to the other side by a sequence of manipulations. It usually pays to start with the more complicated side and try to reduce it to the simpler side.

Suggestion 2. If Suggestion 1 doesn't seem to work, try to simplify each side of the equation separately. If you can reduce each side to the *same* expression, you can conclude that the original equation is an identity.

Suggestion 3. In carrying out Suggestion 1 or 2, try the following:

(a) Perform indicated additions or subtractions of fractions. (First obtain common denominators, of course.)

(b) Perform indicated multiplications or divisions of expressions.

(c) Simplify fractions by canceling common factors in the numerator and denominator.

(d) See whether you can come closer to your goal by factoring combinations of terms.

(e) Try multiplying both numerator and denominator of a fraction by the same expression.

(f) Try rewriting all trigonometric expressions in terms of sines and cosines.

Examples Prove that each equation is an identity.

1 $(\sin^2 \theta)(1 + \cot^2 \theta) = 1$

Solution We follow Suggestion 1, starting with the left side of the equation (the more complicated side):

$$(\sin^2 \theta)(1 + \cot^2 \theta) = \sin^2 \theta \csc^2 \theta = (\sin \theta \csc \theta)^2 = 1^2 = 1.$$

2 $\cos^4 t - \sin^4 t = 1 - 2\sin^2 t$

Solution Again, we try Suggestion 1. The left side of the equation involves fourth powers, so it is perhaps a bit more complicated than the right side. Therefore, we start with the left side. Trying parts (a)–(f) of Suggestion 3, one at a time, we find that (d) is the first reasonable idea. So we begin by factoring:

$$\begin{aligned}\cos^4 t - \sin^4 t &= (\cos^2 t)^2 - (\sin^2 t)^2 = (\cos^2 t + \sin^2 t)(\cos^2 t - \sin^2 t)\\ &= 1(\cos^2 t - \sin^2 t) = \cos^2 t - \sin^2 t = (1 - \sin^2 t) - \sin^2 t\\ &= 1 - 2\sin^2 t.\end{aligned}$$

3 $2\sec^2 x = \dfrac{1}{1 - \sin x} + \dfrac{1}{1 + \sin x}$

Solution Here we start with the right side, which seems more complicated than the left side, and we apply part (a) of Suggestion 3. Thus,

$$\frac{1}{1-\sin x}+\frac{1}{1+\sin x}=\frac{1(1+\sin x)}{(1-\sin x)(1+\sin x)}+\frac{1(1-\sin x)}{(1+\sin x)(1-\sin x)}$$

$$=\frac{1+\sin x}{1-\sin^2 x}+\frac{1-\sin x}{1-\sin^2 x}=\frac{1+\sin x+1-\sin x}{1-\sin^2 x}$$

$$=\frac{2}{1-\sin^2 x}=\frac{2}{\cos^2 x}=2\sec^2 x.$$

4 $\dfrac{1+\sin(-t)}{\cos(-t)}=\dfrac{\cos(-t)}{1-\sin(-t)}$

Solution Since $\sin(-t) = -\sin t$ and $\cos(-t) = \cos t$, the left side of the equation is equal to $(1 - \sin t)/\cos t$ and the right side is equal to $\cos t/(1 + \sin t)$. Therefore, it will be enough to prove that

$$\frac{1-\sin t}{\cos t}=\frac{\cos t}{1+\sin t}.$$

Notice that it would *not* be correct to "cross multiply" here, since we don't know that the equation is true—indeed, our job is to show that it is true. Thus, we work with both sides separately. Neither side appears to be more complicated than the other, so let's start with the left side. We try part (e) of Suggestion 3. Thus,

$$\frac{1-\sin t}{\cos t}=\frac{(1-\sin t)(1+\sin t)}{\cos t(1+\sin t)}=\frac{1-\sin^2 t}{\cos t(1+\sin t)}=\frac{\cos^2 t}{\cos t(1+\sin t)}$$

$$=\frac{\cos t}{1+\sin t}.$$

Example Show that the trigonometric equation

$$\sin x+\cos x=\tan x+1$$

is not an identity.

Solution If we let $x = 0$, we find that both sides of the equation are defined and that the equation becomes $1 = 1$, which is true. From this, however, we can't conclude that the equation is an identity, since there may be other values of x for which it is false. If we let $x = \pi/2$, the left side of the equation is $\sin(\pi/2) + \cos(\pi/2) = 1 + 0 = 1$, but the right side of the equation is undefined. Again, this is inconclusive, since an identity is only required to be true for values of the variable for which both sides are defined. Suppose we try $x = \pi/4$. Then the left side of the equation is $\sin(\pi/4) + \cos(\pi/4) = (\sqrt{2}/2) + (\sqrt{2}/2) = \sqrt{2}$, whereas the right side of the equation is $\tan(\pi/4) + 1 = 1 + 1 = 2$. Therefore, the equation is not an identity.

PROBLEM SET 2

In problems 1–52, show that each trigonometric equation is an identity.

1. $\sin\theta\sec\theta = \tan\theta$
2. $\cos\alpha\tan\alpha\csc\alpha = 1$
3. $\tan x\cos x = \sin x$
4. $\sin\beta\cot\beta\sec\beta = 1$
5. $\csc(-t)\tan(-t) = \sec t$
6. $\sin(-u) = \sin^2 u\csc(-u)$
7. $\tan\alpha\sin\alpha + \cos\alpha = \sec\alpha$
8. $\dfrac{\sec x\csc x}{\tan x + \cot x} = 1$
9. $\dfrac{\sin\beta}{\csc\beta} + \dfrac{\cos\beta}{\sec\beta} = 1$
10. $2 - \sin^2\theta = 1 + \cos^2\theta$
11. $\cos^2 t(1 + \tan^2 t) = 1$
12. $\sec^2 v(1 - \sin^2 v) = 1$
13. $\sec^2 w\cot^2 w - \cos^2 w\csc^2 w = 1$
14. $\tan^4 u - \sec^4 u = 1 - 2\sec^2 u$
15. $\sin^2\theta\cot^2\theta + \cos^2\theta\tan^2\theta = 1$
16. $\cot^2\gamma - \cos^2\gamma = \cot^2\gamma\cos^2\gamma$
17. $\sin^2 v + \tan^2 v + \cos^2 v = \sec^2 v$
18. $2\csc\beta - \cot\beta\cos\beta = \sin\beta + \csc\beta$
19. $\sin^2 x + \cos^2 x(1 - \tan^2 x) = \cos^2 x$
20. $\sin^4 t - \cos^4 t + 2\sin^2 t\cot^2 t = 1$
21. $\dfrac{\tan\theta}{1 + \tan^2\theta} = \dfrac{\sin\theta}{\sec\theta}$
22. $\dfrac{\cos^2 s}{\sin s} + \dfrac{1}{\csc s} = \csc s$
23. $\dfrac{\sin^2 t}{\cos t} + \cos t = \sec t$
24. $\dfrac{1}{\tan\theta + \cot\theta} = \sin\theta\cos\theta$
25. $\dfrac{\sin^3 t}{\cos t} + \sin t\cos t = \tan t$
26. $\dfrac{\csc x - \sec x}{\csc x + \sec x} = \dfrac{\cot x - 1}{\cot x + 1}$
27. $\dfrac{\sin\beta + \cos\beta}{\sin\beta - \cos\beta} = \dfrac{\sec\beta + \csc\beta}{\sec\beta - \csc\beta}$
28. $\left(\dfrac{1 + \csc t}{\csc t}\right)^2\sec^2 t = \dfrac{1 + \sin t}{1 - \sin t}$
29. $\dfrac{\sin x\cos x}{1 - 2\sin^2 x} = \dfrac{\tan x}{1 - \tan^2 x}$
30. $\dfrac{(1 - \cot y)^2}{\csc^2 y} + 2\sin y\cos y = 1$
31. $\dfrac{\tan u\sin u}{\tan u - \sin u} = \dfrac{\sin u}{1 - \cos u}$
32. $(\sec\gamma - \tan\gamma)^2 = \dfrac{1 - \sin\gamma}{1 + \sin\gamma}$
33. $\dfrac{1 - \cot(-\alpha)}{1 - \tan(-\alpha)} = \cot\alpha$
34. $\dfrac{1}{\csc x - \cot x} = \dfrac{2}{\sin x} - \dfrac{1}{\csc x + \cot x}$
35. $(\cot\beta + \csc\beta)^2 = \dfrac{\sec\beta + 1}{\sec\beta - 1}$
36. $\dfrac{\csc^2 t + \sec^2 t}{\csc t\sec t} = \cot t + \tan t$
37. $\dfrac{\cos\alpha}{1 + \cos\alpha\tan\alpha} = \dfrac{1 - \cos\alpha\tan\alpha}{\cos\alpha}$
38. $\dfrac{\cot\beta - \csc\beta + 1}{\cot\beta + \csc\beta - 1} = \dfrac{\sin\beta}{1 + \cos\beta}$
39. $\dfrac{\sin\theta}{\cot\theta + \csc\theta} - \dfrac{\sin\theta}{\cot\theta - \csc\theta} = 2$
40. $\dfrac{\tan x - \tan y}{1 + \tan x\tan y} = \dfrac{\cot y - \cot x}{1 + \cot x\cot y}$
41. $\dfrac{1}{\sin^2 t} + \dfrac{1}{\cos^2 t} = \dfrac{1}{\sin^2 t - \sin^4 t}$
42. $\cos^6\theta - \sin^6\theta = (2\cos^2\theta - 1)(1 - \sin^2\theta\cos^2\theta)$
43. $\dfrac{\cos(-\alpha)}{1 + \tan(-\alpha)} - \dfrac{\sin(-\alpha)}{1 + \cot(-\alpha)} = \sin\alpha + \cos\alpha$
44. $(1 + \tan\beta + \sec\beta)(1 + \cot\beta - \csc\beta) = 2$
45. $\dfrac{\sec u}{\csc u(1 + \sec u)} + \dfrac{1 + \cos u}{\sin u} = 2\csc u$
46. $\dfrac{\sec^4 x + \tan^4 x}{\sec^2 x\tan^2 x} = \dfrac{\cos^4 x}{\sin^2 x} + 2$
47. $(1 + \tan\beta + \cot\beta)(\cos\beta - \sin\beta) = \dfrac{\csc\beta}{\sec^2\beta} - \dfrac{\sec\beta}{\csc^2\beta}$
48. $\sqrt{\dfrac{\sec\gamma - \tan\gamma}{\sec\gamma + \tan\gamma}} = \dfrac{|\cos\gamma|}{1 + \sin\gamma}$

49. $(1 - \cot w)^2(1 + \cot w)^2 + 4\cot^2 w = \csc^4 w$

50. $\dfrac{\cot\alpha + \csc\alpha}{\sin\alpha + \cot(-\alpha) + \csc(-\alpha)} + \sec\alpha = 0$

51. $(1 + \sin\omega t + \cos\omega t)^2 = 2(1 + \sin\omega t)(1 + \cos\omega t)$

52. $\dfrac{\cot x}{1 - \tan x} + \dfrac{\tan x}{1 - \cot x} = 1 + \dfrac{\sec x}{\sin x}$

In problems 53–59, show that the given trigonometric equation is *not* a trigonometric identity.

53. $\sin\theta - \sec\theta = \tan\theta - 1$

54. $(\sin x + \cos x)^2 = 1$

55. $\dfrac{\sin t + \tan t}{\cos t + \tan t} = \tan t$

56. $\cos(\gamma + \pi) = \cos\gamma$

57. $\sqrt{1 + \sin^2 u} = 1 + \sin u$

[C] **58.** $\ln(\sin x) = \sin(\ln x)$

[C] **59.** $\sin(t^2) = \sin^2 t$

60. Give an example of a trigonometric equation that is true for three different values of the variable, but isn't an identity.

---

3 Trigonometric Formulas for Sums and Differences

The fundamental trigonometric identities considered in Section 1 express relationships among trigonometric functions of a single variable. In this section we develop trigonometric identities involving sums or differences of *two* variables.

Since $75^\circ = 30^\circ + 45^\circ$, you might ask whether $\cos 75^\circ$ can be found by calculating $\cos 30^\circ + \cos 45^\circ$. The answer is *no*; $\cos 75^\circ < 1$, whereas

$$\cos 30^\circ + \cos 45^\circ = \frac{\sqrt{3}}{2} + \frac{\sqrt{2}}{2} = \frac{\sqrt{3} + \sqrt{2}}{2} > 1.$$

The correct formula for the cosine of the sum of two angles is given by the following theorem.

Theorem 1 **Cosine of a Sum**

$$\cos(\alpha + \beta) = \cos\alpha\cos\beta - \sin\alpha\sin\beta$$

To see why the formula in Theorem 1 is correct, consider the unit circles in Figures 1 (a) and (b). In Figure 1(a), angle AOB is $\alpha + \beta$. Triangle COD in Figure 1(b) is obtained by rotating triangle AOB through the angle $-\alpha$. In Figure 1(a), we have $A = (\cos(\alpha + \beta), \sin(\alpha + \beta))$ and $B = (1,0)$, so, by the distance formula,

$$\begin{aligned}
|\overline{AB}|^2 &= [\cos(\alpha + \beta) - 1]^2 + [\sin(\alpha + \beta) - 0]^2 \\
&= \cos^2(\alpha + \beta) - 2\cos(\alpha + \beta) + 1 + \sin^2(\alpha + \beta) \\
&= [\cos^2(\alpha + \beta) + \sin^2(\alpha + \beta)] - 2\cos(\alpha + \beta) + 1 \\
&= 1 - 2\cos(\alpha + \beta) + 1 = 2 - 2\cos(\alpha + \beta).
\end{aligned}$$

Figure 1

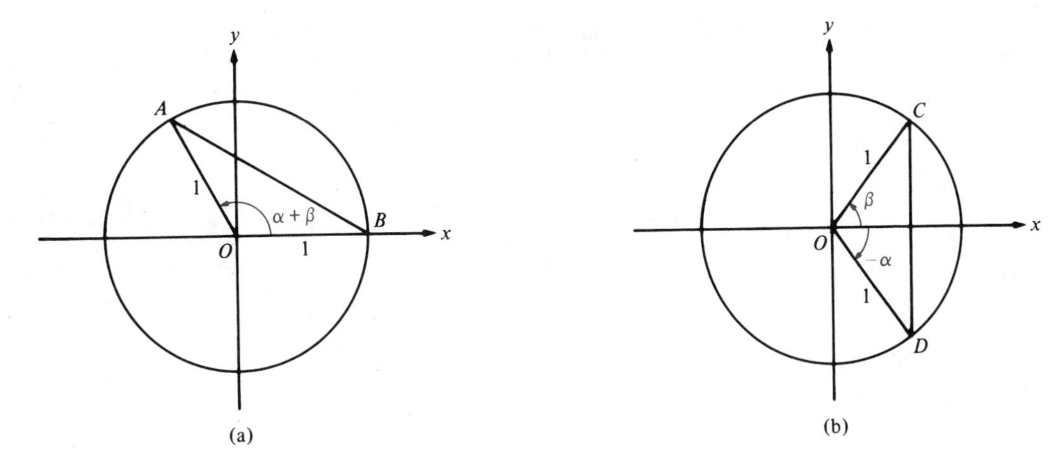

In Figure 1(b),

$$C = (\cos\beta, \sin\beta) \quad \text{and} \quad D = (\cos(-\alpha), \sin(-\alpha)) = (\cos\alpha, -\sin\alpha),$$

so, by the distance formula again,

$$\begin{aligned} |\overline{CD}|^2 &= (\cos\beta - \cos\alpha)^2 + [\sin\beta - (-\sin\alpha)]^2 = (\cos\beta - \cos\alpha)^2 + (\sin\beta + \sin\alpha)^2 \\ &= \cos^2\beta - 2\cos\beta\cos\alpha + \cos^2\alpha + \sin^2\beta + 2\sin\beta\sin\alpha + \sin^2\alpha \\ &= (\cos^2\beta + \sin^2\beta) + (\cos^2\alpha + \sin^2\alpha) - 2\cos\beta\cos\alpha + 2\sin\beta\sin\alpha \\ &= 1 + 1 - 2(\cos\beta\cos\alpha - \sin\beta\sin\alpha) \\ &= 2 - 2(\cos\alpha\cos\beta - \sin\alpha\sin\beta). \end{aligned}$$

Because triangle AOB is congruent to triangle COD, it follows that

$$|\overline{AB}|^2 = |\overline{CD}|^2.$$

Hence,

$$\begin{aligned} 2 - 2\cos(\alpha + \beta) &= 2 - 2(\cos\alpha\cos\beta - \sin\alpha\sin\beta) \\ -2\cos(\alpha + \beta) &= -2(\cos\alpha\cos\beta - \sin\alpha\sin\beta) \\ \cos(\alpha + \beta) &= \cos\alpha\cos\beta - \sin\alpha\sin\beta, \end{aligned}$$

which is the formula given in Theorem 1.

Example Find the exact value of $\cos 75°$.

Solution By Theorem 1,

$$\begin{aligned} \cos 75° &= \cos(30° + 45°) = \cos 30° \cos 45° - \sin 30° \sin 45° \\ &= \frac{\sqrt{3}}{2} \cdot \frac{\sqrt{2}}{2} - \frac{1}{2} \cdot \frac{\sqrt{2}}{2} = \frac{(\sqrt{3} - 1)\sqrt{2}}{4}. \end{aligned}$$

Theorem 2 **Cosine of a Difference**

$$\cos(\alpha - \beta) = \cos\alpha\cos\beta + \sin\alpha\sin\beta$$

Proof In Theorem 1, replace β by $-\beta$ to obtain

$$\cos(\alpha - \beta) = \cos[\alpha + (-\beta)] = \cos\alpha\cos(-\beta) - \sin\alpha\sin(-\beta)$$
$$= \cos\alpha\cos\beta - \sin\alpha(-\sin\beta) = \cos\alpha\cos\beta + \sin\alpha\sin\beta.$$

Example Find the exact value of $\cos 15°$.

Solution By Theorem 2,

$$\cos 15° = \cos(45° - 30°) = \cos 45°\cos 30° + \sin 45°\sin 30°$$
$$= \frac{\sqrt{2}}{2}\cdot\frac{\sqrt{3}}{2} + \frac{\sqrt{2}}{2}\cdot\frac{1}{2} = \frac{\sqrt{2}(\sqrt{3}+1)}{4}.$$

In Section 2 of Chapter 2, we discussed the function-cofunction relationships for trigonometric functions of *acute angles*. The following theorem shows that these relationships hold for arbitrary angles.

Theorem 3 **Cofunction Theorem**

If θ is a real number or an angle measured in radians, then

(i) $\cos\left(\frac{\pi}{2} - \theta\right) = \sin\theta$ (ii) $\sin\left(\frac{\pi}{2} - \theta\right) = \cos\theta$

(iii) $\cot\left(\frac{\pi}{2} - \theta\right) = \tan\theta$ (iv) $\tan\left(\frac{\pi}{2} - \theta\right) = \cot\theta$

(v) $\csc\left(\frac{\pi}{2} - \theta\right) = \sec\theta$ (vi) $\sec\left(\frac{\pi}{2} - \theta\right) = \csc\theta$

Proof We prove (i), (ii), and (iii) here and leave the remaining proofs as an exercise (Problem 49).

(i) By Theorem 2,

$$\cos\left(\frac{\pi}{2} - \theta\right) = \cos\frac{\pi}{2}\cos\theta + \sin\frac{\pi}{2}\sin\theta = 0\cdot\cos\theta + 1\cdot\sin\theta = \sin\theta.$$

(ii) In part (i), written from right to left, we replace θ by $(\pi/2) - \theta$ to obtain

$$\sin\left(\frac{\pi}{2} - \theta\right) = \cos\left[\frac{\pi}{2} - \left(\frac{\pi}{2} - \theta\right)\right] = \cos\theta.$$

(iii) By parts (i) and (ii),

$$\cot\left(\frac{\pi}{2} - \theta\right) = \frac{\cos\left(\frac{\pi}{2} - \theta\right)}{\sin\left(\frac{\pi}{2} - \theta\right)} = \frac{\sin\theta}{\cos\theta} = \tan\theta.$$

Naturally, if we measure an angle α in degrees instead of radians, the relationships expressed in Theorem 3 are written as

$$\cos(90^\circ - \alpha) = \sin\alpha, \quad \sin(90^\circ - \alpha) = \cos\alpha,$$

and so forth.

Now, by combining Theorem 2 and Theorem 3, we can prove the following important theorem.

Theorem 4 **Sine of a Sum**

$$\sin(\alpha + \beta) = \sin\alpha\cos\beta + \sin\beta\cos\alpha$$

Proof By parts (i) and (ii) of Theorem 3,

$$\sin(\alpha + \beta) = \cos\left[\frac{\pi}{2} - (\alpha + \beta)\right] = \cos\left(\frac{\pi}{2} - \alpha - \beta\right) = \cos\left[\left(\frac{\pi}{2} - \alpha\right) - \beta\right]$$

$$= \cos\left(\frac{\pi}{2} - \alpha\right)\cos\beta + \sin\left(\frac{\pi}{2} - \alpha\right)\sin\beta = \sin\alpha\cos\beta + \cos\alpha\sin\beta$$

$$= \sin\alpha\cos\beta + \sin\beta\cos\alpha.$$

Example Find the exact value of $\sin\frac{7\pi}{12}$.

Solution By Theorem 4, we have

$$\sin\frac{7\pi}{12} = \sin\left(\frac{\pi}{3} + \frac{\pi}{4}\right) = \sin\frac{\pi}{3}\cos\frac{\pi}{4} + \sin\frac{\pi}{4}\cos\frac{\pi}{3}$$

$$= \frac{\sqrt{3}}{2}\cdot\frac{\sqrt{2}}{2} + \frac{\sqrt{2}}{2}\cdot\frac{1}{2} = \frac{\sqrt{2}(\sqrt{3} + 1)}{4}.$$

The following theorem follows directly from Theorem 4 and the even–odd identities.

Theorem 5 **Sine of a Difference**

$$\sin(\alpha - \beta) = \sin\alpha\cos\beta - \sin\beta\cos\alpha$$

Proof Replacing β by $-\beta$ in Theorem 4, we have

$$\sin(\alpha - \beta) = \sin[\alpha + (-\beta)] = \sin\alpha\cos(-\beta) + \sin(-\beta)\cos\alpha$$
$$= \sin\alpha\cos\beta - \sin\beta\cos\alpha.$$

The four identities

$$\sin(\alpha + \beta) = \sin\alpha\cos\beta + \sin\beta\cos\alpha \qquad \sin(\alpha - \beta) = \sin\alpha\cos\beta - \sin\beta\cos\alpha$$
$$\cos(\alpha + \beta) = \cos\alpha\cos\beta - \sin\alpha\sin\beta \qquad \cos(\alpha - \beta) = \cos\alpha\cos\beta + \sin\alpha\sin\beta$$

are used so often that they should be memorized. We recommend that you memorize the first identity in the form: *The sine of the sum of two angles is equal to the sine of the first times the cosine of the second plus the sine of the second times the cosine of the first.* The remaining three identities can be stated in a similar manner.

Example 1 Simplify each expression:

(a) $\sin(90^\circ + \theta)$ (b) $\cos\left(\dfrac{\pi}{2} + x\right)$

Solution (a) $\sin(90^\circ + \theta) = \sin 90^\circ \cos\theta + \sin\theta\cos 90^\circ = 1 \cdot \cos\theta + (\sin\theta)\cdot 0 = \cos\theta$

(b) $\cos\left(\dfrac{\pi}{2} + x\right) = \cos\dfrac{\pi}{2}\cos x - \sin\dfrac{\pi}{2}\sin x = 0 \cdot \cos x - 1 \cdot \sin x = -\sin x$

Example 2 Suppose that α and β are angles in standard position; α is in quadrant I, $\sin\alpha = \frac{3}{5}$, β is in quadrant II, and $\cos\beta = -\frac{5}{13}$. Find (a) $\cos\alpha$, (b) $\sin\beta$, (c) $\sin(\alpha + \beta)$, (d) $\cos(\alpha + \beta)$, and (e) the quadrant containing angle $\alpha + \beta$ in standard position.

Solution (a) Since α is in quadrant I, $\cos\alpha$ is positive, and

$$\cos\alpha = \sqrt{1 - \sin^2\alpha} = \sqrt{1 - (\tfrac{3}{5})^2} = \sqrt{\tfrac{16}{25}} = \tfrac{4}{5}$$

(b) Since β is in quadrant II, $\sin\beta$ is positive and

$$\sin\beta = \sqrt{1 - \cos^2\beta} = \sqrt{1 - (-\tfrac{5}{13})^2} = \sqrt{\tfrac{144}{169}} = \tfrac{12}{13}$$

(c) $\sin(\alpha + \beta) = \sin\alpha\cos\beta + \sin\beta\cos\alpha = \frac{3}{5}(-\frac{5}{13}) + \frac{12}{13}\cdot\frac{4}{5} = \frac{33}{65}$

(d) $\cos(\alpha + \beta) = \cos\alpha\cos\beta - \sin\alpha\sin\beta = \frac{4}{5}(-\frac{5}{13}) - \frac{3}{5}\cdot\frac{12}{13} = -\frac{56}{65}$

(e) Since $\sin(\alpha + \beta)$ is positive and $\cos(\alpha + \beta)$ is negative, it follows that $\alpha + \beta$ is a quadrant II angle.

Example 3 Simplify each expression:

(a) $\cos 25^\circ \cos 20^\circ - \sin 25^\circ \sin 20^\circ$ (b) $\sin\dfrac{2\pi}{5}\cos\dfrac{3\pi}{5} + \sin\dfrac{3\pi}{5}\cos\dfrac{2\pi}{5}$

Solution (a) Reading the formula for the cosine of a sum from right to left, we have

$$\cos 25^\circ \cos 20^\circ - \sin 25^\circ \sin 20^\circ = \cos(25^\circ + 20^\circ) = \cos 45^\circ = \sqrt{2}/2.$$

(b) Reading the formula for the sine of a sum from right to left, we have

$$\sin\frac{2\pi}{5}\cos\frac{3\pi}{5} + \sin\frac{3\pi}{5}\cos\frac{2\pi}{5} = \sin\left(\frac{2\pi}{5} + \frac{3\pi}{5}\right) = \sin\pi = 0.$$

Example 4 As we noticed in Section 5 of Chapter 2, the graphs of the sine and cosine functions suggest that

$$\sin(x + \pi) = -\sin x \quad \text{and} \quad \cos(x + \pi) = -\cos x.$$

Prove these identities.

Solution

$$\begin{aligned}\sin(x + \pi) &= \sin x \cos\pi + \sin\pi\cos x = (\sin x)(-1) + 0 \cdot \cos x \\ &= -\sin x \\ \cos(x + \pi) &= \cos x\cos\pi - \sin x\sin\pi = (\cos x)(-1) - (\sin x)\cdot 0 \\ &= -\cos x\end{aligned}$$

Example 5 Prove the trigonometric identity

$$\cos(x + y)\cos(x - y) = \cos^2 x - \sin^2 y.$$

Solution Beginning with the left side and applying the formulas for the cosine of a sum and for the cosine of a difference, we have

$$\begin{aligned}\cos(x + y)\cos(x - y) &= (\cos x\cos y - \sin x\sin y)(\cos x\cos y + \sin x\sin y) \\ &= (\cos x\cos y)^2 - (\sin x\sin y)^2 \\ &= \cos^2 x\cos^2 y - \sin^2 x\sin^2 y \\ &= \cos^2 x(1 - \sin^2 y) - (1 - \cos^2 x)\sin^2 y \\ &= \cos^2 x - \cos^2 x\sin^2 y - \sin^2 y + \cos^2 x\sin^2 y \\ &= \cos^2 x - \sin^2 y.\end{aligned}$$

Do you see what gave us the idea to substitute $1 - \sin^2 y$ for $\cos^2 y$ and $1 - \cos^2 x$ for $\sin^2 x$ in the fourth step of the calculation above? [Hint: We wanted the expression $\cos^2 x - \sin^2 y$ to show up.]

Using the formulas for the sine and cosine of a sum, we can derive a useful formula for the tangent of a sum.

Theorem 6 **Tangent of a Sum**

If $\cos\alpha \neq 0$, $\cos\beta \neq 0$, and $\cos(\alpha + \beta) \neq 0$, then

$$\tan(\alpha + \beta) = \frac{\tan\alpha + \tan\beta}{1 - \tan\alpha\tan\beta}.$$

Proof

$$\tan(\alpha + \beta) = \frac{\sin(\alpha + \beta)}{\cos(\alpha + \beta)} = \frac{\sin\alpha\cos\beta + \sin\beta\cos\alpha}{\cos\alpha\cos\beta - \sin\alpha\sin\beta}$$

Dividing numerator and denominator of the fraction on the right side of the equation above by $\cos\alpha\cos\beta$, we get

$$\tan(\alpha + \beta) = \frac{\dfrac{\sin\alpha\cos\beta}{\cos\alpha\cos\beta} + \dfrac{\sin\beta\cos\alpha}{\cos\alpha\cos\beta}}{\dfrac{\cos\alpha\cos\beta}{\cos\alpha\cos\beta} - \dfrac{\sin\alpha\sin\beta}{\cos\alpha\cos\beta}} = \frac{\dfrac{\sin\alpha}{\cos\alpha} + \dfrac{\sin\beta}{\cos\beta}}{1 - \left(\dfrac{\sin\alpha}{\cos\alpha}\right)\left(\dfrac{\sin\beta}{\cos\beta}\right)}$$

$$= \frac{\tan\alpha + \tan\beta}{1 - \tan\alpha\tan\beta}.$$

Example 1 Assuming that $\cos\alpha \neq 0$, $\cos\beta \neq 0$, and $\cos(\alpha - \beta) \neq 0$, show that

$$\tan(\alpha - \beta) = \frac{\tan\alpha - \tan\beta}{1 + \tan\alpha\tan\beta}.$$

Solution In Theorem 6, we replace β by $-\beta$ to obtain

$$\tan(\alpha - \beta) = \tan[\alpha + (-\beta)] = \frac{\tan\alpha + \tan(-\beta)}{1 - \tan\alpha\tan(-\beta)} = \frac{\tan\alpha - \tan\beta}{1 + \tan\alpha\tan\beta}.$$

Example 2 Simplify the expression $\dfrac{\tan 43^\circ + \tan 17^\circ}{1 - \tan 43^\circ \tan 17^\circ}$.

Solution Reading the formula in Theorem 6 from right to left, we have

$$\frac{\tan 43^\circ + \tan 17^\circ}{1 - \tan 43^\circ \tan 17^\circ} = \tan(43^\circ + 17^\circ) = \tan 60^\circ = \sqrt{3}.$$

PROBLEM SET 3

In problems 1–12, find the exact value of the indicated trigonometric function.

1. $\sin 105^\circ$ [Hint: $105^\circ = 60^\circ + 45^\circ$.]
2. $\cos 105^\circ$
3. $\sin 195^\circ$ [Hint: $195^\circ = 150^\circ + 45^\circ$.]
4. $\tan 195^\circ$
5. $\cos\dfrac{17\pi}{12}$ $\left[\text{Hint: } \dfrac{17\pi}{12} = \dfrac{5\pi}{4} + \dfrac{\pi}{6}.\right]$
6. $\cot\dfrac{17\pi}{12}$
7. $\sin\dfrac{11\pi}{12}$ $\left[\text{Hint: } \dfrac{11\pi}{12} = \dfrac{7\pi}{6} - \dfrac{\pi}{4}.\right]$
8. $\sec\dfrac{11\pi}{12}$
9. $\tan\dfrac{13\pi}{12}$ $\left[\text{Hint: } \dfrac{13\pi}{12} = \dfrac{3\pi}{4} + \dfrac{\pi}{3}.\right]$
10. $\cot\dfrac{13\pi}{12}$
11. $\tan\dfrac{19\pi}{12}$ $\left[\text{Hint: } \dfrac{19\pi}{12} = \dfrac{11\pi}{6} - \dfrac{\pi}{4}.\right]$
12. $\csc\dfrac{19\pi}{12}$

In problems 13–34, simplify each expression.

13. $\sin(180° - \theta)$
14. $\cos(180° + t)$
15. $\cos(\pi - \alpha)$
16. $\sin(270° - x)$
17. $\sin(360° - s)$
18. $\cos(2\pi - \gamma)$
19. $\tan(180° - t)$
20. $\sec(\pi + u)$
21. $\csc\left(\frac{3\pi}{2} + \beta\right)$
22. $\tan(270° + \theta)$ [Caution: Theorem 6 doesn't work here.]
23. $\sin 17° \cos 13° + \sin 13° \cos 17°$
24. $\cos 43° \cos 17° - \sin 43° \sin 17°$
25. $\sin \frac{17\pi}{36} \cos \frac{11\pi}{36} - \sin \frac{11\pi}{36} \cos \frac{17\pi}{36}$
26. $\cos \frac{5\pi}{7} \sin \frac{2\pi}{7} + \cos \frac{2\pi}{7} \sin \frac{5\pi}{7}$
27. $\cos 2x \cos x + \sin 2x \sin x$
28. $\sin(\alpha - \beta)\cos \beta + \cos(\alpha - \beta)\sin \beta$
29. $\cos(\alpha + \beta)\cos \beta + \sin(\alpha + \beta)\sin \beta$
30. $\cos(x + y) - \cos(x - y)$
31. $\dfrac{\tan 42° + \tan 18°}{1 - \tan 42° \tan 18°}$
32. $\dfrac{\tan 50° - \tan 20°}{1 + \tan 50° \tan 20°}$
33. $\dfrac{\tan 8x - \tan 7x}{1 + \tan 8x \tan 7x}$
34. $\tan\left(\frac{\pi}{2} + \theta\right)$

35. Suppose that α and β are angles in standard position; α is in quadrant I, $\sin \alpha = \frac{4}{5}$, β is in quadrant II, and $\sin \beta = \frac{12}{13}$. Find (a) $\cos \alpha$, (b) $\cos \beta$, (c) $\sin(\alpha + \beta)$, (d) $\cos(\alpha + \beta)$, (e) $\sin(\alpha - \beta)$, (f) $\tan(\alpha + \beta)$, and (g) the quadrant containing the angle $\alpha + \beta$ in standard position.
36. Suppose that $\frac{\pi}{2} < s < \pi$, $\frac{\pi}{2} < t < \pi$, $\sin s = \frac{5}{13}$, and $\tan t = -\frac{3}{4}$. Find (a) $\sin(s - t)$, (b) $\cos(s - t)$, (c) $\sin(s + t)$, (d) $\sec(s + t)$, and (e) the quadrant containing the angle θ in standard position if the radian measure of θ is $s + t$.
37. If α and β are positive acute angles, $\sin(\alpha + \beta) = \frac{56}{65}$, and $\sin \beta = \frac{5}{13}$, find $\sin \alpha$. [Hint: $\sin \alpha = \sin((\alpha + \beta) - \beta)$.]
38. Derive a formula for $\tan(\alpha + \beta)$ for the case in which $\cos \alpha = 0$ and $\sin \beta \neq 0$.

In problems 39–48, show that each equation is an identity.

39. $\dfrac{\sin(\alpha + \beta)}{\sin \alpha \cos \beta} = 1 + \cot \alpha \tan \beta$
40. $\tan(\theta + \pi) = \tan \theta$
41. $\sin\left(t - \frac{\pi}{3}\right) + \cos\left(t - \frac{\pi}{6}\right) = \sin t$
42. $\cos(x + y)\cos(x - y) = 1 - \sin^2 x - \sin^2 y$
43. $\tan\left(x + \frac{\pi}{2}\right) = \tan\left(x - \frac{\pi}{2}\right)$
44. $\tan \alpha - \tan \beta = \dfrac{\sin(\alpha - \beta)}{\cos \alpha \cos \beta}$
45. $\cot(x + y) = \dfrac{\cot x \cot y - 1}{\cot x + \cot y}$
46. $\dfrac{\sin(s + t)}{\sin(s - t)} = \dfrac{\tan s + \tan t}{\tan s - \tan t}$
47. $\dfrac{1 - \tan^2\left(\frac{\pi}{4} - \theta\right)}{1 + \tan^2\left(\frac{\pi}{4} - \theta\right)} = 2 \sin \theta \cos \theta$
48. $\tan^2\left(\frac{\pi}{2} - t\right)\sec^2 t - \sin^2\left(\frac{\pi}{2} - t\right)\csc^2 t = 1$
49. Prove parts (iv), (v), and (vi) of Theorem 3.
50. If $x + y = \pi/4$, show that

$$\tan y = \frac{1 - \tan x}{1 + \tan x}.$$

C In problems 51–56, use a calculator to verify the equation for the indicated values of the variables.

51. $\cos(\alpha - \beta) = \cos\alpha\cos\beta + \sin\alpha\sin\beta$ for $\alpha = \frac{3\pi}{5}$, $\beta = \frac{\pi}{5}$

52. $\sin(x + y) = \sin x\cos y + \sin y\cos x$ for $x = 2.31$, $y = 1.07$

53. $\tan(\alpha + \beta) = \dfrac{\tan\alpha + \tan\beta}{1 - \tan\alpha\tan\beta}$ for $\alpha = 22.4°$, $\beta = 17.3°$

54. $\cos(s + t) = \cos s\cos t - \sin s\sin t$ for $s = \frac{5\pi}{11}$, $t = \frac{7\pi}{9}$

55. $\sin(s - t) = \sin s\cos t - \cos s\sin t$ for $s = 0.791$, $t = 1.234$

56. $\tan(\alpha - \beta) = \dfrac{\tan\alpha - \tan\beta}{1 + \tan\alpha\tan\beta}$ for $\alpha = 91.7°$, $\beta = 87.2°$

---

4 Double-Angle and Half-Angle Formulas

In this section we study the important double-angle and half-angle formulas. These identities, which are obtained as special cases of the formulas derived in the last section, enable us to evaluate trigonometric functions of 2θ or $\theta/2$ in terms of trigonometric functions of θ.

The sine of twice an angle is *not* in general the same as twice the sine of the angle. For instance, $\sin 30° = \frac{1}{2}$; $\sin 2(30°) = \sin 60° = \sqrt{3}/2 \neq 2\sin 30°$. The correct formula for the sine of twice an angle is given by the following theorem.

Theorem 1 **Double-Angle Formula for Sine**

$$\sin 2\theta = 2\sin\theta\cos\theta.$$

Proof In the formula

$$\sin(\alpha + \beta) = \sin\alpha\cos\beta + \sin\beta\cos\alpha,$$

let $\alpha = \beta = \theta$, so that

$$\sin(\theta + \theta) = \sin\theta\cos\theta + \sin\theta\cos\theta$$

or

$$\sin 2\theta = 2\sin\theta\cos\theta.$$

Example 1 Verify Theorem 1 for $\theta = 30°$.

Solution We know that $\sin 60° = \sqrt{3}/2$. Using Theorem 1 with $\theta = 30°$, we also get

$$\sin 60° = \sin 2(30°) = 2\sin 30°\cos 30° = 2\left(\frac{1}{2}\right)\left(\frac{\sqrt{3}}{2}\right) = \frac{\sqrt{3}}{2}.$$

Example 2 Write $\sin 100^\circ$ in terms of trigonometric functions of 50°.

Solution

$$\sin 100^\circ = \sin 2(50^\circ) = 2\sin 50^\circ \cos 50^\circ$$

Example 3 Reduce the expression $2\sin 8x \cos 8x$ to the value of a function of $16x$.

Solution Reading Theorem 1 from right to left with $\theta = 8x$, we have

$$2\sin 8x \cos 8x = \sin[2(8x)] = \sin 16x.$$

Example 4 Suppose that α is a quadrant I angle and that $\sin\alpha = \frac{3}{5}$. Find $\sin 2\alpha$.

Solution We are going to use the formula $\sin 2\alpha = 2\sin\alpha\cos\alpha$, but first we must find $\cos\alpha$. Because α is a quadrant I angle, we know that $\cos\alpha$ is positive, so

$$\cos\alpha = \sqrt{1 - \sin^2\alpha} = \sqrt{1 - (\tfrac{3}{5})^2} = \sqrt{\tfrac{16}{25}} = \tfrac{4}{5}.$$

Therefore,

$$\sin 2\alpha = 2\sin\alpha\cos\alpha = 2(\tfrac{3}{5})(\tfrac{4}{5}) = \tfrac{24}{25}.$$

There are three double-angle formulas for the cosine—a basic formula and two alternative versions of it.

Theorem 2 **Double-Angle Formulas for Cosine**

(i) $\cos 2\theta = \cos^2\theta - \sin^2\theta$

(ii) $\cos 2\theta = 2\cos^2\theta - 1$

(iii) $\cos 2\theta = 1 - 2\sin^2\theta$

Proof (i) In the formula

$$\cos(\alpha + \beta) = \cos\alpha\cos\beta - \sin\alpha\sin\beta,$$

let $\alpha = \beta = \theta$, so that

$$\cos(\theta + \theta) = \cos\theta\cos\theta - \sin\theta\sin\theta$$

or

$$\cos 2\theta = \cos^2\theta - \sin^2\theta.$$

(ii) In the basic formula just derived, substitute $1 - \cos^2\theta$ for $\sin^2\theta$ to obtain

$$\cos 2\theta = \cos^2\theta - (1 - \cos^2\theta) = 2\cos^2\theta - 1.$$

(iii) In the basic formula $\cos 2\theta = \cos^2\theta - \sin^2\theta$, substitute $1 - \sin^2\theta$ for $\cos^2\theta$ to obtain

$$\cos 2\theta = 1 - \sin^2\theta - \sin^2\theta = 1 - 2\sin^2\theta.$$

Example 1 Write $\cos(\pi/4)$ in terms of trigonometric functions of $\pi/8$.

Solution By Theorem 2, with $\theta = \pi/8$,

$$\cos\frac{\pi}{4} = \cos 2\left(\frac{\pi}{8}\right) = \cos^2\frac{\pi}{8} - \sin^2\frac{\pi}{8}.$$

Alternatively, we could have written

$$\cos\frac{\pi}{4} = 2\cos^2\frac{\pi}{8} - 1 \quad \text{or} \quad \cos\frac{\pi}{4} = 1 - 2\sin^2\frac{\pi}{8}.$$

Example 2 Reduce the expression $1 - 2\sin^2 5\theta$ to the value of a function of 10θ.

Solution Reading part (iii) of Theorem 2 from right to left and replacing θ by 5θ, we have

$$1 - 2\sin^2 5\theta = \cos 10\theta.$$

Example 3 If $\sin\beta = -\frac{7}{10}$, find $\cos 2\beta$.

Solution By part (iii) of Theorem 2 with $\theta = \beta$, we have

$$\cos 2\beta = 1 - 2\sin^2\beta = 1 - 2(-\tfrac{7}{10})^2 = \tfrac{50}{50} - \tfrac{49}{50} = \tfrac{1}{50}.$$

Example 4 Express $\cos 4x$ in terms of $\sin x$.

Solution By part (ii) of Theorem 2 with $\theta = 2x$,

$$\cos 4x = \cos 2(2x) = 2\cos^2 2x - 1 = 2(\cos 2x)^2 - 1.$$

Therefore, by part (iii) of Theorem 2 with $\theta = x$,

$$\begin{aligned}\cos 4x &= 2(1 - 2\sin^2 x)^2 - 1 = 2(1 - 4\sin^2 x + 4\sin^4 x) - 1 \\ &= 1 - 8\sin^2 x + 8\sin^4 x.\end{aligned}$$

If we solve the equation $\cos 2\theta = 2\cos^2\theta - 1$ in part (ii) of Theorem 2 for $\cos^2\theta$, we obtain the identity

$$\cos^2\theta = \frac{1 + \cos 2\theta}{2}.$$

Similarly, the equation $\cos 2\theta = 1 - 2\sin^2\theta$ in part (iii) of Theorem 2 can be solved for $\sin^2\theta$, and we have

$$\sin^2\theta = \frac{1 - \cos 2\theta}{2}.$$

The formulas above for $\cos^2\theta$ and $\sin^2\theta$ are used extensively in calculus, and they provide the basis for the half-angle formulas to be derived shortly. We record them for future use in the following theorem.

Theorem 3 **Formulas for $\cos^2\theta$ and $\sin^2\theta$**

$$\cos^2\theta = \frac{1+\cos 2\theta}{2} \quad \text{and} \quad \sin^2\theta = \frac{1-\cos 2\theta}{2}$$

Example 1 Prove the identity

$$\sin^4\theta = \tfrac{3}{8} - \tfrac{1}{2}\cos 2\theta + \tfrac{1}{8}\cos 4\theta.$$

Solution Using the second equation of Theorem 3, we can write

$$\sin^4\theta = (\sin^2\theta)^2 = \left(\frac{1-\cos 2\theta}{2}\right)^2 = \tfrac{1}{4}(1 - 2\cos 2\theta + \cos^2 2\theta).$$

Now, using the first equation of Theorem 3 with θ replaced by 2θ, we have

$$\sin^4\theta = \tfrac{1}{4}(1 - 2\cos 2\theta + \cos^2 2\theta) = \tfrac{1}{4}\left[1 - 2\cos 2\theta + \left(\frac{1+\cos 4\theta}{2}\right)\right]$$
$$= \tfrac{1}{4} - \tfrac{1}{2}\cos 2\theta + \tfrac{1}{8} + \tfrac{1}{8}\cos 4\theta = \tfrac{3}{8} - \tfrac{1}{2}\cos 2\theta + \tfrac{1}{8}\cos 4\theta.$$

If we put $\alpha = \beta = \theta$ in Theorem 6 of Section 3, we obtain the double-angle formula for the tangent:

$$\tan 2\theta = \frac{2\tan\theta}{1-\tan^2\theta}.$$

A formula for $\tan^2\theta$ can be obtained as follows:

$$\tan^2\theta = \frac{\sin^2\theta}{\cos^2\theta} = \frac{\left(\dfrac{1-\cos 2\theta}{2}\right)}{\left(\dfrac{1+\cos 2\theta}{2}\right)}$$

so that

$$\tan^2\theta = \frac{1-\cos 2\theta}{1+\cos 2\theta}.$$

If you let $\theta = \alpha/2$ in Theorem 3 and in the preceding formula for $\tan^2\theta$, you will obtain the formulas in the following theorem (Problem 34).

Theorem 4 **Half-Angle Formulas**

(i) $\cos^2\dfrac{\alpha}{2} = \dfrac{1+\cos\alpha}{2}$

(ii) $\sin^2\dfrac{\alpha}{2} = \dfrac{1-\cos\alpha}{2}$

(iii) $\tan^2\dfrac{\alpha}{2} = \dfrac{1-\cos\alpha}{1+\cos\alpha}$, for $\cos\alpha \neq -1$.

The equations in Theorem 4 can be written in the form

$$\cos\frac{\alpha}{2} = \pm\sqrt{\frac{1+\cos\alpha}{2}}, \quad \sin\frac{\alpha}{2} = \pm\sqrt{\frac{1-\cos\alpha}{2}}, \quad \text{and} \quad \tan\frac{\alpha}{2} = \pm\sqrt{\frac{1-\cos\alpha}{1+\cos\alpha}}.$$

In using these formulas, you can determine the algebraic sign (+ or −) by considering which quadrant contains the angle $\alpha/2$ in standard position.

Example 1 Find the exact values of (a) $\sin 22.5^\circ$ and (b) $\cos(\pi/12)$.

Solution (a) Because 22.5° is a quadrant I angle, we have

$$\sin 22.5^\circ = \sin\frac{45^\circ}{2} = \sqrt{\frac{1-\cos 45^\circ}{2}} = \sqrt{\frac{1-(\sqrt{2}/2)}{2}} = \sqrt{\frac{2-\sqrt{2}}{4}} = \frac{\sqrt{2-\sqrt{2}}}{2}$$

(b) Because $\pi/12$ is a quadrant I angle, we have

$$\cos\frac{\pi}{12} = \cos\frac{\pi/6}{2} = \sqrt{\frac{1+\cos(\pi/6)}{2}} = \sqrt{\frac{1+(\sqrt{3}/2)}{2}} = \sqrt{\frac{2+\sqrt{3}}{4}} = \frac{\sqrt{2+\sqrt{3}}}{2}$$

Example 2 Suppose that $\frac{3\pi}{2} < \theta < 2\pi$ and $\cos\theta = \frac{5}{13}$. Find (a) $\sin\frac{\theta}{2}$ and (b) $\tan\frac{\theta}{2}$.

Solution Because $3\pi/2 < \theta < 2\pi$, we have $\frac{1}{2}(3\pi/2) < \frac{1}{2}\theta < \frac{1}{2}(2\pi)$, or $3\pi/4 < \theta/2 < \pi$. Therefore, $\theta/2$ is a quadrant II angle, and so $\sin(\theta/2) > 0$ and $\tan(\theta/2) < 0$.

$$\text{(a)}\ \sin\frac{\theta}{2} = \sqrt{\frac{1-\cos\theta}{2}} = \sqrt{\frac{1-(5/13)}{2}} = \sqrt{\frac{1}{2}\left(\frac{8}{13}\right)} = \sqrt{\frac{4}{13}} = \frac{2\sqrt{13}}{13}$$

$$\text{(b)}\ \tan\frac{\theta}{2} = -\sqrt{\frac{1-\cos\theta}{1+\cos\theta}} = -\sqrt{\frac{1-(5/13)}{1+(5/13)}} = -\sqrt{\frac{13-5}{13+5}}$$

$$= -\sqrt{\frac{8}{18}} = -\sqrt{\frac{4}{9}} = -\frac{2}{3}$$

Example 3 Let $z = \tan(\theta/2)$. Show that:

(a) $\cos\theta = \dfrac{1-z^2}{1+z^2}$ (b) $\sin\theta = \dfrac{2z}{1+z^2}$ (c) $\tan\dfrac{\theta}{2} = \dfrac{1-\cos\theta}{\sin\theta}$

Solution (a) Using part (iii) of Theorem 4, we have

$$\frac{1-z^2}{1+z^2} = \frac{1-\tan^2\frac{\theta}{2}}{1+\tan^2\frac{\theta}{2}} = \frac{1-\left(\frac{1-\cos\theta}{1+\cos\theta}\right)}{1+\left(\frac{1-\cos\theta}{1+\cos\theta}\right)}$$

$$= \frac{1+\cos\theta-(1-\cos\theta)}{1+\cos\theta+(1-\cos\theta)} = \frac{2\cos\theta}{2} = \cos\theta.$$

(b) Using the fundamental identities, we have

$$\frac{2z}{1+z^2} = \frac{2\tan\frac{\theta}{2}}{1+\tan^2\frac{\theta}{2}} = \frac{2\tan\frac{\theta}{2}}{\sec^2\frac{\theta}{2}} = 2\tan\frac{\theta}{2}\cos^2\frac{\theta}{2}$$

$$= 2\frac{\sin\frac{\theta}{2}}{\cos\frac{\theta}{2}}\cos^2\frac{\theta}{2} = 2\sin\frac{\theta}{2}\cos\frac{\theta}{2}.$$

But, by Theorem 1,

$$2\sin\frac{\theta}{2}\cos\frac{\theta}{2} = \sin\left[2\left(\frac{\theta}{2}\right)\right] = \sin\theta.$$

Therefore,

$$\frac{2z}{1+z^2} = \sin\theta.$$

(c) Using the results of (a) and (b), we have

$$\frac{1-\cos\theta}{\sin\theta} = \frac{1-\left(\frac{1-z^2}{1+z^2}\right)}{\left(\frac{2z}{1+z^2}\right)} = \frac{(1+z^2)-(1-z^2)}{2z} = \frac{2z^2}{2z} = z = \tan\frac{\theta}{2}.$$

Example 4 Find the exact value of $\tan\frac{\pi}{8}$.

Solution By part (c) of Example 3 above,

$$\tan\frac{\pi}{8} = \tan\left[\frac{1}{2}\left(\frac{\pi}{4}\right)\right] = \frac{1-\cos\frac{\pi}{4}}{\sin\frac{\pi}{4}} = \frac{1-\frac{\sqrt{2}}{2}}{\frac{\sqrt{2}}{2}} = \frac{2-\sqrt{2}}{\sqrt{2}} = \frac{2}{\sqrt{2}} - 1 = \sqrt{2}-1.$$

Example 5 If $z = \tan\frac{\theta}{2}$, write $\frac{1}{1-\cos\theta}$ in terms of z.

Solution By part (a) of Example 3,

$$\frac{1}{1-\cos\theta} = \frac{1}{1-\left(\frac{1-z^2}{1+z^2}\right)} = \frac{1+z^2}{(1+z^2)-(1-z^2)} = \frac{1+z^2}{2z^2}.$$

PROBLEM SET 4

In problems 1–6, use the double-angle formulas to write the given expression in terms of the values of trigonometric functions of an angle half as large.

1. $\sin 76^\circ$ **2.** $\tan 62^\circ$ **3.** $\cos 144^\circ$ **4.** $\cos 4x$ **5.** $\sin \frac{2\pi}{9}$ **6.** $\tan \frac{\pi}{7}$

In problems 7–12, use the double-angle formulas to reduce the given expression to the value of a single trigonometric function of an angle twice as large.

7. $2 \sin 29^\circ \cos 29^\circ$ **8.** $\cos^2 4\alpha - \sin^2 4\alpha$ **9.** $2\cos^2 \frac{5\theta}{2} - 1$

10. $\frac{\tan 5t}{1 - \tan^2 5t}$ **11.** $1 - 2\sin^2 \frac{\pi}{17}$ **12.** $2 \sin\left(\frac{x+y}{2}\right) \cos\left(\frac{x+y}{2}\right)$

In problems 13–18, use the given information to find (a) $\sin 2\theta$, (b) $\cos 2\theta$, and (c) $\tan 2\theta$.

13. $\sin \theta = \frac{4}{5}$, θ in quadrant I **14.** $\csc \theta = \frac{5}{3}$, θ in quadrant II

15. $\cos \theta = -\frac{5}{13}$, θ in quadrant III **16.** $\sec \theta = -\frac{17}{8}$, θ not in quadrant II

17. $\tan \theta = -\frac{12}{5}$, θ in quadrant II **18.** $\cot \theta = \frac{1}{2}$, θ in quadrant III

In problems 19–26, rewrite each expression in a form involving only a single trigonometric function.

19. $\frac{\sin 2x}{2 \sin x}$ **20.** $(\sin 2\alpha + \cos 2\alpha)^2 - 2 \sin 4\alpha$ **21.** $(\sin t - \cos t)^2 + 3 \sin 2t$

22. $\frac{1 - \cos 2u}{\sin 2u}$ **23.** $\sin x \cos^3 x + \sin^3 x \cos x$ **24.** $\frac{\sin 2\beta}{1 + \cos 2\beta}$

25. $\frac{\tan 2\theta - \sin 2\theta}{2 \tan 2\theta}$ **26.** $\frac{\sin 2x}{1 - \cos 2x}$

In problems 27–33, (a) use the half-angle formulas to write the given expression in terms of the value of a trigonometric function of an angle twice as large, and (b) find the exact numerical value of each expression.

27. $\cos 15^\circ$ **28.** $\sin 67.5^\circ$ **29.** $\cos \frac{5\pi}{8}$ **30.** $\tan \frac{\pi}{12}$

31. $\tan 157.5^\circ$ **32.** $\cos \frac{11\pi}{12}$ **33.** $\sin 202.5^\circ$

34. Prove Theorem 4.

In problems 35–38, use the given information to find (a) $\sin(\theta/2)$ (b) $\cos(\theta/2)$, and (c) $\tan(\theta/2)$.

35. $\cos \theta = \frac{7}{25}$, $0 < \theta < 90^\circ$ **36.** $\cos \theta = -\frac{4}{5}$, θ in quadrant II

37. $\cos \theta = -\frac{5}{13}$, $\pi < \theta < \frac{3\pi}{2}$ **38.** $\cot \theta = -\frac{15}{8}$, θ in quadrant IV

In problems 39–44, reduce the given expression to the value of a single trigonometric function of an angle half as large.

39. $-\sqrt{\frac{1 + \cos 250^\circ}{2}}$ **40.** $-\sqrt{\frac{1 - \cos 400^\circ}{2}}$ **41.** $\frac{1 - \cos 6x}{\sin 6x}$

42. $\dfrac{\sin 10\theta}{1 - \cos 10\theta}$ **43.** $\sqrt{\dfrac{1 + \cos(2\pi/5)}{2}}$ **44.** $\sqrt{\dfrac{1 - \cos(11\pi/15)}{2}}$

In problems 45–55, prove that each trigonometric equation is an identity.

45. $\cos^4\theta - \sin^4\theta = \cos 2\theta$

46. $2\csc 2x = \sec x \csc x$

47. $\sin 4t = 2(\sin 2t)(1 - 2\sin^2 t)$

48. $\dfrac{\sin^2 2u}{\sin^2 u} = 4\cos^2 u$

49. $\sin 3\theta = (\sin\theta)(3 - 4\sin^2\theta)$ [Hint: $3\theta = \theta + 2\theta$.]

50. $\cos 3t = 4\cos^3 t - 3\cos t$

51. $\dfrac{\sec^2 x}{2 - \sec^2 x} = \sec 2x$

52. $\sin^2 t\cos^2 t = \frac{1}{8}(1 - \cos 4t)$

53. $\dfrac{\cot\alpha - 1}{\cot\alpha + 1} = \dfrac{\cos 2\alpha}{1 + \sin 2\alpha}$

54. $\cos^4 x = \frac{3}{8} + \frac{1}{2}\cos 2x + \frac{1}{8}\cos 4x$

55. $\csc\theta - \cot\theta = \tan\dfrac{\theta}{2}$

56. Show that $\cos 20^\circ$ is a root of the equation $8x^3 - 6x - 1 = 0$. [Hint: Use problem 50.]

In problems 57–60, write the given expression in terms of z, where $z = \tan(\theta/2)$. (Use the results of Example 3, page 136.)

57. $\sec\theta$ **58.** $\csc\theta$ **59.** $\cot\theta$ **60.** $\dfrac{\sin\theta}{2 - \cos\theta}$

61. Using Theorem 3, show that, if b and c are positive constants, the quantity $y = b\cos^2 ct$ oscillates in accordance with the simple harmonic model $y = a\cos(\omega t - \phi) + k$, with $a = b/2$, $\omega = 2c$, $\phi = 0$, and $k = b/2$.

---

5 Product Formulas, Sum Formulas, and Graphs of Combinations of Sine and Cosine Functions

In applied mathematics, it is often necessary to rewrite a product of sine or cosine functions as a sum or difference of such functions, or vice versa. This can be done by using the identities in Theorems 1 and 2.

Theorem 1 **Product Formulas**

(i) $\sin\alpha\cos\beta = \frac{1}{2}[\sin(\alpha + \beta) + \sin(\alpha - \beta)]$

(ii) $\cos\alpha\sin\beta = \frac{1}{2}[\sin(\alpha + \beta) - \sin(\alpha - \beta)]$

(iii) $\cos\alpha\cos\beta = \frac{1}{2}[\cos(\alpha + \beta) + \cos(\alpha - \beta)]$

(iv) $\sin\alpha\sin\beta = \frac{1}{2}[\cos(\alpha - \beta) - \cos(\alpha + \beta)]$

Proof We prove (i) and (iv) here and leave (ii) and (iii) as exercises (Problem 22).

(i) Using the formulas

$$\sin(\alpha + \beta) = \sin\alpha\cos\beta + \sin\beta\cos\alpha$$
$$\sin(\alpha - \beta) = \sin\alpha\cos\beta - \sin\beta\cos\alpha,$$

and adding the second equation to the first, we obtain

$$\sin(\alpha + \beta) + \sin(\alpha - \beta) = 2\sin\alpha\cos\beta.$$

It follows that

$$\sin\alpha\cos\beta = \tfrac{1}{2}[\sin(\alpha + \beta) + \sin(\alpha - \beta)].$$

(iv) Using the formulas

$$\cos(\alpha - \beta) = \cos\alpha\cos\beta + \sin\alpha\sin\beta$$
$$\cos(\alpha + \beta) = \cos\alpha\cos\beta - \sin\alpha\sin\beta,$$

and subtracting the second equation from the first, we obtain

$$\cos(\alpha - \beta) - \cos(\alpha + \beta) = 2\sin\alpha\sin\beta.$$

It follows that

$$\sin\alpha\sin\beta = \tfrac{1}{2}[\cos(\alpha - \beta) - \cos(\alpha + \beta)].$$

Example Express each product as a sum or difference:

(a) $\sin 5\theta\cos 3\theta$ (b) $\sin 3t\sin t$

Solution (a) By part (i) of Theorem 1 with $\alpha = 5\theta$ and $\beta = 3\theta$,

$$\sin 5\theta\cos 3\theta = \tfrac{1}{2}[\sin(5\theta + 3\theta) + \sin(5\theta - 3\theta)] = \tfrac{1}{2}\sin 8\theta + \tfrac{1}{2}\sin 2\theta.$$

(b) By part (iv) of Theorem 1 with $\alpha = 3t$ and $\beta = t$,

$$\sin 3t\sin t = \tfrac{1}{2}[\cos(3t - t) - \cos(3t + t)] = \tfrac{1}{2}\cos 2t - \tfrac{1}{2}\cos 4t.$$

Theorem 2 **Sum Formulas**

(i) $\sin\gamma + \sin\theta = 2\sin\left(\frac{\gamma + \theta}{2}\right)\cos\left(\frac{\gamma - \theta}{2}\right)$

(ii) $\sin\gamma - \sin\theta = 2\cos\left(\frac{\gamma + \theta}{2}\right)\sin\left(\frac{\gamma - \theta}{2}\right)$

(iii) $\cos\gamma + \cos\theta = 2\cos\left(\frac{\gamma + \theta}{2}\right)\cos\left(\frac{\gamma - \theta}{2}\right)$

(iv) $\cos\gamma - \cos\theta = -2\sin\left(\frac{\gamma + \theta}{2}\right)\sin\left(\frac{\gamma - \theta}{2}\right)$

Proof We prove (i) here and leave (ii), (iii), and (iv) as exercises (Problem 24). In part (i) of Theorem 1, let

$$\alpha = \frac{\gamma + \theta}{2} \quad \text{and} \quad \beta = \frac{\gamma - \theta}{2}.$$

Thus,

$$\alpha + \beta = \frac{\gamma + \theta}{2} + \frac{\gamma - \theta}{2} = \frac{2\gamma}{2} = \gamma$$

and

$$\alpha - \beta = \frac{\gamma + \theta}{2} - \frac{\gamma - \theta}{2} = \frac{2\theta}{2} = \theta.$$

It follows that

$$\sin\left(\frac{\gamma + \theta}{2}\right)\cos\left(\frac{\gamma - \theta}{2}\right) = \tfrac{1}{2}[\sin\gamma + \sin\theta],$$

or

$$\sin\gamma + \sin\theta = 2\sin\left(\frac{\gamma + \theta}{2}\right)\cos\left(\frac{\gamma - \theta}{2}\right).$$

Example 1 Express $\sin 10x + \sin 4x$ as a product.

Solution Using part (i) of Theorem 2 with $\gamma = 10x$ and $\theta = 4x$, we have

$$\sin 10x + \sin 4x = 2\sin\left(\frac{10x + 4x}{2}\right)\cos\left(\frac{10x - 4x}{2}\right) = 2\sin 7x\cos 3x.$$

Example 2 Prove that the equation

$$\frac{\sin(x + h) - \sin x}{h} = \cos\left(x + \frac{h}{2}\right)\frac{\sin(h/2)}{h/2}$$

is an identity.

Solution Using part (ii) of Theorem 2 with $\gamma = x + h$ and $\theta = x$, we have

$$\begin{aligned}
\frac{\sin(x + h) - \sin x}{h} &= \frac{2\cos\left[\dfrac{(x + h) + x}{2}\right]\sin\left[\dfrac{(x + h) - x}{2}\right]}{h} \\
&= \frac{2\cos\left(\dfrac{2x + h}{2}\right)\sin\dfrac{h}{2}}{h} \\
&= \cos\left(x + \frac{h}{2}\right)\frac{\sin(h/2)}{h/2}.
\end{aligned}$$

5.1 Graphs of Combinations of Sine and Cosine Functions

We have already graphed the sum of algebraic functions in Section 6 of Chapter 1 by adding ordinates. The same method works for graphing sums of trigonometric functions.

Example Ⓒ Sketch $y = \sin x + 2\cos 2x$ by adding ordinates.

Solution We begin by sketching the graphs of

$$y_1 = \sin x \quad \text{and} \quad y_2 = 2\cos 2x$$

on the same coordinate system (Figure 1). Then we add the ordinates y_1 and y_2 for selected values of x to obtain values of

$$y = \sin x + 2\cos 2x.$$

(A calculator is useful here.) By plotting the resulting points and connecting them with a smooth curve, we obtain the desired graph (Figure 1).

Figure 1

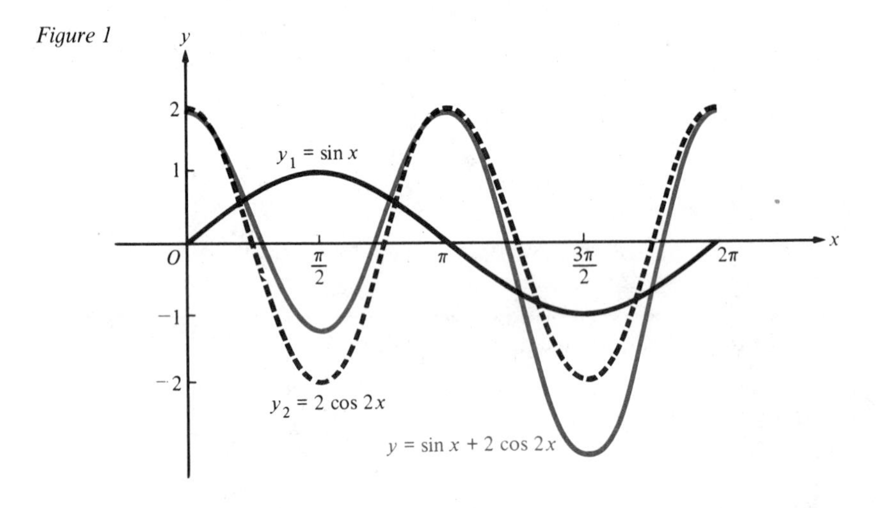

In sketching the graphs of certain combinations of sines and cosines, you will find the following theorem very helpful.

Theorem 3 **Reduction of $A\cos\theta + B\sin\theta$ to the Form $a\cos(\theta - \phi)$**

Suppose that A and B are constants; let $a = \sqrt{A^2 + B^2}$, and let ϕ be an angle in standard position with the point (A, B) on its terminal side. Then $A = a\cos\phi$, $B = a\sin\phi$, and, for all values of θ,

$$A\cos\theta + B\sin\theta = a\cos(\theta - \phi).$$

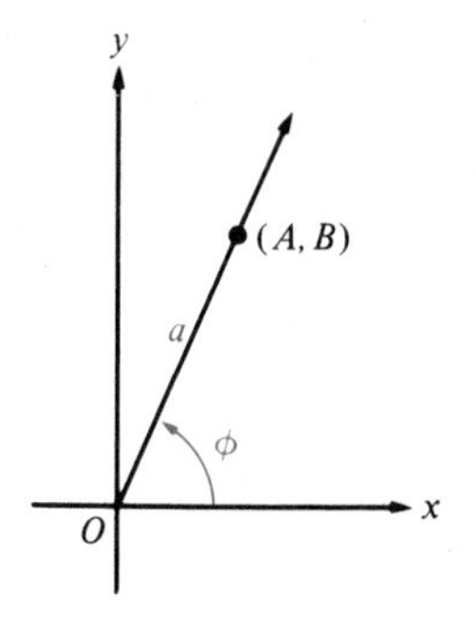

Figure 2

Proof Notice that $a = \sqrt{A^2 + B^2}$ is the distance between (A, B) and $(0, 0)$ (Figure 2). Hence, by the definitions of $\sin\phi$ and $\cos\phi$,

$$\sin\phi = \frac{B}{a} \quad \text{and} \quad \cos\phi = \frac{A}{a}.$$

It follows that

$$B = a\sin\phi \quad \text{and} \quad A = a\cos\phi.$$

Therefore,

$$\begin{aligned} A\cos\theta + B\sin\theta &= (a\cos\phi)\cos\theta + (a\sin\phi)\sin\theta \\ &= a(\cos\theta\cos\phi + \sin\theta\sin\phi) = a\cos(\theta - \phi). \end{aligned}$$

Example Sketch the graph of $y = \sqrt{3}\cos\pi t + \sin\pi t$.

Solution We could use the method of adding ordinates; but, since the same quantity πt appears in both terms, we prefer to use Theorem 3 with $\theta = \pi t$, $A = \sqrt{3}$, and $B = 1$. Then $a = \sqrt{A^2 + B^2} = \sqrt{4} = 2$ and the equations $A = a\cos\phi$ and $B = a\sin\phi$ become $\sqrt{3} = 2\cos\phi$ and $1 = 2\sin\phi$, or

$$\cos\phi = \frac{\sqrt{3}}{2} \quad \text{and} \quad \sin\phi = \frac{1}{2}.$$

Because both $\cos\phi$ and $\sin\phi$ are positive, ϕ is a quadrant I angle. Evidently, $\phi = \pi/6$. Therefore, by Theorem 3, we can rewrite the original equation in the form $y = a\cos(\theta - \phi)$, that is,

$$y = 2\cos\left(\pi t - \frac{\pi}{6}\right).$$

According to the discussion in Section 7 of Chapter 2, the graph is a simple harmonic curve with amplitude $a = 2$, angular frequency $\omega = \pi$, phase angle $\phi = \pi/6$, phase shift $S = \phi/\omega = (\pi/6)/\pi = \frac{1}{6}$, period $T = 2\pi/\omega = 2\pi/\pi = 2$, and vertical shift $k = 0$ (Figure 3).

Figure 3

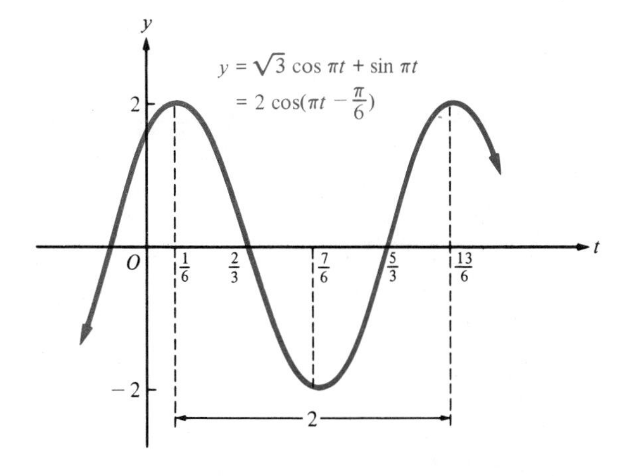

To sketch the graph of a product of sine and cosine functions, you can use Theorem 1 to rewrite the product as a sum or difference, and then use the method of adding or subtracting ordinates.

Example Ⓒ Sketch the graph of

$$y = 2\sin\frac{3x}{2}\sin\frac{x}{2}.$$

Solution Using the product formula

$$\sin\alpha\sin\beta = \tfrac{1}{2}[\cos(\alpha - \beta) - \cos(\alpha + \beta)]$$

in part (iv) of Theorem 1 with $\alpha = 3x/2$ and $\beta = x/2$, we find that

$$2\sin\frac{3x}{2}\sin\frac{x}{2} = \cos\left(\frac{3x}{2} - \frac{x}{2}\right) - \cos\left(\frac{3x}{2} + \frac{x}{2}\right) = \cos x - \cos 2x.$$

Therefore, the original equation can be rewritten as

$$y = \cos x - \cos 2x.$$

In Figure 4, we have sketched graphs of $y_1 = \cos x$ and $y_2 = \cos 2x$ on the same coordinate system. Then, subtracting the ordinate y_2 from the ordinate y_1 for selected values of x, we obtain corresponding values of $y = \cos x - \cos 2x$. (Here a calculator proves to be quite useful.) By plotting the resulting points and connecting them with a smooth curve, we obtain the required graph (Figure 4).

Figure 4

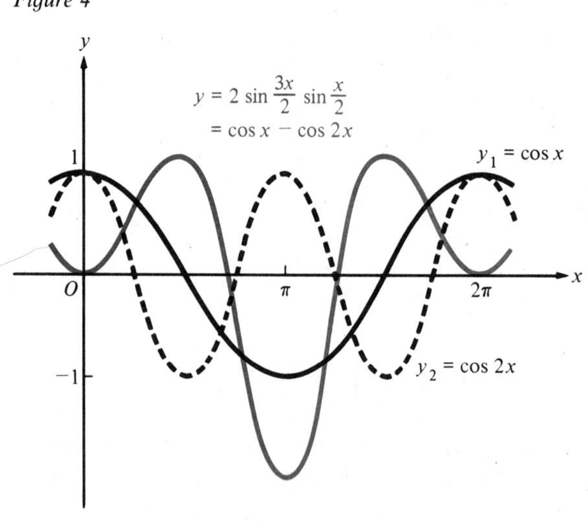

Many natural variations result from combinations of two or more simple harmonic oscillations with different frequencies, amplitudes, and phase angles. These variations may no longer be simple harmonic oscillations. For instance, pulsations in the brightness or magnitude of a variable star may be caused by a combination of factors, such as periodic expansion and contraction, variable temperature, and periodic eclipses by a companion star. The resulting pulsations, although they may occur in a regular pattern such as that shown in Figure 5, may not be simple harmonic oscillations.

Figure 5

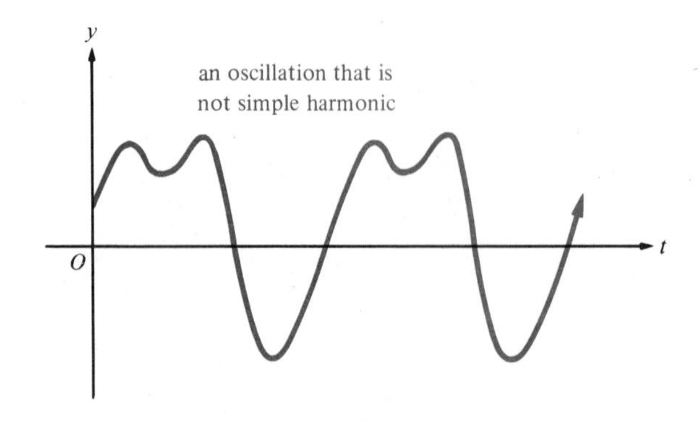

When two simple harmonic oscillations (for instance, musical tones) combine by addition, the resulting oscillation is called a **superposition** of the original oscillations. Thus, a superposition of the oscillations represented by

$$y_1 = a_1 \cos(\omega_1 t - \phi_1) + k_1 \quad \text{and} \quad y_2 = a_2 \cos(\omega_2 t - \phi_2) + k_2$$

is represented by

$$y = y_1 + y_2.$$

The graph of such a superposition can be sketched by adding ordinates. This graph is often **periodic,** in that it repeats itself over intervals of length T (called the **period**).

It can be shown that the superposition of two simple harmonic oscillations with periods T_1 and T_2 will be a periodic oscillation only if T_1/T_2 is a rational number. If

$$T_1/T_2 = n/m,$$

where n and m are positive integers and the fraction n/m is reduced to lowest terms, then the period T of the superposition is given by

$$T = mT_1 = nT_2.$$

Example Find the period T of the oscillation represented by

$$y = 2\cos\left(3\pi t - \frac{\pi}{3}\right) + \frac{1}{2}\cos\left(\frac{4}{5}\pi t + \frac{\pi}{12}\right).$$

Solution The oscillation is a superposition of the simple harmonic oscillations

$$y_1 = 2\cos\left(3\pi t - \frac{\pi}{3}\right) \quad \text{and} \quad y_2 = \frac{1}{2}\cos\left(\frac{4}{5}\pi t + \frac{\pi}{12}\right),$$

which have periods

$$T_1 = \frac{2\pi}{\omega_1} = \frac{2\pi}{3\pi} = \frac{2}{3} \quad \text{and} \quad T_2 = \frac{2\pi}{\omega_2} = \frac{2\pi}{(4/5)\pi} = \frac{5}{2}.$$

Here,

$$\frac{T_1}{T_2} = \frac{2/3}{5/2} = \frac{4}{15},$$

so the period of the superposition is given by

$$T = 15T_1 = 15(\tfrac{2}{3}) = 10$$

[and also by $T = 4T_2 = 4(\tfrac{5}{2}) = 10$].

Even if the superposition of two simple harmonic oscillations is periodic, it usually isn't a simple harmonic oscillation (for instance, see Figure 4). However, *the superposition of simple harmonic oscillations*

$$y_1 = a_1 \cos(\omega t - \phi_1) + k_1 \quad \textit{and} \quad y_2 = a_2 \cos(\omega t - \phi_2) + k_2,$$

which have the same period $T = 2\pi/\omega$, *is also a simple harmonic oscillation of period* T (for instance, see Figure 3). We leave the proof as an exercise (Problem 60).

PROBLEM SET 5

In problems 1–10, use Theorem 1 to express each product as a sum or difference.

1. $\sin 105^\circ \cos 40^\circ$
2. $\cos \frac{5\pi}{12} \sin \frac{\pi}{12}$
3. $2 \cos \frac{5\pi}{8} \cos \frac{\pi}{8}$
4. $10 \sin 160^\circ \sin 42^\circ$
5. $\sin 3\theta \cos 5\theta$
6. $\cos 7y \sin 3y$
7. $\sin 7s \sin 5t$
8. $\cos 3t \cos 5t$
9. $\sin x \cos 3x$
10. $a \cos(\omega_1 t + \phi_1) \cos(\omega_2 t + \phi_2)$

In problems 11–20, use Theorem 2 to express each sum or difference as a product.

11. $\sin 80^\circ + \sin 20^\circ$
12. $\sin 50^\circ - \sin 20^\circ$
13. $\cos \frac{3\pi}{8} + \cos \frac{\pi}{8}$
14. $\cos \frac{7\pi}{12} - \cos \frac{5\pi}{12}$
15. $\sin 4\theta + \sin 2\theta$
16. $\sin 4x - \sin 6x$
17. $\cos 2t - \cos 4t$
18. $\cos 3\gamma + \cos 8\gamma$
19. $\sin(\alpha + \beta) - \sin \alpha$
20. $a \cos(\omega_1 t + \phi_1) + a \cos(\omega_2 t + \phi_2)$
21. Use Theorem 2 to simplify each expression; then find its exact numerical value.

    (a) $\dfrac{\cos 75^\circ - \cos 15^\circ}{\sin 75^\circ + \sin 15^\circ}$ (b) $\dfrac{\sin 80^\circ + \sin 10^\circ}{\cos 80^\circ + \cos 10^\circ}$
22. Prove parts (ii) and (iii) of Theorem 1.
23. Show that $\sin 50^\circ + \sin 10^\circ = \sin 70^\circ$.
24. Prove parts (ii), (iii), and (iv) of Theorem 2.

In problems 25–38, prove that each equation is a trigonometric identity.

25. $\dfrac{\sin 5\alpha + \sin \alpha}{\cos \alpha - \cos 5\alpha} = \cot 2\alpha$
26. $\dfrac{\cos \theta + \cos 9\theta}{\sin \theta + \sin 9\theta} = \cot 5\theta$
27. $2 \sin 3s \cos 2t = \sin(3s + 2t) + \sin(3s - 2t)$
28. $\sin(x + y)\sin(x - y) = \cos^2 y - \cos^2 x$
29. $2 \sin\left(\frac{\pi}{4} + x\right)\sin\left(\frac{\pi}{4} - x\right) = \cos 2x$
30. $(\sin \alpha)(\sin 3\alpha + \sin \alpha) = \cos^2 \alpha - \cos \alpha \cos 3\alpha$
31. $\dfrac{\sin 4t - \sin 2t}{\sin 4t + \sin 2t} = \dfrac{\tan t}{\tan 3t}$
32. $\dfrac{2 \sin 3x}{\cos 2x} - \dfrac{1}{\sin x} = -\cos 4x \sec 2x \csc x$
33. $\dfrac{\cos 5s + \cos 2s}{\sin 5s + \sin 2s} = \cot \dfrac{7s}{2}$
34. $\dfrac{\sin x + \sin y}{\sin x - \sin y} = \dfrac{\tan \frac{1}{2}(x + y)}{\tan \frac{1}{2}(x - y)}$
35. $\dfrac{\sin x - \sin y}{\cos x - \cos y} = -\cot\left(\dfrac{x + y}{2}\right)$
36. $\dfrac{\cos x + \cos y}{\cos y - \cos x} = \cot\left(\dfrac{x + y}{2}\right)\cot\left(\dfrac{x - y}{2}\right)$
37. $\dfrac{\cos(x + h) - \cos x}{h} = -\sin\left(x + \dfrac{h}{2}\right)\dfrac{\sin(h/2)}{h/2}$
38. $\dfrac{\cos 4t + \cos 3t + \cos 2t}{\sin 4t + \sin 3t + \sin 2t} = \cot 3t$

Ⓒ In problems 39–44, use addition or subtraction of ordinates to sketch each graph.

39. $y = \sin x + \cos x$
40. $y = \sin 2x + 2 \cos x$
41. $y = \sin x - \cos 2x$
42. $y = 3 \sin 2\pi x + 2 \sin 4\pi x$
43. $y = \sin 2t + \sin 3t$
44. $y = 2 \cos t - \sin \frac{1}{2}t$

In problems 45–48, use Theorem 3 to rewrite each equation in the form $y = a \cos(\omega t - \phi)$, and then sketch the graph.

45. $y = \sin 2t + \cos 2t$
46. $y = -\sqrt{3} \cos\left(\frac{\pi}{2} t\right) + \sin\left(\frac{\pi}{2} t\right)$
47. $y = \cos 2\pi t - \sqrt{3} \sin 2\pi t$
48. $y = 3 \cos(2t - 1) - 3 \sin(2t - 1)$

C In problems 49–54, use Theorem 1 to rewrite the product as a sum or difference, and then sketch the graph by adding or subtracting ordinates.

49. $y = \cos 4x \sin 2x$ **50.** $y = 3\cos 3t \cos 7t$ **51.** $y = 2\sin 3t \sin t$

52. $y = -\sin\frac{3x}{2}\cos\frac{x}{2}$ **53.** $y = -\cos \pi x \cos 2\pi x$ **54.** $y = 5\sin\frac{4t}{3}\cos\frac{2t}{3}$

In problems 55–58, find the period T of the oscillation represented by each equation.

55. $y = 3\cos(2\pi t - \pi) + 2\cos\left(\frac{\pi}{2}t - \frac{\pi}{2}\right) + 4$ **56.** $y = 2\cos\left(\frac{3}{2}t + 1\right) + 4\cos\left(\frac{2}{3}t - 2\right) - 5$

57. $y = \cos\frac{\sqrt{2}}{2}t + \cos(3\sqrt{2}t - 1)$ **58.** $y = \sqrt{3}\cos\left(\frac{4\pi}{3}t - \frac{\pi}{7}\right) + 2\sin\left(\frac{3\pi}{2}t + \frac{\pi}{5}\right)$

C **59.** Suppose you sketch the graph of

$$y = 19\sin x - 10\sin 2x$$

on the interval $0 \le x \le \pi$ by plotting points corresponding to

$$x = 0, \frac{\pi}{6}, \frac{\pi}{4}, \frac{\pi}{3}, \frac{\pi}{2}, \frac{2\pi}{3}, \frac{3\pi}{4}, \frac{5\pi}{6}, \text{ and } \pi.$$

If you connect these points with a smooth curve, you will obtain the graph shown in Figure 6. However, this is *not* the correct graph. Find the error, correct it, and redraw the graph.

Figure 6

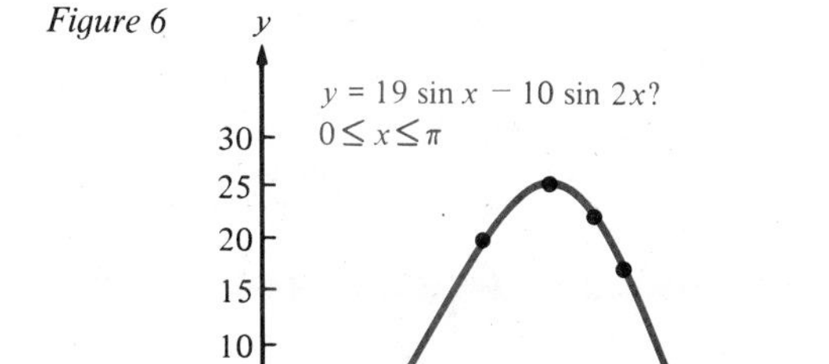

60. Assume $y_1 = a_1\cos(\omega t - \phi_1) + k_1$ and $y_2 = a_2\cos(\omega t - \phi_2) + k_2$. Let $\theta = \omega t - \phi_1$, $A = a_1 + a_2\cos(\phi_2 - \phi_1)$, $B = a_2\sin(\phi_2 - \phi_1)$, $k = k_1 + k_2$, and $y = y_1 + y_2$.
(a) Show that $\omega t - \phi_2 = \theta - (\phi_2 - \phi_1)$.
(b) Use the result in part (a) to show that

$$y_2 = a_2\cos(\phi_2 - \phi_1)\cos\theta + a_2\sin(\phi_2 - \phi_1)\sin\theta + k_2.$$

(c) Use the result in part (b) to show that

$$y = A\cos\theta + B\sin\theta + k.$$

(d) Use the result in part (c) together with Theorem 3 to rewrite y in the form

$$y = a\cos(\theta - \phi) + k.$$

(e) Conclude that $y = a\cos(\omega t - \phi_0) + k$, where $\phi_0 = \phi_1 + \phi$.

C **61.** Even when sketched carefully and correctly, the graph of $y = 19\sin x - 10\sin 2x$ (see problem 59) appears to have a maximum at $x = 2\pi/3$. Use a calculator to check values of y for values of x close to $2\pi/3$ to see if this is really so.

62. A 50 kilohertz radio frequency oscillation with the equation $y_1 = 10\sin(\pi \times 10^5)t$ is amplitude modulated by an audio frequency tone with the equation $y_2 = \sin(2\pi \times 10^4)t$ to produce a signal with the equation $y = y_2 y_1$. Sketch three graphs over the interval $0 \le t \le \frac{1}{5000}$ to show (a) the radio frequency oscillation, (b) the audio frequency tone, and (c) the amplitude modulated signal.

6 Inverse Trigonometric Functions

In Sections 2, 3, and 4 of Chapter 2, we learned how to find the values (or the approximate values) of the trigonometric functions of known angles. In many applications of trigonometry, the problem is the other way around—to find unknown angles when information is given about the values of trigonometric functions of the angles. The **inverse trigonometric functions** can be used to solve such problems.

Recall from Section 6 of Chapter 1 that by reflecting the graph of an invertible function f across the line $y = x$ we obtain the graph of the inverse function f^{-1} (Figure 1). However, in order for f to be invertible, no horizontal line can intersect its graph more than once.

Figure 1

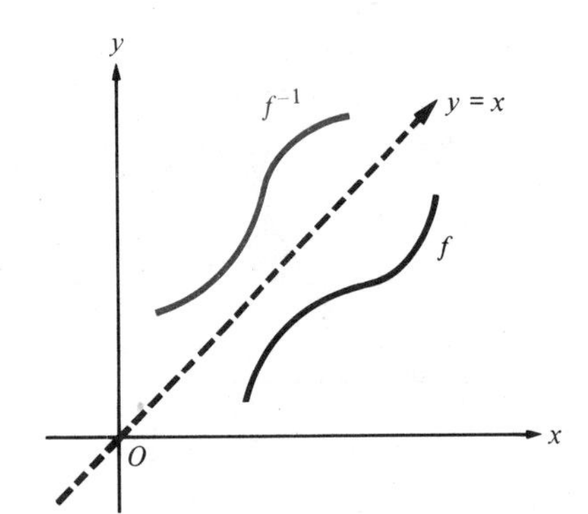

Figure 2

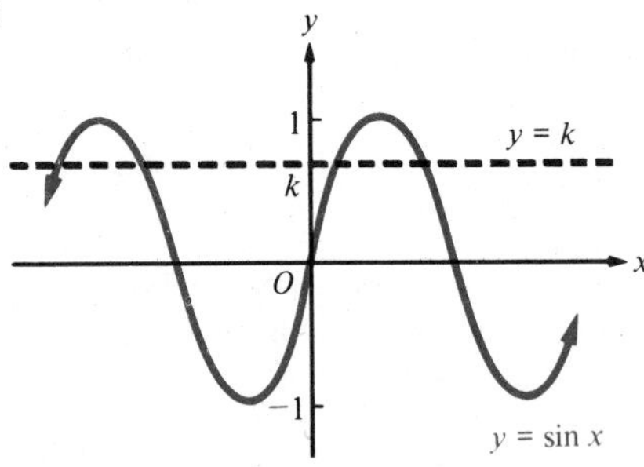

Because the sine function is periodic, any horizontal line $y = k$ with $-1 \leq k \leq 1$ intersects the graph of $y = \sin x$ repeatedly (Figure 2), and therefore the sine function is *not* invertible. But consider the portion of the graph of $y = \sin x$ between $x = -\pi/2$ and $x = \pi/2$ (Figure 3a). No horizontal line intersects this curve more than once, so its reflection across the line $y = x$ (Figure 3b) is the graph of a function. This function is often called the "inverse sine" and denoted by $\sin^{-1}$.

Figure 3

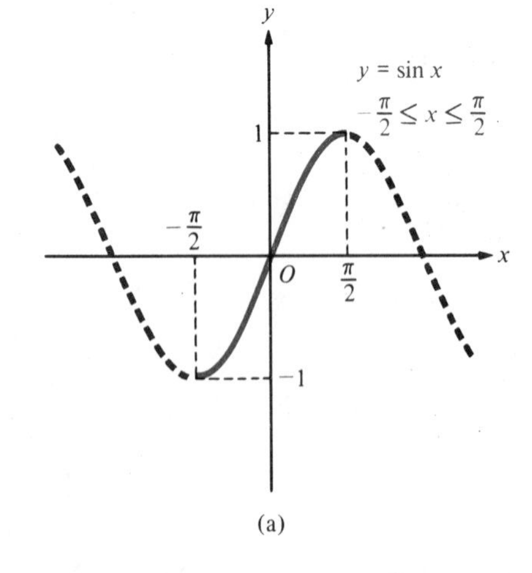

(a)

(b)

Of course, the terminology "inverse sine function" and the notation $\sin^{-1}$ aren't really correct because the sine function (defined on the set of all real numbers) has no inverse. Because of this, some people use the notation Sine (with a *capital* S) for the function in Figure 3a and Sin^{-1} for its inverse in Figure 3b. Another source of difficulty is the possible confusion of $\sin^{-1} x$ with $(\sin x)^{-1} = 1/\sin x = \csc x$. To avoid such confusion, many people use the notation "$\arcsin x$," meaning the arc (or angle) whose sine is x, rather than $\sin^{-1} x$. Because both the notations $\sin^{-1} x$ and $\arcsin x$ are in common use in mathematics and its applications, we shall use them interchangeably. Thus, we make the following definition.

Definition 1 **The Inverse Sine or Arcsine**

The *inverse sine* or *arcsine* function, denoted by $\sin^{-1}$ or arcsin, is defined by

$$\sin^{-1} x = y \quad \text{if and only if} \quad x = \sin y \quad \text{and} \quad -\frac{\pi}{2} \le y \le \frac{\pi}{2}.$$

In words, $\sin^{-1} x$ is the angle (or number) between $-\pi/2$ and $\pi/2$ whose sine is x. The graph of

$$y = \sin^{-1} x \quad \text{or} \quad y = \arcsin x$$

(Figure 4) shows that $\sin^{-1}$ is an increasing function with domain $[-1, 1]$ and range $[-\pi/2, \pi/2]$. The graph appears to be symmetric about the origin, indicating that $\sin^{-1}$ is an odd function; that is,

$$\sin^{-1}(-x) = -\sin^{-1} x.$$

Figure 4

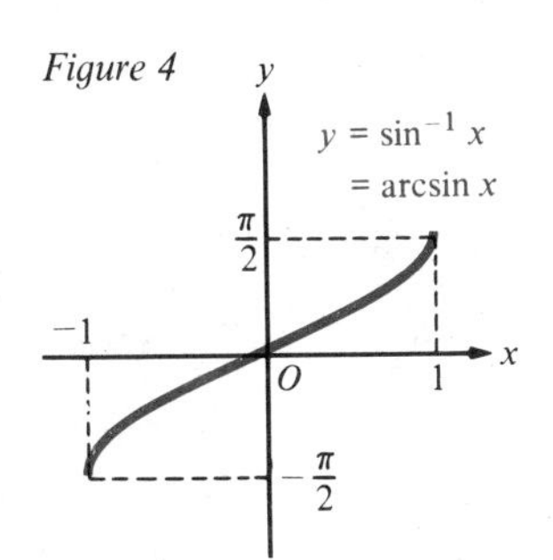

A scientific calculator can be used to find approximate values of $\sin^{-1} x$ (or $\arcsin x$). (If a calculator isn't available, you can find these values by reading Table 2 of Appendix Section 3 "backwards.") Some calculators have a $\sin^{-1}$ key; some have a key marked arcsin; and some have an INV key which must be pressed before the SIN key to give the inverse sine. Again, you must be careful to put the calculator in radian mode if you want $\sin^{-1} x$ in radians—otherwise, you will get the angle in *degrees* (between $-90°$ and $90°$) whose sine is x.

Example Find each value:

(a) $\sin^{-1} \frac{1}{2}$ (b) $\arcsin(-\sqrt{3}/2)$ Ⓒ (c) $\sin^{-1} 0.7321$

Solution (a) Here a calculator isn't needed because we know that $\pi/6$ radian $(30°)$ is an angle whose sine is $\frac{1}{2}$. Thus, because $-\pi/2 \le \pi/6 \le \pi/2$, we have $\sin^{-1} \frac{1}{2} = \pi/6$.

(b) Again we don't need a calculator because we know that $-\pi/3$ radians $(-60°)$ is an angle whose sine is $-\sqrt{3}/2$. Thus, because $-\pi/2 \le -\pi/3 \le \pi/2$, we have $\arcsin(-\sqrt{3}/2) = -\pi/3$.

(c) Using a 10-digit calculator in radian mode, we obtain

$$\sin^{-1} 0.7321 = 0.821399673.$$

(In degree mode, we get $\sin^{-1} 0.7321 = 47.06273457°$.)

The remaining five trigonometric functions are also periodic, and therefore, like the sine function, they are not invertible. However, by restricting these functions to suitable intervals we can define corresponding inverses, just as we did for the sine function. The appropriate portions of the graphs of the cosine and tangent functions are shown in Figure 5. Thus, we make the following definitions.

Figure 5

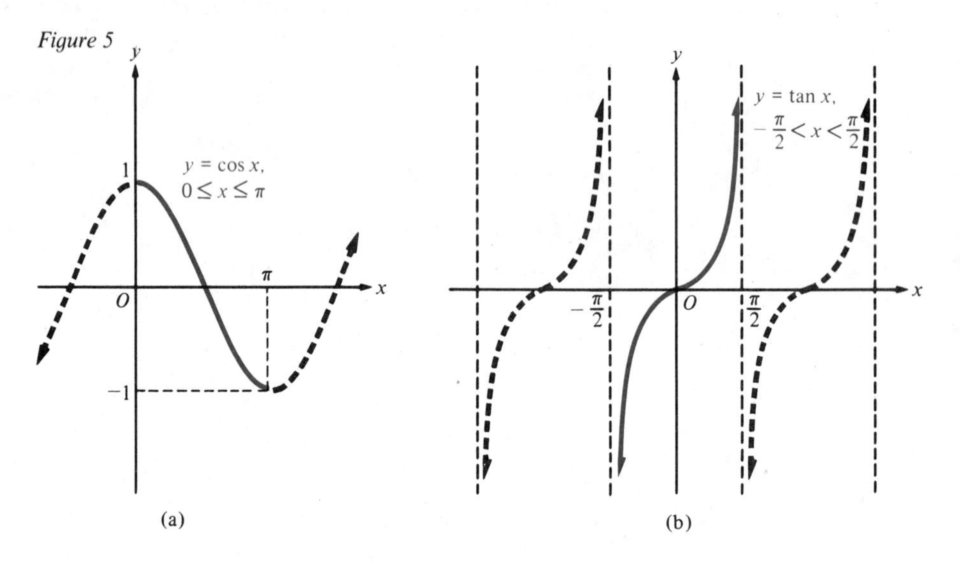

Definition 2 **The Inverse Cosine or Arccosine**

The *inverse cosine* or *arccosine* function, denoted by $\cos^{-1}$ or arccos, is defined by

$$\cos^{-1} x = y \quad \text{if and only if} \quad x = \cos y \quad \text{and} \quad 0 \le y \le \pi.$$

Definition 3 **The Inverse Tangent or Arctangent**

The inverse tangent or arctangent function, denoted by $\tan^{-1}$ or arctan, is defined by

$$\tan^{-1} x = y \quad \text{if and only if} \quad x = \tan y \quad \text{and} \quad -\frac{\pi}{2} < y < \frac{\pi}{2}.$$

Figure 6

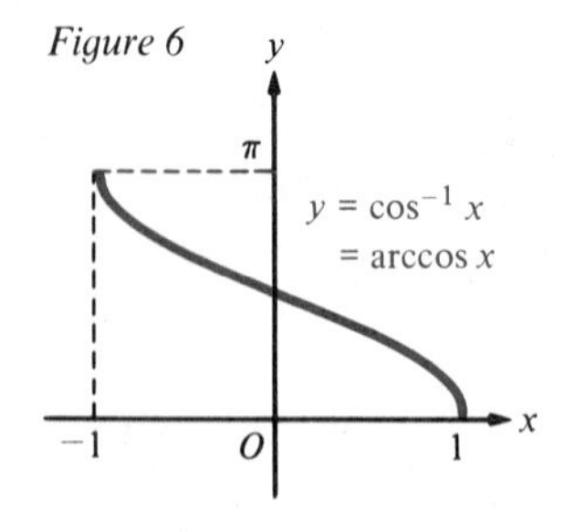

In words, $\cos^{-1} x$ is the angle (or number) between 0 and π whose cosine is x. Likewise, $\tan^{-1} x$ is the angle (or number) between $-\pi/2$ and $\pi/2$ whose tangent is x.

The graph of $y = \cos^{-1} x$, obtained by reflecting the curve in Figure 5a across the line $y = x$, shows that $\cos^{-1}$ (or arccos) is a decreasing function with domain $[-1, 1]$ and range $[0, \pi]$ (Figure 6). Notice that the graph of $\cos^{-1}$ is *not* symmetric about the origin nor about the y axis. Thus, in spite of the fact that cosine is an even function, $\cos^{-1}$ is not.

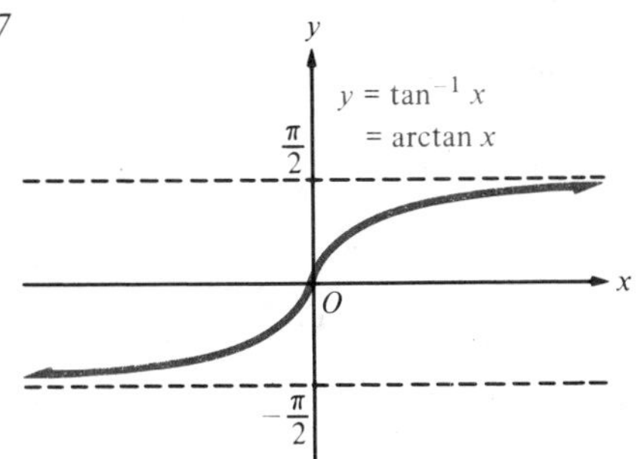

Figure 7

The graph of $y = \tan^{-1} x$, obtained by reflecting the curve in Figure 5b across the line $y = x$, shows that $\tan^{-1}$ (or arctan) is an increasing function with domain $\mathbb{R}$ and range $(-\pi/2, \pi/2)$ (Figure 7). Notice that the lines $y = -\pi/2$ and $y = \pi/2$ are horizontal asymptotes. The graph appears to be symmetric about the origin, indicating that $\tan^{-1}$ is an odd function; that is,

$$\tan^{-1}(-x) = -\tan^{-1} x$$

(Problem 39).

Example Find each value:

(a) $\cos^{-1}(-\sqrt{2}/2)$

Ⓒ (b) arccos 0.6675

(c) $\arctan\sqrt{3}$

Ⓒ (d) $\tan^{-1}(-2.498)$

Solution (a) If $\cos^{-1}(-\sqrt{2}/2) = y$, then $-\sqrt{2}/2 = \cos y$ and $0 \leq y \leq \pi$. Therefore, $y = 3\pi/4$.

(b) Using a 10-digit calculator in radian mode, we find that

$$\arccos 0.6675 = 0.839950077.$$

(c) If $\arctan\sqrt{3} = y$, then $\sqrt{3} = \tan y$ and $-\pi/2 < y < \pi/2$. Therefore, $y = \pi/3$.

(d) Using a 10-digit calculator in radian mode, we find that

$$\tan^{-1}(-2.498) = -1.190013897.$$

If you enter a number x between -1 and 1 in a calculator, take $\sin^{-1} x$, and then take the sine of the result, you will get x back again. Thus,

$$x \xmapsto{\sin^{-1}} y \xmapsto{\sin} x.$$

Do you see why? Therefore, we have the identity

$$\sin(\sin^{-1} x) = x \quad \text{for} \quad -1 \leq x \leq 1.$$

However, if you try the same thing the other way around, first taking the sine and then taking the inverse sine, you might not get your original number back. For instance, starting with $x = 2$, we get (on a 10-digit calculator in radian mode)

$$2 \xmapsto{\sin} 0.909297427 \xmapsto{\sin^{-1}} 1.141592654.$$

The reason is simply that $\sin^{-1} 0.909297427$ is the number *between* $-\pi/2$ *and* $\pi/2$ whose sine is 0.909297427, and 2 does not lie between $-\pi/2$ and $\pi/2$. However, if you start with a number *between* $-\pi/2$ *and* $\pi/2$, take the sine, and then take the inverse sine, you will get your original number back. In other words,

$$\sin^{-1}(\sin x) = x \quad \text{for} \quad -\pi/2 \leq x \leq \pi/2.$$

Similar rules apply to cos and $\cos^{-1}$, and to tan and $\tan^{-1}$ (Problems 45 and 47).

In calculus, it's sometimes necessary to find exact values of expressions such as $\sin(\tan^{-1}\frac{2}{3})$. The following example shows how this can be done.

Example Find the exact value of $\sin(\tan^{-1}\frac{2}{3})$.

Solution We begin by sketching an acute angle θ in a right triangle such that $\tan^{-1}\frac{2}{3} = \theta$; that is, $\frac{2}{3} = \tan\theta = \text{opp}/\text{adj}$. This is accomplished simply by letting opp = 2 units and adj = 3 units (Figure 8). By the Pythagorean theorem,

$$\text{hyp} = \sqrt{\text{opp}^2 + \text{adj}^2} = \sqrt{2^2 + 3^2} = \sqrt{13}.$$

Figure 8

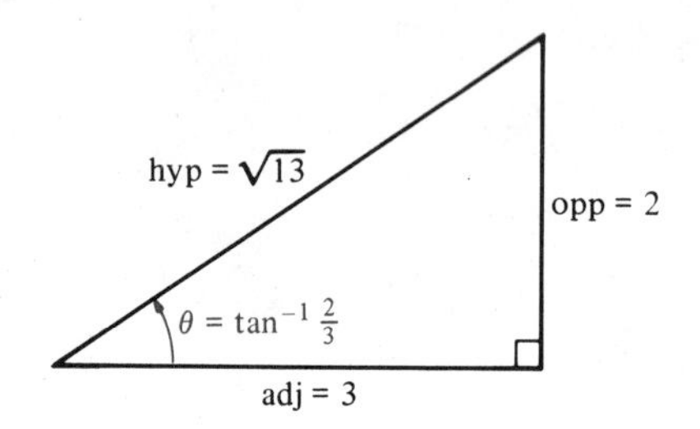

Therefore,

$$\sin(\tan^{-1}\tfrac{2}{3}) = \sin\theta = \frac{\text{opp}}{\text{hyp}} = \frac{2}{\sqrt{13}} = \frac{2\sqrt{13}}{13}.$$

Another method for finding exact values of expressions involving inverse trigonometric functions is to use identities such as the one developed in the following example.

Example Show that $\cos(\tan^{-1}x) = \dfrac{1}{\sqrt{1 + x^2}}$.

Solution Let $\tan^{-1}x = y$, so that $x = \tan y$ and $-\pi/2 < y < \pi/2$. Then

$$1 + x^2 = 1 + \tan^2 y = \sec^2 y.$$

Because $-\pi/2 < y < \pi/2$, it follows that $\sec y > 0$; hence,

$$\sqrt{1 + x^2} = \sec y.$$

Therefore,

$$\frac{1}{\sqrt{1 + x^2}} = \frac{1}{\sec y} = \cos y = \cos(\tan^{-1}x).$$

The remaining inverse trigonometric functions $\cot^{-1}$, $\sec^{-1}$, and $\csc^{-1}$ (that is, arccot, arcsec, and arccsc, respectively) aren't used as often as $\sin^{-1}$, $\cos^{-1}$, and $\tan^{-1}$, so you won't find them on the keys of many scientific calculators. (See Problems 48–51.)

PROBLEM SET 6

In problems 1–12, evaluate each expression without using a calculator or tables.

1. $\sin^{-1} 1$
2. $\arcsin \frac{\sqrt{3}}{2}$
3. $\arcsin\left(-\frac{\sqrt{2}}{2}\right)$
4. $\cos^{-1}\left(-\frac{1}{2}\right)$
5. $\arccos 1$
6. $\cos^{-1} \frac{\sqrt{3}}{2}$
7. $\sin^{-1} \frac{\sqrt{2}}{2}$
8. $\cos^{-1} 0$
9. $\arccos \frac{1}{2}$
10. $\arctan 1$
11. $\tan^{-1}(-1)$
12. $\tan^{-1} \frac{\sqrt{3}}{3}$

Ⓒ In problems 13–24, use a calculator (or Appendix Table 2, Section 3) to evaluate each expression. Give answers in *radians*.

13. $\arcsin 0.6442$
14. $\arccos 0.6675$
15. $\cos^{-1} 0.9051$
16. $\tan^{-1} 0.2500$
17. $\arctan 2$
18. $\sin^{-1}(-0.5495)$
19. $\tan^{-1}(-3.224)$
20. $\cos^{-1}(-\frac{1}{8})$
21. $\arcsin(-0.5505)$
22. $\arccos\left(-\frac{5}{11}\right)$
23. $\sin^{-1} \frac{\sqrt{5}}{4}$
24. $\tan^{-1}\left(-\frac{\sqrt{7}}{3}\right)$

In problems 25–36, find the exact value of each expression without using a calculator or tables.

25. $\sin\left(\sin^{-1} \frac{3}{4}\right)$
26. $\tan(\tan^{-1} 3)$
27. $\sin^{-1}\left(\sin \frac{\pi}{6}\right)$
28. $\tan^{-1}\left(\tan \frac{5\pi}{4}\right)$
29. $\sin\left(\tan^{-1} \frac{4}{3}\right)$
30. $\tan\left(\sin^{-1} \frac{\sqrt{5}}{5}\right)$
31. $\cos\left(\sin^{-1} \frac{\sqrt{10}}{10}\right)$
32. $\cos[\arctan(-2)]$
33. $\tan\left(\sin^{-1} \frac{4}{5}\right)$
34. $\sin(2 \arcsin \frac{2}{3})$
35. $\sec(\cos^{-1} \frac{7}{10})$
36. $\cos(2 \sin^{-1} \frac{1}{8})$

In problems 37–44, show that the given equation is an identity.

37. $\sin^{-1}(-x) = -\sin^{-1} x$
38. $\cos^{-1} x = \frac{\pi}{2} - \sin^{-1} x$
39. $\arctan(-x) = -\arctan x$
40. $\cos^{-1}(-x) = \pi - \cos^{-1} x$
41. $\tan(\sin^{-1} x) = \frac{x}{\sqrt{1 - x^2}}$
42. $\sin(\arctan x) = \frac{x}{\sqrt{1 + x^2}}$
43. $\sin(2 \arcsin x) = 2x\sqrt{1 - x^2}$
44. $\tan(\frac{1}{2} \arccos x) = \sqrt{\frac{1 - x}{1 + x}}$

45. For what values of x is it true that
(a) $\cos(\cos^{-1} x) = x$? (b) $\cos^{-1}(\cos x) = x$?

Ⓒ 46. To the nearest hundredth of a degree, find the two vertex angles α and β of a 3–4–5 right triangle (Figure 9).

47. For what values of x is it true that
(a) $\tan(\arctan x) = x$? (b) $\arctan(\tan x) = x$?

48. The **inverse cotangent** function, $\cot^{-1}$ or arccot, is defined by $\cot^{-1} x = y$ if and only if $x = \cot y$ and $0 < y < \pi$. (a) Sketch the graph of $y = \cot^{-1} x$. (b) Show that

$$\cot^{-1} x = \frac{\pi}{2} - \tan^{-1} x.$$

Figure 9

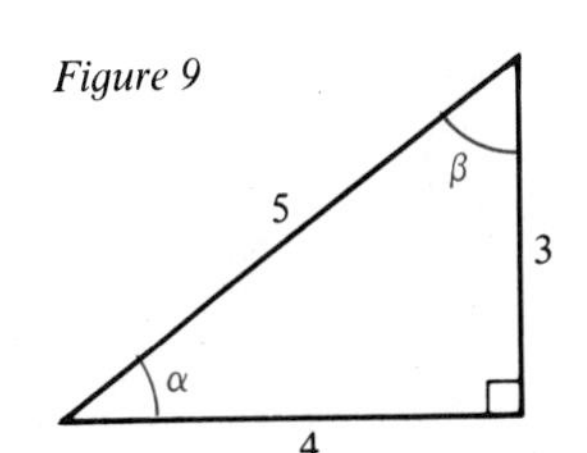

49. One popular definition of the **inverse secant** function, $\sec^{-1}$ or arcsec, is $\sec^{-1} x = y$ if and only if $x = \sec y$ and $0 \le y \le \pi$ with $y \ne \pi/2$. (a) Sketch the graph of $y = \sec^{-1} x$. (b) Show that

$$\sec^{-1} x = \cos^{-1}\left(\frac{1}{x}\right) \quad \text{for} \quad |x| \ge 1.$$

50. An alternative definition of the inverse secant function (see problem 49) used in some calculus textbooks is $\sec^{-1} x = y$ if and only if $x = \sec y$ and either $0 \le y < \pi/2$ or else $\pi \le y < 3\pi/2$. Sketch the graph of $\sec^{-1}$ according to this definition.

51. One popular definition of the **inverse cosecant** function, $\csc^{-1}$ or arccsc, is $\csc^{-1} x = y$ if and only if $x = \csc y$ and $-\pi/2 \le y \le \pi/2$ with $y \ne 0$. (a) Sketch the graph of $y = \csc^{-1} x$. (b) Show that

$$\csc^{-1} x = \sin^{-1}\left(\frac{1}{x}\right) \quad \text{for} \quad |x| \ge 1.$$

7 Trigonometric Equations

In Section 1, we mentioned that a trigonometric equation that is not an identity is called a **conditional** equation. To **solve** a conditional trigonometric equation means to find all values of the unknown for which the equation is true; these values are called the **solutions** of the equation. If a side condition such as $0^\circ \le \theta < 360^\circ$ is given along with a trigonometric equation, we understand that the solutions consist of all values of θ that satisfy *both* the equation and the side condition.

An equation, such as $\sin\theta = \sqrt{3}/2$, whose left side is a trigonometric function of the unknown and whose right side is a constant, is called a **simple** trigonometric equation. A side condition such as $0^\circ \le \theta < 360^\circ$ may or may not be involved. Even if there is no side condition, we usually *begin* by solving the equation for nonnegative values of the unknown that are less than 360° or 2π radians. The complete set of solutions is then obtained by adding all integer multiples of 360° or 2π radians to the values thus obtained. (Why?)

Examples Solve each simple trigonometric equation. Use degree or radian measure as indicated.

1 $\quad \sin\theta = \dfrac{\sqrt{3}}{2}, \quad 0^\circ \le \theta < 360^\circ$

Solution Recall that $\sin 60^\circ = \sqrt{3}/2$. Therefore, $\theta = 60^\circ$ is the only solution in quadrant I. (Why?) Because the values of $\sin\theta$ are negative in quadrants III and IV, the only possible remaining solution between 0° and 360° must lie in quadrant II, and its reference angle must be $\theta_R = 60^\circ$ (Figure 1). Hence, $\theta = 180^\circ - 60^\circ = 120^\circ$. Therefore, the solutions are

$$\theta = 60^\circ \quad \text{and} \quad \theta = 120^\circ.$$

Figure 1

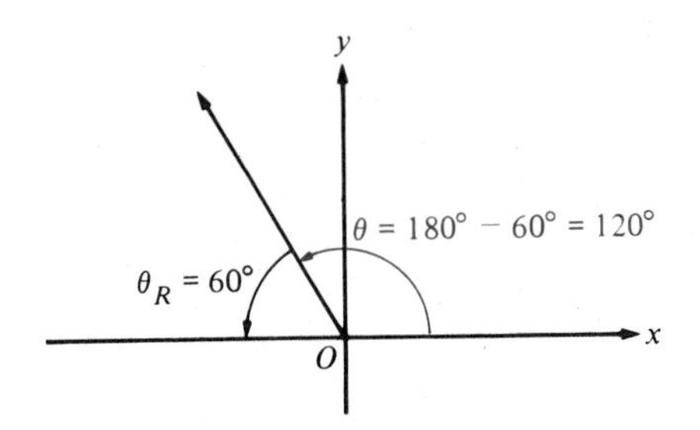

2 $\sec\alpha = -\sqrt{2}, \quad 0° \le \alpha < 360°$

Solution Recall that the values of $\sec\alpha$ are negative only in quadrants II and III, and that $\sec 45° = \sqrt{2}$. Therefore, by constructing angles in quadrants II and III with 45° reference angles (Figure 2), we find the solutions

$$\alpha = 180° - 45° = 135° \quad \text{and} \quad \alpha = 180° + 45° = 225°.$$

Figure 2

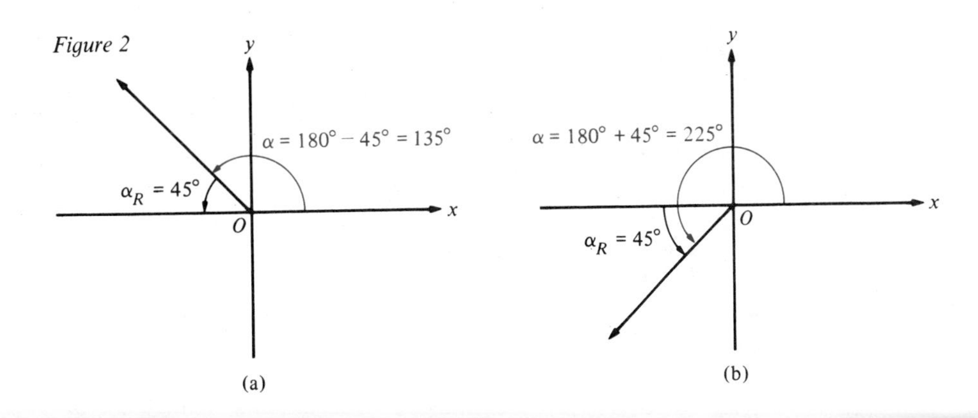

3 $\sec t = -\sqrt{2}, \quad t$ in radians

Solution Here there is no side condition. Nevertheless, we *begin* by solving the equation with the side condition $0 \le t < 2\pi$. Proceeding as in Example 2, but measuring angles in radians, we find the two solutions

$$t = \pi - \frac{\pi}{4} = \frac{3\pi}{4} \quad \text{and} \quad t = \pi + \frac{\pi}{4} = \frac{5\pi}{4}.$$

Therefore, the complete set of solutions consists of all real numbers

$$t = \frac{3\pi}{4} + 2\pi k \quad \text{and} \quad t = \frac{5\pi}{4} + 2\pi k,$$

where k denotes an arbitrary integer.

Ⓒ **4** $\cot x = -0.8333, \quad 0 \le x < 2\pi$

Solution We begin by rewriting the equation as

$$\tan x = 1/(-0.8333)$$

so that we can use the $\tan^{-1}$ (or arctan) key on a scientific calculator. If we simply calculate $\tan^{-1}[1/(-0.8333)]$, we get -0.876077723, which is *not* a solution because it doesn't satisfy the condition $0 \le x < 2\pi$. Recall that the values of $\tan x$ are negative in quadrants II and IV. With the calculator in radian mode, we enter 0.8333, take the reciprocal, and then take the inverse tangent to obtain the reference angle

$$\theta_R = \tan^{-1}(1/0.8333) = 0.876077723$$

Figure 3

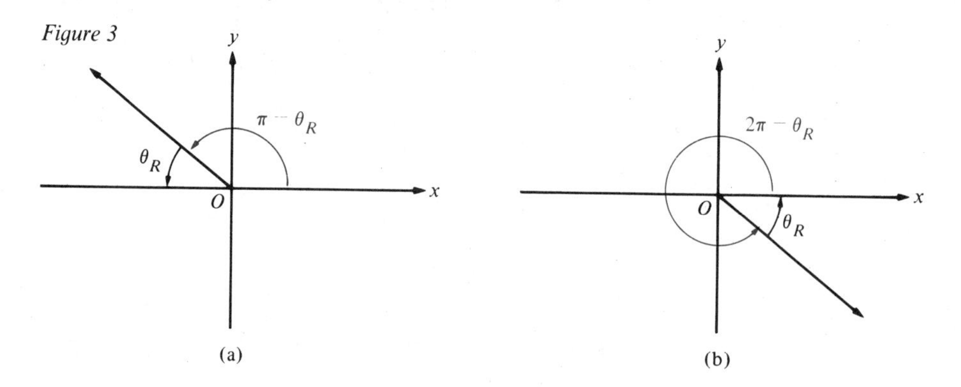

(Figure 3). Therefore, the solution in quadrant II is $x = \pi - \theta_R$ and the solution in quadrant IV is $x = 2\pi - \theta_R$; that is,

$$x = \pi - 0.876077723 = 2.265514931$$

and

$$x = 2\pi - 0.876077723 = 5.407107585.$$

Figure 4

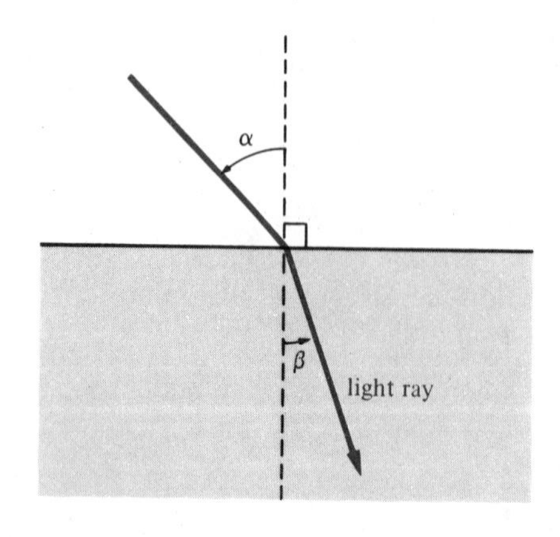

Snell's law of refraction, which was discovered around 1620 by the Dutch physicist Willebrord Snell (1591–1626), says that a light ray is bent (refracted) as it passes from a first medium into a second medium according to the equation

$$\frac{\sin \alpha}{\sin \beta} = \mu,$$

where α is the **angle of incidence** (Figure 4), β is the **angle of refraction** (Figure 4), and μ (the Greek letter mu) is a constant called the **index of refraction** of the second medium with respect to the first.

Example Ⓒ The index of refraction of flint glass with respect to air is $\mu = 1.650$. Determine the angle of refraction β of a ray of light that strikes a block of flint glass with an angle of incidence $\alpha = 35^\circ$.

Solution From Snell's law $\dfrac{\sin\alpha}{\sin\beta} = \mu$, we obtain the simple trigonometric equation

$$\sin\beta = \frac{\sin\alpha}{\mu} = \frac{\sin 35^\circ}{1.650}.$$

From the geometry of the problem (Figure 4), we have the side condition $0 < \beta < 90^\circ$. Using a calculator in degree mode, we find that

$$\beta = \sin^{-1}\left(\frac{\sin 35^\circ}{1.650}\right) = 20.34^\circ$$

to the nearest hundredth of a degree.

Many of the techniques used for solving trigonometric equations are similar to those used for solving algebraic equations. The basic idea is to reduce the equation to one or more simple trigonometric equations, and then use the method illustrated above.

Examples Solve each trigonometric equation subject to the side condition $0^\circ \le \theta < 360^\circ$ or $0 \le t < 2\pi$.

1 $\cos 3t = \sqrt{2}/2$

Solution Let $x = 3t$, and rewrite the equation as the simple trigonometric equation $\cos x = \sqrt{2}/2$. The side condition $0 \le t < 2\pi$ is equivalent to $0 \le 3t < 6\pi$ or $0 \le x < 6\pi$. The solutions of $\cos x = \sqrt{2}/2$ for $0 \le x < 2\pi$, obtained using the method described previously, are $x = \pi/4$ and $x = 7\pi/4$. Adding 2π to these, we obtain two more solutions,

$$x = \frac{\pi}{4} + 2\pi = \frac{9\pi}{4} \quad \text{and} \quad x = \frac{7\pi}{4} + 2\pi = \frac{15\pi}{4},$$

both of which satisfy the condition $2\pi \le x < 4\pi$. Finally, adding 2π again, we obtain two more solutions,

$$x = \frac{9\pi}{4} + 2\pi = \frac{17\pi}{4} \quad \text{and} \quad x = \frac{15\pi}{4} + 2\pi = \frac{23\pi}{4},$$

both of which satisfy the condition $4\pi \le x < 6\pi$. Thus, the solutions for $0 \le x < 6\pi$ are

$$3t = x = \frac{\pi}{4}, \frac{7\pi}{4}, \frac{9\pi}{4}, \frac{15\pi}{4}, \frac{17\pi}{4}, \text{ and } \frac{23\pi}{4},$$

and so the values of t are

$$t = \frac{x}{3} = \frac{\pi}{12}, \frac{7\pi}{12}, \frac{3\pi}{4}, \frac{5\pi}{4}, \frac{17\pi}{12}, \text{ and } \frac{23\pi}{12}.$$

2 $\cos\theta \cot\theta = \cos\theta$

Solution We rewrite the equation as

$$\cos\theta\cot\theta - \cos\theta = 0 \quad \text{or} \quad \cos\theta(\cot\theta - 1) = 0,$$

so that, with the side condition $0^\circ \le \theta < 360^\circ$, we have:

$\cos\theta = 0$	$\cot\theta - 1 = 0$
$\theta = 90^\circ, 270^\circ$	$\cot\theta = 1$
	$\theta = 45^\circ, 225^\circ$

Therefore, the solutions are

$$\theta = 45^\circ, 90^\circ, 225^\circ, \text{and } 270^\circ.$$

(Note: We were careful *not* to divide both sides of the original equation by the variable factor $\cos\theta$. Had we done so, we would have lost the solutions $\theta = 90^\circ$ and $\theta = 270^\circ$.)

3 $2\cos^2 t - \sin t - 1 = 0.$

Solution Using the Pythagorean identity $\cos^2 t + \sin^2 t = 1$, we rewrite the equation as

$$2(1 - \sin^2 t) - \sin t - 1 = 0 \quad \text{or} \quad -2\sin^2 t - \sin t + 1 = 0,$$

that is,

$$2\sin^2 t + \sin t - 1 = 0 \quad \text{or} \quad (2\sin t - 1)(\sin t + 1) = 0.$$

Setting each factor equal to zero and keeping in mind the side condition $0 \le t < 2\pi$, we have:

$2\sin t - 1 = 0$	$\sin t + 1 = 0$
$\sin t = \frac{1}{2}$	$\sin t = -1$
$t = \frac{\pi}{6}, \frac{5\pi}{6}$	$t = \frac{3\pi}{2}$

Therefore, the solutions are

$$t = \frac{\pi}{6}, \frac{5\pi}{6}, \text{and } \frac{3\pi}{2}.$$

4 $\tan\theta - \sec\theta = 1$

Solution We rewrite the equation as $\sec\theta = \tan\theta - 1$, and square both sides. (Note: When proving a trigonometric *identity*, we are not permitted to square both sides. However, in solving a *conditional* equation, we may square both sides *provided that we later check for extraneous roots.*) We have

$$\sec^2\theta = \tan^2\theta - 2\tan\theta + 1.$$

Using the Pythagorean identity $1 + \tan^2 \theta = \sec^2 \theta$, we can rewrite the last equation as

$$1 + \tan^2 \theta = \tan^2 \theta - 2 \tan \theta + 1 \quad \text{or} \quad 0 = -2 \tan \theta,$$

that is,

$$\tan \theta = 0.$$

Therefore, with the side condition $0° \leq \theta < 360°$, we have

$$\theta = 0° \quad \text{or} \quad \theta = 180°.$$

We must now check our solutions. Substituting $\theta = 0°$ in the original equation, we get

$$\tan 0° - \sec 0° = 1 \quad \text{or} \quad 0 - 1 = 1,$$

which is *false*. Therefore, $\theta = 0°$ isn't a solution. Substituting $\theta = 180°$ in the original equation, we get

$$\tan 180° - \sec 180° = 1 \quad \text{or} \quad 0 - (-1) = 1,$$

which is *true*. Thus, $\theta = 180°$ is the only solution.

5 Ⓒ $5 \tan^2 t + 2 \tan t - 7 = 0$ (Round off to four decimal places.)

Solution Factoring the left side of the equation, we have

$$(5 \tan t + 7)(\tan t - 1) = 0.$$

Setting the first factor equal to zero, we obtain the simple trigonometric equation

$$\tan t = -\tfrac{7}{5} = -1.4.$$

Using a calculator (as in Example 4 on page 156), we find that

$$t \approx 2.1910 \quad \text{or} \quad t \approx 5.3326.$$

Setting the second factor equal to zero, we obtain the simple trigonometric equation

$$\tan t = 1,$$

which has the solutions

$$t = \frac{\pi}{4} \approx 0.7854 \quad \text{and} \quad t = \frac{5\pi}{4} \approx 3.9270.$$

Hence, rounded off to four decimal places, the solutions are

$$t = 0.7854, 2.1910, 3.9270, \text{ and } 5.3326.$$

Ecologists often use the following equations as a **simple predator–prey model:**

$$x = a_1 \cos \omega t + k_1$$

$$y = a_2 \cos(\omega t - \phi) + k_2,$$

where x is the number of prey at time t and y is the number of predators at time t.

Here ω, a_1, a_2, k_1, k_2, and ϕ are constants determined by the particular predator–prey relationship. Notice that when $t = 0$, the prey population has its maximum value $x = a_1 + k_1$.

That x and y oscillate in a simple harmonic fashion is explained as follows: As the predators begin consuming the prey, the predator population increases and the prey population decreases. After a while, there aren't enough prey left to support the increased number of predators, and the predators begin to die off. Fewer predators means that more prey survive, and the prey population begins to increase. These oscillations continue indefinitely, unless something disturbs the ecological balance.

Example In a certain habitat, the simple predator–prey model for owls (the predator) and field mice (the prey) is determined by the values $\omega = 2$, $a_1 = 500$, $a_2 = 10$, $k_1 = 2000$, $k_2 = 50$, and $\phi = \pi/4$, where t is the time in years measured from an instant when the prey population had its maximum value $x = 2500$. (a) Find the values of $t > 0$ for which the owl population is $y = 55$. [C] (b) Find the field mouse population x when $y = 55$.

Solution (a) Substitute $y = 55$, $a_2 = 10$, $\omega = 2$, $\phi = \pi/4$, and $k_2 = 50$ in the equation $y = a_2 \cos(\omega t - \phi) + k_2$ to obtain

$$55 = 10\cos\left(2t - \frac{\pi}{4}\right) + 50;$$

that is,

$$\cos\left(2t - \frac{\pi}{4}\right) = \frac{1}{2}.$$

It follows that

$$2t - \frac{\pi}{4} = \frac{\pi}{3} + 2\pi n \quad \text{or} \quad 2t - \frac{\pi}{4} = \frac{5\pi}{3} + 2\pi n,$$

where n is an integer. Thus,

$$t = \frac{7\pi}{24} + \pi n \quad \text{or} \quad t = \frac{23\pi}{24} + \pi n.$$

Since $t > 0$, then n must be a *nonnegative* integer.

(b) When $t = \dfrac{7\pi}{24} + \pi n$,

$$\begin{aligned} x &= 500\cos 2\left(\frac{7\pi}{24} + \pi n\right) + 2000 \\ &= 500\cos\left(\frac{7\pi}{12} + 2\pi n\right) + 2000 \\ &= 500\cos\frac{7\pi}{12} + 2000 \approx 1871. \end{aligned}$$

When $t = \dfrac{23\pi}{24} + \pi n$,

$$x = 500\cos 2\left(\frac{23\pi}{24} + \pi n\right) + 2000 = 500\cos\left(\frac{23\pi}{12} + 2\pi n\right) + 2000$$

$$= 500\cos\frac{23\pi}{12} + 2000 \approx 2483.$$

Thus, when there are 55 owls in the habitat, the model predicts that there are either 1871 or 2483 field mice.

PROBLEM SET 7

In problems 1–16, solve each simple trigonometric equation. Use degree or radian measure as indicated. Do not use a calculator or tables.

1. $\cos\theta = 0$, $0^\circ \le \theta < 360^\circ$
2. $\cos x = 1$, $0 \le x < 2\pi$
3. $\cot\theta = 1$, $0^\circ \le \theta < 360^\circ$
4. $\tan s = 1$, $0 \le s < 2\pi$
5. $\sin t = \frac{1}{2}$, $0 \le t < 2\pi$
6. $\sec\alpha = 2$, $0^\circ \le \alpha < 360^\circ$
7. $\tan x = -\sqrt{3}$, $0 \le x < 2\pi$
8. $\csc t = -2$, $0 \le t < 2\pi$
9. $\sin t = -1$, $0 \le t < 2\pi$
10. $\sec\theta = -2\sqrt{3}/3$, $0^\circ \le \theta < 720^\circ$
11. $\tan\theta = 0$, θ in degrees
12. $\sin\theta = 0$, $0^\circ \le \theta < 720^\circ$
13. $\sec t = \sqrt{2}$, t in radians
14. $\sec x = 2$, x in radians
15. $\tan x = \sqrt{3}$, x in radians
16. $\csc\beta = 2\sqrt{3}/3$, β in degrees

Ⓒ In problems 17–22, use a calculator to solve each simple trigonometric equation. Use degree or radian measure as indicated and round off all answers to four decimal places.

17. $\sin\theta = \frac{3}{4}$, $0^\circ \le \theta < 360^\circ$
18. $\csc x = 4.5201$, $x > 0$ in radians
19. $\cos t = \frac{2}{3}$, $0 \le t < 2\pi$
20. $\tan\beta = \frac{5}{4}$, $0^\circ \le \beta < 720^\circ$
21. $\cot x = 0.2884$, x in radians
22. $\sec\alpha = 1.5763$, α in degrees

In problems 23–46, solve each trigonometric equation with the side condition $0^\circ \le \theta < 360^\circ$ or $0 \le t < 2\pi$. Do not use a calculator or tables.

23. $\sin 2\theta = \frac{1}{2}$
24. $\sec 2t = 2$
25. $\tan 3\theta = \sqrt{3}$
26. $\csc 3t = -2$
27. $\sin\frac{1}{2}t = \sqrt{3}/2$
28. $\sec\frac{1}{2}\theta = 2\sqrt{3}/3$
29. $\cos\frac{1}{3}t = 0$
30. $\tan\frac{1}{3}\theta = 1$
31. $2\sin\theta + 1 = 0$
32. $\sin t\cos t = 0$
33. $\sin^2 t = \frac{1}{2}\sin t$
34. $\sin(\theta + 40^\circ) = 1$
35. $3\tan^2\theta - 1 = 0$
36. $2\cos\left(t - \dfrac{\pi}{6}\right) = -\sqrt{2}$
37. $\sec^2 t - 2 = 0$
38. $4\sin t\cos t = 1$
39. $\tan t\sin t = \sqrt{3}\sin t$
40. $\cos\theta + \sqrt{3}\sin\theta = -1$
41. $\tan^2\theta - \tan\theta = 0$
42. $\sin\theta + \cos\theta = 1$
43. $\cot^2 t + \csc^2 t = 3$
44. $\tan^4 t - 2\sec^2 t + 3 = 0$
45. $(\sin\theta - 1)(2\cos\theta + \sqrt{3}) = 0$
46. $\sec^2 t + \csc^2 t = \sec^2 t\csc^2 t$

Ⓒ In problems 47–50, solve each trigonometric equation with the side condition $0^\circ \le \theta < 360^\circ$ or $0 \le t < 2\pi$. Use a calculator and round off all answers to four decimal places.

47. $7\sin^2 t - 10\sin t + 3 = 0$
48. $\cot^2\theta - 2\cot\theta - 3 = 0$
49. $3\sin^2\theta + \sqrt{3}\cos\theta = 3$
50. $8\sec^2 t + 2\tan t - 9 = 0$

51. An oscillating signal voltage is given by the equation $E = 75\cos(\omega t - \phi)$ millivolts, with angular frequency $\omega = 120\pi$, phase angle $\phi = \pi/2$, and time t in seconds. This voltage is applied to an oscilloscope, and the triggering circuit of the oscilloscope starts the sweep when E reaches the value 35 millivolts. Find the smallest positive value of t for which triggering occurs.

52. An irrigation ditch has a cross section in the shape of an isosceles trapezoid that is wider at the top than at the bottom (Figure 5). The bottom and the equal sides of the trapezoid are each 2 meters long. If θ is the acute angle between the horizontal and the side of the ditch, (a) show that $A = 4\sin\theta(1 + \cos\theta)$ gives the cross-sectional area of the ditch, and (b) solve for θ if $A = 3\sqrt{3}$ square meters.

Figure 5

53. The index of refraction of water with respect to air is $\mu = 1.333$. Use Snell's law (page 156) to determine the angle of refraction β of a ray of light that strikes the surface of a tank of water at an angle of incidence $\alpha = 15^\circ$.

54. In optics, it is shown that if the index of refraction of a first medium with respect to a second is μ, then the index of refraction of the second medium with respect to the first is $1/\mu$. The index of refraction of flint glass with respect to air is $\mu = 1.650$. Figure 6 shows a light ray initiating in a block of flint glass and emerging into the air. The *critical angle* α_C for flint glass is the value of α for which $\beta = 90^\circ$. (For $\alpha \geq \alpha_C$ the ray is totally reflected back into the glass when it strikes the surface.) Find the critical angle for flint glass.

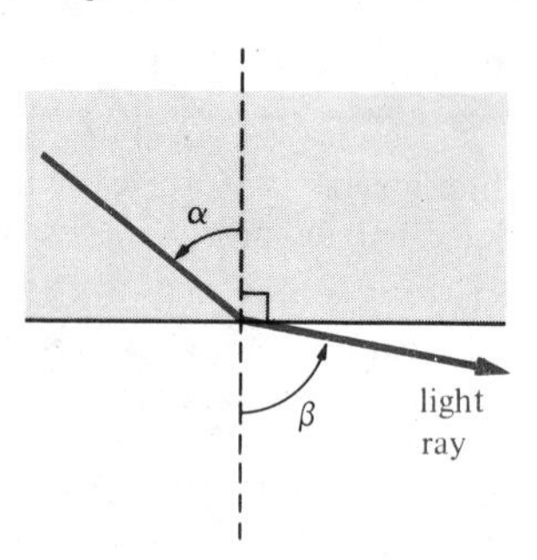

Figure 6

55. Suppose that in a certain tropical habitat, the simple predator–prey model (page 159) for boa constrictors (the predator) and wild pigs (the prey) is determined by the values $\omega = \frac{1}{2}$, $a_1 = 10$, $a_2 = 4$, $k_1 = 40$, $k_2 = 10$, and $\phi = \pi/3$, where t is the time in years measured from an instant when the prey population had its maximum value $x = 50$. (a) Find the values of t for which the boa population is $y = 8$. (b) Find the wild pig population x when $y = 8$.

REVIEW PROBLEM SET

In problems 1–10, simplify each expression.

1. $\dfrac{\sin(-\theta)}{\cos(-\theta)}$

2. $\dfrac{-\sin(-\alpha)}{-\cos(-\alpha)}$

3. $\csc x - \cos x \cot x$

4. $\sec\theta - \sin\theta\tan\theta$

5. $\csc^2 t \tan^2 t - 1$

6. $(\cot x + 1)^2 - \csc^2 x$

7. $\dfrac{\sec^2 u + 2\tan u}{1 + \tan u}$

8. $\dfrac{\sec\beta}{\cot\beta + \tan\beta}$

9. $\dfrac{\sin^2\theta + 2\cos^2\theta}{\sin\theta\cos\theta} - 2\cot\theta$

10. $\dfrac{1}{\csc y - \cot y} - \dfrac{1}{\csc y + \cot y}$

In problems 11 and 12, rewrite each radical expression as a trigonometric expression containing no radical by making the indicated trigonometric substitution.

11. $\sqrt{(25 - x^2)^3}$, $\quad x = 5\sin\theta, \quad -\dfrac{\pi}{2} \leq \theta \leq \dfrac{\pi}{2}$

12. $\dfrac{x}{\sqrt{x^2 - 4}}$, $\quad x = 2\sec t, \quad 0 \leq t < \dfrac{\pi}{2}$

In problems 13–26, prove that each equation is an identity.

13. $\dfrac{\sin\theta}{\csc\theta} - 1 = \dfrac{-1}{\sec^2\theta}$

14. $\dfrac{\csc\alpha}{\cot\alpha + \tan\alpha} = \cos\alpha$

15. $\dfrac{1 - \cot^2 t}{\tan^2 t - 1} = \cot^2 t$

16. $\dfrac{1}{\tan\beta} + \dfrac{\sin\beta}{1 + \cos\beta} = \csc\beta$

17. $\dfrac{1 + \tan x}{1 + \cot x} = \dfrac{\sec x}{\csc x}$

18. $\dfrac{\sin y + \tan y}{\cot y + \csc y} = \dfrac{\sin y}{\cot y}$

19. $\dfrac{\sin\beta + \cos\beta}{\sec\beta + \csc\beta} = \dfrac{\sin\beta}{\sec\beta}$

20. $\dfrac{1 - (\sin t - \cos t)^2}{\sin t} = 2\cos t$

21. $\dfrac{1 - \tan\theta}{1 + \tan\theta} = \dfrac{\cot\theta - 1}{\cot\theta + 1}$

22. $(\sec u - \tan u)^2 = \dfrac{1 - \sin u}{1 + \sin u}$

23. $\dfrac{\cos s}{\sec s - \tan s} = 1 + \sin s$

24. $\sqrt{\dfrac{\sec x - 1}{\sec x + 1}} = \dfrac{|\sin x|}{1 + \cos x}$

25. $\dfrac{\sin t}{\cot t} - \dfrac{\cos t}{\sec t} = \dfrac{\tan t - \cos t \cot t}{\csc t}$

26. $(\csc\omega + \sec\omega)^2 = \dfrac{\sec^2\omega + 2\tan\omega}{\sin^2\omega}$

In problems 27–36, simplify each expression.

27. $\cos(360^\circ - \theta)$

28. $\tan(2\pi - \beta)$

29. $\sin(270^\circ + \alpha)$

30. $\cos(270^\circ - \phi)$

31. $\sin(2\pi + t)$

32. $\cot\left(\dfrac{3\pi}{2} + x\right)$

33. $\sin 37^\circ \cos 23^\circ + \cos 37^\circ \sin 23^\circ$

34. $\dfrac{\tan(\pi/5) + \tan(\pi/20)}{1 - \tan(\pi/5)\tan(\pi/20)}$

35. $\sin x \cos y - \sin\left(x + \dfrac{\pi}{2}\right)\sin(-y)$

36. $\cos(\pi - t) - \tan t \cos\left(\dfrac{\pi}{2} - t\right)$

37. Use the fact that $\dfrac{7\pi}{12} = \dfrac{\pi}{4} + \dfrac{\pi}{3}$ to find the exact numerical value of
(a) $\sin\dfrac{7\pi}{12}$ (b) $\cos\dfrac{7\pi}{12}$ (c) $\tan\dfrac{7\pi}{12}$.

38. Use the fact that $75^\circ = 120^\circ - 45^\circ$ to find the exact numerical value of
(a) $\sin 75^\circ$ (b) $\cos 75^\circ$ (c) $\sec 75^\circ$.

In problems 39–48, assume that α is in quadrant IV, $\cos\alpha = \frac{3}{5}$, β is in quadrant I, $\sin\beta = \frac{8}{17}$, γ is in quadrant II, $\cos\gamma = -\frac{24}{25}$, θ is in quadrant II, and $\sin\theta = \frac{5}{13}$. Find the exact numerical value of each expression.

39. $\sin(\alpha + \beta)$

40. $\cos(\gamma + \theta)$

41. $\sin(\beta + \theta)$

42. $\sin(\alpha - \gamma)$

43. $\cos(\beta - \gamma)$

44. $\sin(\beta - \gamma)$

45. $\tan(\beta - \gamma)$

46. $\sec(\beta - \gamma)$

47. $\sin(\theta - \gamma)$

48. $\cos(\beta - \theta)$

In problems 49–56, prove that each equation is an identity.

49. $\sin(90^\circ + \theta) = \sin(90^\circ - \theta)$

50. $\sin\left(t + \dfrac{\pi}{6}\right) + \cos\left(t + \dfrac{\pi}{3}\right) = \cos t$

51. $\dfrac{\sin(s + t)}{\cos s \cos t} = \tan s + \tan t$

52. $\sin(\beta + 30^\circ) + \cos(60^\circ - \beta) = 2\sin(\beta + 30^\circ)$

53. $\dfrac{\cos(x - y)}{\cos x \sin y} = \tan x + \cot y$

54. $\tan(\alpha + 135^\circ) = \dfrac{\tan\alpha - 1}{\tan\alpha + 1}$

55. $\sin(\alpha - \beta)\cos\beta + \cos(\alpha - \beta)\sin\beta = \sin\alpha$

56. $\tan(\alpha + \beta)\tan(\alpha - \beta) = \dfrac{\tan^2\alpha - \tan^2\beta}{1 - \tan^2\alpha\tan^2\beta}$

57. If $\sec x = \frac{25}{7}$ and $0 < x < \pi/2$, find the exact numerical value of
(a) $\sin 2x$ (b) $\cos 2x$ (c) $\cos(x/2)$ (d) $\tan(x/2)$.

58. If $\tan\theta = \frac{5}{12}$ and $180° < \theta < 270°$, find the exact numerical value of
(a) $\sin(\theta/2)$ (b) $\cos(\theta/2)$ (c) $\cos 2\theta$, (d) $\tan 2\theta$.

In problems 59–68, simplify each expression.

59. $\cos^2 2x - \sin^2 2x$

60. $1 - 2\sin^2\dfrac{t}{2}$

61. $2\sin\dfrac{t}{2}\cos\dfrac{t}{2}$

62. $\cos^4 2\theta - \sin^4 2\theta$

63. $2\sin^2\dfrac{\theta}{2} + \cos\theta$

64. $\dfrac{\sin 4\pi t}{4\sin\pi t\cos\pi t}$

65. $\dfrac{\tan\omega t}{1 - \tan^2\omega t}$

66. $\dfrac{2\tan(t/2)}{1 - \tan^2(t/2)}$

67. $\dfrac{\cos^2(v/2) - \cos v}{\sin^2(v/2)}$

68. $2\sin 2x\cos^3 2x + 2\sin^3 2x\cos 2x$

In problems 69–84, prove that each equation is an identity.

69. $\dfrac{\sin 2\theta + \sin\theta}{\cos 2\theta + \cos\theta + 1} = \tan\theta$

70. $\dfrac{\sec x - 1}{2\sec x} = \sin^2\dfrac{x}{2}$

71. $\tan t + \cot t = 2\csc 2t$

72. $\csc w - \cot w = \tan\dfrac{w}{2}$

73. $\dfrac{\cos^2\dfrac{x}{2} - \cos x}{\sin^2\dfrac{x}{2}} = 1$

74. $\dfrac{\tan\theta - \sin\theta}{\sin\theta\sec\theta} = \tan\dfrac{\theta}{2}\sin\theta$

75. $8\sin^2\dfrac{\theta}{2}\cos^2\dfrac{\theta}{2} = 1 - \cos 2\theta$

76. $\cos^4\dfrac{x}{2} + \sin^4\dfrac{x}{2} = 1 - \dfrac{1}{2}\sin^2 x$

77. $\dfrac{\tan x + \sin x}{2\tan x} = \cos^2\dfrac{x}{2}$

78. $\dfrac{\tan\alpha - \sin\alpha}{2\tan\alpha} = \sin^2\dfrac{\alpha}{2}$

79. $\sin 4t = 4\sin t\cos t(1 - 2\sin^2 t)$

80. $\cos 4w = 8\cos^4 w - 8\cos^2 w + 1$

81. $4\sin\beta\cos^2\dfrac{\beta}{2} = \sin 2\beta + 2\sin\beta$

82. $\dfrac{1 - \cos x - \tan^2\dfrac{x}{2}}{\sin^2\dfrac{x}{2}} = \dfrac{2\cos x}{1 + \cos x}$

83. $\dfrac{\sin 3x}{\sin x} + \dfrac{\cos 3x}{\cos x} = \sin 4x\sec x\csc x$

84. $\dfrac{\cos 3u}{\sin u} - \dfrac{\sin 3u}{\cos u} = \dfrac{\cos 4u}{\sin u\cos u}$

In problems 85–88, express each product as a sum or difference.

85. $\sin\dfrac{3x}{2}\cos\dfrac{x}{2}$

86. $\cos 4\beta\sin 2\beta$

87. $\sin 37.5°\sin 7.5°$

88. $\sin 75°\cos 15°$

In problems 89–92, rewrite each expression as a product.

89. $\sin 55° + \sin 5°$

90. $\cos\dfrac{5\pi}{3} + \cos\dfrac{\pi}{12}$

91. $\sin 4\beta - \sin\beta$

92. $\sin 11t + \sin 5t$

In problems 93–98, prove that each equation is an identity.

93. $\dfrac{\cos 5x + \cos 3x}{\sin 5x - \sin 3x} = \cot x$

94. $\dfrac{2\sin 2\theta\cos\theta - \sin\theta}{\cos\theta - 2\sin 2\theta\sin\theta} = \tan 3\theta$

95. $4\sin 3t\cos 3t\sin t = \cos 5t - \cos 7t$

96. $4\sin 4\alpha\cos 2\alpha\sin\alpha = \cos\alpha - \cos 3\alpha + \cos 5\alpha - \cos 7\alpha$

97. $\dfrac{\sin\theta + \sin 3\theta + \sin 5\theta}{\cos\theta + \cos 3\theta + \cos 5\theta} = \tan 3\theta$

98. $2\sin\left(\dfrac{\alpha-\beta}{2} + \dfrac{\pi}{4}\right)\cos\left(\dfrac{\alpha+\beta}{2} - \dfrac{\pi}{4}\right) = \sin\alpha + \cos\beta$

Ⓒ In problems 99–102, use addition or subtraction of ordinates to sketch each graph. (A calculator will be helpful.)

99. $y = 2\sin 3x + 3\cos 2x$

100. $y = 0.12\sin \pi t + 0.6\sin 2\pi t$

101. $y = \sin t - \sin\dfrac{t}{3}$

102. $y = \sin x + \tan x$

Ⓒ **103.** The variations in air pressure caused at a certain point by simultaneously playing the pure tone "concert A" and the pure tone one octave higher are given by

$$y = 0.12\sin 880\pi t + 0.06\sin 1760\pi t,$$

where t is the time in seconds. Sketch the graph showing these air-pressure variations.

Ⓒ **104.** Believers in biorhythms sometimes plot a *composite* biorhythm curve by adding all three ordinates (physical, emotional, and intellectual) on a biorhythm chart. Using the data in problem 39 on page 110, sketch a composite biorhythm curve over a 62-day period.

Ⓒ **105.** Use the product formula to rewrite the quantity y given by $y = 4\sin t\cos(4t/3)$ as a superposition of two simple harmonic oscillations. Then plot a graph of y as a function of t using the method of addition of ordinates.

106. The equation for a radio frequency carrier wave with frequency v_r, which is amplitude modulated by a pure audio tone with frequency v_a, is

$$y = a(1 + m\sin 2\pi v_a t)\sin 2\pi v_r t,$$

where a and m are constants. Rewrite this equation to show that y is a superposition of simple harmonic oscillations.

In problems 107–110, use Theorem 3 on page 142 to rewrite the given equation in the form $y = a\cos(\omega t - \phi)$ and then sketch the graph.

107. $y = 2\sin\left(\dfrac{\pi}{3}t - \pi\right) - 2\cos\left(\dfrac{\pi}{3}t - \pi\right)$

Ⓒ **108.** $y = 4\cos\left(\dfrac{\pi}{2}t - \dfrac{\pi}{3}\right) - 3\sin\left(\dfrac{\pi}{2}t - \dfrac{\pi}{3}\right)$

109. $y = \sqrt{3}\cos(2t - 3) - \sin(2t - 3)$

Ⓒ **110.** $y = 2\cos(3t - 1) - 3\sin(3t - 1)$

In problems 111 and 112, find the period T of the oscillation represented by each equation.

111. $y = 2\cos\left(\dfrac{\pi}{3}t - \dfrac{\pi}{4}\right) + \sqrt{3}\cos\left(\dfrac{\pi}{5}t - \dfrac{\pi}{3}\right) + 2$

112. $y = \sqrt{2}\cos\left(\dfrac{\sqrt{3}}{2}t - \sqrt{5}\right) - \sin\left(\dfrac{\sqrt{3}}{3}t - \sqrt{3}\right) - 1$

113. Write the oscillation represented by $y = 2\cos\pi t\cos 3\pi t$ as a superposition of two simple harmonic oscillations and determine its period T.

114. Figure 1 illustrates the phenomenon of *beats* or *hetrodyning* which occurs when two sounds of slightly different frequencies v_1 and v_2 are superposed. If the sounds are represented by $y_1 = a\cos(2\pi v_1 t)$ and $y_2 = a\cos(2\pi v_2 t)$, show that the superposition $y = y_1 + y_2$ can be regarded as a sound of frequency $(v_1 + v_2)/2$ amplitude modulated by an oscillation of frequency $|v_1 - v_2|/2$. Explain why one hears "beats" with a frequency of $|v_1 - v_2|$.

Figure 1

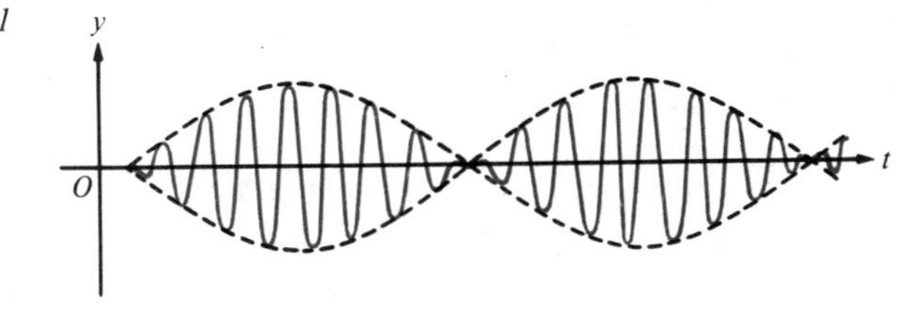

In problems 115–118, evaluate each expression without the use of a calculator or tables.

115. $\sin^{-1}(-\frac{1}{2})$ **116.** $\arccos(-\sqrt{3}/2)$ **117.** $\arctan\sqrt{3}$ **118.** $\operatorname{arcsec}\sqrt{2}$

C In problems 119–128, use a calculator (or Appendix Table 2, Section 3) to evaluate each expression. Give answers in radians.

119. $\sin^{-1}0.3750$ **120.** $\arccos(-0.3901)$ **121.** $\cos^{-1}0.9273$ **122.** $\tan^{-1}57.29$
123. $\arctan 1.425$ **124.** $\cos^{-1}\frac{1}{8}$ **125.** $\arcsin(-\frac{3}{4})$ **126.** $\tan^{-1}8$
127. $\cos^{-1}(-\frac{10}{11})$ **128.** $\arctan 1.007$

In problems 129–134, find the exact value of each expression without using a calculator or tables.

129. $\cos(\tan^{-1}\frac{4}{3})$ **130.** $\sin[\arctan(-\frac{5}{12})]$ **131.** $\sin[\sin^{-1}(-\frac{12}{13})]$
132. $\tan[\arccos(-\frac{3}{5})]$ **133.** $\arcsin[\sin(19\pi/14)]$ **134.** $\sin(\sin^{-1}\frac{2}{3} + \sin^{-1}\frac{3}{4})$

In problems 135–138, show that the given equation is an identity.

135. $\sin(\cos^{-1}x) = \sqrt{1 - x^2}$ **136.** $\sin(\frac{1}{2}\arccos x) = \sqrt{\dfrac{1 - x}{2}}$

137. $\tan(\arccos x) = \dfrac{\sqrt{1 - x^2}}{x}$ **138.** $\tan(\tan^{-1}x + \tan^{-1}y) = \dfrac{x + y}{1 - xy}$

139. A picture a meters high hangs on a wall so that its bottom is b meters above the eye level of an observer. If the observer stands x meters from the wall (Figure 2), show that the angle θ subtended by the picture is given by

$$\theta = \arctan\frac{a + b}{x} - \arctan\frac{b}{x}.$$

Figure 2

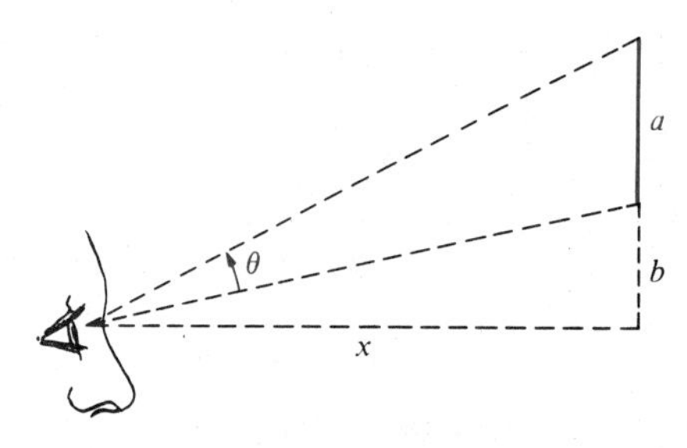

140. Two parallel lines, each at a distance a from the center of a circle of radius r, cut off the region of area A shown in Figure 3. Show that

$$A = 2a\sqrt{r^2 - a^2} + 2r^2\arcsin\frac{a}{r}.$$

Figure 3

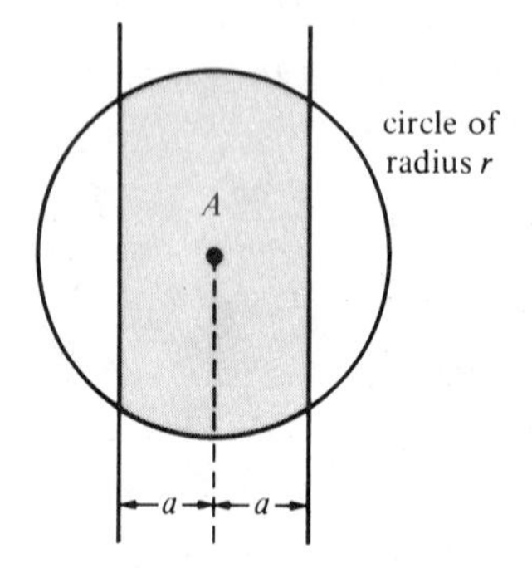

In problems 141–152, solve each trigonometric equation with the side condition $0^\circ \leq \theta < 360^\circ$ or $0 \leq t < 2\pi$. Do not use a calculator or tables.

141. $\tan\theta = \dfrac{\sqrt{3}}{3}$

142. $4\sin^2 t - 3 = 0$

143. $4\cos^2 t - 3 = 0$

144. $\tan^2\theta + (1 - \sqrt{3})\tan\theta - \sqrt{3} = 0$

145. $2\sin^2\theta + \sqrt{2}\sin\theta = 0$

146. $\tan t = \sec t + 1$

147. $\cot^2 t + 3\csc t + 3 = 0$

148. $\cot\theta = \csc\theta - 1$

149. $\tan 4\theta = \sqrt{3}$

150. $\sin 3t \cos t = \sin t \cos 3t + 1$

151. $\sin 3\theta + \sin\theta = 0$

152. $\cos\dfrac{t}{4} = \dfrac{\sqrt{3}}{2}$

C In problems 153–156, solve each trigonometric equation with the side condition $0^\circ \leq \theta < 360^\circ$ or $0 \leq t < 2\pi$. Use a calculator and round off all answers to four decimal places.

153. $\cos 2\theta = \frac{3}{5}$

154. $\cot\dfrac{\theta}{2} = -2.017$

155. $6\sin^2 t + \sin t - 2 = 0$

156. $2\sec^2 t + \tan t - 3 = 0$

157. The index of refraction of a diamond with respect to air is $\mu = 2.417$. Use Snell's law (page 156) to determine the angle of refraction β of a ray of light that strikes the surface of a diamond at an angle of incidence $\alpha = 20^\circ$.

158. A ray of light passing through a plate of material with parallel faces is displaced, but not deviated (Figure 4); that is, the emerging ray is parallel to the ingoing ray. If d is the amount of displacement, t the thickness of the plate, α the angle of incidence, and β the angle of refraction, show that

$$\beta = \arctan\left(\tan\alpha - \frac{d}{t}\sec\alpha\right).$$

Figure 4

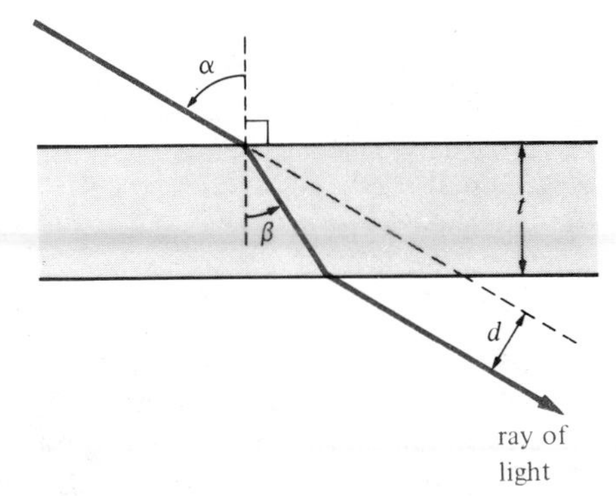

4 Applications of Trigonometry

In this chapter, we study some of the applications of the trigonometric functions to geometry. We begin by "solving triangles," that is, finding certain parts of triangles when other parts are known. To do this, we derive and apply the *law of sines* and the *law of cosines.* The chapter also includes an introduction to vectors and a discussion of polar coordinates.

1 Right Triangles

Figure 1

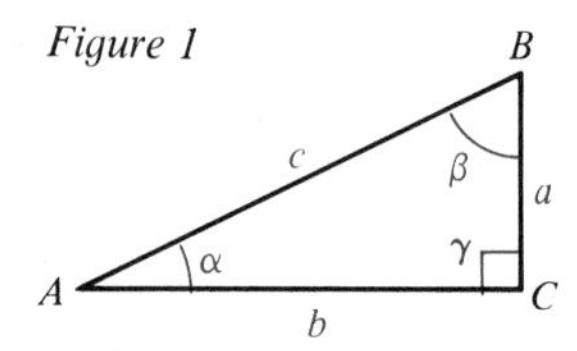

In the right triangle ACB shown in Figure 1, the angles are denoted by α at vertex A, β at vertex B, and γ at vertex C. The lengths of the sides opposite angles α, β, and γ are denoted by a, b, and c. Note that angles α and β are complementary acute angles, that angle γ is a right angle, and that c is the hypotenuse of right triangle ACB. Therefore,

$$\sin\alpha = \cos\beta = \frac{a}{c} \qquad \csc\alpha = \sec\beta = \frac{c}{a}$$

$$\cos\alpha = \sin\beta = \frac{b}{c} \qquad \sec\alpha = \csc\beta = \frac{c}{b}$$

$$\tan\alpha = \cot\beta = \frac{a}{b} \qquad \cot\alpha = \tan\beta = \frac{b}{a}.$$

If the lengths of two sides of a right triangle are given, or if one side and an acute angle are given, then these formulas can be used to solve for the remaining angles and sides of the triangle. This procedure is called **solving the right triangle.**

In solving a right triangle, it is necessary to use a calculator or a table of trigonometric functions, unless the special angles 30°, 45°, or 60° are involved. You should always keep in mind that solutions obtained using either a calculator or tables are *approximations.* In this section, we round off all angles to the nearest hundredth of a degree, and all side lengths to four significant digits.

Examples Ⓒ Assume that right triangle ACB is labeled as in Figure 1. In each case, solve the triangle and sketch it.

1 $b = 31, \quad \alpha = 43.33^\circ$

Solution We must find a, c, and β. Because α and β are complementary,

$$\beta = 90^\circ - \alpha = 90^\circ - 43.33^\circ = 46.67^\circ.$$

To find a, we notice that

$$\tan\alpha = \frac{a}{b}, \quad \text{so} \quad a = b\tan\alpha.$$

Using a calculator, we find that

$$\tan\alpha = \tan 43.33^\circ = 0.943341386.$$

Therefore, rounded off to four significant digits,

$$a = b\tan\alpha = 31\tan 43.33^\circ = 29.24.$$

(Note that it isn't necessary to write down the intermediate result $\tan 43.33^\circ = 0.943341386$. We have done it here so you can check your calculator work.) To find c, we notice that

$$\cos\alpha = \frac{b}{c}, \quad \text{so} \quad c = \frac{b}{\cos\alpha}.$$

Therefore, rounded off to four significant digits,

$$c = \frac{b}{\cos\alpha} = \frac{31}{\cos 43.33^\circ} = \frac{31}{0.727413564} = 42.62.$$

(Again, if you are using your calculator efficiently, it won't be necessary to write down the fraction $31/0.727413564$. You should be able to calculate $31/\cos 43.33^\circ$ directly with only a few key strokes.) Notice that we could have used the Pythagorean theorem $c^2 = a^2 + b^2$ to solve for c:

$$c = \sqrt{a^2 + b^2} = \sqrt{(29.24)^2 + 31^2}$$
$$= 42.61.$$

(The discrepancy in the last decimal place was caused by using the rounded off value for a. The result $c = 42.62$ is actually more accurate.) The triangle is shown in Figure 2.

Figure 2

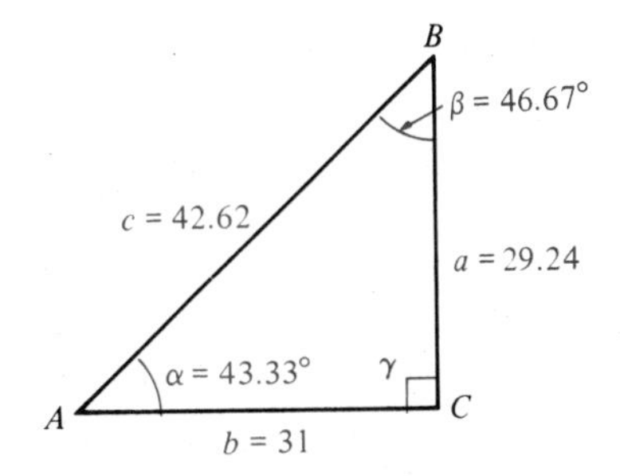

2 $a = 4, \quad b = 3$

Solution By the Pythagorean theorem,

$$c = \sqrt{a^2 + b^2} = \sqrt{16 + 9} = 5.$$

Now we have

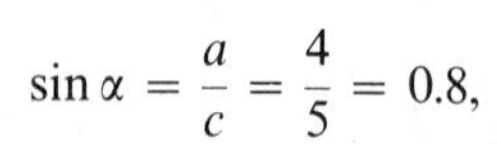

$$\sin\alpha = \frac{a}{c} = \frac{4}{5} = 0.8,$$

so

$$\alpha = \sin^{-1} 0.8 = 53.13^\circ.$$

It follows that

$$\beta = 90^\circ - \alpha = 90^\circ - 53.13^\circ = 36.87^\circ.$$

The triangle is shown in Figure 3.

Figure 3

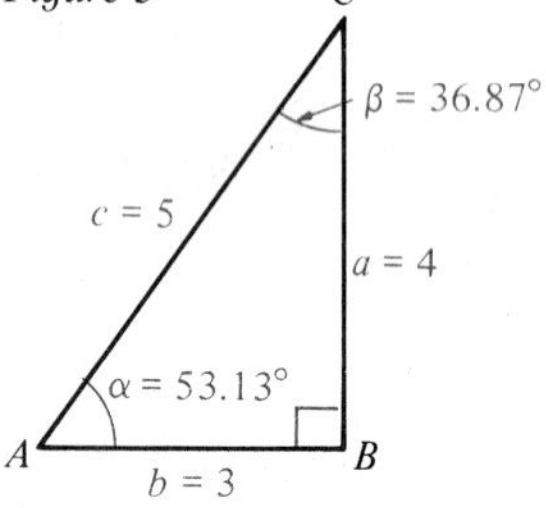

In many applications of trigonometry, we represent the relevant features of a real-world situation with triangles, and then solve the triangles for their unknown parts.

Example Ⓒ A railway track rises 300 feet per mile (5280 feet). Find the angle α at which the track is inclined from the horizontal (Figure 4).

Solution From Figure 4,

$$\tan\alpha = \tfrac{300}{5280},$$

so

$$\alpha = \tan^{-1} \tfrac{300}{5280} = 3.25^\circ.$$

Figure 4

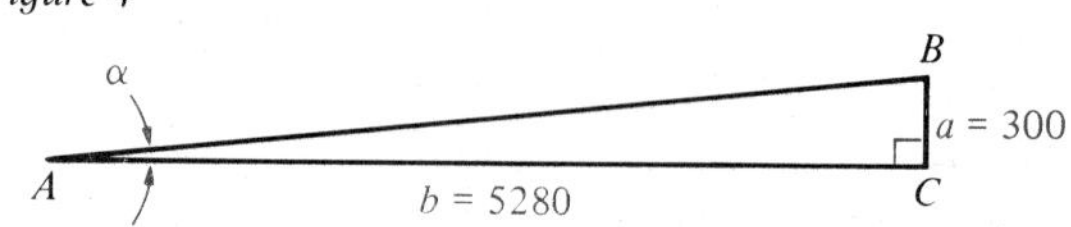

The acute angle formed by a horizontal line and an observer's line of sight to any object above the horizontal is called an **angle of elevation** (Figure 5a). Similarly, the acute angle formed by a horizontal line and an observer's line of sight to any object below the horizontal is called an **angle of depression** (Figure 5b).

Figure 5

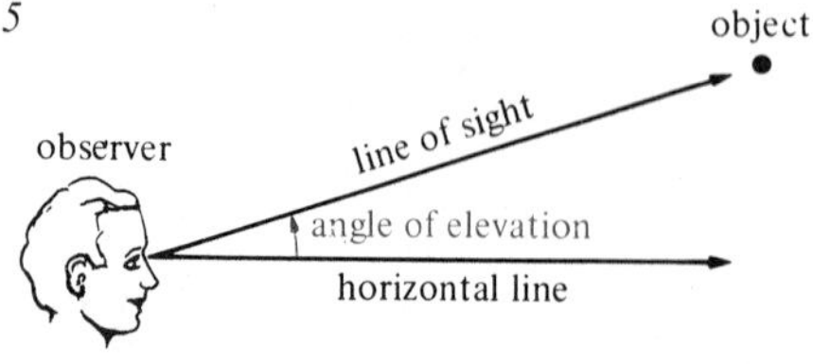

(a)

(b)

Example 1 Ⓒ From a point on level ground 75 meters from the base of a television transmitting tower, the angle of elevation of the top of the tower is 68.17°. Find the height h of the tower (Figure 6).

Solution In Figure 6, $\tan 68.17^\circ = \frac{h}{75}$, so,

$$h = 75 \tan 68.17^\circ = 187.2 \text{ meters.}$$

Figure 6

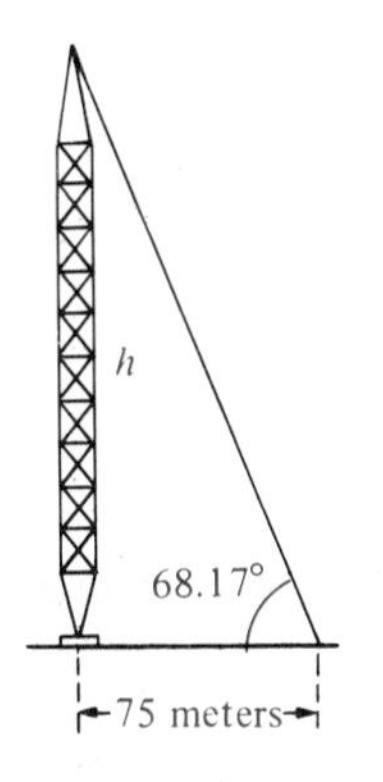

Figure 7

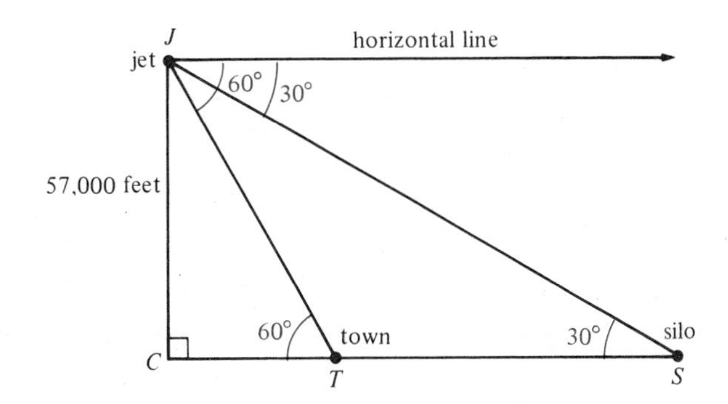

Example 2 Ⓒ A high-altitude military-reconnaissance jet photographs a missile silo under construction near a small town. The jet is at an altitude of 57,000 feet, and the angles of depression of the town and silo are 60° and 30°, respectively. Assuming that the jet, the silo, and the town lie in the same vertical plane, find the distance between the town and the silo (Figure 7).

Solution In Figure 7, we want to find $|\overline{TS}|$, which is $|\overline{CS}| - |\overline{CT}|$. From elementary geometry, we know that alternate interior angles are equal, so angle $CTJ = 60^\circ$ and angle $CSJ = 30^\circ$. In right triangle TCJ,

$$\tan 60^\circ = \frac{|\overline{CJ}|}{|\overline{CT}|} = \frac{57{,}000}{|\overline{CT}|},$$

so

$$|\overline{CT}| = \frac{57{,}000}{\tan 60^\circ} = \frac{57{,}000}{\sqrt{3}} = \frac{57{,}000\sqrt{3}}{3} = 19{,}000\sqrt{3}.$$

In right triangle SCJ,

$$\tan 30^\circ = \frac{|\overline{CJ}|}{|\overline{CS}|} = \frac{57{,}000}{|\overline{CS}|},$$

so

$$|\overline{CS}| = \frac{57{,}000}{\tan 30^\circ} = \frac{57{,}000}{\sqrt{3}/3} = 57{,}000\sqrt{3}.$$

Hence,

$$|\overline{TS}| = |\overline{CS}| - |\overline{CT}| = 57{,}000\sqrt{3} - 19{,}000\sqrt{3} = 38{,}000\sqrt{3} \approx 65{,}820 \text{ feet.}$$

In some applications of trigonometry, especially in surveying and navigation, the **direction** or **bearing** of a point Q as viewed from a point P is defined to be the positive angle the ray from P through Q makes with a north–south line through P. Such an angle is specified as being measured east or west from north or south. For instance, in Figure 8, the bearing from P to Q is 61° east of north, or N61°E. Similarly, the bearing from P to R is N28°W; the bearing from P to T is S15°W; and the bearing from P to V is S40°E.

Figure 8

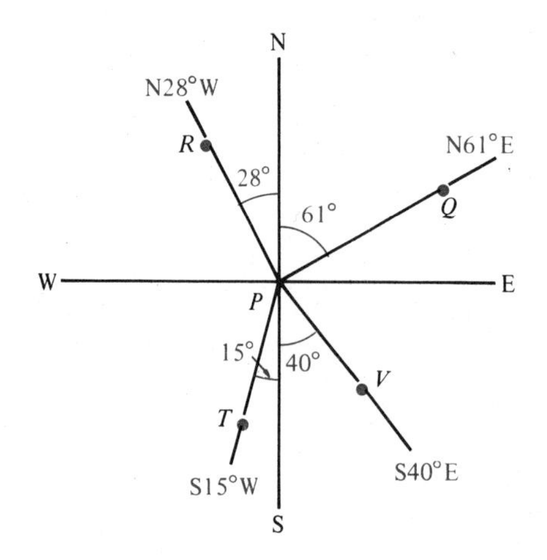

Figure 9

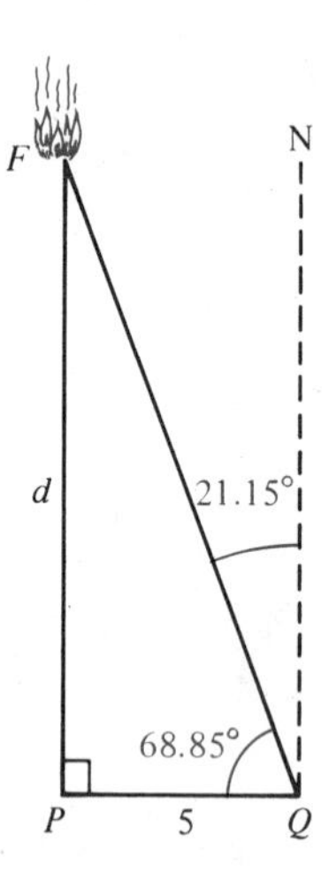

Example Ⓒ From an observation point P, a forest ranger sights a fire F directly to the north. Another ranger at point Q, 5 kilometers due east of P, sights the same fire at a bearing N21.15°W (Figure 9). Find the distance d between P and the fire.

Solution In Figure 9, angle PQF is complementary to the 21.15° angle; therefore,

$$\text{angle } PQF = 90^\circ - 21.15^\circ = 68.85^\circ.$$

Also,

$$\tan 68.85^\circ = \frac{d}{5},$$

so

$$d = 5 \tan 68.85^\circ = 12.92.$$

Therefore, the fire is approximately 12.92 kilometers due north of observation point P.

C PROBLEM SET 1

In problems 1–18, assume that the right triangle ACB is labeled as in Figure 10. In each case solve the triangle and sketch it. Round off all angles to the nearest hundredth of a degree and all side lengths to four significant digits.

Figure 10

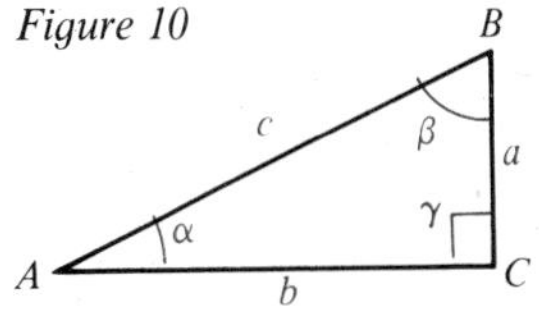

1. $a = 5,\quad b = 4$
2. $b = 91,\quad \beta = 30°$
3. $a = 10,\quad c = 15$
4. $c = 100,\quad \beta = 41°$
5. $b = 1700,\quad \alpha = 31.23°$
6. $c = 10^4,\quad \beta = 17.81°$
7. $a = 7.132,\quad c = 9.209$
8. $b = 0.01523,\quad \beta = 11.3°$
9. $a = 3.32,\quad b = 4331$
10. $a = 113.5,\quad c = 217.0$
11. $b = 2570,\quad \beta = 15.45°$
12. $a = 50,\quad b = 120$
13. $b = 5.673 \times 10^{-3}$ meter, $\alpha = 67°45'$
14. $a = 8.141 \times 10^5$ meters, $\alpha = 1.10°$
15. $b = 30.73$ miles, $c = 77.17$ miles
16. $a = 9.200 \times 10^7$ kilometers, $\alpha = 51°33'$
17. $a = 4.932 \times 10^{-6}$ meter, $b = 4.101 \times 10^{-6}$ meter
18. $c = 1.410 \times 10^{15}$ kilometers, $\alpha = 62°13'$

19. A 30-foot ladder leaning against a vertical wall just reaches a window sill. If the ladder makes an angle of 47° with the level ground, how high is the window sill? (Round off your answer to the nearest foot.)
20. A guy wire 8 meters long helps support a CB base antenna mounted on top of a flat roof. If the wire makes an angle of 49.5° with the horizontal roof, how far above the roof is it attached to the antenna? (Round off your answer to two decimal places.)
21. A kite string makes an angle of 28° with the level ground, and 73 meters of string is out. How high is the kite? (Round off your answer to the nearest meter.)
22. An engineer is designing an access ramp for an elevated expressway that is 40 feet above level ground. The ramp must be straight and cannot be inclined more than 15° from the horizontal. Because of existing structures near the expressway, the horizontal distance between the beginning of the ramp and the expressway cannot exceed 150 feet. Can the engineer design such a ramp?
23. A jetliner is climbing so that its path is a straight line that makes an angle of 8.5° with the horizontal. How many meters does the jetliner rise while traveling 300 meters along its path? (Round off your answer to the nearest meter.)
24. A portion of a tunnel under a river is straight for 150 meters and descends 10 meters in this distance. (a) To the nearest hundredth of a degree, what angle does this part of the tunnel make with the horizontal? (b) What is the *horizontal* distance, to the nearest 0.1 meter, between the ends of this part of the tunnel?
25. A monument 22 meters high casts a shadow 31 meters long. Find the angle of elevation of the sun to the nearest tenth of a degree.
26. To measure the height of a cloud cover at night, a spotlight is aimed straight upward from the ground. The resulting spot of light on the clouds is viewed from a point on the level ground 850 meters from the spotlight, and the angle of elevation is measured at 61.8°. Find the height of the cloud cover to the nearest meter.
27. A wrecking company has contracted to knock down an old brick chimney. In order to determine in advance just where the chimney will fall, it is necessary to find its height. The top of the chimney is viewed from a point 150 meters from its base, and the angle of elevation is measured at 33.25°. Find the height of the chimney, rounded off to the nearest 0.1 meter.
28. A rectangular panel to collect solar energy rests on flat ground and is tilted toward the sun. The edge resting on the ground is 3.217 meters long, and the upper edge is 1.574 meters above the ground. The panel is located near Chicago and its latitude is 41°50′. Solar engineers recommend that the angle between the panel and the ground be equal to the latitude of its location. Assuming that this recommendation has been followed, find the surface area of the panel. (Round off your answer to three decimal places.)

29. A lifeguard is seated on a high platform so that her eyes are 7 meters above sea level. Suddenly she spots the dorsal fin of a great white shark at a 4° angle of depression. Estimate, to the nearest meter, the horizontal distance between the platform and the shark.
30. A customs officer located on a straight shoreline observes a smuggler's motorboat making directly for the closest point on shore, an abandoned lighthouse. The angle between the shoreline and the officer's line of sight to the motorboat is 34.6°, the motorboat is traveling at 18 knots (nautical miles per hour), and it is 6 nautical miles from the lighthouse. The officer immediately departs by car on a straight road along the shoreline, hoping to reach the lighthouse 10 minutes before the smugglers in order to apprehend them as they land. How fast must the officer drive? (One nautical mile is 6076 feet, whereas one statute, or ordinary, mile is 5280 feet.)
31. Biologists studying the migration of birds are following a migrating flock in a light plane. The birds are flying at a constant altitude of 1200 feet and the plane is following at a constant altitude of 1700 feet. The biologists must maintain a distance of at least 600 feet between the plane and the flock in order to avoid disturbing the birds; therefore, they must monitor the angle of depression of the flock from the plane. Find the maximum allowable angle of depression, rounded off to the nearest degree.
32. An observer on a bluff 100 meters above the surface of a Scottish loch sees the head of an aquatic monster at an angle of depression of 18.45°. The monster, swimming directly away from the observer, immediately submerges. Five minutes later, the monster's head reappears, now at an angle of depression of 14.05°. How fast is the monster swimming?
33. A nature photographer using a telephoto lens photographs a rare bird roosting on a high branch of a tree at an angle of elevation of 22.5°. The distance between the lens and the bird is 330 feet. In order to obtain a more detailed photograph of the bird, the photographer cautiously moves closer to the base of the tree. The angle of elevation of the bird is now 51.25°. Find the new distance between the photographer's lens and the bird.
34. The angle of elevation of the top of a tower from a point 100 meters from the top is measured to be α degrees. Thus, the height h of the tower is calculated to be $h = 100 \sin \alpha$ meters. Suppose, however, that a 0.1° error has been made, and that the true angle of elevation is $\alpha + 0.1^\circ$. Then the true height of the tower is $100 \sin(\alpha + 0.1^\circ)$, and the error E in the calculated value of h is given by $E = 100 \sin(\alpha + 0.1^\circ) - 100 \sin \alpha$, the true value minus the calculated value. Using a calculator, find E if (a) $\alpha = 20^\circ$, (b) $\alpha = 40^\circ$, and (c) $\alpha = 60^\circ$. (Round off your answer to two decimal places.)
35. A lifeguard at station A sights a swimmer directly south of him. Another lifeguard at station B, 230 feet directly east of A, sights the same swimmer at a bearing of S49.5°W. How far is the swimmer from station A?
36. Refer to problem 34. (a) Sketch a graph of E as a function of α for $0 < \alpha < 90^\circ$. (b) Does the error E increase or decrease as α increases? (c) If you know that your measurement of α is subject to an error of as much as 0.1°, to how many decimal places should you round off your calculated value of h?
37. An airplane A is in the air at a position due west of an airport, and another airplane B is 15 miles south of A. From B, the bearing of the airport is N58.5°E. How far is airplane A from the airport?
38. To find the width $|\overline{PQ}|$ of a lake, a surveyor measures 3000 feet from P in the direction of N$41^\circ 40'$W to locate point R. The surveyor then determines that the bearing of Q from R is N$71^\circ 30'$E. Find the width of the lake if the point Q is located so that angle QPR is a right angle.
39. At 1:00 P.M. a ship is 24 nautical miles directly east of a lighthouse. The ship is sailing due north at 32 knots (nautical miles per hour). What is the bearing of the lighthouse from the ship at 3:00 P.M.?

2 The Law of Sines

In this section and the next, we shall be studying relationships among the three angles α, β, and γ and the opposite sides a, b, and c of a *general* triangle ABC (Figure 1). The following theorem, called the *law of sines*, relates the lengths of the three sides to the sines of the three vertex angles.

Figure 1

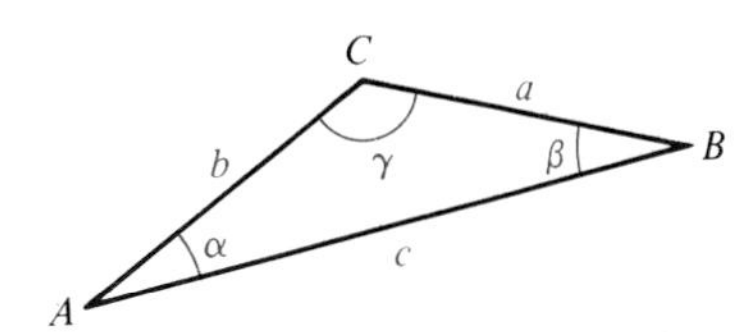

Theorem 1 **The Law of Sines**

> In the general triangle ABC in Figure 1,
>
> $$\frac{\sin\alpha}{a} = \frac{\sin\beta}{b} = \frac{\sin\gamma}{c}$$

Proof We show that $\dfrac{\sin\alpha}{a} = \dfrac{\sin\beta}{b}$ and leave the similar proof that $\dfrac{\sin\beta}{b} = \dfrac{\sin\gamma}{c}$ as an exercise (Problem 38). Drop a perpendicular $\overline{CD}$ from vertex C to the straight line ℓ containing the vertices A and B. Figure 2 shows the case in which both angles α and β are acute, so that D lies *between* A and B on ℓ. [If either α or β is not acute, D falls outside of the segment $\overline{AB}$ on ℓ, and the following argument has to be modified slightly (Problem 37).] In right triangles ADC and BDC, we have

$$\sin\alpha = \frac{|\overline{CD}|}{b} \quad \text{and} \quad \sin\beta = \frac{|\overline{CD}|}{a},$$

so

$$b\sin\alpha = |\overline{CD}| \quad \text{and} \quad a\sin\beta = |\overline{CD}|.$$

Consequently,

$$b\sin\alpha = a\sin\beta.$$

Dividing both sides of the last equation by ab, we obtain

$$\frac{\sin\alpha}{a} = \frac{\sin\beta}{b}.$$

Figure 2

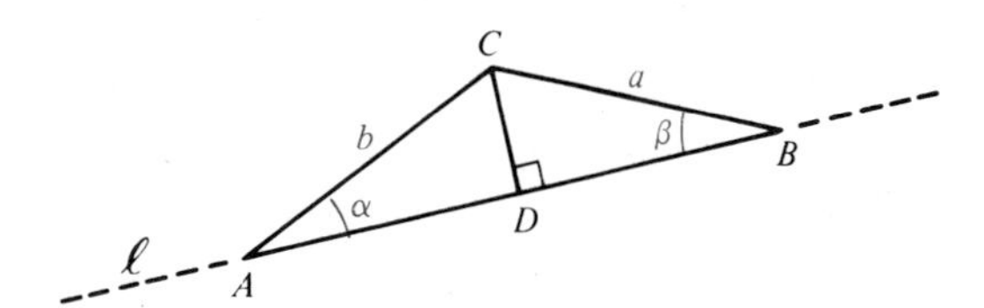

Notice that the law of sines (Theorem 1) can be written in the alternative form

$$\frac{a}{\sin \alpha} = \frac{b}{\sin \beta} = \frac{c}{\sin \gamma}.$$

If two angles of triangle ABC are given, then the third angle can be found by using the relationship

$$\alpha + \beta + \gamma = 180°;$$

hence, the three denominators $\sin \alpha$, $\sin \beta$, and $\sin \gamma$ can be found by using a calculator (or a table of sines). Now, if any one of the sides a, b, or c is also given, then the equations

$$\frac{a}{\sin \alpha} = \frac{b}{\sin \beta} = \frac{c}{\sin \gamma}$$

can be solved for the remaining two sides.

The following examples illustrate the procedure for solving a triangle when *two angles and one side* are given (or can be determined from information provided). Unless otherwise indicated, we shall round off all angles to the nearest hundredth of a degree, and all side lengths to four significant digits.

Example 1 In triangle ABC (Figure 3), suppose that $\alpha = 41°$, $\beta = 77°$, and $a = 74$. Solve for γ, b, and c.

Figure 3

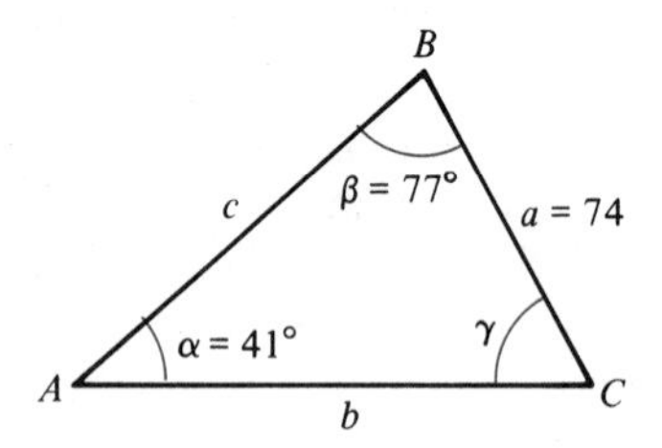

Solution

$$\gamma = 180° - \alpha - \beta = 180° - 41° - 77° = 62°$$

By the law of sines,

$$\frac{a}{\sin 41°} = \frac{b}{\sin 77°} = \frac{c}{\sin 62°};$$

hence, since $a = 74$,

$$b = \frac{a \sin 77°}{\sin 41°} = \frac{74 \sin 77°}{\sin 41°} = 109.9$$

and

$$c = \frac{a \sin 62°}{\sin 41°} = \frac{74 \sin 62°}{\sin 41°} = 99.59.$$

Example 2 A statue of height 70 meters ($\overline{BC}$ in Figure 4) stands atop a hill of height h. From a point A at ground level, the angle of elevation of the base B of the statue is 20.75° and the angle of elevation of the top C of the statue is 28.30°. Find the height h.

Figure 4

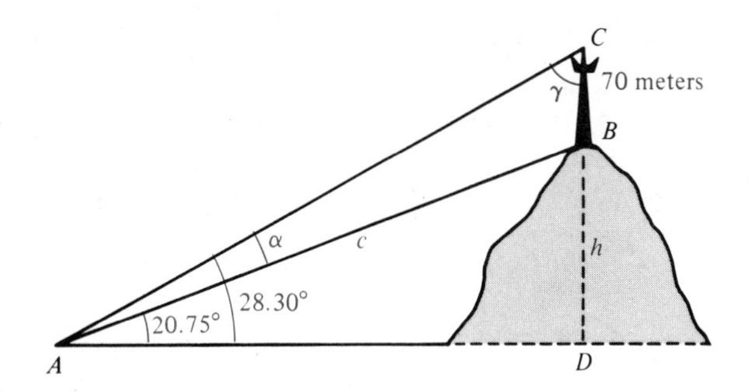

Solution In right triangle ADB, we have

$$\sin 20.75^\circ = h/c$$

so

$$h = c \sin 20.75^\circ.$$

Our plan is to find c and then use the last equation to find h. We know the length of one side $\overline{BC}$ in triangle ABC, so we can use the law of sines to find $c = |\overline{AB}|$, provided that we can find two of the vertex angles in this triangle. Angle α is easily found:

$$\alpha = 28.30^\circ - 20.75^\circ = 7.55^\circ.$$

In right triangle ADC, angle γ is complementary to angle DAC; hence,

$$\gamma = 90^\circ - \text{angle } DAC = 90^\circ - 28.30^\circ = 61.70^\circ.$$

Now we can apply the law of sines to triangle ABC to obtain

$$\frac{c}{\sin \gamma} = \frac{70}{\sin \alpha}$$

so that

$$c = \frac{70 \sin \gamma}{\sin \alpha} = \frac{70 \sin 61.70^\circ}{\sin 7.55^\circ},$$

and it follows that

$$h = c \sin 20.75^\circ = \left(\frac{70 \sin 61.70^\circ}{\sin 7.55^\circ}\right) \sin 20.75^\circ = 166.2 \text{ meters}.$$

Example 3 Solve the triangle ABC if $\alpha = 82.17^\circ$, $\gamma = 103.50^\circ$, and $b = 615$.

Solution We have

$$\beta = 180^\circ - (\alpha + \gamma) = 180^\circ - 185.67^\circ = -5.67^\circ,$$

which is impossible because we cannot have a triangle with a negative vertex angle. We conclude that there is no triangle satisfying the given conditions.

More generally, if the specifications for a triangle require that the sum of two vertex angles exceeds 180°, then no such triangle will exist.

2.1 The Ambiguous Case

Because there are several possibilities, the situation in which you are given the *lengths of two sides* of a triangle and the *angle opposite one of them* is called the **ambiguous case.** For instance, suppose you are given side a, side b, and angle α in triangle ABC. You might try to construct triangle ABC from this information by drawing a line segment $\overline{AC}$ of length b and a ray ℓ that starts at A and makes an angle α with $\overline{AC}$ (Figure 5). To find the remaining vertex B, you could use a compass to draw an arc of a circle of radius a with center C. If the arc intersects the ray ℓ at point B, then ABC is the desired triangle. (Why?)

Figure 5

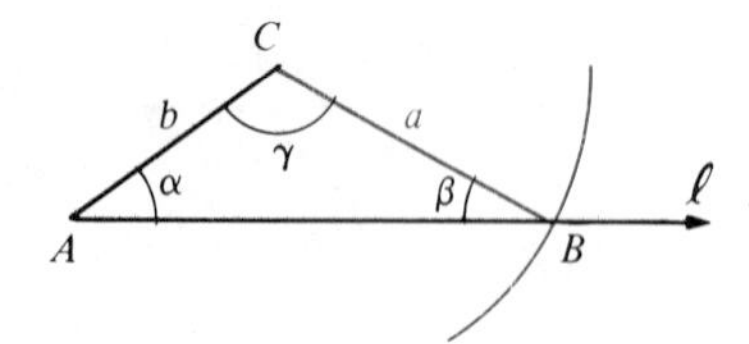

As Figure 6 illustrates, there are actually *three* possibilities if you try to construct triangle ABC by the method above: The circle does not intersect the ray ℓ at all and there is **no triangle** $\boldsymbol{ABC}$ (Figure 6a); it intersects the ray ℓ in exactly one point B and there is just **one triangle** $\boldsymbol{ABC}$ (Figures 5 and 6b); or it intersects the ray ℓ in two points B_1 and B_2 and there are **two triangles** $\boldsymbol{AB_1C}$ and $\boldsymbol{AB_2C}$ that satisfy the given conditions (Figure 6c).

Figure 6

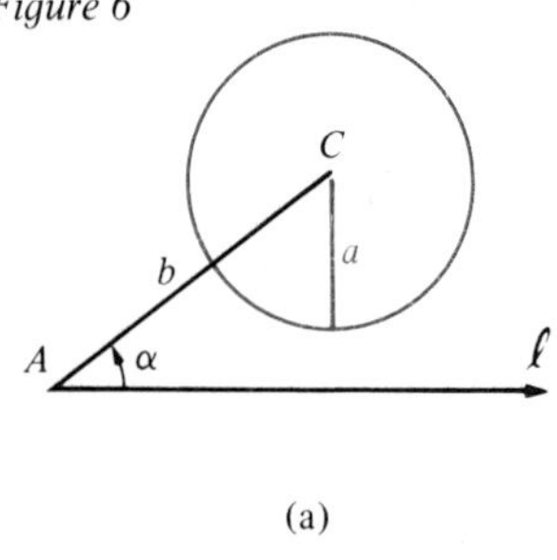

(a)

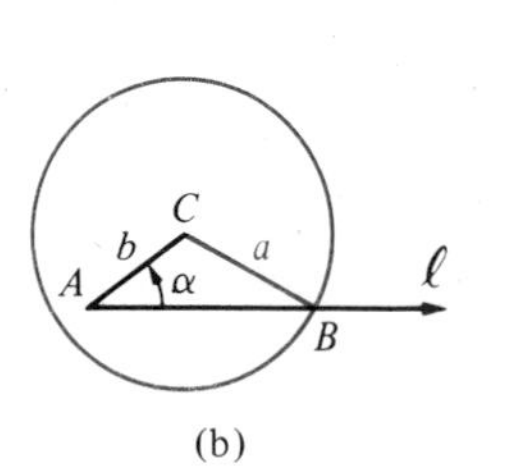

(b)

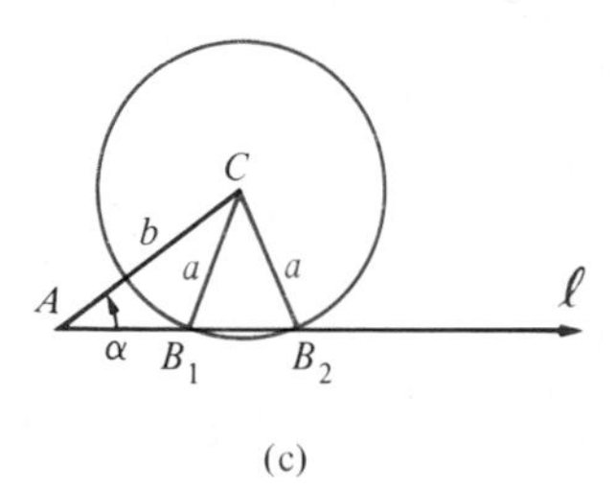

(c)

In the ambiguous case, you can use a calculator (or Appendix Table IIA) to solve the triangle ABC without bothering to draw any diagrams at all. Just use the law of sines,

$$\frac{\sin\alpha}{a} = \frac{\sin\beta}{b},$$

to evaluate $\sin\beta$:

$$\sin\beta = b\,\frac{\sin\alpha}{a}, \qquad 0 < \beta < 180^\circ.$$

Recall that the sine of an angle is never greater than 1; hence, if $b\,\dfrac{\sin\alpha}{a} > 1$, this trigonometric equation has no solution, and *no triangle satisfies the given conditions.* If $b\,\dfrac{\sin\alpha}{a} = 1$, the equation has only one solution, $\beta = 90^\circ$. If $b\,\dfrac{\sin\alpha}{a} < 1$, the trigonometric equation has two solutions (Problem 39), namely

$$\beta_1 = \sin^{-1}\left(b\,\frac{\sin\alpha}{a}\right) \quad \text{and} \quad \beta_2 = 180^\circ - \beta_1.$$

Once you have determined β (or β_1 and β_2), you know two angles and two sides of the triangle (or triangles), and you can solve the triangle by using the methods previously explained. Even if there are two solutions β_1 and β_2 of the trigonometric equation for β, it is possible that only one of these solutions corresponds to an actual triangle satisfying the given conditions (see Figure 6b).

Examples Solve each triangle by using the method for the ambiguous case.

1 © $a = 95, \quad b = 117, \quad \alpha = 65°$

Solution Here,

$$b\frac{\sin\alpha}{a} = 117\frac{\sin 65°}{95} = 1.116 > 1,$$

so there is *no triangle* satisfying the given conditions.

2 $a = 10, \quad b = 20, \quad \alpha = 30°$

Solution Because

$$\sin\beta = b\frac{\sin\alpha}{a} = 20\frac{\sin 30°}{10} = 20\frac{1/2}{10} = 1,$$

it follows that $\beta = 90°$, and the triangle is a right triangle with hypotenuse $b = 20$ (Figure 7). Therefore,

$$\gamma = 90° - \alpha = 90° - 30° = 60°.$$

Figure 7

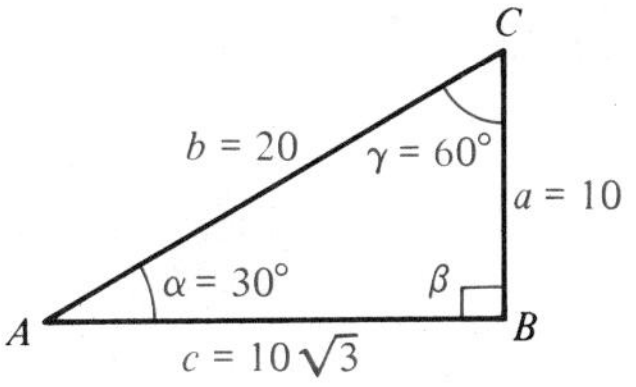

Also, by the Pythagorean theorem,

$$c^2 + a^2 = b^2,$$

so

$$\begin{aligned} c &= \sqrt{b^2 - a^2} \\ &= \sqrt{20^2 - 10^2} \\ &= \sqrt{300} = 10\sqrt{3}. \end{aligned}$$

3 © $a = 10, \quad b = 11, \quad \alpha = 57°$

Solution Here

$$b\frac{\sin\alpha}{a} = 11\frac{\sin 57°}{10} = 0.9225 < 1.$$

Thus,

$$\beta_1 = \sin^{-1}\left(b\frac{\sin\alpha}{a}\right) = \sin^{-1}\left(11\frac{\sin 57°}{10}\right) = 67.30°$$

and

$$\beta_2 = 180° - \beta_1 = 180° - 67.30° = 112.70°.$$

Figure 8

(a)

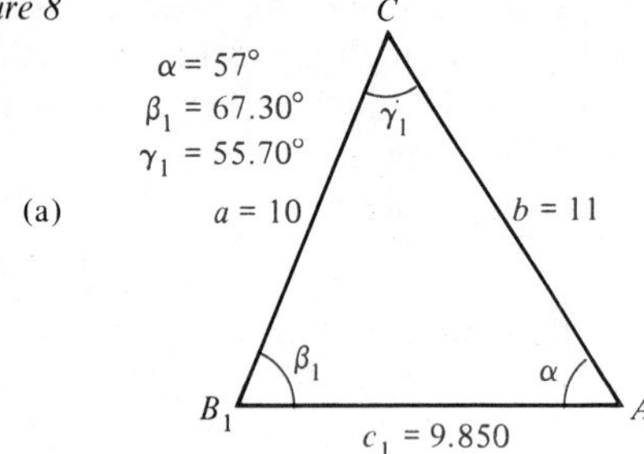

(b)

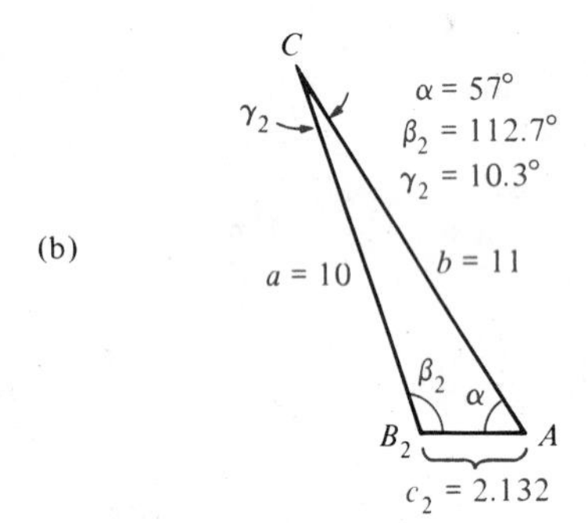

The solution $\beta_1 = 67.30°$ leads to a triangle AB_1C with

$$\gamma_1 = 180° - (\alpha + \beta_1)$$
$$= 180° - (57° + 67.30°) = 55.70°$$

and because $\dfrac{c_1}{\sin \gamma_1} = \dfrac{a}{\sin \alpha}$

$$c_1 = \frac{a \sin \gamma_1}{\sin \alpha} = \frac{10 \sin 55.70°}{\sin 57°} = 9.850$$

(Figure 8a). The solution $\beta_2 = 112.70°$ leads to a second triangle AB_2C with

$$\gamma_2 = 180° - (\alpha + \beta_2)$$
$$= 180° - (57° + 112.70°) = 10.30°$$

and because $\dfrac{c_2}{\sin \gamma_2} = \dfrac{a}{\sin \alpha}$

$$c_2 = \frac{a \sin \gamma_2}{\sin \alpha} = \frac{10 \sin 10.3°}{\sin 57°} = 2.132$$

(Figure 8b).

4 [C] $a = 70, \quad b = 65, \quad \alpha = 40°$

Solution Here

$$b \frac{\sin \alpha}{a} = 65 \frac{\sin 40°}{70} = 0.5969 < 1.$$

Thus,

$$\beta_1 = \sin^{-1}\left(b \frac{\sin \alpha}{a}\right) = \sin^{-1}\left(65 \frac{\sin 40°}{70}\right) = 36.65°$$

and

$$\beta_2 = 180° - \beta_1 = 143.35°.$$

Because it is impossible to have an angle $\beta_2 = 143.35°$ in a triangle that already has an angle of 40° (143.35° + 40° > 180°), the solution $\beta_2 = 143.35°$ does not lead to an actual triangle. However, the solution $\beta = \beta_1 = 36.65°$ leads to

$$\gamma = 180° - (\alpha + \beta) = 180° - (40° + 36.65°) = 103.35°$$

Figure 9

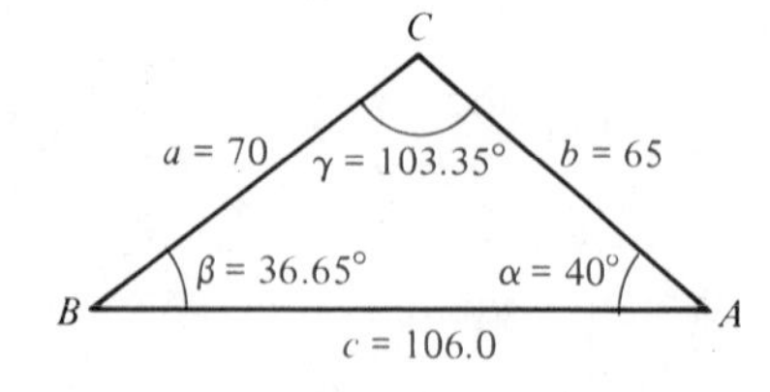

and because $\dfrac{c}{\sin \gamma} = \dfrac{a}{\sin \alpha}$

$$c = \frac{a \sin \gamma}{\sin \alpha} = \frac{70 \sin 103.35°}{\sin 40°} = 106.0$$

(Figure 9).

PROBLEM SET 2

C In problems 1–8, use the law of sines to solve each triangle ABC. Round off angles to the nearest hundredth of a degree and side lengths to four significant digits.

1. $a = 32, \beta = 57°, \gamma = 38°$
2. $b = 16, \alpha = 120°, \gamma = 30°$
3. $b = 24, \beta = 38°, \gamma = 21°$
4. $a = 42.9, \alpha = 32°, \beta = 81.5°$
5. $b = 47.3, \alpha = 48.31°, \gamma = 57.82°$
6. $a = 120, \alpha = 132°, \beta = 61°$
7. $a = 95, \alpha = 132°, \beta = 24.45°$
8. $b = 67.5, \alpha = 61.25°, \beta = 65.75°$

C In problems 9–16, solve each triangle ABC by using the method for the ambiguous case. Round off angles to the nearest hundredth of a degree and side lengths to four significant digits. Be certain to find all possible triangles that satisfy the given conditions.

9. $a = 50, b = 30, \alpha = 45°$
10. $a = 40, b = 70, \alpha = 30°$
11. $a = 31, b = 33, \alpha = 60°$
12. $a = 10, b = 26, \beta = 110°$
13. $a = 27, b = 52, \alpha = 70°$
14. $b = 4.5, c = 9, \gamma = 60°$
15. $a = 65.52, b = 55.51, \alpha = 111.5°$
16. $a = 12.41, b = 81.69, \beta = 36.67°$

C In problems 17–24, solve each triangle ABC. Round off angles to the nearest hundredth of a degree and side lengths to four significant digits. In the ambiguous cases, be certain to find all possible triangles that satisfy the given conditions.

17. $b = 52.05$ feet, $\alpha = 42.85°, \beta = 61.50°$
18. $b = 194.5$ meters, $\alpha = 82.25°, \beta = 69.45°$
19. $a = 110$ kilometers, $b = 100$ kilometers, $\alpha = 59.33°$
20. $a = 8.143 \times 10^6$ meters, $b = 1.271 \times 10^7$ meters, $\alpha = 34.65°$
21. $a = 2090$ meters, $b = 5579$ meters, $\alpha = 22°$
22. $a = 2.011 \times 10^{-10}$ meter, $b = 1.509 \times 10^{-10}$ meter, $\beta = 30.2°$
23. $a = 10.07$ kilometers, $b = 15.15$ kilometers, $\alpha = 67.67°$
24. $c = 0.4351$ meter, $\beta = 48.50°, \gamma = 96.25°$

C **25.** A surveyor lays out a baseline segment $\overline{AB}$ of length 346 meters. An inaccessible point C forms the third vertex of a triangle ABC. If the surveyor determines that angle $CAB = 62°40'$ and that angle $CBA = 54°30'$, find $|\overline{AC}|$ and $|\overline{BC}|$.

C **26.** The crankshaft $\overline{OA}$ of an engine is 2 inches long and the connecting rod $\overline{AP}$ is 8 inches long (Figure 10). If angle APO is 10° at a certain instant, find angle AOP at this instant. [Caution: There may be more than one solution.]

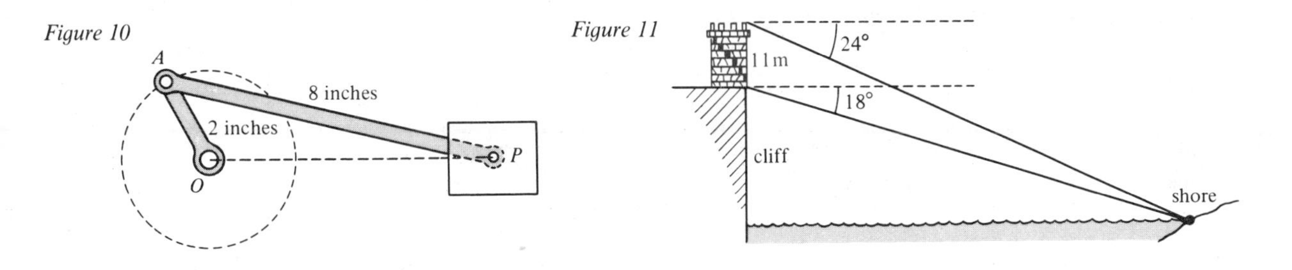

Figure 10

Figure 11

C **27.** An observation tower stands on the edge of a vertical cliff rising directly above a river (Figure 11). From the top of the tower, the angle of depression of a point on the opposite shore of the river is 24°; from the base of the tower, the angle of depression of the same point is 18°. If the top of the tower is 11 meters above its base, how wide is the river?

C **28.** Plans for an oil pipeline require that it be diverted, as shown in Figure 12, around the mating grounds of a caribou herd. Suppose that angle $BAC = 22°$, angle $ABC = 48°$, and $|\overline{AB}| = 7$ kilometers. If the pipeline costs $200,000 per kilometer, find the additional cost of the pipeline because of the diversion.

Figure 12

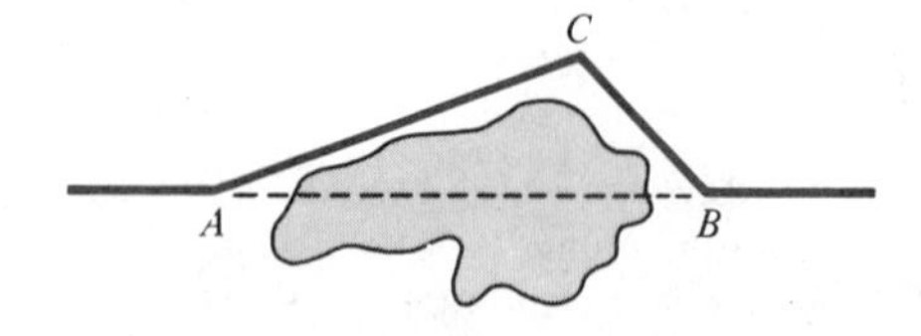

C **29.** A nature photographer spots a rare northern shrike perched in a tree at an angle of elevation of 14°. Using the range finder on her camera, she determines that the shrike is 103 meters from her lens. She creeps toward the bird until its angle of elevation is 20°. Find the distance between the lens and the bird in the photographer's new position.

C **30.** A commercial jet is flying at constant speed on a level course that will carry it directly over Albuquerque, New Mexico. Through a hole in the cloud cover, the captain observes the lights of the city at an angle of depression of 7°. Three minutes later, the captain catches a second glimpse of the city, now at an angle of depression of 13°. In how many *more* minutes will the jet be directly over the city?

C **31.** A sports parachutist is sighted simultaneously by two observers who are 5 kilometers apart. The observed angles of elevation are 26.5° and 18.2°. Assuming that the parachutist and the two observers lie in the same vertical plane, how high above the ground is the parachutist?

C **32.** Riding at constant speed across the desert on his faithful camel Clyde, Lawrence sights the top of a tall palm tree. He assumes that the palm tree is located at an oasis, and heads directly toward it. When first sighted, the top of the palm tree was at an angle of elevation of 4°. Twenty minutes later, the angle of elevation has increased to 9°. In how many *more* minutes will Lawrence reach the palm tree?

C **33.** A CB transmitter is illegally transmitting at excessive power from point C, and it is under surveillance by two FCC agents situated 3.8 kilometers apart at points A and B. Using direction finders, the agents find that angle $CAB = 40.1°$ and angle $CBA = 31.8°$. How far is the illegal transmitter from the agent at point A?

C **34.** Two Coast Guard stations A and B are located along a straight coastline. Station A is 20 miles due north of station B. A supertanker in distress is observed from station A at a bearing of S9°10′E, and from station B at a bearing of S37°40′E. Find the distance from the supertanker to the nearest point on the coastline.

C **35.** Two joggers running at the same constant speed set out from the same point C. They run along separate straight paths toward a straight highway. The first jogger arrives at point A along the highway after running for 70 minutes. Some time later, the second jogger arrives at point B along the highway. If angle CAB is 105° and angle CBA is 50°, how many minutes were required by the second jogger to run from C to B?

36. In triangle ABC, use the law of sines and the fact that $\sin\gamma = \sin[180° - (\alpha + \beta)]$ to prove that $c = b\cos\alpha + a\cos\beta$.

37. In the proof of Theorem 1, suppose that angle α is not acute. Redraw Figure 2 to illustrate this situation and then show that $\dfrac{\sin\alpha}{a} = \dfrac{\sin\beta}{b}$ is still true. [Hint: $\sin(180° - \alpha) = \sin\alpha$.]

38. Complete the proof of Theorem 1 by showing that $\dfrac{\sin\beta}{b} = \dfrac{\sin\gamma}{c}$.

39. If $0 < \beta_1 < 90°$ and $\beta_2 = 180° - \beta_1$, show that $90° < \beta_2 < 180°$ and $\sin\beta_1 = \sin\beta_2$.

40. Use the law of sines to establish the **law of tangents:** For any triangle ABC,

$$\frac{a-b}{a+b} = \frac{\tan\left(\frac{\alpha-\beta}{2}\right)}{\tan\left(\frac{\alpha+\beta}{2}\right)}.$$

Begin by proving each of the following:

(a) $\frac{a-b}{b} = \frac{\sin\alpha - \sin\beta}{\sin\beta}$ (b) $\frac{a+b}{b} = \frac{\sin\alpha + \sin\beta}{\sin\beta}$ (c) $\frac{a-b}{a+b} = \frac{\sin\alpha - \sin\beta}{\sin\alpha + \sin\beta}$

[C] In problems 41–43, use the law of tangents in problem 40 to solve triangle ABC.

41. $a = 9, \beta = 73°, \alpha = 69°$ **42.** $b = 27, \beta = 33°, \gamma = 77°$ **43.** $a = 15, \gamma = 67°, \alpha = 81°$

44. The identity $(a - b)\cos\frac{\gamma}{2} = c\sin\left(\frac{\alpha-\beta}{2}\right)$ is called **Mollweide's formula.** It is often used to check the solution of a triangle ABC because it contains all six parts of the triangle. If, when values are substituted in Mollweide's formula, the result is not a true statement, then an error has been made in solving the triangle. Prove Mollweide's formula.

[C] **45.** Use Mollweide's formula to check your solutions of the triangles in problems 41 and 43.

46. (a) Show that the area $\mathscr{A}$ of triangle ABC is given by the formula $\mathscr{A} = \frac{1}{2}ab\sin\gamma$. [C] (b) Use the formula in part (a) to find the area of triangle ABC if $a = 12.7$ meters, $b = 8.91$ meters, and $\gamma = 46.5°$.

3 The Law of Cosines

The law of sines is not effective for solving a triangle when *two sides and the angle between them are given,* or when *all three sides are given.* However, the following *law of cosines* can be used in these cases.

Theorem 1 **The Law of Cosines**

In triangle ABC (Figure 1):

(i) $c^2 = a^2 + b^2 - 2ab\cos\gamma$

(ii) $a^2 = b^2 + c^2 - 2bc\cos\alpha$

(iii) $b^2 = a^2 + c^2 - 2ac\cos\beta$

Figure 1

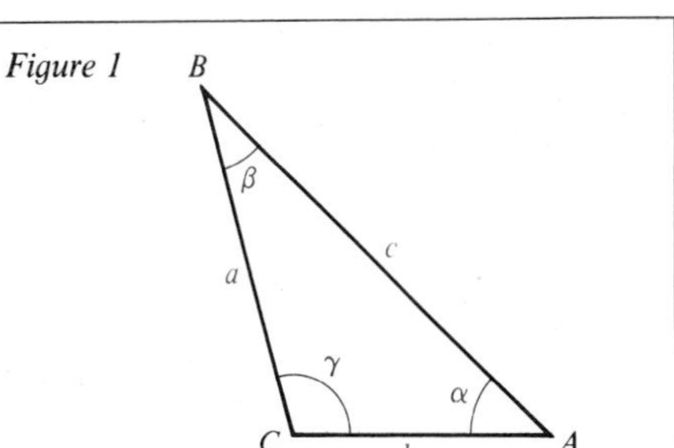

Proof We prove (i) here; (ii) and (iii) are proved in exactly the same way. To begin with, we place triangle ABC on a cartesian coordinate system with vertex C at the origin and vertex A on the positive x axis (Figure 2). Then vertex B has coordinates $B = (a\cos\gamma, a\sin\gamma)$. (Why?) Because $A = (b, 0)$, it follows from the distance formula that

Figure 2

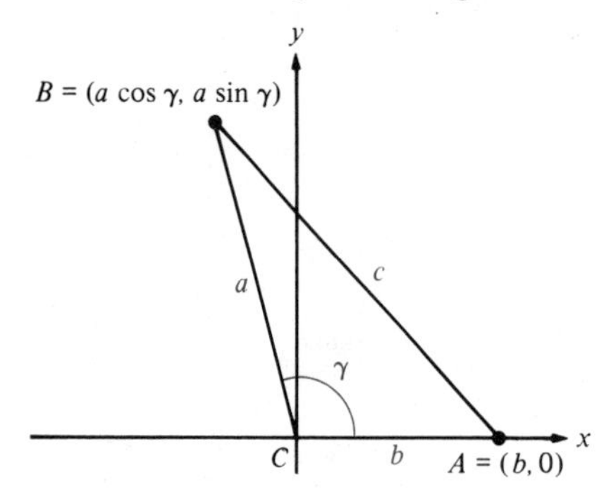

$$\begin{aligned} c^2 &= (a\cos\gamma - b)^2 + (a\sin\gamma - 0)^2 \\ &= a^2\cos^2\gamma - 2ab\cos\gamma + b^2 + a^2\sin^2\gamma \\ &= a^2(\cos^2\gamma + \sin^2\gamma) + b^2 - 2ab\cos\gamma \\ &= a^2 + b^2 - 2ab\cos\gamma. \end{aligned}$$

Rather than memorizing the three formulas in Theorem 1, we recommend that you learn the following statement, which summarizes all three formulas.

The Law of Cosines

> The square of the length of any side of a triangle is equal to the sum of the squares of the lengths of the other two sides minus twice the product of the lengths of these two sides and the cosine of the angle between them.

It's interesting to notice that, when the angle mentioned above is 90°, the law of cosines reduces to the Pythagorean theorem (Problem 35).

Example 1 Ⓒ In triangle ABC (Figure 3), find c to four significant digits if $a = 5$, $b = 10$, and $\gamma = 37.85°$.

Solution By the law of cosines,

$$\begin{aligned} c^2 &= a^2 + b^2 - 2ab\cos\gamma = 5^2 + 10^2 - 2(5)(10)\cos 37.85° \\ &= 25 + 100 - 100\cos 37.85° = 46.04. \end{aligned}$$

Therefore,

$$c = \sqrt{46.04} = 6.785.$$

Figure 3

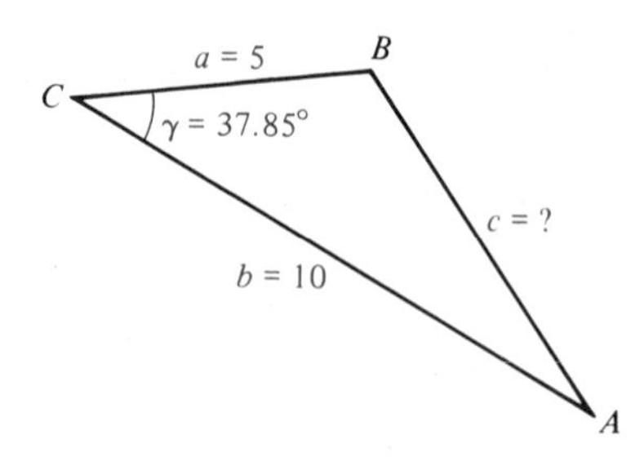

Example 2 Ⓒ In triangle ABC (Figure 4), find β to the nearest hundredth of a degree if $a = 3$, $b = 4$, and $c = 2$.

Figure 4

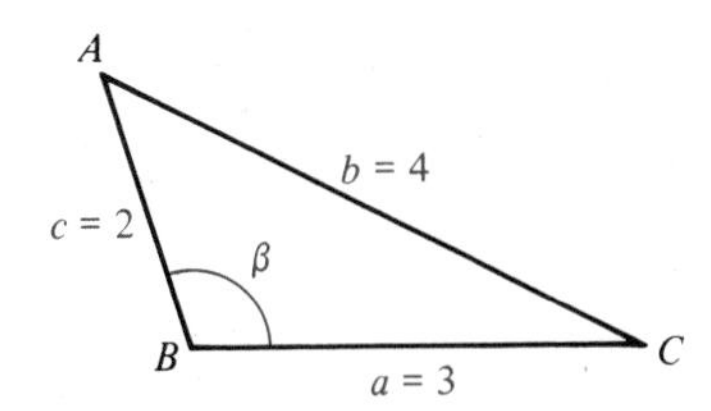

Solution We begin by writing the law of cosines in the form that contains $\cos\beta$:

$$b^2 = a^2 + c^2 - 2ac\cos\beta.$$

Solving this equation for $\cos\beta$, we obtain

$$\cos\beta = \frac{a^2 + c^2 - b^2}{2ac}.$$

Therefore,

$$\beta = \cos^{-1}\frac{a^2 + c^2 - b^2}{2ac} = \cos^{-1}\frac{3^2 + 2^2 - 4^2}{2(3)(2)}$$

$$= \cos^{-1}\frac{-3}{12} = \cos^{-1}\left(-\frac{1}{4}\right) = 104.48^\circ.$$

Example 3 Ⓒ In order to find the horizontal distance between two towers A and B that are separated by a grove of trees, a point C that is readily accessible from both A and B is chosen (Figure 5), and the following measurements are made: $|\overline{CA}| = 341$ meters, $|\overline{CB}| = 549$ meters, angle $ACB = 115.5^\circ$. Find $|\overline{AB}|$ to four significant digits.

Figure 5

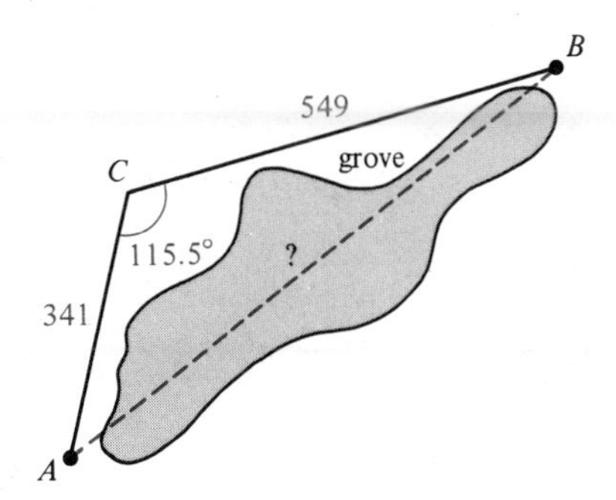

Solution By the law of cosines,

$$|\overline{AB}|^2 = |\overline{CA}|^2 + |\overline{CB}|^2 - 2|\overline{CA}||\overline{CB}|\cos(\text{angle } ACB)$$

$$= 341^2 + 549^2 - 2(341)(549)\cos 115.5^\circ;$$

hence,

$$|\overline{AB}| = \sqrt{341^2 + 549^2 - 2(341)(549)\cos 115.5^\circ} = 760.8 \text{ meters}.$$

The law of cosines can be used to establish a number of interesting formulas involving the geometry of triangles (see Problems 36–42). One of the most useful of these is **Hero's** (or Heron's) **area formula:**

$$\mathscr{A} = \sqrt{s(s-a)(s-b)(s-c)} \qquad \text{where } s = \tfrac{1}{2}(a+b+c),$$

which gives the area $\mathscr{A}$ of a triangle with sides of lengths a, b, and c (Problem 40). The quantity s, which is one-half of the perimeter $a + b + c$ of the triangle, is called the *semiperimeter.*

Example Ⓒ A cornfield that is bounded by three straight roads has side lengths $a = 600$ feet, $b = 400$ feet, and $c = 500$ feet (Figure 6). Find the area $\mathscr{A}$ of the cornfield.

Solution We use Hero's formula. The semiperimeter of the cornfield is

Figure 6

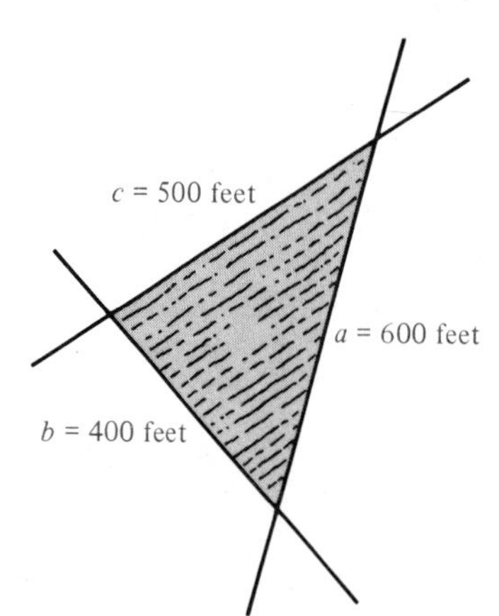

$$\begin{aligned} s &= \tfrac{1}{2}(a+b+c) \\ &= \tfrac{1}{2}(600+400+500) \\ &= 750. \end{aligned}$$

Therefore,

$$\begin{aligned} \mathscr{A} &= \sqrt{s(s-a)(s-b)(s-c)} \\ &= \sqrt{750(750-600)(750-400)(750-500)} \\ &= \sqrt{750(150)(350)(250)} = \sqrt{9{,}843{,}750{,}000} \\ &\approx 99{,}216 \text{ square feet.} \end{aligned}$$

PROBLEM SET 3

Ⓒ In problems 1–12, use the law of cosines to find the specified unknown part of the triangle ABC (Figure 1). Round off angles to the nearest hundredth of a degree and side lengths to four significant digits.

1. Find c if $a = 3$, $b = 10$, and $\gamma = 60°$.
2. Find a if $b = 68$, $c = 14$, and $\alpha = 24.5°$.
3. Find b if $a = 23$, $c = 31$, and $\beta = 112°$.
4. Find a if $b = 3.2$, $c = 2.4$, and $\alpha = 117°$.
5. Find a if $c = 5.78$, $b = 4.78$, and $\alpha = 35.25°$.
6. Find β if $a = 200$, $b = 50$, and $c = 177$.
7. Find γ if $a = 2$, $b = 3$, and $c = 4$.
8. Find α if $a = 2\sqrt{61}$, $b = 8$, and $c = 10$.
9. Find β if $a = 1240$, $b = 876$, and $c = 918$.
10. Find γ if $a = 0.64$, $b = 0.27$, and $c = 0.49$.
11. Find α if $a = 189$, $b = 214$, and $c = 325$.
12. Find β if $a = 189$, $b = 214$, and $c = 325$.

Ⓒ In problems 13–18, use the law of cosines or the law of sines (or both) to solve each triangle ABC. Round off angles to the nearest hundredth of a degree and side lengths to four significant digits. In the ambiguous cases, be certain to find all possible triangles that satisfy the given conditions.

13. $a = 327$, $b = 251$, $\gamma = 72.45°$
14. $a = 100$, $b = 150$, $c = 300$
15. $a = 312$, $b = 490$, $\alpha = 33.75°$
16. $a = 87$, $c = 124$, $\alpha = 35.33°$
17. $a = 321$, $b = 456$, $c = 654$
18. $a = 59$, $b = 22$, $\beta = 117°20'$

[C] **19.** To find the distance between two points A and B, a surveyor chooses a point C that is 52 meters from A and 64 meters from B. If angle ACB is $72°20'$, find the distance between A and B. Round off your answer to four significant digits.

[C] **20.** A tracking antenna A on the surface of the earth is following a spacecraft B (Figure 7). The tracking antenna is aimed 36° above the horizon, and the distance from the antenna to the spacecraft is 6521 miles. If the radius of the earth is 3959 miles, how high is the spacecraft above the surface of the earth?

Figure 7

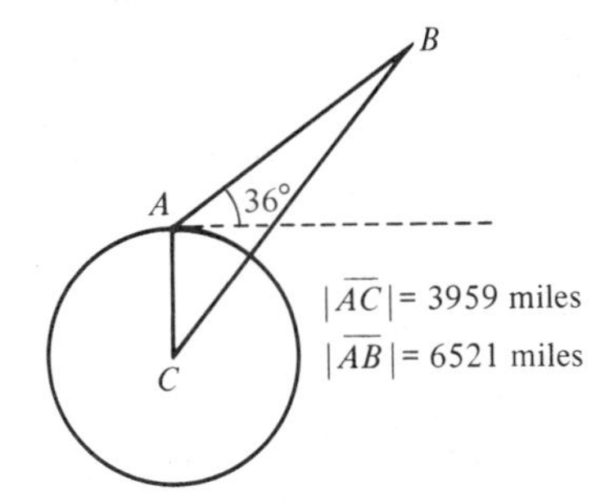

[C] **21.** Engineers are planning to construct a straight tunnel from point A on one side of a hill to point B on the other side. From a third point C, the distances $|\overline{CA}| = 837$ meters and $|\overline{CB}| = 1164$ meters are measured. If angle ACB is measured as 44.5°, find the length of the tunnel when it is completed. Round off your answer to four significant digits.

[C] **22.** A molecule of ammonia (NH_3) consists of three hydrogen atoms and one nitrogen atom, which form the vertices of a pyramid with four triangular faces. In one of the faces containing the nitrogen atom (N) and two hydrogen atoms (H), the distance between the two H atoms is 1.628 angstroms and the distance from the N atom to either H atom is 1.014 angstroms (Figure 8). [One angstrom (abbreviated Å) is 10^{-10} meter.] Find the angle formed by the two line segments joining the H atoms to the N atom.

Figure 8

N
1.014 Å   1.014 Å
H   H
1.628 Å

[C] **23.** Two jet fighters leave an air base at the same time and fly at the same speed along straight courses, forming an angle of 135.65° with each other. After the jets have each flown 402 kilometers, how far apart are they?

[C] **24.** A ship leaves a harbor at noon and sails S52°W at 22 knots (nautical miles per hour) until 2:30 P.M. At that time it changes its course and sails N22°W at a reduced speed of 18 knots until 5:00 P.M. At 5:00 P.M., how far is the ship from the harbor and what is its bearing from the harbor?

[C] **25.** Two straight roads cross at an angle of 75°. A bus on one road is 5 miles from the intersection and moving away from it at 50 miles per hour. At the same instant, a truck on the other road is 10 miles from the intersection and moving away from it at the rate of 55 miles per hour (Figure 9). If the bus and truck maintain constant speeds, what is the distance between them after 45 minutes?

Figure 9

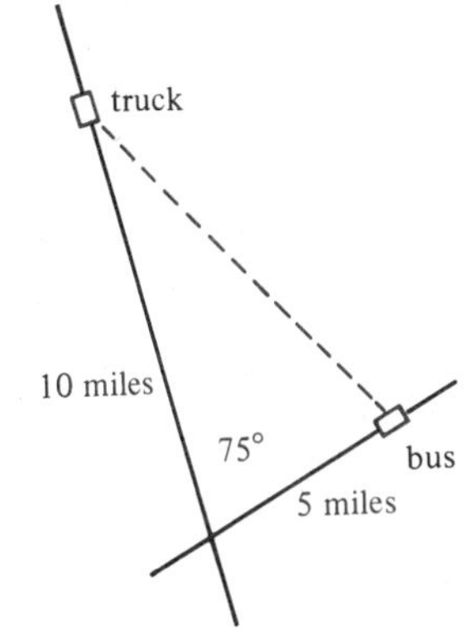

[C] **26.** The navigator of an oil tanker plots a straight-line course from point A to point B. Because of an error, the tanker proceeds from point A along a slightly different straight-line course. After sailing for 5 hours on the wrong course, the error is discovered and the tanker is turned through an angle of 23° so that it is now heading directly toward point B. After 3 more hours, the tanker reaches point B. Assuming that the tanker was proceeding at a constant speed, calculate the time lost because of the error.

[C] In problems 27–30, use Hero's area formula to find the area of a triangle with the given side lengths a, b, and c. Round off all answers to four significant digits.

27. $a = 5$ centimeters, $b = 8$ centimeters, $c = 7$ centimeters

28. $a = 7.301$ kilometers, $b = 11.15$ kilometers, $c = 15.22$ kilometers

29. $a = 13.03$ feet, $b = 15.77$ feet, $c = 17.20$ feet
30. $a = 1.110 \times 10^5$ meters, $b = 9.812 \times 10^4$ meters, $c = 1.728 \times 10^5$ meters

Ⓒ 31. Offshore drilling rights are granted to RIPCO oil company in a triangular region bounded by sides of lengths 33 kilometers, 28 kilometers, and 7 kilometers. Find the area of this region.

Ⓒ 32. A bird sanctuary along the straight shoreline of a lake is bounded by straight line segments $\overline{AB}$, $\overline{BC}$, and $\overline{CD}$ as shown in Figure 10. Suppose that $|\overline{AB}| = 320$ meters, $|\overline{BC}| = 200$ meters, $|\overline{CD}| = 360$ meters, angle $ABC = 146°$, and angle $BCD = 77°$. Find (a) the area of the bird sanctuary and (b) the distance $|\overline{AD}|$ along the shoreline. [Hint: Break the quadrilateral $ABCD$ into two triangles—for instance, ABC and ACD—then solve these triangles.]

Figure 10

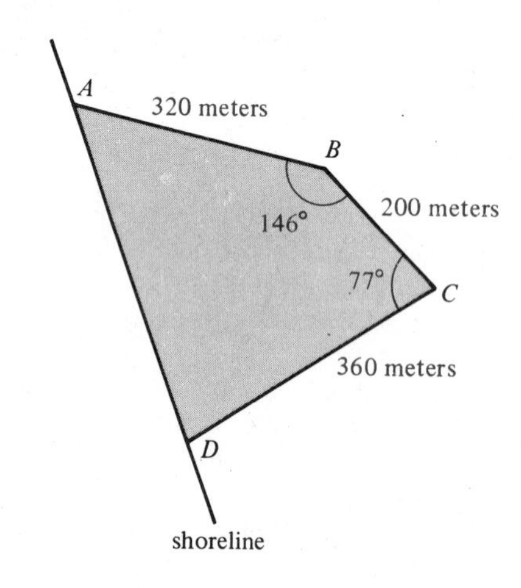

Ⓒ 33. The Great Pyramid of Cheops at Al Giza, Egypt, has four congruent triangular faces. Each face is an isosceles triangle with equal sides of length 219 meters and base 230 meters (Figure 11). Find the *total* surface area of the pyramid.

Ⓒ 34. In problem 33, find (a) the height h of the pyramid and (b) the volume V of the pyramid. [Hint: The volume of a pyramid is one-third the height times the area of the base.]

Figure 11

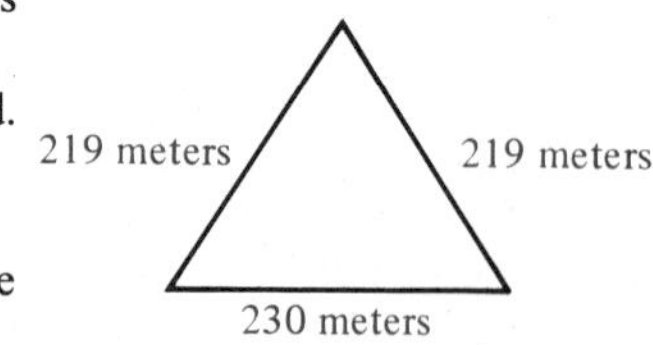

35. Show that the Pythagorean theorem is a special case of the law of cosines.
36. Let $s = \frac{1}{2}(a + b + c)$ be the semiperimeter of triangle ABC (Figure 12). Combine the law of cosines

$$a^2 = b^2 + c^2 - 2bc\cos\alpha$$

and the half-angle formula

$$\cos^2\frac{\alpha}{2} = \frac{1 + \cos\alpha}{2}$$

to obtain the formula

$$\cos\frac{\alpha}{2} = \sqrt{\frac{s(s-a)}{bc}}.$$

37. For every triangle ABC (Figure 12), show that

$$\frac{\cos\alpha}{a} + \frac{\cos\beta}{b} + \frac{\cos\gamma}{c} = \frac{a^2 + b^2 + c^2}{2abc}.$$

Figure 12

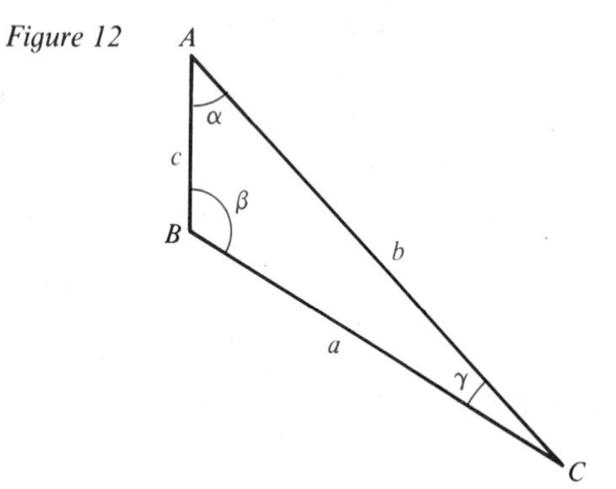

38. Proceeding as in problem 36, but using the half-angle formula

$$\sin^2 \frac{\alpha}{2} = \frac{1 - \cos \alpha}{2},$$

show that

$$\sin \frac{\alpha}{2} = \sqrt{\frac{(s-b)(s-c)}{bc}}.$$

39. For every triangle ABC (Figure 12), show that γ is an acute angle if and only if $a^2 + b^2 > c^2$.

40. In Figure 13, the area $\mathscr{A}$ of triangle ABC is given by $\mathscr{A} = \frac{1}{2}hb$.

(a) Show that $\mathscr{A} = \frac{1}{2}bc \sin \alpha$.

(b) Using the double-angle formula, show that

$$\sin \alpha = 2 \sin \frac{\alpha}{2} \cos \frac{\alpha}{2}.$$

(c) Combine (a) and (b) to obtain

$$\mathscr{A} = bc \sin \frac{\alpha}{2} \cos \frac{\alpha}{2}.$$

(d) Combine (c), problem 36, and problem 38 to obtain Hero's formula

$$\mathscr{A} = \sqrt{s(s-a)(s-b)(s-c)}.$$

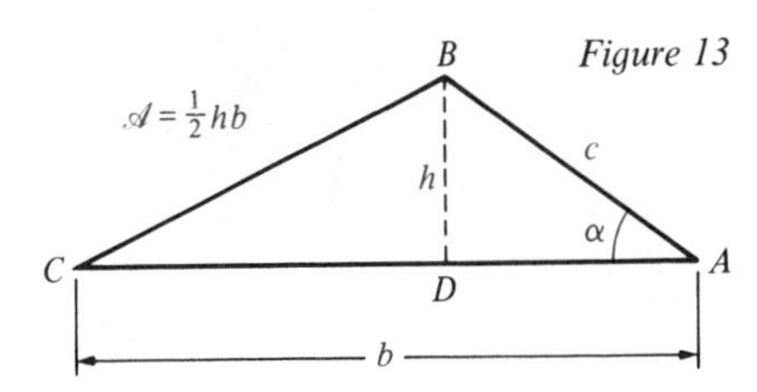

Figure 13

41. Suppose you are given three positive numbers a, b, and c. Under what conditions will there exist a triangle ABC with side lengths a, b, and c? [Hint: Under what conditions could you form a triangle with three sticks of wood of lengths a, b, and c?]

42. The center O of the circle inscribed in a triangle ABC always lies at the intersection of the bisectors of the three vertex angles (Figure 14). If $s = \frac{1}{2}(a + b + c)$ is the semiperimeter of triangle ABC, show that the radius r of the inscribed circle is given by

$$r = \sqrt{\frac{(s-a)(s-b)(s-c)}{s}}.$$

[Hint: Use Hero's formula and the fact that the area of triangle ABC is the sum of the areas of triangles AOB, BOC, and COA.]

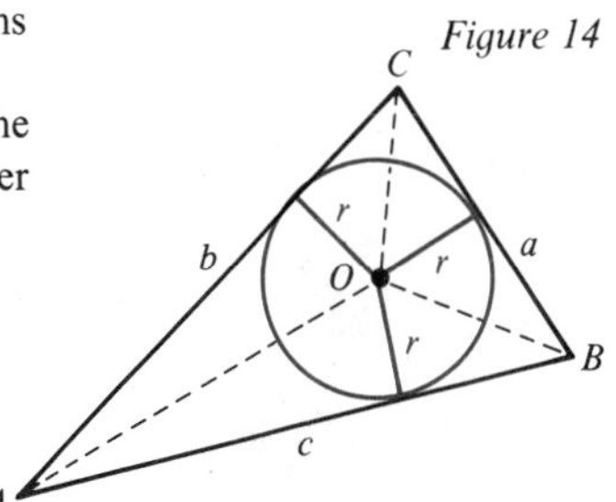

Figure 14

4 Vectors

Many applications of mathematics involve force, velocity, acceleration, or other quantities that have both a *magnitude* and a *direction*. For instance, if you push on a stone, the force you exert has both a magnitude (say, 5 pounds) and a direction (say, due east). You can represent this force by an arrow that has a length (5 units) corresponding to the magnitude and that points in the appropriate direction (due east) (Figure 1).

Figure 1

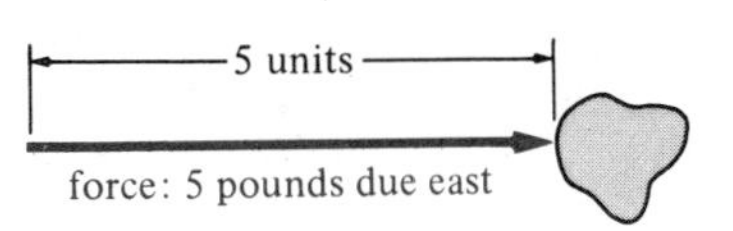

Figure 2

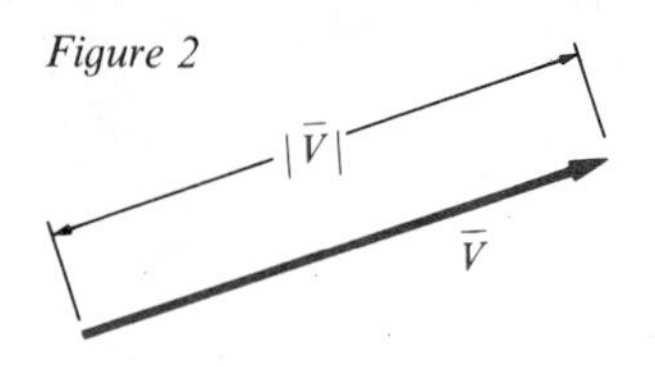

More generally, a **vector** can be thought of as an arrow, that is, a line segment with a direction specified by an arrowhead. We use letters with bars, such as $\bar{A}$, $\bar{B}$, $\bar{C}$, and so forth, to denote vectors. If $\bar{V}$ is a vector, we denote the **length** or **magnitude** of $\bar{V}$ by $|\bar{V}|$ (Figure 2). A vector $\bar{V}$ with tail end or **initial point** A and head end or **terminal point** B may be written as

$$\bar{V} = \overrightarrow{AB}$$

Figure 3

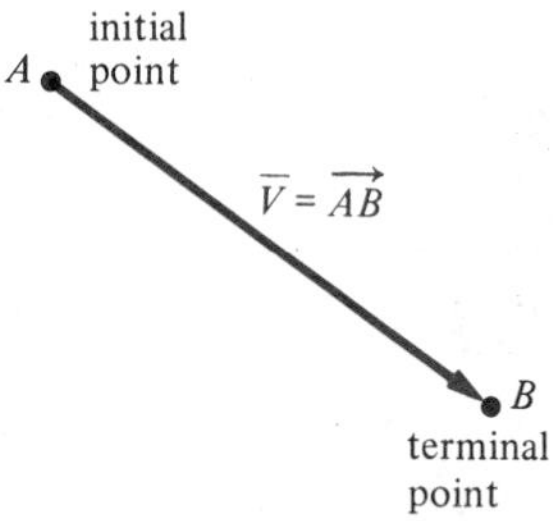

(Figure 3). This notation tells us that we are dealing with a *vector* $\overrightarrow{AB}$ rather than a line segment $\overline{AB}$, and it shows the direction of the vector—from the initial point A to the terminal point B. Notice that the magnitude

$$|\bar{V}| = |\overrightarrow{AB}| = |\overline{AB}|$$

is just the distance between A and B.

Let's agree to say that two vectors $\overrightarrow{AB}$ and $\overrightarrow{CD}$ are **equal** and to write

$$\overrightarrow{AB} = \overrightarrow{CD}$$

if they are parallel and have the same direction and magnitude (Figure 4). In other words, you may move a vector around freely if you keep it parallel to its original position. In particular, you can always move a vector $\bar{R}$ in the xy plane so that its tail end (initial point) is at the origin O (Figure 5). Then $\bar{R}$ will extend from O to its head end (terminal point) $P = (x, y)$, and we have

$$\bar{R} = \overrightarrow{OP}.$$

Figure 4

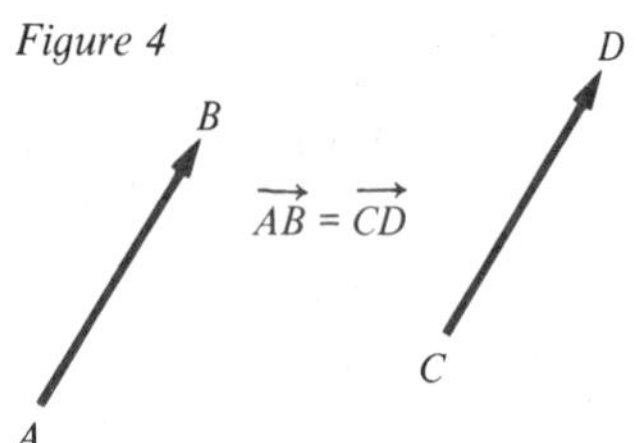

Figure 5

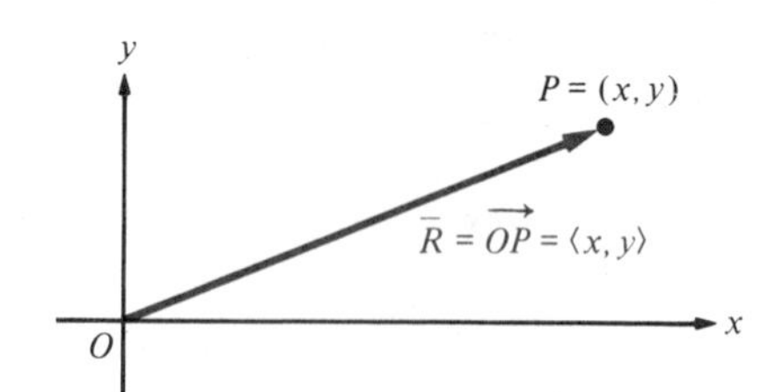

We say that $\bar{R} = \overrightarrow{OP}$ is the **position vector** of the point $P = (x, y)$; we call $\boldsymbol{x}$ and $\boldsymbol{y}$ the **components** of the vector $\bar{R}$; and we write

$$\bar{R} = \langle x, y\rangle.$$

The special symbol $\langle x, y\rangle$ is used to avoid confusing the *vector* $\langle x, y\rangle$ and the *point* $P = (x, y)$.

The **zero vector,** denoted by $\bar{0}$, is defined by $\bar{0} = \langle 0, 0\rangle$. You can think of $\bar{0}$ as an arrow whose length has shrunk to 0, so that its initial and terminal points coincide. Not only is $\bar{0}$ the only vector whose magnitude is 0, but it is the only vector whose direction is indeterminate.

Let $\bar{R}$ be a nonzero vector in the xy plane. Move $\bar{R}$, if necessary, so that its initial point is at the origin O. If $\bar{R}$ falls along the terminal side of an angle θ in standard position, we say that θ is a **direction angle** for $\bar{R}$ (Figure 6). Notice that any angle coterminal with θ is also a direction angle for $\bar{R}$. The vector $\bar{R}$ can be

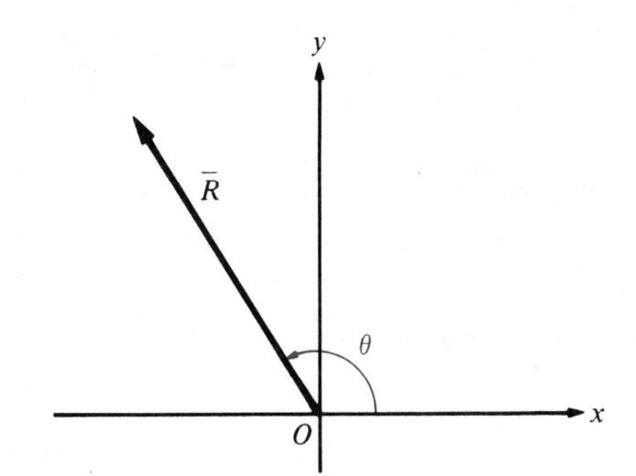

Figure 6

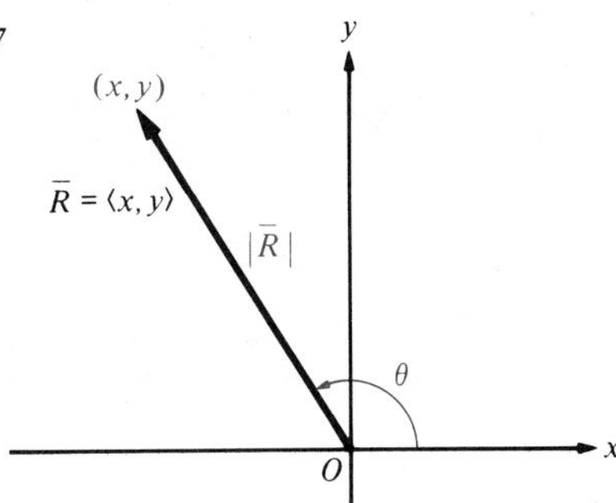

Figure 7

specified either by giving its components $\bar{R} = \langle x, y\rangle$ or by giving its magnitude (length) $|\bar{R}|$ and its direction angle θ (Figure 7). The quantities x, y, $|\bar{R}|$, and θ are related as follows:

> (i) $|\bar{R}| = \sqrt{x^2 + y^2}$
>
> (ii) $\cos\theta = \dfrac{x}{|\bar{R}|}$ or $x = |\bar{R}|\cos\theta$
>
> (iii) $\sin\theta = \dfrac{y}{|\bar{R}|}$ or $y = |\bar{R}|\sin\theta$
>
> (iv) $\tan\theta = \dfrac{y}{x}$ if $x \neq 0$.

(See Definition 1 on page 75.) These equations can be used to find $|\bar{R}|$ and θ in terms of x and y or vice versa.

Example 1 Ⓒ Sketch each vector, find its magnitude, and find its smallest positive direction angle θ.

(a) $\bar{A} = \langle 4, 3\rangle$ (b) $\bar{B} = \langle 2\sqrt{3}, -1\rangle$

Solution The two vectors, which extend from the origin to the points $(4, 3)$ and $(2\sqrt{3}, -1)$, are shown in Figure 8.

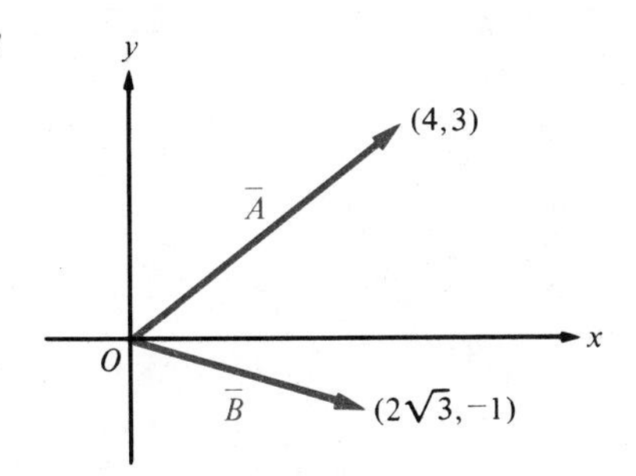

Figure 8

(a) By equation (i) above,

$$|\bar{A}| = \sqrt{4^2 + 3^2} = \sqrt{25} = 5 \text{ units.}$$

By equation (ii) above,

$$\cos\theta = \frac{4}{|\bar{A}|} = \frac{4}{5} = 0.8.$$

Because the point $(4, 3)$ is in quadrant I, it follows that θ is an acute angle; hence,

$$\theta = \cos^{-1} 0.8 \approx 36.87^\circ.$$

(b) $|\bar{B}| = \sqrt{(2\sqrt{3})^2 + (-1)^2} = \sqrt{13}$ units. Because the point $(2\sqrt{3}, -1)$ is in quadrant IV, it follows that $270° < \theta < 360°$. Also,

$$\cos\theta = \frac{2\sqrt{3}}{\sqrt{13}} = 2\sqrt{3/13},$$

so the reference angle θ_R corresponding to θ is given by

$$\theta_R = \cos^{-1} 2\sqrt{3/13} \approx 16.10°.$$

Therefore,

$$\theta = 360° - \theta_R \approx 360° - 16.10° = 343.90°.$$

Example 2 Find the components of the vector $\bar{V}$ if $|\bar{V}| = 3$ and a direction angle for $\bar{V}$ is $\theta = 150°$.

Solution We use equations (ii) and (iii) in the forms $x = |\bar{V}|\cos\theta$ and $y = |\bar{V}|\sin\theta$. Thus, we have

$$x = |\bar{V}|\cos\theta = 3\cos 150° = 3\left(-\frac{\sqrt{3}}{2}\right) = -\frac{3\sqrt{3}}{2}$$

and

$$y = |\bar{V}|\sin\theta = 3\sin 150° = 3\left(\frac{1}{2}\right) = \frac{3}{2}$$

so

$$\bar{V} = \left\langle -\frac{3\sqrt{3}}{2}, \frac{3}{2} \right\rangle.$$

As we mentioned earlier, forces can be represented by vectors: the length $|\bar{F}|$ of a force vector $\bar{F}$ is the magnitude of the force, while the direction of $\bar{F}$ indicates the direction of the force. Laboratory experiments have revealed the following: If two forces $\bar{F}_1$ and $\bar{F}_2$ act simultaneously on a particle, the effect is the same as if a single force $\bar{F}$, called the **resultant,** were acting on the particle. The resultant $\bar{F}$ forms a diagonal of the parallelogram in which $\bar{F}_1$ and $\bar{F}_2$ are adjacent edges (Figure 9). The resultant vector $\bar{F}$ is also called the **sum** of the two vectors $\bar{F}_1$ and $\bar{F}_2$, and we write

$$\bar{F} = \bar{F}_1 + \bar{F}_2.$$

Figure 9

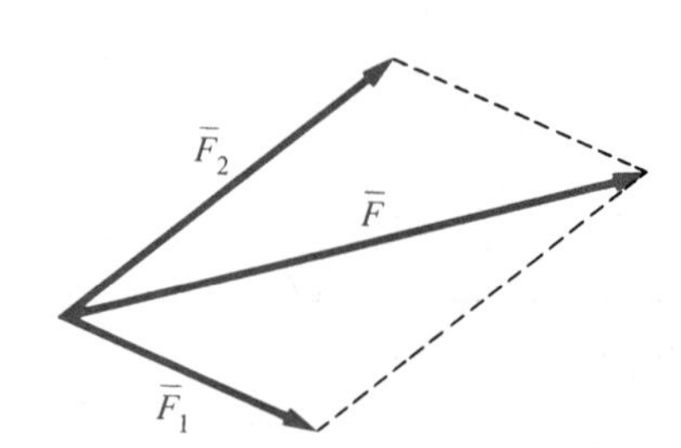

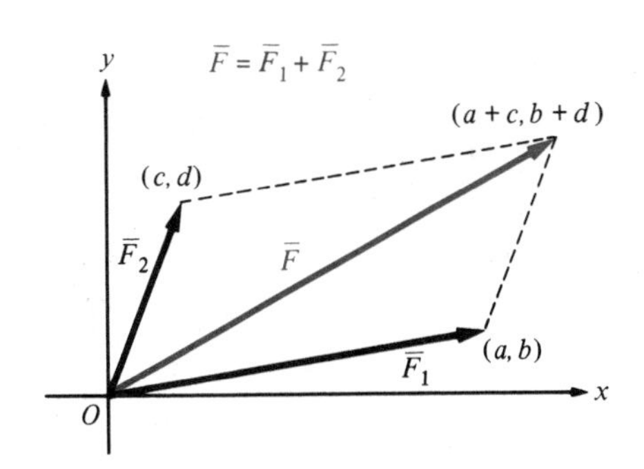

Figure 10

By applying the **parallelogram rule,** illustrated in Figure 9, you can form the sum $\bar{F} = \bar{F}_1 + \bar{F}_2$ of any two vectors, whether they represent forces or not. (Also, see Problem 31.) By examining Figure 10, you can see that *the components of the sum of two vectors are the sums of their corresponding components*, that is,

$$\langle a, b\rangle + \langle c, d\rangle = \langle a + c, b + d\rangle.$$

(See Problem 35).

Example If $\bar{A} = \langle 2, -7\rangle$ and $\bar{B} = \langle\sqrt{3}, 5\rangle$, find the components of $\bar{A} + \bar{B}$.

Solution

$$\bar{A} + \bar{B} = \langle 2, -7\rangle + \langle\sqrt{3}, 5\rangle = \langle 2 + \sqrt{3}, -7 + 5\rangle = \langle 2 + \sqrt{3}, -2\rangle.$$

To *double* a force $\bar{F}$ means to double its magnitude and leave its direction the same. Thus, $2\bar{F}$ is a vector twice as long as $\bar{F}$, and pointing in the same direction (Figure 11).

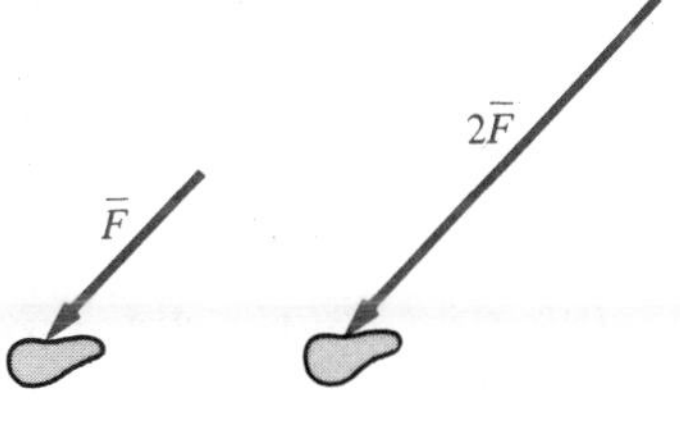

Figure 11

More generally, if s is a real number and $\bar{F}$ is a vector, we define $s\bar{F}$ to be a vector $|s|$ times as long as $\bar{F}$, and in the same direction if s is positive, or in the opposite direction if s is negative. Naturally, if either $s = 0$ or $\bar{F} = \bar{0}$ (or both), we define $s\bar{F} = \bar{0}$. In a product $s\bar{F}$, the real number s is often called a **scalar** (because it corresponds to a point on a number scale). Thus, the vector $s\bar{F}$ is referred to as a **scalar multiple** of the vector $\bar{F}$. Notice that

$$|s\bar{F}| = |s||\bar{F}|.$$

When a vector is multiplied by a scalar, the components of the vector are multiplied by the scalar. Thus, if $\bar{F} = \langle a, b\rangle$, then $s\bar{F} = \langle sa, sb\rangle$, that is,

$$s\langle a, b\rangle = \langle sa, sb\rangle.$$

Figure 12

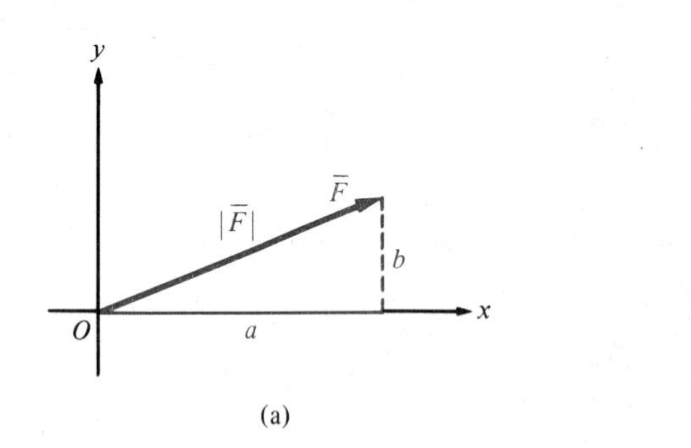

(a)

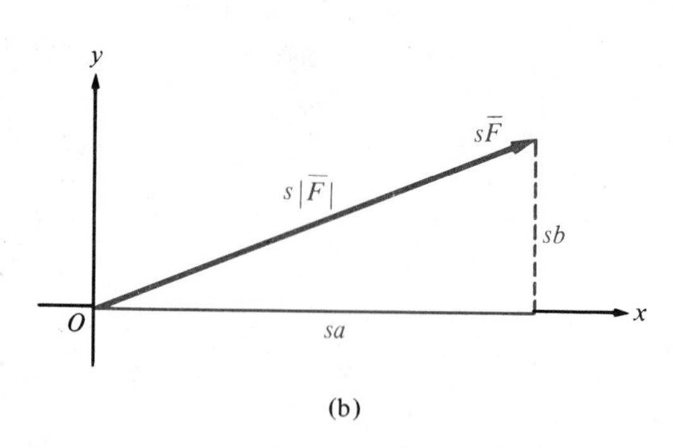

(b)

You can confirm this for the case in which s is positive and $\bar{F} = \langle a, b \rangle$ lies in quadrant I by considering the similar triangles in Figure 12a and b. The remaining cases are left as an exercise (Problem 34).

Example Let $A = \langle -2, \sqrt{3} \rangle$ and $\bar{B} = \langle 1, 4 \rangle$. Find the components of
(a) $5\bar{A}$ (b) $-3\bar{B}$ (c) $2\bar{A} + 3\bar{B}$.

Solution
(a) $5\bar{A} = 5\langle -2, \sqrt{3} \rangle = \langle 5(-2), 5\sqrt{3} \rangle = \langle -10, 5\sqrt{3} \rangle$
(b) $-3\bar{B} = -3\langle 1, 4 \rangle = \langle -3(1), -3(4) \rangle = \langle -3, -12 \rangle$
(c) $2\bar{A} + 3\bar{B} = 2\langle -2, \sqrt{3} \rangle + 3\langle 1, 4 \rangle = \langle 2(-2), 2\sqrt{3} \rangle + \langle 3(1), 3(4) \rangle$
$= \langle -4, 2\sqrt{3} \rangle + \langle 3, 12 \rangle = \langle -4 + 3, 2\sqrt{3} + 12 \rangle$
$= \langle -1, 2\sqrt{3} + 12 \rangle$

By the definition of multiplication of a vector by a scalar, the vector $(-1)\bar{F}$ has the same magnitude as $\bar{F}$, but points in the opposite direction. For simplicity, $(-1)\bar{F}$ is often written as $-\bar{F}$ (Figure 13). If $\bar{F} = \langle a, b \rangle$, then $-\bar{F} = \langle -a, -b \rangle$, that is,

$$-\langle a, b \rangle = \langle -a, -b \rangle$$

(Problem 36). The **difference** of two vectors $\bar{A}$ and $\bar{B}$ is defined by

$$\bar{A} - \bar{B} = \bar{A} + (-\bar{B}).$$

In other words, to **subtract** a vector $\bar{B}$ from a vector $\bar{A}$, you add $-\bar{B}$ to $\bar{A}$ (see Problem 33.) Evidently, if $\bar{A} = \langle a, b \rangle$ and $\bar{B} = \langle c, d \rangle$, then $\bar{A} - \bar{B} = \langle a - c, b - d \rangle$, that is,

$$\langle a, b \rangle - \langle c, d \rangle = \langle a - c, b - d \rangle$$

(Problem 37).

Figure 13

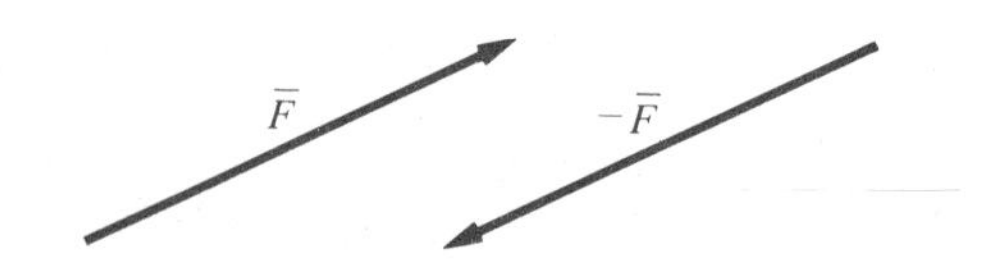

Example If $\bar{A} = \langle 4, -1 \rangle$ and $\bar{B} = \langle 7, \frac{2}{3} \rangle$, find the components of

(a) $-\bar{A}$
(b) $\bar{A} - \bar{B}$
(c) $2\bar{A} - 3\bar{B}$.

Solution

(a) $-\bar{A} = -\langle 4, -1 \rangle = \langle -4, -(-1) \rangle = \langle -4, 1 \rangle$

(b) $\bar{A} - \bar{B} = \langle 4, -1 \rangle - \langle 7, \frac{2}{3} \rangle = \langle 4 - 7, -1 - \frac{2}{3} \rangle = \langle -3, -\frac{5}{3} \rangle$

(c) $2\bar{A} - 3\bar{B} = 2\langle 4, -1 \rangle - 3\langle 7, \frac{2}{3} \rangle = \langle 8, -2 \rangle - \langle 21, 2 \rangle$
$= \langle 8 - 21, -2 - 2 \rangle = \langle -13, -4 \rangle$

Figure 14

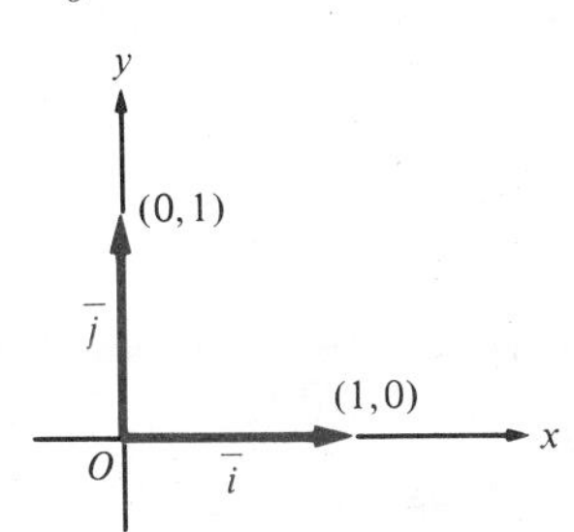

A popular alternative way to represent vectors in the xy plane is based on the two special vectors $\bar{i} = \langle 1, 0 \rangle$ and $\bar{j} = \langle 0, 1 \rangle$ (Figure 14). Notice that $\bar{i}$ and $\bar{j}$, which are called the **standard unit basis vectors,** have length 1 and point in the directions of the positive x and y axes, respectively. If $\bar{R} = \langle x, y \rangle$, then

$$\bar{R} = \langle x, y \rangle = \langle x, 0 \rangle + \langle 0, y \rangle = x\langle 1, 0 \rangle + y\langle 0, 1 \rangle = x\bar{i} + y\bar{j}.$$

In other words, $\bar{R} = \langle x, y \rangle$ and $\bar{R} = x\bar{i} + y\bar{j}$ are just two different ways of saying that $\bar{R}$ is the vector with components x and y. Using the standard unit basis vectors, we can write the following rules for addition, subtraction, and multiplication of vectors by scalars:

(i) $(a\bar{i} + b\bar{j}) + (c\bar{i} + d\bar{j}) = (a + c)\bar{i} + (b + d)\bar{j}$

(ii) $(a\bar{i} + b\bar{j}) - (c\bar{i} + d\bar{j}) = (a - c)\bar{i} + (b - d)\bar{j}$

(iii) $s(a\bar{i} + b\bar{j}) = (sa)\bar{i} + (sb)\bar{j}$.

Example If $\bar{A} = 2\bar{i} + 5\bar{j}$ and $\bar{B} = 3\bar{i} - \bar{j}$, find

(a) $\bar{A} + \bar{B}$ (b) $\bar{A} - \bar{B}$ (c) $\frac{4}{5}\bar{A}$ (d) $3\bar{A} - 2\bar{B}$

Solution

(a) $\bar{A} + \bar{B} = (2\bar{i} + 5\bar{j}) + (3\bar{i} - \bar{j}) = (2 + 3)\bar{i} + (5 - 1)\bar{j} = 5\bar{i} + 4\bar{j}$

(b) $\bar{A} - \bar{B} = (2\bar{i} + 5\bar{j}) - (3\bar{i} - \bar{j}) = (2 - 3)\bar{i} + (5 + 1)\bar{j} = -\bar{i} + 6\bar{j}$

(c) $\frac{4}{5}\bar{A} = \frac{4}{5}(2\bar{i} + 5\bar{j}) = [\frac{4}{5}(2)]\bar{i} + [\frac{4}{5}(5)]\bar{j} = \frac{8}{5}\bar{i} + 4\bar{j}$

(d) $3\bar{A} - 2\bar{B} = 3(2\bar{i} + 5\bar{j}) - 2(3\bar{i} - \bar{j}) = (6\bar{i} + 15\bar{j}) - (6\bar{i} - 2\bar{j})$
$= (6 - 6)\bar{i} + [15 - (-2)]\bar{j} = 17\bar{j}$

PROBLEM SET 4

In problems 1–12, sketch each vector, find its magnitude, and find its smallest nonnegative direction angle θ.

1. $\bar{R} = \langle -4, 4 \rangle$
© 2. $\bar{A} = \langle 6, 8 \rangle$
© 3. $\bar{B} = \langle 4, 3 \rangle$
4. $\bar{D} = \langle 5, 0 \rangle$
5. $\bar{C} = \langle 1, -\sqrt{3} \rangle$
6. $\bar{F} = \langle 0, 5 \rangle$
© 7. $\bar{H} = \langle \frac{3}{8}, \frac{1}{2} \rangle$
© 8. $\bar{K} = \langle 3.71, -4.88 \rangle$
© 9. $\bar{M} = \langle -\sqrt{5}, \sqrt{11} \rangle$
10. $\bar{N} = \langle \pi, \pi\sqrt{3} \rangle$
© 11. $\bar{P} = 3\bar{i} + 4\bar{j}$
© 12. $\bar{Q} = 2\sqrt{5}\bar{i} - 4\bar{j}$

In problems 13–20, the magnitude $|\bar{V}|$ and a direction angle θ of a vector $\bar{V}$ are given. In each case, find the components of $\bar{V}$.

13. $|\bar{V}| = 5, \theta = 45^\circ$
14. $|\bar{V}| = 3, \theta = 180^\circ$
15. $|\bar{V}| = 4\sqrt{2}, \theta = 135^\circ$
16. $|\bar{V}| = 9, \theta = -\dfrac{5\pi}{3}$
17. $|\bar{V}| = 2, \theta = -\dfrac{5\pi}{6}$
Ⓒ **18.** $|\bar{V}| = 4, \theta = 82^\circ$
Ⓒ **19.** $|\bar{V}| = 7.03, \theta = 75.25^\circ$
Ⓒ **20.** $|\bar{V}| = 8.12, \theta = 247^\circ 10'$

In problems 21–30, find (a) $\bar{A} + \bar{B}$, (b) $\bar{A} - \bar{B}$, (c) $3\bar{A} + 4\bar{B}$, and (d) $2\bar{A} - 3\bar{B}$.

21. $\bar{A} = \langle 3,4 \rangle, \bar{B} = \langle -1,4 \rangle$
22. $\bar{A} = \langle -2,6 \rangle, \bar{B} = \langle 3, -5 \rangle$
23. $\bar{A} = \langle -3,2 \rangle, \bar{B} = \langle \frac{5}{2}, \frac{7}{3} \rangle$
24. $\bar{A} = \langle 3, -3 \rangle, \bar{B} = \langle -5,0 \rangle$
25. $\bar{A} = 2\langle 1, -3 \rangle, \bar{B} = 3\langle 3, -5 \rangle$
26. $\bar{A} = -\bar{i} - \bar{j}, \bar{B} = 9\bar{j}$
27. $\bar{A} = 2\bar{i} + 5\bar{j}, \bar{B} = -3\bar{i} + 7\bar{j}$
28. $\bar{A} = \frac{7}{4}\bar{i} - \bar{j}, \bar{B} = \bar{i} - \frac{11}{3}\bar{j}$
29. $\bar{A} = 6\bar{i} - 5\bar{j}, \bar{B} = 2(2\bar{i} - 3\bar{j})$
30. $\bar{A} = 3.07\bar{i} + 0.33\bar{j}, \bar{B} = 7.65\bar{i} - 3.44\bar{j}$

31. The sum $\bar{A} + \bar{B}$ of two vectors can be obtained geometrically by applying the following **head-to-tail rule:** Move $\bar{B}$ if necessary so that its tail end coincides with the head end of $\bar{A}$ (Figure 15). Then $\bar{A} + \bar{B}$ is the vector whose tail end is the tail end of $\bar{A}$ and whose head end is the head end of $\bar{B}$. Show that the head-to-tail rule follows from the parallelogram rule.

Figure 15

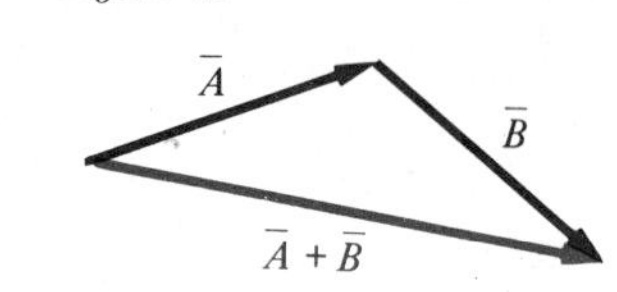

32. Using the head-to-tail rule (problem 31), give a geometric justification of the following equation, which is known as the **associative law of vector addition:**

$$\bar{A} + (\bar{B} + \bar{C}) = (\bar{A} + \bar{B}) + \bar{C}.$$

33. Justify the **triangle rule** for subtraction of vectors: Move $\bar{B}$ if necessary so that its tail end coincides with the tail end of $\bar{A}$ (Figure 16). Then $\bar{A} - \bar{B}$ is the vector whose tail end is the head end of $\bar{B}$ and whose head end is the head end of $\bar{A}$.

Figure 16

$\bar{A}$ $\quad$ $\bar{A} - \bar{B}$ $\quad$ $\bar{B}$

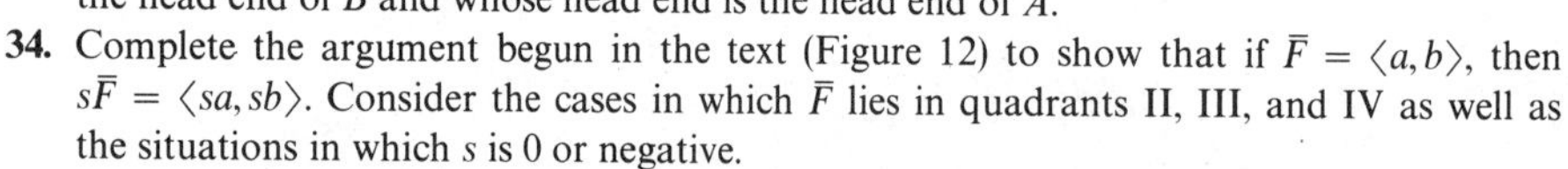

34. Complete the argument begun in the text (Figure 12) to show that if $\bar{F} = \langle a,b \rangle$, then $s\bar{F} = \langle sa, sb \rangle$. Consider the cases in which $\bar{F}$ lies in quadrants II, III, and IV as well as the situations in which s is 0 or negative.

35. Figure 10 shows a special situation in which both vectors $\bar{F}_1$ and $\bar{F}_2$ lie in quadrant I. Check several other configurations to convince yourself that the components of $\bar{F}_1 + \bar{F}_2$ are always equal to the sum of the corresponding components of $\bar{F}_1$ and $\bar{F}_2$.

36. Prove that if $\bar{F} = \langle a,b \rangle$, then $-\bar{F} = \langle -a, -b \rangle$.

37. Prove that if $\bar{A} = \langle a,b \rangle$ and $\bar{B} = \langle c,d \rangle$, then $\bar{A} - \bar{B} = \langle a - c, b - d \rangle$.

38. If $\bar{A} + \bar{X} = \bar{B}$, prove that $\bar{X} = \bar{B} - \bar{A}$.

39. If $\bar{A} + \bar{X} = \bar{A}$, prove that $\bar{X} = \bar{0}$.

40. If $s\bar{F} = \bar{0}$, prove that either $s = 0$ or $\bar{F} = \bar{0}$.

41. Prove that $s(\bar{A} + \bar{B}) = s\bar{A} + s\bar{B}$.

42. If $\bar{A}$ and $\bar{B}$ are two vectors and α is the angle between them, then the **dot product** of $\bar{A}$ and $\bar{B}$ is defined to be the number

$$\bar{A} \cdot \bar{B} = |\bar{A}|\,|\bar{B}| \cos \alpha.$$

Use the law of cosines to show that if $\bar{A} = \langle a,b \rangle$ and $\bar{B} = \langle c,d \rangle$, then

$$\bar{A} \cdot \bar{B} = ac + bd.$$

43. Use the formula obtained in problem 42 to calculate $\bar{A} \cdot \bar{B}$ if (a) $\bar{A} = \langle 5,1 \rangle$ and $\bar{B} = \langle 4, -7 \rangle$; (b) $\bar{A} = \langle -3,2 \rangle$ and $\bar{B} = \langle -4,3 \rangle$; (c) $\bar{A} = \bar{i} + 2\bar{j}$ and $\bar{B} = 3\bar{i} - 5\bar{j}$; (d) $\bar{A} = \bar{i}$ and $\bar{B} = \bar{j}$; and (e) $\bar{A} = \bar{B} = \bar{i}$.

44. Use the result of problem 42 to show that for any three vectors $\bar{A}$, $\bar{B}$, and $\bar{C}$,

$$\bar{A} \cdot (\bar{B} + \bar{C}) = \bar{A} \cdot \bar{B} + \bar{A} \cdot \bar{C}.$$

5 Polar Coordinates

Until now we have specified the position of points in the plane by means of cartesian coordinates; however, in some situations it is simpler and more natural to use a different system called *polar coordinates*. To set up a polar coordinate system in the plane, we choose a fixed point O called the **pole** and a fixed ray with endpoint O. The ray is called the **polar axis** (Figure 1a).

Figure 1

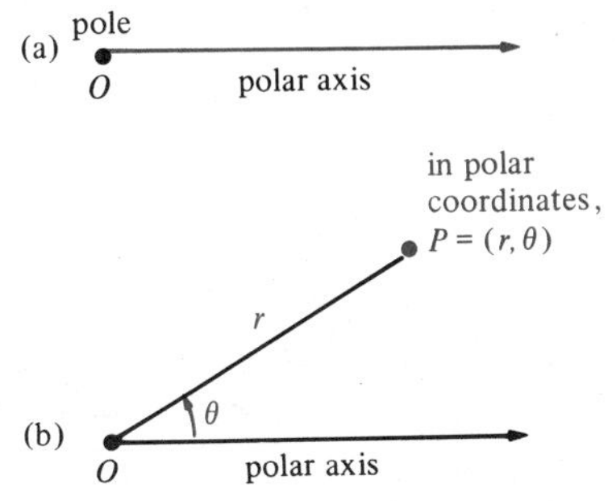

(a) pole O polar axis

(b) O polar axis; θ; r; in polar coordinates, $P = (r, \theta)$

Now let P be any point in the plane and denote by r the distance between P and the pole O, so that $r = |\overline{OP}|$ (Figure 1b). If $P \neq O$, then the ray containing P with endpoint O forms the terminal side of an angle θ with the polar axis as its initial side (Figure 1b). We shall refer to the ordered pair (r, θ) as the **polar coordinates** of the point P, and we write

$$P = (r, \theta).$$

The angle θ can be measured either in degrees or radians.

The polar coordinates (r, θ) locate the point P on a grid formed by concentric circles with center O and rays with endpoint O (Figure 2). The value of r determines a particular circle of radius r; the value of θ designates a specific ray making an angle θ with the polar axis; and P lies at the intersection of the circle and the ray. For instance, the point

$$P = (4, 30^\circ)$$

lies at the intersection of the circle $r = 4$ and the ray $\theta = 30^\circ$.

Figure 2

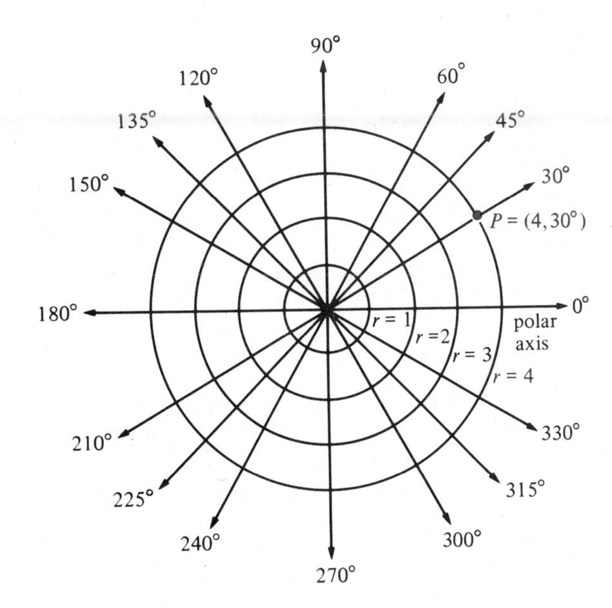

If $r = 0$ in the polar coordinate system, we understand that the point $(r, \theta) = (0, \theta)$ is at the pole O no matter what the angle θ may be. Thus,

$$O = (0, \theta).$$

Figure 3

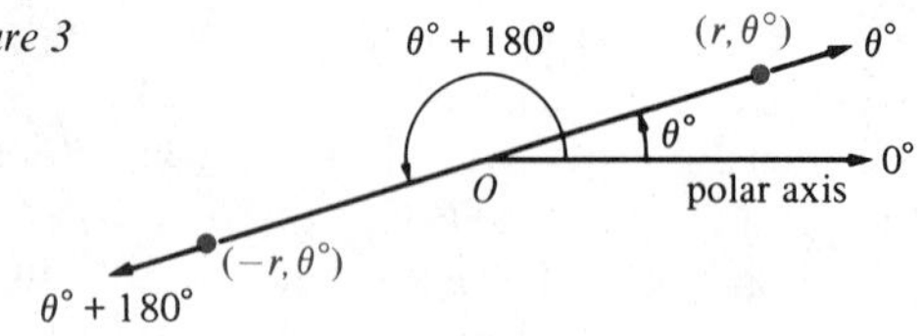

Also, it is convenient to allow r to be negative by making the following definition: The point $(-r, \theta^\circ)$ is located $|r|$ units from the pole O, but on the ray opposite to the θ° ray, that is, on the $\theta^\circ + 180^\circ$ ray (Figure 3). Therefore,

$$(-r, \theta^\circ) = (r, \theta^\circ + 180^\circ),$$

or if θ is measured in radians,

$$(-r, \theta) = (r, \theta + \pi).$$

To "plot the polar point (r, θ)," means to draw a diagram showing the pole O, the polar axis, and the point P whose polar coordinates are (r, θ). You will find it easier to plot polar points if you use polar graph paper.

Example Plot each point in the polar coordinate system.

(a) $(2, 60^\circ)$ (b) $(4, 135^\circ)$ (c) $(-\frac{5}{2}, 30^\circ)$ (d) $\left(2, -\frac{\pi}{4}\right)$

(e) $(2, 0^\circ)$ (f) $(0, 0^\circ)$ (g) $(-4, -45^\circ)$

Solution (a) To plot the polar point $(2, 60^\circ)$, we measure an angle of 60° from the polar axis and then locate the point 2 units from the pole on the terminal side of this angle. The polar points in parts (b) through (g) are plotted in a similar way (Figure 4).

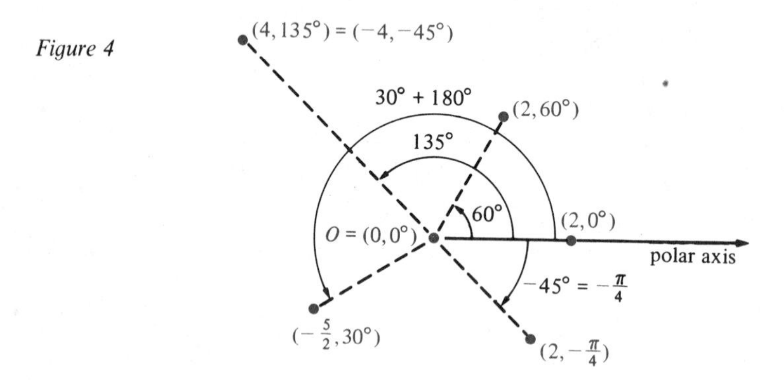

Figure 4

Unlike the cartesian coordinate system, in which a point P has only one representation, a point P has many different representations in the polar coordinate system. Not only do we have $(r, \theta^\circ) = (-r, \theta^\circ + 180^\circ)$ but, because $\pm 360^\circ$ corresponds to one revolution about the pole, we also have

$$(r, \theta^\circ) = (r, \theta^\circ + 360^\circ) = (r, \theta^\circ - 360^\circ).$$

Indeed, if n is any integer,

$$(r, \theta^\circ) = (r, \theta^\circ + 360^\circ \cdot n),$$

or, if angles are measured in radians,

$$(r, \theta) = (r, \theta + 2\pi n).$$

Example Give other polar representations $P = (r, \theta^\circ)$ of the polar point $P = (2, 30^\circ)$ that satisfy each of the following conditions:

(a) $r < 0$ and $0^\circ \leq \theta^\circ < 360^\circ$ (b) $r > 0$ and $-360^\circ < \theta^\circ \leq 0^\circ$

(c) $r < 0$ and $-360^\circ < \theta^\circ \leq 0^\circ$

Solution (a) $P = (2, 30^\circ) = (-2, 30^\circ + 180^\circ) = (-2, 210^\circ)$
(b) $P = (2, 30^\circ) = (2, 30^\circ - 360^\circ) = (2, -330^\circ)$
(c) $P = (-2, 210^\circ) = (-2, 210^\circ - 360^\circ) = (-2, -150^\circ)$

It is sometimes necessary to convert from cartesian to polar coordinates or vice versa. To do this, we place the pole for the polar coordinate system at the origin of the cartesian coordinate system, and position the polar axis along the positive x axis (Figure 5). Now suppose that the point P has cartesian coordinates (x, y) and polar coordinates (r, θ) with $r > 0$. Then $\cos\theta = x/r$ and $\sin\theta = y/r$; hence,

Figure 5

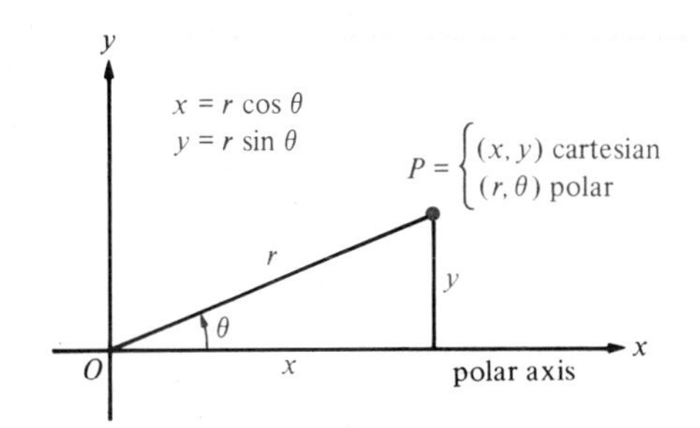

$$x = r\cos\theta \quad \text{and} \quad y = r\sin\theta.$$

It can be shown (Problem 51) that these equations hold in all cases, even when $r \leq 0$.

Example Convert the following polar coordinates to cartesian coordinates:

(a) $(4, 30^\circ)$ (b) $\left(6, -\dfrac{3\pi}{4}\right)$ [c] (c) $(-2, 133.5^\circ)$

Solution We use the conversion equations $x = r\cos\theta$ and $y = r\sin\theta$.

(a) $x = r\cos\theta = 4\cos 30^\circ = 4\left(\dfrac{\sqrt{3}}{2}\right) = 2\sqrt{3}$

$y = r\sin\theta = 4\sin 30^\circ = 4(\tfrac{1}{2}) = 2$

Hence, the cartesian coordinates are $(2\sqrt{3}, 2)$.

(b) $x = r\cos\theta = 6\cos\left(-\dfrac{3\pi}{4}\right) = 6\left(-\dfrac{\sqrt{2}}{2}\right) = -3\sqrt{2}$

$y = r\sin\theta = 6\sin\left(-\dfrac{3\pi}{4}\right) = 6\left(-\dfrac{\sqrt{2}}{2}\right) = -3\sqrt{2}$

Hence, the cartesian coordinates are $(-3\sqrt{2}, -3\sqrt{2})$.

(c) Using a calculator, and rounding off to four significant digits, we have

$$x = r\cos\theta = -2\cos 133.5^\circ = 1.377$$
$$y = r\sin\theta = -2\sin 133.5^\circ = -1.451$$

Hence, the approximate cartesian coordinates are $(1.377, -1.451)$.

From the conversion equations $x = r\cos\theta$ and $y = r\sin\theta$, it follows that

$$\begin{aligned} x^2 + y^2 &= r^2\cos^2\theta + r^2\sin^2\theta \\ &= r^2(\cos^2\theta + \sin^2\theta) \\ &= r^2, \end{aligned}$$

so

$$r = \pm\sqrt{x^2 + y^2}.$$

Also, if $x \neq 0$,

$$\frac{y}{x} = \frac{r\sin\theta}{r\cos\theta} = \frac{\sin\theta}{\cos\theta} = \tan\theta,$$

so

$$\tan\theta = \frac{y}{x} \quad \text{for} \quad x \neq 0.$$

The equations $r = \pm\sqrt{x^2 + y^2}$ and $\tan\theta = y/x$ do not determine r and θ uniquely in terms of x and y, simply because the point P whose cartesian coordinates are (x, y) has an unlimited number of different representations in the polar coordinate system. Thus, in using these equations to find the polar coordinates of P, you must pay attention to the quadrant in which P lies, because this will help you to determine an appropriate value of θ.

Example 1 Convert the following cartesian coordinates to polar coordinates with $r \geq 0$ and $-180^\circ < \theta^\circ \leq 180^\circ$:

(a) $(2, 2)$

Ⓒ (b) $(5.031, -2.758)$

Solution (a) $r = \sqrt{2^2 + 2^2} = \sqrt{8} = 2\sqrt{2}$, and $\tan\theta = 2/2 = 1$. Because the point lies in quadrant I, it follows that $0^\circ < \theta^\circ < 90^\circ$, and so $\theta = 45^\circ$. Hence, the polar coordinates are $(2\sqrt{2}, 45^\circ)$.

(b) Using a calculator, and rounding off to four significant digits, we have

$$r = \sqrt{(5.031)^2 + (-2.758)^2} = 5.737.$$

Here $\tan\theta = -2.758/5.031$, and, because the point lies in quadrant IV, we have $-90^\circ < \theta^\circ < 0^\circ$. Therefore, rounded off to the nearest hundredth of a degree,

$$\theta = \tan^{-1}\frac{-2.758}{5.031} = -28.73^\circ.$$

Hence, the approximate polar coordinates are $(5.737, -28.73^\circ)$.

Example 2 Convert the following cartesian coordinates to polar coordinates with $r \geq 0$ and $-\pi < \theta \leq \pi$.

(a) $(0, -7)$ [C] (b) $(-33.01, 29.77)$

Solution (a) $r = \sqrt{0^2 + (-7)^2} = \sqrt{49} = 7$. Because the point lies on the negative y axis, $\theta = -\pi/2$, and the polar coordinates are $(7, -\pi/2)$.

(b) Using a calculator, and rounding off to four significant digits, we have

$$r = \sqrt{(-33.01)^2 + (29.77)^2} = 44.45.$$

Because the point lies in quadrant II, we have $\pi/2 < \theta < \pi$, and because $\tan\theta = 29.77/(-33.01)$, the corresponding reference angle is

$$\theta_R = \tan^{-1}\frac{29.77}{33.01} = 0.7338$$

(rounded off to four significant digits). Hence,

$$\theta = \pi - \theta_R = 2.408,$$

and the approximate polar coordinates are $(44.45, 2.408)$.

PROBLEM SET 5

In problems 1–20, plot each point in the polar coordinate system.

1. $(6, 30°)$ **2.** $\left(10, \frac{\pi}{3}\right)$ **3.** $(3, 120°)$ **4.** $(0, 25°)$

5. $\left(4, -\frac{\pi}{6}\right)$ **6.** $(-4, 90°)$ **7.** $(-5, 135°)$ **8.** $\left(\frac{5}{2}, -270°\right)$

9. $(0, -45°)$ **10.** $\left(-\frac{7}{3}, \frac{\pi}{6}\right)$ **11.** $\left(\frac{17}{2}, -45°\right)$ **12.** $(5, -315°)$

13. $\left(\frac{5}{3}, -210°\right)$ **14.** $(3.75, -330°)$ **15.** $\left(-4.5, -\frac{\pi}{3}\right)$ **16.** $(-4.75, -30°)$

17. $\left(-2, -\frac{3\pi}{4}\right)$ **18.** $\left(-\frac{1}{2}, -\frac{3\pi}{2}\right)$ **19.** $(4.8, 0°)$ **20.** $(-3.25, 0)$

21. Plot the point $(5, 45°)$ in the polar coordinate system, and then give other polar representations $(r, \theta°)$ of the same point that satisfy each of the following conditions:
(a) $r < 0$ and $0° \leq \theta° < 360°$
(b) $r > 0$ and $-360° < \theta° \leq 0°$
(c) $r < 0$ and $-360° < \theta° \leq 0°$

22. Plot the point $\left(-3, \frac{13\pi}{6}\right)$ in the polar coordinate system, and then give other polar representations (r, θ) of the same point that satisfy each of the following conditions:
(a) $r < 0$ and $0 \leq \theta < 2\pi$
(b) $r > 0$ and $-2\pi < \theta \leq 0$
(c) $r < 0$ and $-2\pi < \theta \leq 0$

In problems 23–34, convert the polar coordinates to cartesian coordinates. If use of a calculator is indicated, round off your answers to four significant digits.

23. $(4, 45°)$
24. $(3, 60°)$
25. $\left(-5, \frac{\pi}{4}\right)$
26. $\left(0, \frac{\pi}{7}\right)$
27. $\left(1, -\frac{\pi}{4}\right)$
28. $\left(-5, \frac{5\pi}{6}\right)$
29. $(-4, 150°)$
Ⓒ **30.** $(-3, 25°)$
Ⓒ 31. $(2.765, -122.73°)$
Ⓒ **32.** $(-6.772, 133.33°)$
Ⓒ 33. $(17.76, -37.07°)$
Ⓒ **34.** $(3.271 \times 10^7, 44.33°)$

In problems 35–44, convert the cartesian coordinates to polar coordinates with $r \geq 0$ and $-180° < \theta° \leq 180°$. If use of a calculator is indicated, round off r to four significant digits, and θ to the nearest hundredth of a degree.

35. $(\sqrt{2}, \sqrt{2})$
36. $(-1, \sqrt{3})$
37. $(1, \sqrt{3})$
38. $(-2, 2)$
39. $(-3, 0)$
Ⓒ **40.** $(4, 3)$
Ⓒ 41. $(-5, 12)$
Ⓒ **42.** $(-2, -3)$
Ⓒ 43. $(13.01, -15.57)$
Ⓒ **44.** $(3.001 \times 10^{-5}, 2.774 \times 10^{-5})$

In problems 45–50, convert the cartesian coordinates to polar coordinates with $r \geq 0$ and $-\pi < \theta \leq \pi$. If use of a calculator is indicated, round off r and θ to four significant digits.

45. $(2\sqrt{3}, -2)$
46. $(-3, -3\sqrt{3}$
47. $(0, -2)$
48. $(4\sqrt{3}, -4)$
Ⓒ 49. $(-72.02, -91.33)$
Ⓒ **50.** $(-2.102 \times 10^7, 9.901 \times 10^6)$

51. Show that the equations

$$x = r\cos\theta \quad \text{and} \quad y = r\sin\theta$$

work in all cases, even when $r \leq 0$, to convert the polar coordinates (r, θ) of a point P into the cartesian coordinates (x, y) of P.

6 Applications of Vectors and Polar Coordinates

In this section, we present a few of the many applications of vectors and polar coordinates. Notice that there is a particularly close connection between vectors and polar coordinates: If $\bar{R}$ is the position vector of a point P, and if θ is a direction angle for $\bar{R}$ (Figure 1), then $(|\bar{R}|, \theta)$ are polar coordinates for P and

$$\bar{R} = \langle |\bar{R}|\cos\theta, |\bar{R}|\sin\theta \rangle.$$

Figure 1

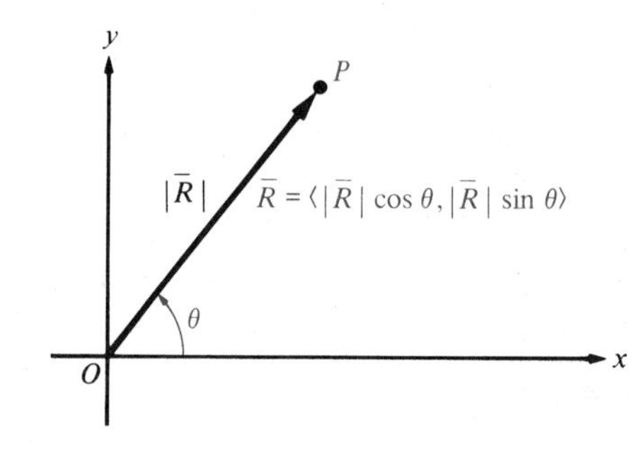

6.1 Applications of Vectors

The following example illustrates how vectors may be used to solve problems involving forces.

Example Ⓒ Two forces $\bar{F}_1$ and $\bar{F}_2$ act on a point. If $|\bar{F}_1| = 30$ pounds, $|\bar{F}_2| = 50$ pounds, and the angle α from $\bar{F}_1$ to $\bar{F}_2$ is 60°, find (a) the magnitude $|\bar{F}|$ of the resultant force $\bar{F}$, and (b) the angle θ from $\bar{F}_1$ to $\bar{F}$.

Solution We begin by setting up an xy coordinate system with the origin O at the point on which both $\bar{F}_1$ and $\bar{F}_2$ act, and with the positive x axis in the direction of $\bar{F}_1$ (Figure 2). Then $\alpha = 60°$ is a direction angle for $\bar{F}_2$; hence, the components of $\bar{F}_2$ are given by

Figure 2

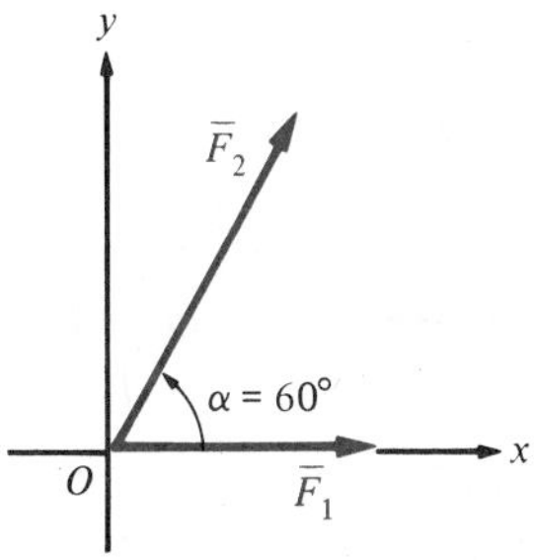

$$\begin{aligned} \bar{F}_2 &= \langle |\bar{F}_2| \cos \alpha, |\bar{F}_2| \sin \alpha \rangle \\ &= \langle 50 \cos 60°, 50 \sin 60° \rangle \\ &= \left\langle \frac{50}{2}, \frac{50\sqrt{3}}{2} \right\rangle = \langle 25, 25\sqrt{3} \rangle. \end{aligned}$$

Since $\bar{F}_1$ lies along the positive x axis and $|\bar{F}_1| = 30$, $\bar{F}_1 = \langle 30, 0 \rangle$. As a consequence,

$$\begin{aligned} \bar{F} = \bar{F}_1 + \bar{F}_2 &= \langle 30, 0 \rangle + \langle 25, 25\sqrt{3} \rangle \\ &= \langle 30 + 25, 0 + 25\sqrt{3} \rangle = \langle 55, 25\sqrt{3} \rangle. \end{aligned}$$

(a) $|\bar{F}| = \sqrt{55^2 + (25\sqrt{3})^2} = \sqrt{3025 + 1875} = \sqrt{4900} = 70$ pounds

Figure 3

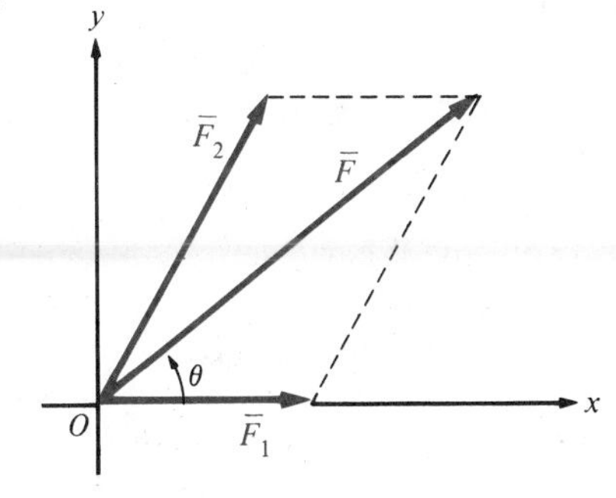

(b) The angle θ from $\bar{F}_1$ to $\bar{F}$ is the direction angle of $\bar{F}$ (Figure 3). Because the components of $\bar{F}$ are both positive, $\bar{F}$ lies in quadrant I, so θ is a quadrant I angle. Therefore, because

$$\cos \theta = \frac{55}{|\bar{F}|} = \frac{55}{70},$$

it follows that

$$\theta = \cos^{-1} \frac{55}{70} \approx 38.21°.$$

Another important application of vectors is in navigation, where the following terminology is commonly used:

1. The **heading** of an airplane is the direction in which it is pointed, and its **air speed** is its speed relative to the air. The vector $\bar{V}_1$ whose magnitude is the air speed and whose direction is the heading represents the **velocity** of the airplane relative to the air (Figure 4).

2. The **course** or **track** of an airplane is the direction in which it is actually moving over the ground, and its **ground speed** is its speed relative to the ground. The vector $\bar{V}$ whose magnitude is the ground speed and whose direction is the course represents the actual *velocity* of the airplane *relative to the ground* (Figure 4).

Figure 4

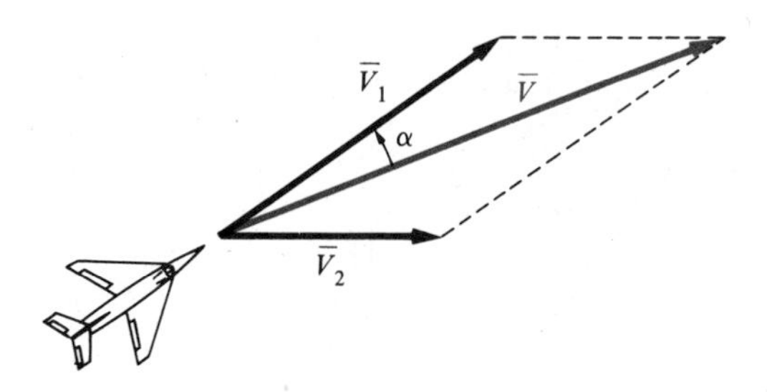

3. The difference $\bar{V}_2 = \bar{V} - \bar{V}_1$ is called the **drift vector.** It represents the *velocity of the air relative to the ground* (Figure 4). Thus,

$$\bar{V} = \bar{V}_1 + \bar{V}_2.$$

The angle α between $\bar{V}$ and $\bar{V}_1$ is called the **drift angle.**

(Similar terminology is used in ship navigation.)

Example Ⓒ A commercial jet is headed N30°E with an air speed of 550 miles per hour. The wind is blowing N155°E at a speed of 50 miles per hour. Find (a) the ground speed, (b) the drift angle, and (c) the course of the jet.

Solution Let $\bar{V}_1$ represent the velocity of the jet relative to the air; let $\bar{V}_2$ represent the velocity of the air relative to the ground; and let $\bar{V}$ represent the velocity of the jet relative to the ground (Figure 5).

Figure 5

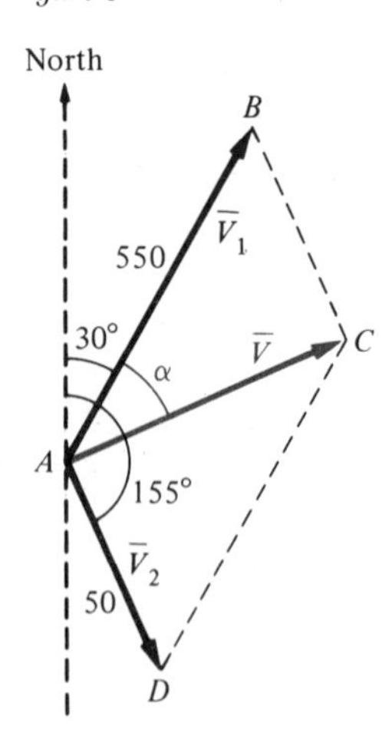

There are two ways to solve this problem: We can set up an xy coordinate system, find the components of $\bar{V}_1$ and $\bar{V}_2$, and then find the components of $\bar{V} = \bar{V}_1 + \bar{V}_2$; or we can use trigonometry to solve the triangles in Figure 5. Because the first method was illustrated in the previous example, we elect to use the second method here. Our plan is to find angle ABC so that we can apply the law of cosines to triangle ABC.

The four interior angles of a parallelogram add up to 360° (why?), and the interior angles at opposite vertices are equal. Hence, in parallelogram $ABCD$,

$$2(\text{angle } ABC) + 2(\text{angle } DAB) = 360^\circ.$$

Therefore,

$$\begin{aligned}\text{angle } ABC &= 180^\circ - (\text{angle } DAB)\\ &= 180^\circ - (155^\circ - 30^\circ)\\ &= 55^\circ.\end{aligned}$$

(a) Applying the law of cosines to triangle ABC, we have

$$|\bar{V}|^2 = |\overline{AC}|^2 = |\overline{AB}|^2 + |\overline{BC}|^2 - 2|\overline{AB}|\,|\overline{BC}|\cos 55^\circ$$
$$= 550^2 + 50^2 - 2(550)(50)\cos 55^\circ$$

so

$$|\bar{V}| = \sqrt{550^2 + 50^2 - 2(550)(50)\cos 55^\circ}$$
$$\approx 522.9 \text{ miles per hour.}$$

(b) Applying the law of sines to triangle ABC, we have

$$\frac{\sin\alpha}{50} = \frac{\sin 55^\circ}{|\bar{V}|},$$

so

$$\sin\alpha = \frac{50\sin 55^\circ}{|\bar{V}|} = \frac{50\sin 55^\circ}{522.9}.$$

Because α is a quadrant I angle, it follows that

$$\alpha = \sin^{-1}\frac{50\sin 55^\circ}{522.9} \approx 4.49^\circ.$$

Hence, rounded off to the nearest hundredth of a degree, the drift angle is 4.49°.
(c) From Figure 5, the vector $\bar{V}$ is $30^\circ + \alpha = 34.49^\circ$ east of due north; hence, the course of the jet is N34.49°E.

6.2 Graphs of Polar Equations

In calculus, polar coordinates are often used to write equations of curves in the plane. An equation, such as $r^2 = 9\cos 2\theta$, that relates the polar coordinates r and θ is called a **polar equation.** Because each point in the plane has a multitude of different polar representations, it is necessary to define the graph of a polar equation with some care: The **graph** of a polar equation consists of all points P in the plane that have at least one polar representation (r, θ) that satisfies the equation.

Example Sketch the graph of each polar equation:

(a) $r = 4$ (b) $r^2 = 16$ (c) $\theta = \frac{\pi}{6}$

Figure 6

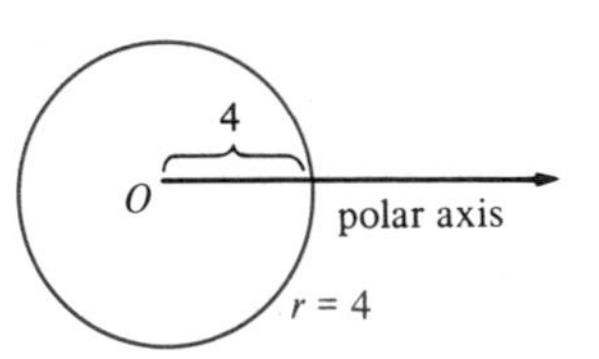

Solution (a) The graph of $r = 4$ is a circle of radius 4 with center at the pole (Figure 6). Notice, for instance, that the point $P = (4, -\pi)$ belongs to the graph in spite of the fact that not all of its representations, such as $P = (-4, 0)$ or $P = (-4, 2\pi)$, satisfy the equation $r = 4$.

(b) The equation $r^2 = 16$ is equivalent to $|r| = 4$, and its graph is the same as the graph of $r = 4$ (Figure 6).

(c) The graph of $\theta = \pi/6$ consists of the entire line through O, making an angle of $\pi/6$ radian (30°) with the polar axis—not just the *ray* as one might at first think (Figure 7). Points of the form $P = \left(r, \dfrac{\pi}{6} + \pi\right)$ belong to the graph because they can be rewritten as $P = (-r, \pi/6)$.

Figure 7

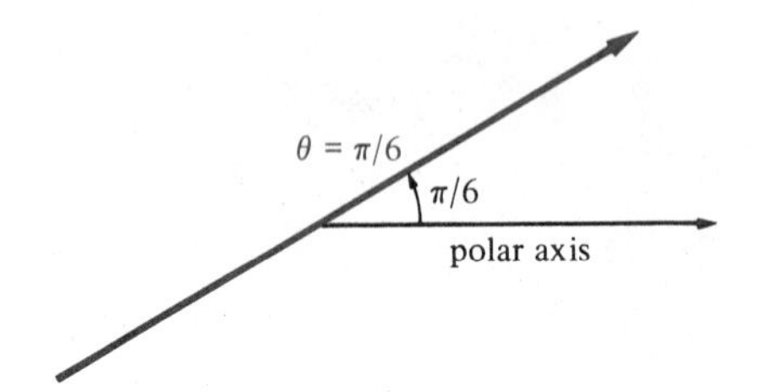

By making the substitutions $x = r\cos\theta$ and $y = r\sin\theta$, you can convert a cartesian equation into a polar equation.

Example Find a polar equation corresponding to the cartesian equation $x^2 + (y + 4)^2 = 16$, and sketch the graph.

Solution In Section 4.1 of Chapter 1, we learned that the graph of an equation of the form $(x - h)^2 + (y - k)^2 = r^2$ is a circle with center (h, k) and radius r. Hence, the graph of $x^2 + (y + 4)^2 = 16$ is a circle of radius 4 with center at the point with cartesian coordinates $(0, -4)$ (Figure 8). Rewriting the equation as

$$x^2 + y^2 + 8y + 16 = 16 \quad \text{or} \quad x^2 + y^2 + 8y = 0,$$

substituting $x = r\cos\theta$ and $y = r\sin\theta$, and using the fact that $x^2 + y^2 = r^2$, we obtain the polar equation

$$r^2 + 8r\sin\theta = 0$$

or

$$r + 8\sin\theta = 0.$$

Figure 8

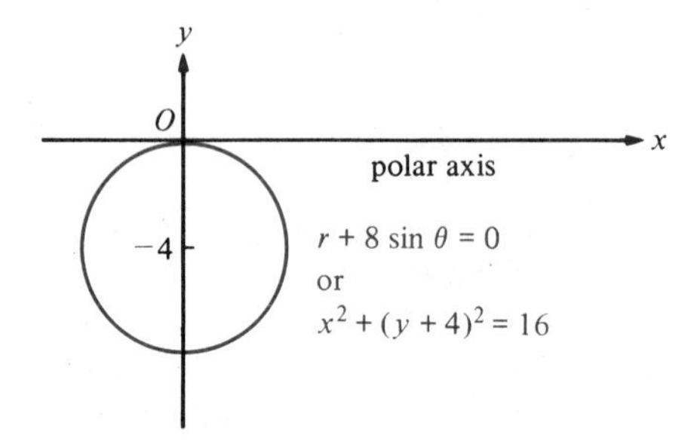

In the last step of the example above, we have divided by r; hence, we may have lost the solution $r = 0$. In this particular case, we have not lost the solution $r = 0$ because, if $r + 8\sin\theta = 0$, then $r = 0$ when $\theta = 0$. More generally, if you multiply a polar equation by r, you *may* introduce an extraneous solution $r = 0$ that doesn't really belong and, conversely, if you divide a polar equation by r, you *may* lose a solution $r = 0$ that really does belong. You must check each case.

Example Find a cartesian equation corresponding to the polar equation $r = 4\tan\theta\sec\theta$, and sketch the graph.

Solution We have

$$r = 4\tan\theta\sec\theta = 4\frac{\sin\theta}{\cos\theta}\cdot\frac{1}{\cos\theta}$$

or

$$r\cos^2\theta = 4\sin\theta.$$

Multiplication by r gives

$$r^2\cos^2\theta = 4r\sin\theta$$

or

$$(r\cos\theta)^2 = 4r\sin\theta.$$

Because $x = r\cos\theta$ and $y = r\sin\theta$, the last equation can be rewritten as

$$x^2 = 4y \quad \text{or} \quad y = \tfrac{1}{4}x^2.$$

By Section 6 of Chapter 1, the graph of $y = \frac{1}{4}x^2$ is a parabola with a vertical axis of symmetry and vertex at $(0, 0)$ (Figure 9). [Notice that multiplication by r did not introduce an extraneous solution $r = 0$ because, if $r\cos^2\theta = 4\sin\theta$, then $r = 0$ when $\theta = 0$.]

Figure 9

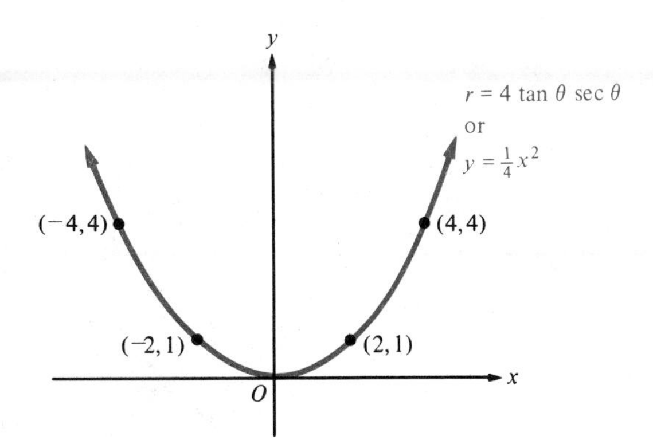

You can also sketch the graph of a polar equation by plotting polar points and connecting them with a smooth curve, in much the same way that you sketch the graph of a cartesian equation.

Example Ⓒ Sketch the graph of $r = 2(1 - \cos\theta)$.

Solution Because $\cos(-\theta) = \cos\theta$, it follows that if the polar point $P = (r, \theta)$ belongs to the graph, so does the polar point $Q = (r, -\theta)$. Therefore, the graph is **symmetric about the polar axis.** We sketch the top half of the graph for $0 \le \theta \le \pi$ using the data in the table below, and then we reflect the top half across the polar axis to obtain the bottom half (Figure 10).

Figure 10

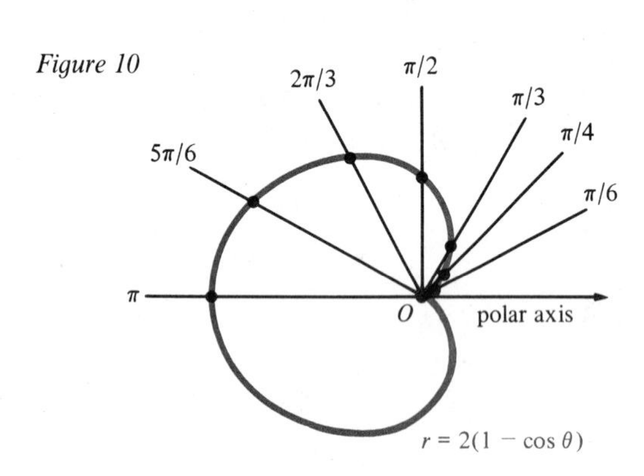

θ	$r = 2(1 - \cos\theta)$
0	0
$\pi/6$	0.27 (approximately)
$\pi/4$	0.59 (approximately)
$\pi/3$	1
$\pi/2$	2
$2\pi/3$	3
$5\pi/6$	3.73 (approximately)
π	4

The curve in Figure 10 is called a *cardioid.*

PROBLEM SET 6

Ⓒ In problems 1–4, assume that the two forces $\bar{F}_1$ and $\bar{F}_2$ act on a point and that α is the angle from $\bar{F}_1$ to $\bar{F}_2$. If $\bar{F}$ is the resultant force, find (a) $|\bar{F}|$ and (b) the angle θ from $\bar{F}_1$ to $\bar{F}$.

1. $|\bar{F}_1| = 25$ pounds, $|\bar{F}_2| = 40$ pounds, $\alpha = 45°$

2. $|\bar{F}_1| = 41$ newtons, $|\bar{F}_2| = 7$ newtons, $\alpha = 27°$

3. $|\bar{F}_1| = 22$ newtons, $|\bar{F}_2| = 33$ newtons, $\alpha = 120°$

4. $|\bar{F}_1| = 3.4$ pounds, $|\bar{F}_2| = 7.3$ pounds, $\alpha = 133°$

Ⓒ **5.** Two children are pulling a third child across the ice on a sled. The first child pulls on her rope with a force of 10 pounds, and the second child pulls on his rope with a force of 7 pounds. If the angle between the ropes is 30°, (a) what is the magnitude of the resultant force, and (b) what angle does the resultant force vector make with the first child's rope?

Ⓒ **6.** Two tugboats are pulling an ocean freighter (Figure 11). The first tugboat exerts a force of 1300 kilograms on a cable making an angle of 25° with the axis of the ship. The second tugboat pulls on a cable making an angle of 32° with the axis of the ship. If the resultant force vector lies directly along the axis of the ship, (a) what force is the second tugboat exerting on the ship, and (b) what is the magnitude of the resultant force vector?

Figure 11

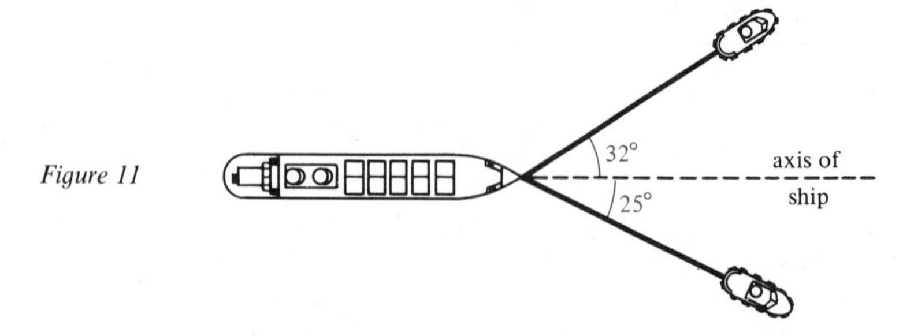

C 7. An airplane is headed N95°E with an air speed of 800 kilometers per hour. The wind is blowing N25°E at a speed of 45 kilometers per hour. Find (a) the ground speed, (b) the drift angle, and (c) the course of the plane.

C 8. A long-distance swimmer can swim 3 kilometers per hour in still water. She is swimming away from and at right angles to a straight shoreline. If a current of 5 kilometers per hour is flowing parallel to the shoreline, find her speed relative to the land and the angle θ between the shoreline and her actual velocity vector.

C 9. A balloon is rising 3 meters per second while a wind is blowing horizontally at 2 meters per second. Find (a) the speed of the balloon, and (b) the angle its track makes with the horizontal.

10. Figure 12 shows the sail $\overline{AB}$ of an iceboat and a particle P of air moving with velocity $\bar{V}_1$ perpendicular to the iceboat's track. Suppose that $\bar{V}_2$ is the velocity of the iceboat and that the sail $\overline{AB}$ makes an angle θ with the iceboat's track. Assume that P is moving at such a speed that it just maintains contact with the sail. (a) Show that $|\bar{V}_1| = |\bar{V}_2| \tan \theta$. (b) Explain why an iceboat can be propelled by a wind at a speed greater than the speed of the wind.

Figure 12

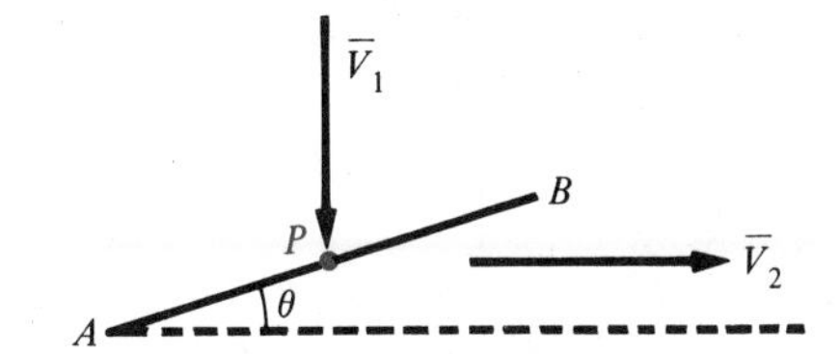

C 11. Three children located at points A, B, and C tug on ropes attached to a ring at point O (Figure 13). The child at A pulls with a force of 4 kilograms in the direction S71°W; the child at B pulls with a force of 12 kilograms in the direction S21°E; and the child at C pulls with a force of 7 kilograms in the direction N82°E. In what direction and with what force should a fourth child pull on a rope attached to the ring in order to counterbalance the other three forces?

Figure 13

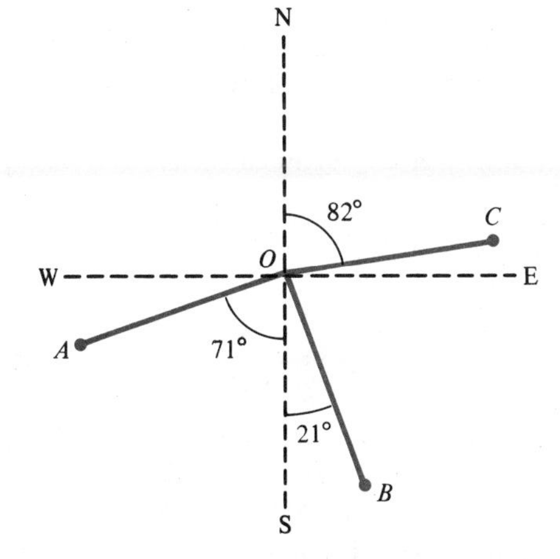

12. Suppose that two forces $\bar{F}_1$ and $\bar{F}_2$ act on a point and that α is the angle from $\bar{F}_1$ to $\bar{F}_2$. Let $\bar{F}$ be the resultant of $\bar{F}_1$ and $\bar{F}_2$ and let θ be the angle from $\bar{F}_1$ to $\bar{F}$.

(a) Show that $|\bar{F}| = (|\bar{F}_1|^2 + |\bar{F}_2|^2 + 2|\bar{F}_1||\bar{F}_2| \cos \alpha)^{1/2}$.

(b) Show that $\theta = \sin^{-1} \dfrac{|\bar{F}_2| \sin \alpha}{|\bar{F}|}$.

In problems 13–18, sketch the graph of each polar equation.

13. $r = 1$
14. $r^2 = 9$
15. $\theta = \frac{\pi}{3}$
16. $\theta^2 = \frac{\pi^2}{16}$
17. $\theta + \frac{\pi}{2} = 0$
18. $\theta^2 - \frac{4\pi}{3}\theta + \frac{7\pi^2}{36} = 0$

In problems 19–24, find a polar equation corresponding to each cartesian equation and sketch the graph. Simplify your answer if possible.

19. $x^2 + y^2 = 25$
20. $y = 6x^2$
21. $3x - 2y = 6$
22. $xy = 1$
23. $(x - 1)^2 + y^2 = 1$
24. $y = \sin x$

In problems 25–30, find a cartesian equation corresponding to each polar equation and sketch the graph. Simplify your answer if possible.

25. $r = 1$
26. $r = 3\cos\theta$
27. $r = \dfrac{1}{\sin\theta - 2\cos\theta}$
28. $\theta = \frac{\pi}{12}$
29. $r\cos\theta = \tan\theta$
30. $r + \frac{1}{r} = 2(\cos\theta + \sin\theta)$

In problems 31–34, use the point-plotting method to sketch the graph of each polar equation.

31. $r = 2(1 - \sin\theta)$ (This is called a *cardioid.*)
32. $r = 3\sin 3\theta$ (This is called a *three-leaved rose.*)
33. $r = \dfrac{6\theta}{\pi}$ (This is called an *Archimedean spiral.*)
34. $r = 3 - 2\cos\theta$ (This is called a *limaçon.*)

REVIEW PROBLEM SET

Ⓒ In problems 1–4, assume that the right triangle ACB is labeled as in Figure 1. In each case, solve the triangle and sketch it. Round off all angles to the nearest hundredth of a degree and all side lengths to four significant digits.

1. $c = 200$ meters, $\alpha = 31.25°$
2. $a = 92.70$ kilometers, $\alpha = 34.45°$
3. $b = 47.33$ microns, $\beta = 56.15°$
4. $b = 1.151 \times 10^4$ meters, $c = 4.703 \times 10^4$ meters

Figure 1

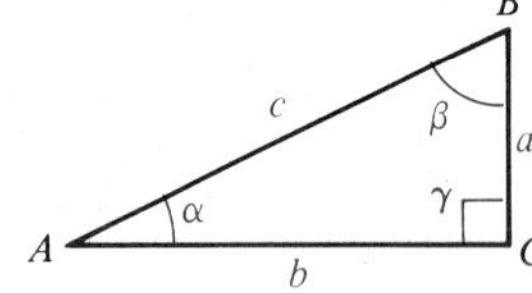

Ⓒ **5.** A straight sidewalk is inclined at an angle of 4.5° from the horizontal. How far must a person walk along this sidewalk to change his or her elevation by 2 meters?

Ⓒ **6.** A ship in distress at night fires a signal rocket straight upward. At an altitude of 0.2 kilometer, the rocket explodes with a brilliant flash that is observed at an angle of elevation of 15.75° by an approaching rescue ship. How far is the rescue ship from the ship in distress?

Ⓒ **7.** A broadcasting antenna tower 200 meters high is to be held vertical by three cables running from a point 10 meters below the top of the tower to concrete anchors sunk in the ground (Figure 2). If the cables are to make angles of 60° with the horizontal, how many meters of cable will be required?

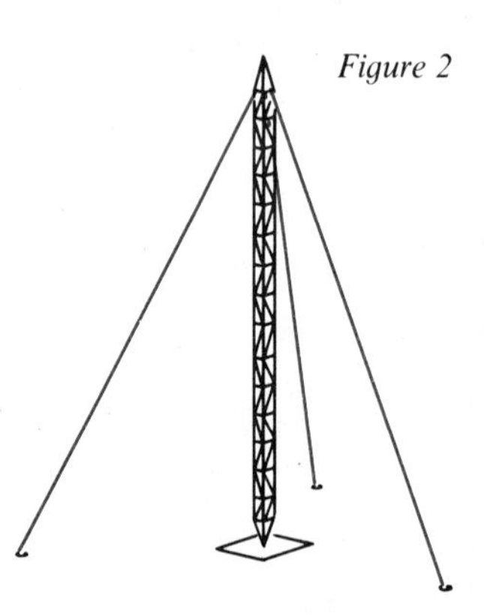

Figure 2

8. Let θ be the angle of depression of the horizon as seen by an astronaut A in a space vehicle h meters above the surface of a planet of radius r meters (Figure 3). Show that

$$h = r(\sec\theta - 1).$$

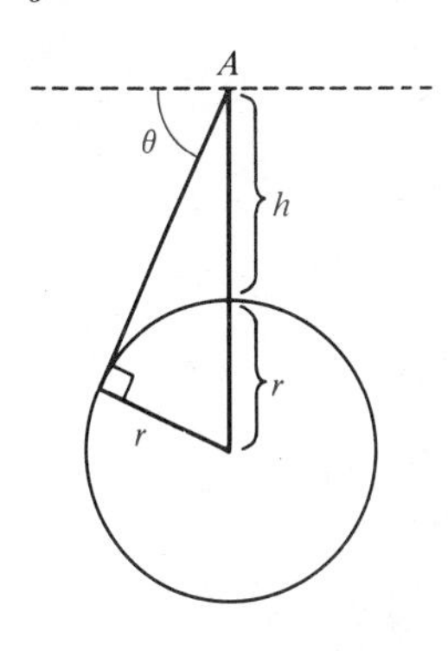

Figure 3

Ⓒ **9.** A certain species of bird is known to fly at an average altitude of 260 meters during migration. An ornithologist observes a migrating flock of these birds flying directly away from her at an angle of elevation of 30°. One minute later, she observes that the angle of elevation of the flock has changed to 20°. Approximately how fast are the birds flying?

Ⓒ **10.** In dealing with alternating current circuits, electrical engineers measure the opposition to current flow by a quantity called **impedance,** whose magnitude Z is measured in ohms. In a circuit driven by an alternating electromotive force (emf) and containing only a resistor and an inductor connected in series, the resistor and inductor both contribute to the impedance. If the resistance is R ohms, the inductance is L henrys, and the driving emf has a frequency of ν hertz, the relationships among Z, R, L, and ν is shown by an **impedance triangle** as in Figure 4. The angle ϕ is called the **phase** and the quantity $2\pi\nu L$ is called the **inductive reactance.** If $R = 4000$ ohms, $L = 2$ henrys, and $\nu = 60$ hertz, find Z and ϕ.

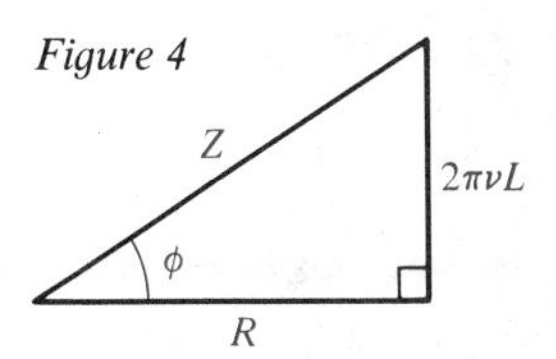

Figure 4

Ⓒ **11.** An electronic echo locator on a commercial fishing boat indicates a school of fish at a slant distance of 575 meters from the boat with an angle of depression of 33.60°. What is the depth of the school of fish?

Ⓒ **12.** A jetliner is flying at an altitude of 10 kilometers. The captain is preparing for a descent to an airport along a straight flight path making an angle of 5° with the horizontal. If the descent will take 12 minutes, what will be the average airspeed of the plane during the descent?

Ⓒ **13.** A ship leaves port and sails S48°10′E for 41 nautical miles. At this point, it turns and sails S41°50′W for 75 nautical miles. Find the bearing of the ship from the port and its distance from the port.

Ⓒ **14.** Two buildings A and B are 2175 feet apart; the angle of depression from the top of B to the top of A is 12.17°; and the angle of depression from the top of B to the bottom of A is 53.35°. Find the heights of the buildings.

Ⓒ In problems 15–20, use the law of sines to find the specified part of each triangle ABC (Figure 5). Round off angles to the nearest hundredth of a degree and side lengths to four significant digits.

15. Find a if $b = 13$ feet, $\alpha = 47°$, and $\gamma = 118°$.
16. Find b if $a = 8.12$ centimeters, $\alpha = 62.45°$, and $\beta = 79.30°$.
17. Find c if $a = 23$ miles, $\alpha = 65°$, and $\beta = 70°$.
18. Find a if $c = 13.70$ kilometers, $\alpha = 62°50'$, and $\beta = 57°30'$.
19. Find γ if $b = 24$ meters, $c = 14$ meters, and $\beta = 38°$.
20. Find α if $a = 5.88$ microns, $c = 12.35$ microns, and $\gamma = 106.55°$.

Figure 5

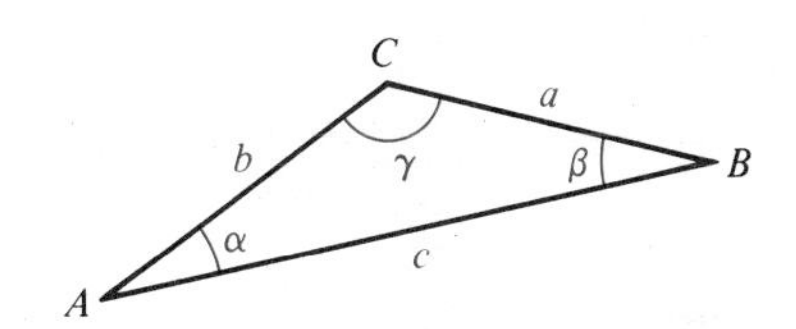

Ⓒ In problems 21–26, solve each triangle ABC (Figure 5) by using the method for the ambiguous case. Round off angles to the nearest hundredth of a degree and side lengths to four significant digits. Be certain to find all possible triangles that satisfy the given conditions.

21. $b = 50\sqrt{3}$ feet, $c = 150$ feet, $\beta = 30°$
22. $a = 4.50$ kilometers, $b = 5.30$ kilometers, $\alpha = 60.33°$
23. $b = 50$ centimeters, $c = 58$ centimeters, $\gamma = 57.25°$

24. $b = 5.94$ angstroms, $c = 7.23$ angstroms, $\beta = 38°$
25. $a = 8.00$ miles, $b = 10.00$ miles, $\alpha = 54°$
26. $a = 40.33$ nautical miles, $b = 42.01$ nautical miles, $\alpha = 110.05°$

Ⓒ In problems 27–32, use the law of cosines to find the specified part of each triangle ABC (Figure 5). Round off angles to the nearest hundredth of a degree and side lengths to four significant digits.

27. Find α if $a = 7$ feet, $b = 8$ feet, and $c = 3$ feet.
28. Find a if $b = 11$ meters, $c = 12$ meters, and $\alpha = 81°$.
29. Find c if $a = 14$ kilometers, $b = 8$ kilometers, and $\gamma = 37°$.
30. Find b if $a = 7$ centimeters, $c = 11$ centimeters, and $\beta = 53.35°$.
31. Find a if $b = 12.30$ inches, $c = 4.85$ inches, and $\alpha = 161.15°$.
32. Find γ if $a = 48.31$ meters, $b = 35.11$ meters, and $c = 63.27$ meters.

Ⓒ In problems 33–38, use an appropriate method to solve each triangle ABC (Figure 5). Round off angles to the nearest hundredth of a degree and side lengths to four significant digits.

33. $a = 49.7$ kilometers, $b = 111$ kilometers, $\gamma = 41.05°$
34. $\alpha = 37.17°$, $\beta = 82.25°$, $a = 7777$ meters
35. $a = 1.12$ microns, $c = 0.98$ micron, $\gamma = 55°$
36. $a = 33.39$ centimeters, $b = 72.27$ centimeters, $c = 106.2$ centimeters
37. $\alpha = 42.45°$, $\beta = 32.15°$, $b = 1.41$ nautical miles
38. $a = 1.80$ millimeters, $b = 1.20$ millimeters, $\alpha = 47°$

Ⓒ **39.** The three sides of a triangle have lengths $a = 27$ meters, $b = 44$ meters, and $c = 65$ meters. Use Hero's formula (page 186) to find the area of the triangle.

Ⓒ **40.** A section of West Germany is called the brown-coal triangle because it contains vast reserves of lignite (brown coal). It is a triangular area with vertices at the industrial cities of Aachen, Cologne, and Düsseldorf. The distance from Aachen to Cologne is 67 kilometers; the distance from Cologne to Düsseldorf is 37 kilometers; and the distance from Düsseldorf to Aachen is 75 kilometers. Find the approximate area of the brown-coal triangle in square kilometers.

Ⓒ **41.** In the western North Atlantic there is a mysterious region, shaped roughly like a triangle with sides of length 925 miles, 850 miles, and 1300 miles, that has been the scene of a number of unexplained disappearances and disasters involving both ships and aircraft. Find the approximate area of this region in square miles.

Ⓒ **42.** Engineers excavating a new subway tunnel in a large city encounter a region of very hard rock (Figure 6). They can either drill the tunnel straight through the rock along path $\overline{AC}$ at a cost of \$5500 per foot, or they can divert the tunnel along path $\overline{AB}$ and then path $\overline{BC}$ at a cost of \$4700 per foot. If angle $BAC = 20°$ and angle $ACB = 10°$, which path would be the cheaper?

Figure 6

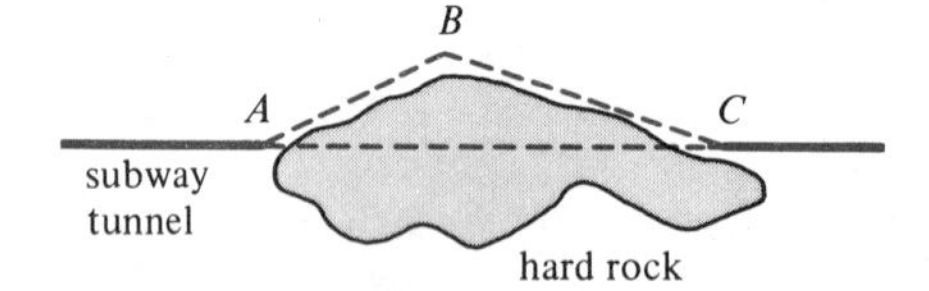

Ⓒ **43.** Two observers are situated on level ground on opposite sides of a tall building. The top of the building is 1600 meters from the first observer at an angle of elevation of 15. The top of the building is 650 meters from the second observer. How far apart are the observers?

Ⓒ **44.** A vertical utility pole is supported on an embankment by a guy wire from 1 meter below the top of the pole to a point 20 meters up the embankment from the foot of the pole (Figure 7). If the embankment makes an angle of 17° with the horizontal and if the guy wire makes an angle of 21° with the horizontal, how tall is the utility pole?

Figure 7

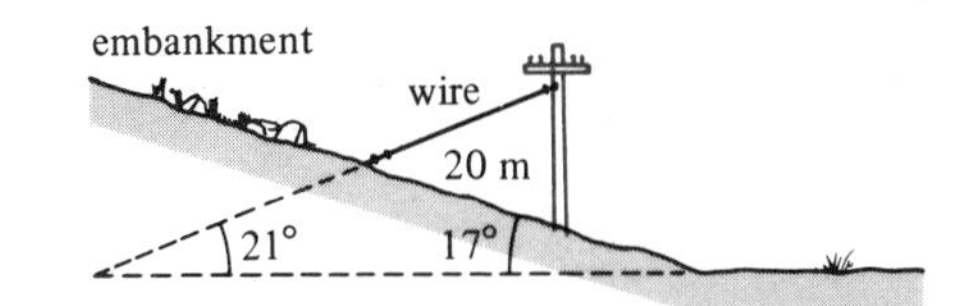

Ⓒ **45.** A CB radio in a person's car can communicate with a base station located at the person's home over a maximum distance of 12 miles. The car leaves the home and travels due west for 5 miles to an interstate highway. The car moves onto the straight highway at precisely 1:00 P.M. traveling N30°E at 55 miles per hour. At what time (to the nearest minute) is communication lost between the car and the base station?

Ⓒ **46.** Apollo objects in the solar system are asteroidlike masses whose orbits intersect the orbit of the earth. Geologists believe that several craterlike formations in Canada and elsewhere were caused by collisions of such objects with the earth. An astronomer observes through a telescope that the angle formed by the lines of sight to two Apollo objects is 43.33°, and that the distances to the two objects are 15,000,000 and 43,000,000 kilometers. How far apart are the two Apollo objects?

Ⓒ **47.** A natural-gas pipeline is to be constructed through a swamp. An engineer obtains the distance across the swamp by establishing points A and B at both ends of it along the path the pipeline is to follow, and finding a third point C outside the swamp at a location that can be seen from both A and B. If $|\overline{CA}| = 1100$ meters, $|\overline{CB}| = 990$ meters, and angle ACB is 75°, what is the distance $|\overline{AB}|$ across the swamp?

Ⓒ **48.** A group of antinuclear activists is conducting a protest march to the construction site of a nuclear power plant. They are marching at constant speed along a flat plane directly toward a huge concrete cooling tower that has already been finished. At 7:00 A.M. the leader of the march observes that the top of the cooling tower is at an angle of elevation of 2°. At 7:30 A.M. the angle of elevation has increased to 5°. Estimate to the nearest minute the time of arrival of the protest group at the construction site.

Ⓒ **49.** A subatomic particle that has been created at point C in a bubble chamber travels in a straight line for 3.5 cm to point B, where it collides with another particle and its path is deflected. It then travels in a straight line from B for 2.3 cm to point A, where it is annihilated. If $|\overline{CA}| = 4.4$ cm, find the deflection angle $\theta = 180° -$ angle CBA.

50. Two small ships are traveling at constant speeds on straight-line courses that intersect. The captain of the first ship measures the bearing of the second ship at the beginning and at the end of a 2-minute interval. The bearing has changed during this interval, so the captain concludes that the ships will not collide. Is this reasoning correct? Why or why not?

In problems 51–62, let $\bar{A} = \langle 2, -1 \rangle$, $\bar{B} = \langle 3, 3 \rangle$, and $\bar{C} = \langle -5, 4 \rangle$. Evaluate each expression.

51. $5\bar{A}$ **52.** $-4\bar{B}$ **53.** $\bar{A} + \bar{B}$ **54.** $\bar{A} - \bar{B}$

55. $2\bar{A} + 3\bar{B}$ **56.** $3\bar{A} + \bar{C}$ **57.** $2\bar{C} - \bar{B}$ **58.** $3\bar{B} + 4\bar{C} - \bar{A}$

59. $|4\bar{A}|$ **60.** $|\bar{A} - \bar{B}|$ **61.** $|\bar{A} + \bar{B} - \bar{C}|$ **62.** $|2\bar{C} - 3\bar{B} + \bar{A}|$

63. Find the components of the vector $\bar{A}$ from the given information about its length $|\bar{A}|$ and its direction angle θ:

(a) $|\bar{A}| = 50, \theta = 30°$ Ⓒ (b) $|\bar{A}| = 70, \theta = 61°$ Ⓒ (c) $|\bar{A}| = 25, \theta = 310°$ (d) $|\bar{A}| = 250, \theta = -\frac{\pi}{6}$

64. Give an example to show that in general $|\bar{A} + \bar{B}| \neq |\bar{A}| + |\bar{B}|$. Are there any cases in which equality does hold?

65. Find the magnitude and the smallest nonnegative direction angle of each vector:
(a) $\langle 1, \sqrt{3} \rangle$ [C] (b) $\langle 12, 15 \rangle$ [C] (c) $\langle -30, 40 \rangle$ [C] (d) $\langle \frac{3}{16}, \frac{1}{4} \rangle$ [C] (e) $3\bar{i} + \sqrt{7}\bar{j}$

66. Give an example to show that the direction angle of the sum of two vectors is not necessarily the sum of their direction angles.

[C] **67.** Suppose that $\bar{F}_1$ and $\bar{F}_2$ represent forces acting on a point, and that α is the angle from $\bar{F}_1$ to $\bar{F}_2$. If $\bar{F}$ denotes the resultant force, find the magnitude $|\bar{F}|$ and the angle θ from $\bar{F}_1$ to $\bar{F}$ in each of the following cases:
(a) $|\bar{F}_1| = 20$ pounds, $|\bar{F}_2| = 18$ pounds, $\alpha = 71^\circ$
(b) $|\bar{F}_1| = 139$ newtons, $|\bar{F}_2| = 156$ newtons, $\alpha = 83^\circ$
(c) $|\bar{F}_1| = 7$ tonnes, $|\bar{F}_2| = 9$ tonnes, $\alpha = 125^\circ$
(d) $|\bar{F}_1| = 300$ dynes, $|\bar{F}_2| = 450$ dynes, $\alpha = 270^\circ$

[C] **68.** Find the angle θ from the position vector $\bar{R}_1 = \langle 5, 3 \rangle$ to the position vector $\bar{R}_2 = \langle -7, 2 \rangle$.

[C] **69.** A cargo jet is headed N45°W with an air speed of 500 knots. The wind is blowing N60°E at a speed of 70 knots. Find (a) the ground speed, (b) the drift angle, and (c) the course of the jet.

[C] **70.** Using ropes, two workers are gently lowering a stone weighing 60 kilograms as shown in Figure 8. The ropes make angles of 60° and 45° with the horizontal. Make a vector diagram showing all three forces acting on the stone, including the force of gravity. Using the fact that all three forces are in equilibrium—that is, their vector sum is $\bar{0}$—find the magnitude of the force exerted by each of the workers on their ropes.

Figure 8

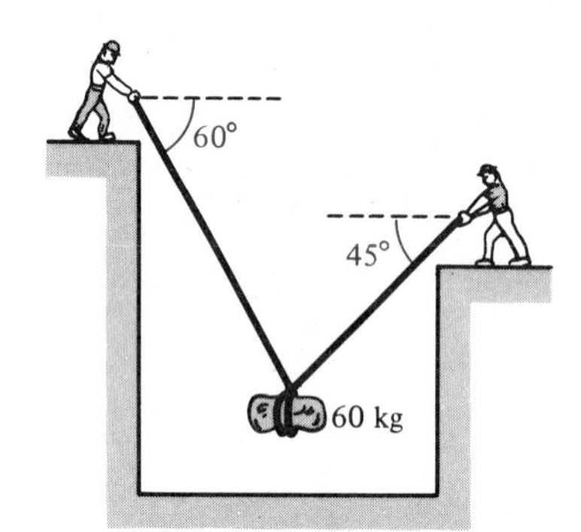

[C] **71.** An oceanographic research vessel leaves Miami on a course for a small island whose bearing is S43°20′E. The vessel is moving through the water at a constant speed of 20 knots, and the captain has charted the course to compensate for the fact that a Gulf Stream current of 7 knots is running due north. Determine the heading of the vessel.

72. If a force represented by a vector $\bar{F}$ acts on a particle and produces a displacement represented by a vector $\bar{D}$, the *work* done is given by $|\bar{F}||\bar{D}|\cos\alpha$, where α is the angle between $\bar{F}$ and $\bar{D}$. Calculate the work done by the force $\bar{F} = \langle -3, 4 \rangle$ in producing the displacement $\bar{D} = \langle 0, 7 \rangle$. [If you use the law of cosines, you won't need a calculator to solve this problem.]

73. Convert the polar coordinates to cartesian coordinates:
(a) $(2, 45^\circ)$ (b) $\left(-2, \frac{\pi}{4}\right)$ [C] (c) $(-1, 280^\circ)$
(d) $(1.5, 0^\circ)$ (e) $(-4, -210^\circ)$ (f) $(6, 315^\circ)$

74. Plot the point $P = (-5, \pi/6)$ in the polar coordinate system and give other polar representations $P = (r, \theta)$ of P that satisfy each of the following conditions:
(a) $r < 0$ and $0 \leq \theta < 2\pi$ (b) $r > 0$ and $-2\pi < \theta \leq 0$
(c) $r < 0$ and $-2\pi < \theta \leq 0$

75. Convert the cartesian coordinates to polar coordinates with $r \geq 0$ and $-\pi < \theta \leq \pi$. If an approximation is necessary, round off the angle θ to the nearest hundredth of a radian.
(a) $(-1, 0)$ (b) $(-7, -7)$ © (c) $(-5, 12)$
© (d) $(3, -4)$ © (e) $(-8, 15)$ © (f) $(-30, -16)$

76. Show that the distance d between the points (r_1, θ_1) and (r_2, θ_2) in the polar coordinate system is given by the formula

$$d = \sqrt{r_1^2 - 2r_1r_2\cos(\theta_1 - \theta_2) + r_2^2}.$$

In problems 77–84, convert each cartesian equation into a corresponding polar equation by making the substitutions $x = r\cos\theta$, $y = r\sin\theta$, and $x^2 + y^2 = r^2$. Simplify your answer if possible.

77. $y = 3x + 1$ **78.** $(x - 1)^2 + (y - 2)^2 = 4$ **79.** $x^2 + y^2 = 4y$
80. $y^2(2 - x) = x^3$ **81.** $y^2 = 6x$ **82.** $(x^2 + y^2)^2 = x^2 - y^2$
83. $4x^2 + y^2 = 4$ **84.** $x^2 - 4y^2 = 4$

In problems 85–88, convert each polar equation into a corresponding cartesian equation.

85. $r = \cos\theta + \sin\theta$ **86.** $r = \sin 2\theta$ **87.** $r^2\cos 2\theta = 2$ **88.** $\theta = \frac{5\pi}{4}$

89. Using the result of problem 87, sketch the graph of $r^2\cos 2\theta = 2$.

90. Using the result in problem 76, write the polar form of the equation of a circle of radius a with center at (r_0, θ_0).

© In problems 91–94, sketch the graph of each polar equation by the point-plotting method.

91. $r = 4(1 + \cos\theta)$ (This is called a *cardioid.*)
92. $r = 2 - 3\sin\theta$ (This is called a *limaçon with a loop.*)
93. $r = \cos 2\theta$ (This is called a *four-leaved rose.*)
94. $r^2 = \cos 2\theta$ (This is called a *lemniscate.*)

5 Complex Numbers and Analytic Geometry

In this chapter, we give a brief introduction to the algebra of complex numbers and to analytic geometry. The trigonometric functions have important roles to play in connection with both of these topics. The chapter includes the complex number system, polar form of complex numbers, De Moivre's theorem, equations of straight lines, conic sections, and translation and rotation of the coordinate axes. Here we can only give the "flavor" of these topics—the interested reader is urged to consult more advanced textbooks for a more detailed and systematic presentation.

1 Complex Numbers

Because the square of a real number is nonnegative, there is no real number whose square is -1. However, mathematicians, who can be highly imaginative people, long ago began to wonder what would happen if there were a "number" i such that $i^2 = -1$. Although i can't be a real number, there are mathematical entities that aren't real numbers and yet behave algebraically like real numbers; for instance, vectors. Leaving aside for now the question of just what i is, let's work with it a bit and see what happens.

If we multiply i by a real number, say, 5, we simply have to write the result as $5i$. More generally, we write the product of i and a real number b as bi or as ib. By further such multiplication, we get nothing new. For instance, if c is a real number and we multiply bi by c, we just get $(cb)i$, which again has the same form—a real number times i. On the other hand, if we multiply bi by ci, we get

$$(bi)(ci) = (bc)i^2 = (bc)(-1) = -bc,$$

a real number! Adding two "numbers" of the form bi and ci produces

$$bi + ci = (b + c)i,$$

another "number" of the same form—nothing new. But if we wish to add a real number, say, 3, to bi, we write the result as $3 + bi$, which is something new! More generally, we write the sum of a real number a and the "number" bi as $a + bi$.

Now, if we suppose that the basic algebraic properties of the real numbers continue to operate for "numbers" of the form $a + bi$, we find that the sum, difference, and product of such "numbers" are again of the same form. Indeed, if a, b, c, and d are real numbers, we have

1. $(a + bi) + (c + di) = a + c + bi + di$
$= (a + c) + (b + d)i$
2. $(a + bi) - (c + di) = a - c + bi - di$
$= (a - c) + (b - d)i$
3. $(a + bi)(c + di) = ac + adi + bic + bidi$
$= ac + adi + bci + bdi^2$
$= ac + adi + bci + bd(-1)$
$= (ac - bd) + (ad + bc)i.$

All very encouraging!

A "number" of the form $a + bi$, where a and b are real numbers, is called a **complex number.** Sums, differences, and products of complex numbers are calculated as in formulas 1, 2, and 3 above. It is understood that $i^2 = -1$ and that the real number 0 has its usual properties with regard to addition and multiplication, so that

$$0 + bi = bi \quad \text{and} \quad 0i = 0.$$

Finally, let's agree that

$$a + bi = c + di \quad \text{means that} \quad a = c \quad \text{and} \quad b = d.$$

In other words, a single equation involving complex numbers represents *two* equations involving real numbers.

The real numbers a and b are called the **real part** and the **imaginary part,** respectively, of the complex number $a + bi$. Thus, to say that two complex numbers are equal is to say that their real parts are equal and their imaginary parts are equal.

The set of all complex numbers, equipped with the algebraic operations of addition, subtraction, and multiplication, is called the **complex number system** and is denoted by the symbol $\mathbb{C}$. Note that a real number a can be regarded as a complex number whose imaginary part is zero, $a = a + 0i$; therefore, the real number system $\mathbb{R}$ forms part of the complex number system $\mathbb{C}$.

It isn't difficult to show that the complex numbers have the same basic algebraic properties as the real numbers. In particular, every nonzero complex number has a reciprocal.

If $a + bi \neq 0$, you can obtain the reciprocal $\dfrac{1}{a + bi}$ by the trick of multiplying the numerator and denominator by the complex number $a - bi$:

$$\frac{1}{a + bi} = \frac{a - bi}{(a + bi)(a - bi)} = \frac{a - bi}{a^2 - (bi)^2} = \frac{a - bi}{a^2 - b^2 i^2}$$

$$= \frac{a - bi}{a^2 - b^2(-1)} = \frac{a - bi}{a^2 + b^2} = \left(\frac{a}{a^2 + b^2}\right) + \left(\frac{-b}{a^2 + b^2}\right)i.$$

In other words, $\frac{a}{a^2 + b^2}$ and $\frac{-b}{a^2 + b^2}$ are the real and imaginary parts of $\frac{1}{a + bi}$. Notice that the denominator $a^2 + b^2$ is a positive real number because of the fact that $a + bi \neq 0$, so a and b cannot both be zero. You can verify by direct calculation that the complex number $\left(\frac{a}{a^2 + b^2}\right) + \left(\frac{-b}{a^2 + b^2}\right)i$ is indeed the reciprocal of the nonzero complex number $a + bi$ (Problem 66).

The complex number $a - bi$ used in the calculation above is called the **complex conjugate** of $a + bi$. Complex conjugates are also useful for dealing with quotients of complex numbers. If $a + bi \neq 0$, the **quotient** $\frac{c + di}{a + bi}$ is defined to be the product of $c + di$ and $1/(a + bi)$. If you multiply the numerator and denominator of a quotient by the complex conjugate of the denominator, you will obtain a fraction with a positive real number in the denominator. This permits you to find the real and imaginary parts of the original quotient.

Example Express each complex number in the form $a + bi$, where a and b are real numbers.

(a) $(3 + 5i) + (6 + 5i)$ (b) $(2 - 3i) - (6 + 4i)$ (c) $(4 + 3i)(2 + 4i)$

(d) $\frac{1}{4 + 3i}$ (e) $\frac{2 - 3i}{1 - 4i}$

Solution

(a) $(3 + 5i) + (6 + 5i) = 3 + 6 + 5i + 5i = 9 + 10i$

(b) $(2 - 3i) - (6 + 4i) = 2 - 6 - 3i - 4i = -4 - 7i = -4 + (-7)i$

(c) $$\begin{aligned}(4 + 3i)(2 + 4i) &= 8 + 16i + 6i + 12i^2\\ &= 8 + 22i + (12)(-1)\\ &= 8 - 12 + 22i = -4 + 22i\end{aligned}$$

(d) Multiplying numerator and denominator by $4 - 3i$, the complex conjugate of $4 + 3i$, we obtain

$$\begin{aligned}\frac{1}{4 + 3i} &= \frac{4 - 3i}{(4 + 3i)(4 - 3i)} = \frac{4 - 3i}{16 - 9i^2}\\ &= \frac{4 - 3i}{16 - 9(-1)} = \frac{4 - 3i}{16 + 9} = \frac{4 - 3i}{25}\\ &= \frac{4}{25} + \frac{-3}{25}i.\end{aligned}$$

(e) Multiplying numerator and denominator by $1 + 4i$, the complex conjugate of $1 - 4i$, we obtain

$$\begin{aligned}\frac{2 - 3i}{1 - 4i} &= \frac{(2 - 3i)(1 + 4i)}{(1 - 4i)(1 + 4i)} = \frac{2 + 8i - 3i - 12i^2}{1 - 16i^2} = \frac{2 + 5i - 12(-1)}{1 - 16(-1)}\\ &= \frac{2 + 12 + 5i}{1 + 16} = \frac{14 + 5i}{17} = \frac{14}{17} + \frac{5}{17}i.\end{aligned}$$

Complex numbers, like real numbers, can be denoted by letters of the alphabet and treated as variables or unknowns. The letters z and w are special favorites for this purpose. If a and b are real numbers and

$$z = a + bi,$$

the complex conjugate of z is often written as

$$\bar{z} = a - bi.$$

Notice that

$$z\bar{z} = (a + bi)(a - bi) = a^2 - (bi)^2$$
$$= a^2 - b^2i^2 = a^2 - b^2(-1) = a^2 + b^2;$$

hence, *$z\bar{z}$ is always a nonnegative real number.* Its principal square root is called the **absolute value** or **modulus** of z and is written as $|z|$. Thus, by definition,

$$|z| = \sqrt{z\bar{z}} = \sqrt{a^2 + b^2}.$$

Therefore,

$$\boxed{z\bar{z} = |z|^2.}$$

Example If $z = 3 + 4i$, find:

(a) $\bar{z}$ (b) $z + \bar{z}$ (c) $z - \bar{z}$ (d) $z\bar{z}$ (e) $|z|$.

Solution

(a) $\bar{z} = 3 - 4i$

(b) $z + \bar{z} = (3 + 4i) + (3 - 4i) = 6$

(c) $z - \bar{z} = (3 + 4i) - (3 - 4i) = 8i$

(d) $z\bar{z} = (3 + 4i)(3 - 4i) = 9 + 16 = 25$

(e) $|z| = \sqrt{z\bar{z}} = \sqrt{25} = 5$

As we have mentioned, a real number a can be regarded as a complex number $a + 0i$ with imaginary part zero. Thus,

$$\bar{a} = a - 0i = a,$$

so *each real number is its own complex conjugate.* In particular,

$$|a| = \sqrt{a\bar{a}} = \sqrt{a^2}.$$

An interesting and useful property of complex conjugation is that it "preserves" all of the algebraic operations; that is, if z and w are complex numbers, then

(i) $\overline{z + w} = \bar{z} + \bar{w}$	(ii) $\overline{z - w} = \bar{z} - \bar{w}$
(iii) $\overline{zw} = \bar{z}\bar{w}$	(iv) $\overline{\left(\dfrac{z}{w}\right)} = \dfrac{\bar{z}}{\bar{w}}$, if $w \neq 0$.

This is easy to show for addition and subtraction (Problem 65). Let's check it for multiplication: Let $z = a + bi$ and let $w = c + di$, where a, b, c, and d are real numbers. Then $\bar{z} = a - bi$ and $\bar{w} = c - di$ and

$$\begin{aligned}\overline{zw} &= \overline{(a + bi)(c + di)} = \overline{(ac - bd) + (ad + bc)i} \\ &= (ac - bd) - (ad + bc)i \\ &= (a - bi)(c - di) = \bar{z}\bar{w}.\end{aligned}$$

We leave it to you to check that the conjugate of a quotient is the quotient of the conjugates (Problem 68).

In Section 2 of Chapter 1, we showed that the absolute value of a product of real numbers is the product of their absolute values. This property extends to complex numbers, as the following calculation shows:

$$|zw| = \sqrt{(zw)\overline{(zw)}} = \sqrt{zw\bar{z}\bar{w}} = \sqrt{z\bar{z}w\bar{w}} = \sqrt{|z|^2|w|^2} = |z|\,|w|.$$

A similar calculation (Problem 69) shows that, for $w \neq 0$,

$$\left|\frac{z}{w}\right| = \frac{|z|}{|w|}.$$

Integral powers of complex numbers are defined just as they are for real numbers: $z^2 = zz$, $z^3 = zzz$, and so on; and, if $z \neq 0$, $z^0 = 1$, $z^{-1} = \dfrac{1}{z}$, $z^{-2} = \dfrac{1}{z^2}$, and so on. Notice that

$$\begin{array}{lll} i^1 = i & i^5 = i & i^9 = i \\ i^2 = -1 & i^6 = -1 & i^{10} = -1 \\ i^3 = -i & i^7 = -i & i^{11} = -i \\ i^4 = 1 & i^8 = 1 & i^{12} = 1 \end{array}$$

and so on. Thus, the positive integer powers of i endlessly repeat the pattern of the first four.

Example Find i^{59}.

Solution We divide 59 by 4 to obtain a quotient of 14 and a remainder of 3. Therefore,

$$i^{59} = i^{(4 \cdot 14)+3} = (i^4)^{14} i^3 = 1^{14}(-i) = -i.$$

Although complex numbers were originally introduced to provide roots for certain algebraic equations (such as $x^2 + 1 = 0$) that have no real solutions, they now have a wide variety of important applications in physics and engineering. For example, in 1893, Charles P. Steinmetz (1865–1923), an American electrical engineer born in Germany, developed a theory of alternating currents based on the complex numbers.

In direct current theory, **Ohm's law**

$$E = IR$$

relates the electromotive force (voltage) E, the current I, and the resistance R. Because of inductive and capacitative effects, voltage and current may be out of phase in alternating current circuits, and the equation $E = IR$ may no longer hold. Steinmetz saw that, by representing voltage and current with *complex numbers* E and I, he could deal algebraically with phase differences. Furthermore, he combined the resistance R, the inductive effect X_L (called **inductive reactance**), and the capacitative effect X_C (called **capacitative reactance**) in a single complex number

$$Z = R + (X_L - X_C)i,$$

called **complex impedance.** Using the complex numbers E, I, and Z, Steinmetz showed that Ohm's law for alternating currents takes the form

$$E = IZ.$$

Today, these ideas of Steinmetz are used routinely by electrical engineers all over the world. It has been said that Steinmetz "generated electricity with the square root of minus one."

PROBLEM SET 1

In problems 1–52, express each complex number in the form $a + bi$, where a and b are real numbers.

1. $(2 + 3i) + (7 - 2i)$
2. $(-1 + 2i) + (3 + 4i)$
3. $(4 + i) + 2(3 - i)$
4. $(4 + 3i) + 3(2 - 5i)$
5. $(5 - 4i) + (7 - 2i)$
6. $(\frac{1}{2} - \frac{2}{3}i) + (\frac{3}{4} + \frac{1}{6}i)$
7. $(3 + 2i) - (5 + 4i)$
8. $(2 - 3i) - i$
9. $3(1 + 2i) - 4(2 + i)$
10. $(\frac{1}{2} - \frac{4}{3}i) - \frac{1}{6}(5 + 7i)$
11. $2(5 + 4i) - 3(7 + 4i)$
12. $i - \frac{1}{2}(1 + 5i)$
13. $-(-3 + 5i) - (4 + 9i)$
14. $(\sqrt{2} + \sqrt{3}i) - \left(\frac{\sqrt{2}}{2} - \frac{\sqrt{3}}{3}i\right)$
15. $(2 + i)(1 + 5i)$
16. $(4 + 3i)(-1 + 2i)$
17. $(7 + 4i)(3 + 6i)$
18. $(7 - 6i)(-5 - i)$
19. $(3 - 2i)(-3 + i)$
20. $i(3 + 7i)$
21. $(-7 + 3i)(-3 + 2i)$
22. $(\frac{2}{3} + \frac{3}{5}i)(\frac{1}{2} - \frac{1}{3}i)$
23. $-8i(5 + 8i)$
24. $(\sqrt{2} + \sqrt{3}i)(\sqrt{2} - \sqrt{3}i)$
25. $(-4i)(-5i)$
26. $(-2i)(3i)(-4i)$
27. $(\frac{1}{2} + \frac{1}{3}i)(\frac{1}{2} - \frac{1}{3}i)$
28. $(\sqrt{2} + \sqrt{3}i)^2$
29. i^{21}
30. i^{41}
31. i^{201}
32. $(1 + i)^3$
33. $(4 + 2i)^2$
34. $(1 - i)^4$
35. $[\frac{1}{2} + (\sqrt{3}/2)i]^2$
36. $(\cos\theta + i\sin\theta)^2$
37. $1/(2 + 3i)$
38. $1/(3 - 4i)$
39. $3/(7 + 2i)$
40. $-4i/(6 - i)$
41. $(3 - 4i)/(4 + 2i)$
42. $(\pi + 4i)/(2 - i)$
43. $(7 + 2i)/(3 - 5i)$
44. $1/i$
45. $(4 + i)/[(3 - 2i) + (4 - 3i)]$
46. $\dfrac{2 + 3i}{4 - 3i} + \dfrac{3 + 5i}{1 - 2i}$
47. $\dfrac{2 - 6i}{3 + i} - \dfrac{4 + i}{3 + i}$
48. $\dfrac{(1 - i)(2 + i)}{(2 - 3i)(3 - 4i)}$
49. $(3 - 2i)/(3 + i) + (4i)/(3 - 7i)$
50. $(3i^3 - 5)/(1 + i^5)$
51. $(3 - 2i)/[(2 + i)(5 + 2i)]$
52. $[(3 - i^7)/(i^9 - 3)]^2$

In problems 53–64, calculate:

(a) $\bar{z}$ (b) $z + \bar{z}$ (c) $z - \bar{z}$ (d) $z\bar{z}$ (e) $|z|$.

53. $z = 2 + i$ **54.** $z = i$ **55.** $z = -i$

56. $z = (1 + i)^2$ **57.** $z = (1 + i)/(1 - i)$ **58.** $z = -3i^5$

59. $z = -12 + 5i$ **60.** $z = 7i^{101}$ **61.** $z = (4 - 3i)/(2 + 4i)$

62. $z = (2 + i)(3 - 7i)$ **63.** $z = 5$ **64.** $z = i + i^2 + i^3$

65. Show that (a) $\overline{z + w} = \bar{z} + \bar{w}$ and (b) $\overline{z - w} = \bar{z} - \bar{w}$.

66. If a and b are real numbers and $a + bi \neq 0$, show by direct calculation that

$$(a + bi)\left[\left(\frac{a}{a^2 + b^2}\right) + \left(\frac{-b}{a^2 + b^2}\right)i\right] = 1.$$

67. If $z \neq 0$, show that $\dfrac{1}{z} = \dfrac{\bar{z}}{|z|^2}$.

68. If $w \neq 0$, show that $\overline{\left(\dfrac{z}{w}\right)} = \dfrac{\bar{z}}{\bar{w}}$.

69. If $w \neq 0$, show that $\left|\dfrac{z}{w}\right| = \dfrac{|z|}{|w|}$.

70. Prove the **triangle inequality**

$$|z + w| \leq |z| + |w|.$$

Hint: Begin by proving that

$$|z + w|^2 \leq (|z| + |w|)^2$$

by using the fact that

$$|z + w|^2 = (z + w)\overline{(z + w)}.$$

2 Polar Form and De Moivre's Theorem

We begin this section by giving an answer to the question (temporarily put aside in Section 1) of just what complex numbers are. Although there are several possible representations of the complex numbers, their interpretation as geometric points in the plane has the most intuitive appeal. According to this scheme, if a and b are real numbers, then the complex number $a + bi$ is represented by the point (a, b) in the complex plane (Figure 1). When each point in a coordinate plane is made to correspond to a complex number in this way, we refer to the plane as the **complex plane.** This correspondence is so compelling that it is natural to *identify* a complex number with its corresponding geometric point, and we shall do so in what follows. Under this identification, there's nothing at all "imaginary" about i or any other complex number; nevertheless, the word "imaginary" continues to be used for historical reasons.

Figure 1

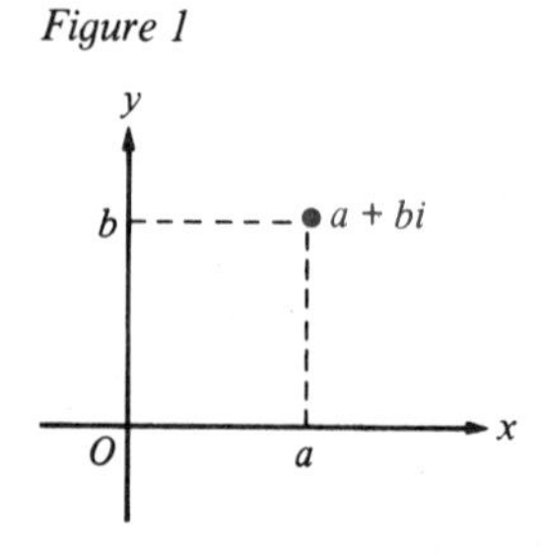

In the complex plane, the x axis is called the **real axis** because its points correspond to real numbers and the y axis is called the **imaginary axis** because its points correspond to multiples of i by real numbers (Figure 2). Notice that the absolute value or modulus $|x + yi| = \sqrt{x^2 + y^2}$ is just the distance $r = \sqrt{x^2 + y^2}$ between the point $x + yi$ and the origin (Figure 2). An angle θ in standard position that contains the point $x + yi$ on its terminal side is called an **argument** of $x + yi$. In other words, the absolute value $r = |x + yi|$ and an argument θ of $x + yi$ are just the polar coordinates (r, θ) of the point $x + yi$. Therefore,

$$x = r\cos\theta$$

and

$$y = r\sin\theta$$

(Chapter 4, Section 5).

Figure 2

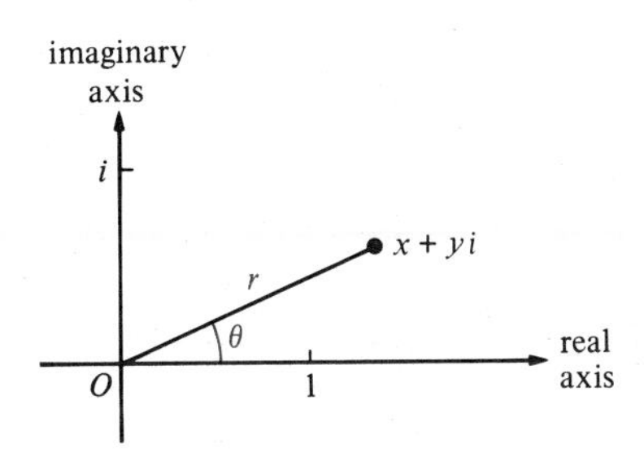

It follows from the discussion above that a complex number z with real part x, imaginary part y, absolute value $r = |z|$, and argument θ can be written either in **cartesian form** (also called **rectangular form**)

$$z = x + yi$$

or in **polar form**

$$\begin{aligned} z &= r\cos\theta + (r\sin\theta)i \\ &= r(\cos\theta + i\sin\theta). \end{aligned}$$

Example 1 Plot the point $z = 4(\cos 60^\circ + i\sin 60^\circ)$ in the complex plane and rewrite z in cartesian form.

Solution The point z is plotted in Figure 3. In cartesian form

$$\begin{aligned} z &= 4\cos 60^\circ + (4\sin 60^\circ)i \\ &= 4(\tfrac{1}{2}) + [4(\sqrt{3}/2)]\,i \\ &= 2 + 2\sqrt{3}i. \end{aligned}$$

Figure 3

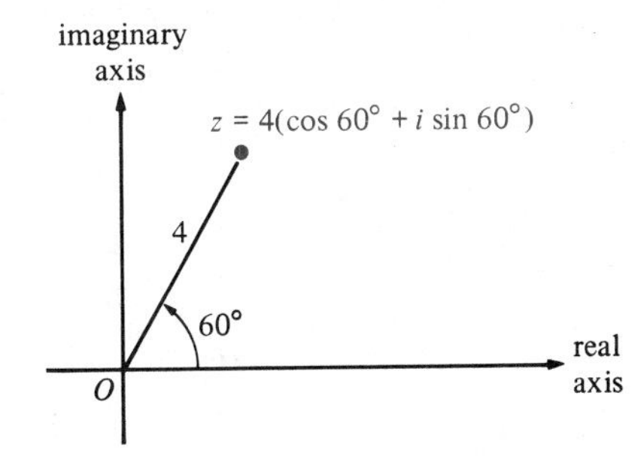

Example 2 Plot the point $z = \sqrt{2} - \sqrt{2}i$ in the complex plane and rewrite z in polar form.

Solution The point z is plotted in Figure 4. The real and imaginary parts of z are $x = \sqrt{2}$ and $y = -\sqrt{2}$, respectively. It follows that

$$r = \sqrt{x^2 + y^2} = \sqrt{(\sqrt{2})^2 + (-\sqrt{2})^2} = \sqrt{2 + 2} = 2,$$

so

$$\cos\theta = \frac{x}{r} = \frac{\sqrt{2}}{2} \quad \text{and} \quad \sin\theta = \frac{y}{r} = -\frac{\sqrt{2}}{2};$$

hence, $\theta = 7\pi/4$. Therefore,

$$z = 2\left(\cos\frac{7\pi}{4} + i\sin\frac{7\pi}{4}\right).$$

Figure 4

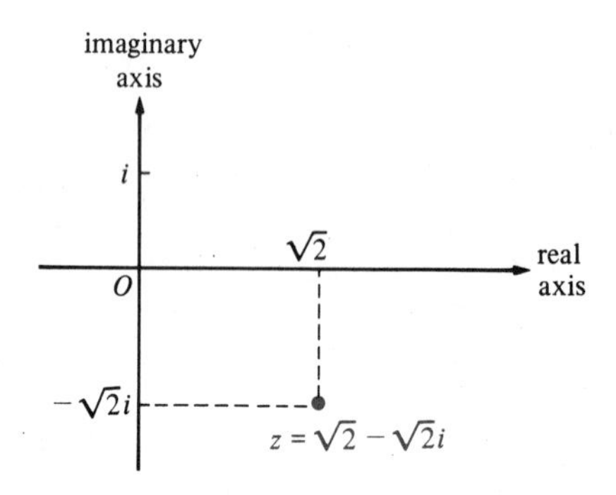

Although complex numbers are most easily added or subtracted when they are written in cartesian form, multiplication and division are more easily carried out when the numbers are expressed in polar form. Indeed, we have the following theorem.

Theorem 1 **Multiplication and Division in Polar Form**

Let $z_1 = r_1(\cos\theta_1 + i\sin\theta_1)$ and $z_2 = r_2(\cos\theta_2 + i\sin\theta_2)$. Then

(i) $z_1 z_2 = r_1 r_2[\cos(\theta_1 + \theta_2) + i\sin(\theta_1 + \theta_2)]$,

(ii) $\dfrac{z_1}{z_2} = \dfrac{r_1}{r_2}[\cos(\theta_1 - \theta_2) + i\sin(\theta_1 - \theta_2)]$, for $z_2 \neq 0$.

Proof (i)
$$\begin{aligned} z_1 z_2 &= r_1(\cos\theta_1 + i\sin\theta_1)r_2(\cos\theta_2 + i\sin\theta_2) \\ &= r_1 r_2(\cos\theta_1 + i\sin\theta_1)(\cos\theta_2 + i\sin\theta_2) \\ &= r_1 r_2[(\cos\theta_1\cos\theta_2 - \sin\theta_1\sin\theta_2) + i(\sin\theta_1\cos\theta_2 + \sin\theta_2\cos\theta_1)] \\ &= r_1 r_2[\cos(\theta_1 + \theta_2) + i\sin(\theta_1 + \theta_2)]. \end{aligned}$$

(ii) Left as an exercise (Problem 73).

In words, part (i) of Theorem 1 says that the modulus (absolute value) of the product of two complex numbers is the product of their moduli (absolute values), while an argument of the product is given by the sum of their arguments. Part (ii) can be expressed analogously (Problem 69).

Example If $z = 6\left(\cos\frac{\pi}{2} + i\sin\frac{\pi}{2}\right)$ and $w = 2\left(\cos\frac{\pi}{6} + i\sin\frac{\pi}{6}\right)$, find (a) zw and (b) $\frac{z}{w}$.

Solution Using Theorem 1, we have

(a) $zw = (6)(2)\left[\cos\left(\frac{\pi}{2} + \frac{\pi}{6}\right) + i\sin\left(\frac{\pi}{2} + \frac{\pi}{6}\right)\right] = 12\left(\cos\frac{2\pi}{3} + i\sin\frac{2\pi}{3}\right)$

(b) $\frac{z}{w} = \frac{6}{2}\left[\cos\left(\frac{\pi}{2} - \frac{\pi}{6}\right) + i\sin\left(\frac{\pi}{2} - \frac{\pi}{6}\right)\right] = 3\left(\cos\frac{\pi}{3} + i\sin\frac{\pi}{3}\right)$.

2.1 Powers and Roots of Complex Numbers

If $z = r(\cos\theta + i\sin\theta)$, then, by part (i) of Theorem 1,

$$z^2 = r^2(\cos 2\theta + i\sin 2\theta).$$

Therefore,

$$z^3 = z^2 z = [r^2(\cos 2\theta + i\sin 2\theta)][r(\cos\theta + i\sin\theta)]$$
$$= r^3(\cos 3\theta + i\sin 3\theta).$$

A similar computation shows that

$$z^4 = r^4(\cos 4\theta + i\sin 4\theta).$$

The pattern emerging here is quite clear and is expressed formally in the following theorem named in honor of Abraham De Moivre (1667–1754), a Frenchman turned Englishman.

Theorem 2 **De Moivre's Theorem**

If $z = r(\cos\theta + i\sin\theta)$, and n is any positive integer, then

$$z^n = r^n(\cos n\theta + i\sin n\theta).$$

In words, De Moivre's theorem says that to raise a complex number to a positive integer power n, raise its modulus (absolute value) to the power n and multiply its argument by n.

Examples Use De Moivre's theorem to evaluate each expression.

1 $[2(\cos 10^\circ + i\sin 10^\circ)]^3$

Solution

$$\begin{aligned}[2(\cos 10^\circ + i\sin 10^\circ)]^3 &= 2^3[\cos 3(10^\circ) + i\sin 3(10^\circ)] \\ &= 8(\cos 30^\circ + i\sin 30^\circ) \\ &= 8\left[\frac{\sqrt{3}}{2} + i\left(\frac{1}{2}\right)\right] \\ &= 4\sqrt{3} + 4i.\end{aligned}$$

2 $(-1 + \sqrt{3}i)^5$

Solution We begin by rewriting $-1 + \sqrt{3}i$ in polar form

$$-1 + \sqrt{3}i = 2\left(\cos\frac{2\pi}{3} + i\sin\frac{2\pi}{3}\right).$$

Therefore,

$$\begin{aligned}(-1 + \sqrt{3}i)^5 &= \left[2\left(\cos\frac{2\pi}{3} + i\sin\frac{2\pi}{3}\right)\right]^5 \\ &= 2^5\left(\cos\frac{10\pi}{3} + i\sin\frac{10\pi}{3}\right) \\ &= 32\left[\cos\left(\frac{4\pi}{3} + 2\pi\right) + i\sin\left(\frac{4\pi}{3} + 2\pi\right)\right] \\ &= 32\left(\cos\frac{4\pi}{3} + i\sin\frac{4\pi}{3}\right) \\ &= 32\left[-\frac{1}{2} + i\left(-\frac{\sqrt{3}}{2}\right)\right] \\ &= -16 - 16\sqrt{3}i.\end{aligned}$$

If w is a complex number and n is a positive integer, then any complex number z such that $z^n = w$ is called a **complex nth root** of w. For $n = 2$ and for $n = 3$, the complex nth roots of w are the **complex square roots** and **complex cube roots,** respectively, of w. By using the formula in the following theorem, we can find all complex nth roots of any nonzero complex number w.

Theorem 3 **Complex nth Roots**

> Let $w = R(\cos\phi + i\sin\phi)$ be a nonzero complex number in polar form and suppose that n is a positive integer. Then there are exactly n different complex nth roots of w, namely,
>
> $$z_0, z_1, z_2, \ldots, z_{n-1},$$
>
> where
>
> $$z_k = \sqrt[n]{R}\left[\cos\left(\frac{\phi}{n} + \frac{2\pi k}{n}\right) + i\sin\left(\frac{\phi}{n} + \frac{2\pi k}{n}\right)\right]$$
>
> for $k = 0, 1, 2, \ldots, n - 1$.

That each of the complex numbers $z_0, z_1, z_2, \ldots, z_{n-1}$ in Theorem 3 satisfies the equation $z^n = w$ is a consequence of De Moivre's theorem (Problem 70). We omit the proof that these complex numbers are different from each other and that they comprise all of the nth roots of w.

Example Find the complex fourth roots of $w = -2 - 2\sqrt{3}i$.

Solution We begin by rewriting w in polar form

$$w = 4\left(\cos\frac{4\pi}{3} + i\sin\frac{4\pi}{3}\right).$$

Then we use Theorem 3 with $R = 4$, $\phi = 4\pi/3$, $n = 4$ to obtain the four complex fourth roots

$$z_0 = \sqrt[4]{4}\left[\cos\left(\frac{4\pi/3}{4} + 0\right) + i\sin\left(\frac{4\pi/3}{4} + 0\right)\right] = \sqrt{2}\left(\cos\frac{\pi}{3} + i\sin\frac{\pi}{3}\right)$$

$$z_1 = \sqrt[4]{4}\left[\cos\left(\frac{4\pi/3}{4} + \frac{2\pi}{4}\right) + i\sin\left(\frac{4\pi/3}{4} + \frac{2\pi}{4}\right)\right] = \sqrt{2}\left(\cos\frac{5\pi}{6} + i\sin\frac{5\pi}{6}\right)$$

$$z_2 = \sqrt[4]{4}\left[\cos\left(\frac{4\pi/3}{4} + \frac{4\pi}{4}\right) + i\sin\left(\frac{4\pi/3}{4} + \frac{4\pi}{4}\right)\right] = \sqrt{2}\left(\cos\frac{4\pi}{3} + i\sin\frac{4\pi}{3}\right)$$

$$z_3 = \sqrt[4]{4}\left[\cos\left(\frac{4\pi/3}{4} + \frac{6\pi}{4}\right) + i\sin\left(\frac{4\pi/3}{4} + \frac{6\pi}{4}\right)\right] = \sqrt{2}\left(\cos\frac{11\pi}{6} + i\sin\frac{11\pi}{6}\right).$$

If desired, these roots can be rewritten in cartesian form as

$$z_0 = \frac{\sqrt{2}}{2} + \frac{\sqrt{6}}{2}i \qquad z_1 = -\frac{\sqrt{6}}{2} + \frac{\sqrt{2}}{2}i \qquad z_2 = -\frac{\sqrt{2}}{2} - \frac{\sqrt{6}}{2}i \qquad z_3 = \frac{\sqrt{6}}{2} - \frac{\sqrt{2}}{2}i.$$

2.2 Complex Roots of Quadratic Equations

If $a \neq 0$ and $b^2 - 4ac \geq 0$, the roots of the quadratic equation $ax^2 + bx + c = 0$ are given by the quadratic formula (Chapter 1, page 7). By using complex numbers, we can extend the quadratic formula to the case in which $b^2 - 4ac < 0$, and even to the case in which the coefficients a, b, and c are complex numbers. Indeed, working with complex numbers, we find that the same proof (by completing the square) applies; however, because the expression $\sqrt{b^2 - 4ac}$ refers to the principal square root of $b^2 - 4ac$, it cannot be used legitimately unless $b^2 - 4ac$ is a nonnegative real number. Nevertheless, with the understanding that $\pm\sqrt{b^2 - 4ac}$ stands for the two complex square roots of $b^2 - 4ac$, it is customary to write the quadratic formula in the form

$$x = \frac{-b \pm \sqrt{b^2 - 4ac}}{2a}.$$

Examples Solve each equation by using the quadratic formula.

1 $x^2 - 2x + 5 = 0$

Solution We use the quadratic formula with $a = 1$, $b = -2$, and $c = 5$. Thus,

$$\begin{aligned} x &= \frac{-(-2) \pm \sqrt{(-2)^2 - 4(1)(5)}}{2(1)} \\ &= \frac{2 \pm \sqrt{-16}}{2} = \frac{2 \pm \sqrt{16(-1)}}{2} \\ &= \frac{2 \pm \sqrt{16}\sqrt{-1}}{2} = \frac{2 \pm 4i}{2} \\ &= 1 \pm 2i. \end{aligned}$$

Therefore, the two roots are the complex numbers

$$x = 1 + 2i \quad \text{and} \quad x = 1 - 2i.$$

2 $z^2 + (4 + 2i)z + (3 + 3i) = 0$

Solution Here $a = 1$, $b = 4 + 2i$, and $c = 3 + 3i$. Thus,

$$\begin{aligned} b^2 - 4ac &= (4 + 2i)^2 - 4(3 + 3i) \\ &= 16 + 16i + 4i^2 - 12 - 12i \\ &= 16 + 16i - 4 - 12 - 12i = 4i. \end{aligned}$$

We intend to use Theorem 3 to find $\pm\sqrt{b^2 - 4ac} = \pm\sqrt{4i}$; that is, the two complex square roots of $4i$. Rewriting $4i$ in polar form, we have

$$4i = 4(\cos 90^\circ + i \sin 90^\circ).$$

Therefore, by Theorem 3, the square roots of $4i$ are

$$\pm 2(\cos 45^\circ + i \sin 45^\circ) = \pm 2\left(\frac{\sqrt{2}}{2} + i\frac{\sqrt{2}}{2}\right) = \pm(\sqrt{2} + \sqrt{2}i).$$

Hence,

$$\begin{aligned} z = \frac{-b \pm \sqrt{b^2 - 4ac}}{2a} &= \frac{-(4 + 2i) \pm \sqrt{4i}}{2(1)} \\ &= \frac{-(4 + 2i) \pm (\sqrt{2} + \sqrt{2}i)}{2} \\ &= \frac{-4 \pm \sqrt{2}}{2} + \frac{-2 \pm \sqrt{2}}{2} i. \end{aligned}$$

In other words, the two roots are the complex numbers

$$z = \frac{-4 + \sqrt{2}}{2} + \frac{-2 + \sqrt{2}}{2} i \quad \text{and} \quad z = \frac{-4 - \sqrt{2}}{2} + \frac{-2 - \sqrt{2}}{2} i.$$

PROBLEM SET 2

In problems 1–8, plot each point z in the complex plane and rewrite z in cartesian form.

1. $z = 4(\cos 30^\circ + i\sin 30^\circ)$
2. $z = 6\left(\cos\frac{\pi}{4} + i\sin\frac{\pi}{4}\right)$
3. $z = 7\left(\cos\frac{3\pi}{4} + i\sin\frac{3\pi}{4}\right)$
4. $z = 8(\cos 270^\circ + i\sin 270^\circ)$
5. $6(\cos 300^\circ + i\sin 300^\circ)$
6. $z = 6\left(\cos\frac{7\pi}{6} + i\sin\frac{7\pi}{6}\right)$
7. $z = 2[\cos(-11\pi/6) + i\sin(-11\pi/6)]$
8. $z = 12(\cos 240^\circ + i\sin 240^\circ)$

In problems 9–18, plot each point z in the complex plane and rewrite z in polar form.

9. $z = -1 + i$
10. $z = 3 - 3i$
11. $z = 3i$
12. $z = 4$
13. $z = 1 + \sqrt{3}i$
14. $z = -2 - \sqrt{3}i$
15. $z = -\sqrt{3} - i$
16. $z = \sqrt{3} - i$
17. $z = -i$
18. $z = -5$

In problems 19–26, use Theorem 1 to find (a) zw and (b) $\frac{z}{w}$. Leave the results in polar form.

19. $z = 4(\cos 70^\circ + i\sin 70^\circ), \quad w = 2(\cos 40^\circ + i\sin 40^\circ)$
20. $z = 8(\cos 80^\circ + i\sin 80^\circ), \quad w = 4(\cos 20^\circ + i\sin 20^\circ)$
21. $z = 14\left(\cos\frac{3\pi}{2} + i\sin\frac{3\pi}{2}\right), \quad w = 7\left(\cos\frac{5\pi}{4} + i\sin\frac{5\pi}{4}\right)$
22. $z = 15\left(\cos\frac{4\pi}{3} + i\sin\frac{4\pi}{3}\right), \quad w = 5\left[\cos\left(-\frac{\pi}{3}\right) + i\sin\left(-\frac{\pi}{3}\right)\right]$
23. $z = 6(\cos 90^\circ + i\sin 90^\circ), \quad w = 3(\cos 45^\circ + i\sin 45^\circ)$
24. $z = 8(\cos 85^\circ + i\sin 85^\circ), \quad w = 4(\cos 55^\circ + i\sin 55^\circ)$
25. $z = 1 + i, \quad w = 1 + \sqrt{3}i$
26. $z = \sqrt{3} + i, \quad w = 1 - i$

In problems 27–38, use De Moivre's theorem to evaluate each expression. Write the result in cartesian form.

27. $[2(\cos 9^\circ + i\sin 9^\circ)]^5$
28. $[\sqrt{2}(\cos 15^\circ + i\sin 15^\circ)]^{10}$
29. $\left[\sqrt{3}\left(\cos\frac{\pi}{10} + i\sin\frac{\pi}{10}\right)\right]^{10}$
30. $\left[2\left(\cos\frac{2\pi}{9} + i\sin\frac{2\pi}{9}\right)\right]^6$
31. $\left[2\left(\cos\frac{\pi}{12} + i\sin\frac{\pi}{12}\right)\right]^9$
32. $\left(\cos\frac{2\pi}{5} + i\sin\frac{2\pi}{5}\right)^{15}$
33. $(\sqrt{3} - i)^5$
34. $(-1 - \sqrt{3}i)^8$
35. $(-1 + i)^{12}$
36. $(1 + i)^{16}$
37. $\left(\frac{1 + i}{\sqrt{2}}\right)^7$
38. $(2 + 2i)^{11}$

In problems 39–48, use Theorem 3 to find all of the indicated complex nth roots.

39. The square roots of $\frac{9}{2} - \frac{9\sqrt{3}}{2}i$
40. The cube roots of $-8i$
41. The cube roots of $-1 + i$
42. The fourth roots of $-2 + 2\sqrt{3}i$
43. The fourth roots of $8 - 8\sqrt{3}i$
44. The fifth roots of 1
45. The fourth roots of -1
46. The sixth roots of $-64i$
47. The fifth roots of $-16\sqrt{2} - 16\sqrt{2}i$
48. The fifth roots of $-\sqrt{3} - i$

In problems 49–62, use the quadratic formula to find all complex roots of each equation.

49. $6x^2 + x + 3 = 0$
50. $10x^2 + 2x + 8 = 0$
51. $5x^2 - 4x + 2 = 0$
52. $4z^2 + 4z + 5 = 0$
53. $z^2 + z + 1 = 0$
54. $w^2 - 2w + \frac{3}{2} = 0$
55. $4w^2 - 2w + 1 = 0$
56. $z^2 - 6z + 13 = 0$
57. $z^2 + 8i = 0$
58. $w^2 + iw + 1 = 0$
59. $z^2 + iz - 1 = 0$
60. $iw^2 + 2w + i = 0$
61. $z^2 - (3 + i)z + 2 + 2i = 0$
62. $\frac{\sqrt{3}}{8}z^2 + \frac{\sqrt{2}}{2}z + i = 0$

63. (a) Show that De Moivre's theorem (Theorem 2) holds for $n = 0$ and for $n = -1$, provided that $z \neq 0$.
(b) With the understanding that $z^{-n} = 1/z^n$ for $z \neq 0$, show that De Moivre's theorem holds for all integer values of n, provided that $z \neq 0$.

64. Suppose that $z = a + bi$, where a and b are real numbers, and let $r = |z|$.
(a) Show that $r + a \geq 0$.
(b) Show that $r - a \geq 0$.
(c) Let $x = \sqrt{\dfrac{r+a}{2}}$ and let $y = \pm\sqrt{\dfrac{r-a}{2}}$, where the algebraic sign is chosen so that xy and b have the same algebraic sign. Show that $x + yi$ is a square root of z.
(d) Show that $-x - yi$ is a square root of z.

65. If n is a positive integer, the complex nth roots of 1 are called the ***n*th roots of unity.** Use Theorem 3 to find all nth roots of unity for (a) $n = 2$, (b) $n = 3$, and (c) $n = 4$.

66. If n is a positive integer, show that (a) the product of two nth roots of unity (problem 65) is again an nth root of unity, (b) the reciprocal of an nth root of unity is again an nth root of unity, and (c) all nth roots of unity lie on the unit circle in the complex plane.

67. If z and w are points in the complex plane, give a geometric interpretation of (a) $\bar{z}$, (b) $-z$, (c) $-\bar{z}$, (d) iz, and (e) $|z - w|$.

68. Let z be a fixed nonzero complex number and let n be a positive integer. (a) Show that the complex nth roots of z lie on a circle in the complex plane, the center of the circle being the origin and the radius being $\sqrt[n]{|z|}$. (b) Show that the complex nth roots of z are equally spaced around the circle in part (a).

69. Express part (ii) of Theorem 1 in words.

70. Using De Moivre's theorem, show that $z_0, z_1, \ldots, z_{n-1}$ in Theorem 3 are nth roots of w.

71. If $z = r(\cos\theta + i\sin\theta)$, show that $z^4 = r^4(\cos 4\theta + i\sin 4\theta)$ by direct use of Theorem 1.

72. If z_1 and z_2 are the complex roots of the equation $az^2 + bz + c = 0$, show that

(a) $z_1 + z_2 = -\dfrac{b}{a}$ (b) $z_1 z_2 = \dfrac{c}{a}$.

73. Prove part (ii) of Theorem 1.

3 Straight Lines and Their Equations

Perhaps the simplest curves in the plane are straight lines (called simply lines) and circles. In Section 4.1 of Chapter 1, we derived equations for circles by using the distance formula. In this section, we derive equations for lines by using the ideas of inclination and slope.

Suppose that L is a line in the xy plane. If L is not parallel to the x axis, then the smallest counterclockwise angle α from the positive x axis to L is called the **inclination** of L. Figure 1a shows a case in which $0° < \alpha < 90°$ and Figure 1b shows a case in which $90° < \alpha < 180°$. Of course, if L is perpendicular to the x axis, we have $\alpha = 90°$. If L is parallel to the x axis, we define $\alpha = 0°$. Therefore, in any case,

$$0° \leq \alpha < 180°.$$

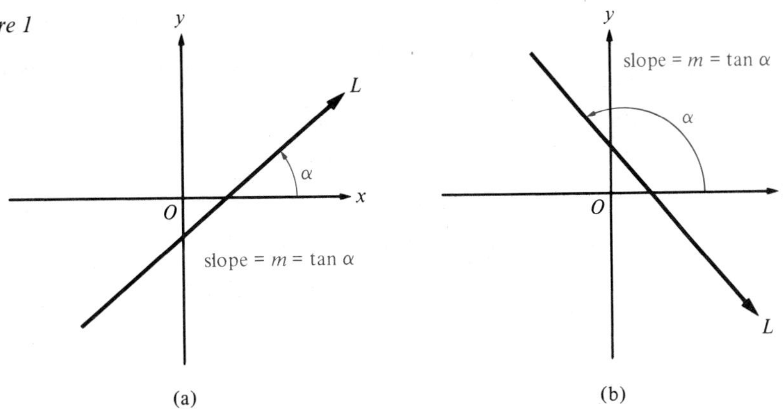

Figure 1

If the line L is not perpendicular to the x axis, we define the **slope** of L to be the tangent of the inclination of L. The slope of a line is traditionally denoted by the letter m; hence, if α is the inclination of L,

$$m = \tan \alpha.$$

The slope of a vertical line is undefined. If L slants upward from left to right, then $0° < \alpha < 90°$, so the slope m of L is positive (Figure 1a). If L slants downward from left to right, then $90° < \alpha < 180°$, so the slope m of L is negative (Figure 1b). A line L is horizontal if and only if its slope is zero.

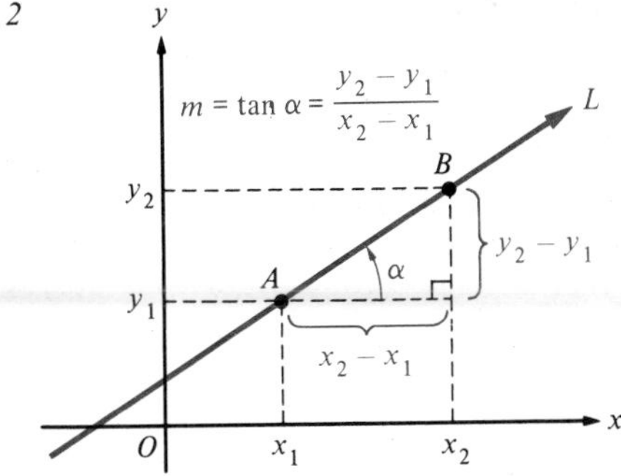

Figure 2

Let $A = (x_1, y_1)$ and $B = (x_2, y_2)$ be two different points on the line L (Figure 2). If B is above and to the right of A as in Figure 2, you can see that the slope of L is given by

$$m = \tan \alpha = \frac{y_2 - y_1}{x_2 - x_1}.$$

Even if B is not above and to the right of A as in Figure 2, it isn't difficult to see that the slope m of L is given by the same formula (Problem 54). Therefore, we have the following theorem.

Theorem 1 **The Slope Formula**

Let L be a line in the xy plane and suppose that α is the inclination of L. If $A = (x_1, y_1)$ and $B = (x_2, y_2)$ are any two different points on L, then the slope m of L is given by

$$m = \tan \alpha = \frac{y_2 - y_1}{x_2 - x_1}.$$

Example 1 Sketch the line in the xy plane that contains the two points $A = (-3, -2)$ and $B = (4, 1)$ and use the slope formula to find the slope m of L.

Solution The line L is sketched in Figure 3. Using the slope formula, we have

$$m = \tan \alpha = \frac{y_2 - y_1}{x_2 - x_1} = \frac{1 - (-2)}{4 - (-3)} = \frac{1 + 2}{4 + 3} = \frac{3}{7}.$$

Figure 3

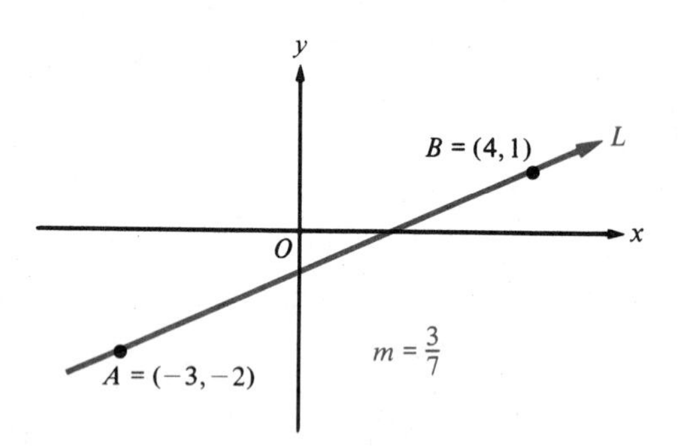

Example 2 Sketch the line in the xy plane that contains the point $P = (1, 2)$ and has slope

(a) $m = \frac{2}{3}$ (b) $m = -\frac{2}{3}$.

Solution (a) The condition $m = \frac{2}{3}$ means that, for every 3 units we move to the right from a point on the line, we must move up 2 units to get back to the line. Thus, if we start at the point $P = (1, 2)$, move 3 units to the right and 2 units up, we arrive at the point $Q = (1 + 3, 2 + 2) = (4, 4)$ on the line. We simply plot the two points P and Q, and use a straightedge to draw the line through these two points (Figure 4a).

(b) The condition $m = -\frac{2}{3}$ means that, for every 3 units we move to the right from a point on the line, we must move down 2 units to get back to the line. If we start at the point $P = (1, 2)$, move 3 units to the right and 2 units down, we arrive at the point $Q = (1 + 3, 2 - 2) = (4, 0)$ on the line. Plotting $P = (1, 2)$ and $Q = (4, 0)$, and using a straightedge, we draw the desired line (Figure 4b).

Figure 4

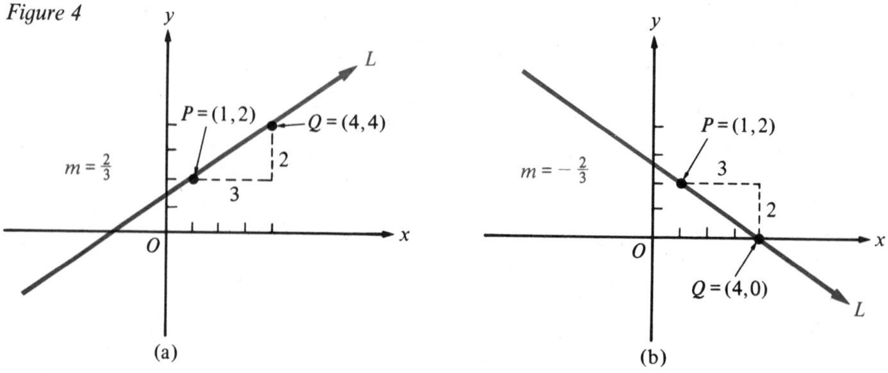

(a)

(b)

Because two parallel lines in the xy plane have the same inclination, it follows that they have the same slope (Figure 5). Conversely, it is easy to show that two distinct lines with the same slope are necessarily parallel (Problem 56), and we have the following theorem.

Figure 5

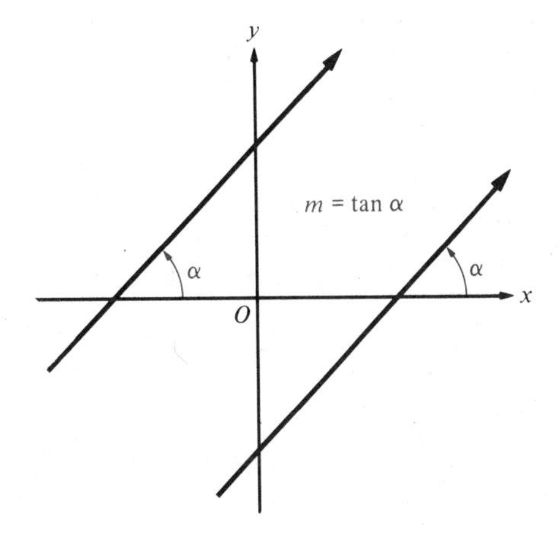

Theorem 2 **Parallelism Condition**

> Two distinct nonvertical straight lines in the xy plane are parallel if and only if they have the same slope.

Example Sketch the line in the xy plane that contains the point $P = (3,4)$ and is parallel to the line segment $\overline{AB}$, where $A = (-1,2)$ and $B = (4,-5)$.

Solution By the slope formula, the line containing the points A and B has slope

$$m = \frac{y_2 - y_1}{x_2 - x_1}$$

$$= \frac{-5 - 2}{4 - (-1)} = \frac{-7}{5}.$$

Figure 6

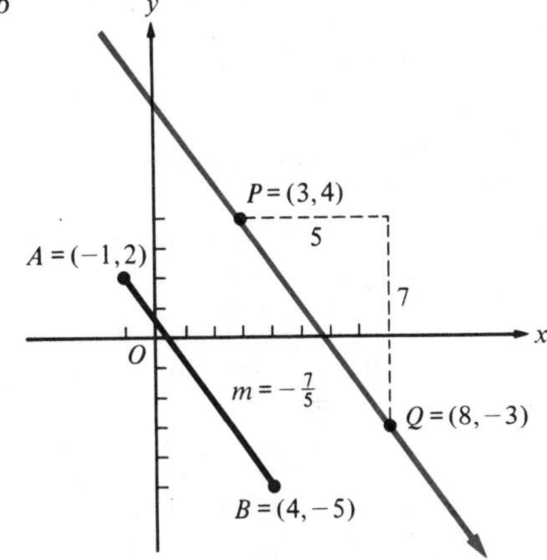

Therefore, by the parallelism condition (Theorem 2), the desired line also has slope $m = -\frac{7}{5}$. Starting at the point $P = (3,4)$, we move 5 units to the right and 7 units down to the point $Q = (3 + 5, 4 - 7) = (8, -3)$. Using a straightedge, we draw the desired line through P and Q (Figure 6).

The following theorem gives a condition for two lines to be perpendicular.

Theorem 3 **Perpendicularity Condition**

> Two nonvertical lines in the xy plane are perpendicular if and only if the slope of one of the lines is the negative of the reciprocal of the slope of the other line.

Proof Let the two lines L_1 and L_2 have slopes m_1 and m_2, respectively. The condition that the slope of either one of the lines is the negative of the reciprocal of the slope of the other can be written as $m_1 m_2 = -1$. Neither the angle between the lines nor their slopes are affected if the origin O of the coordinate system is placed at the point where they intersect (Figure 7). Starting at O on L_1, we move 1 unit to the right and $|m_1|$ units vertically to arrive at the point $A = (1, m_1)$ on L_1. Likewise, the point $B = (1, m_2)$ belongs to L_2. By the Pythagorean theorem and its converse, triangle AOB is a right triangle if and only if

$$|\overline{AB}|^2 = |\overline{OA}|^2 + |\overline{OB}|^2.$$

Figure 7

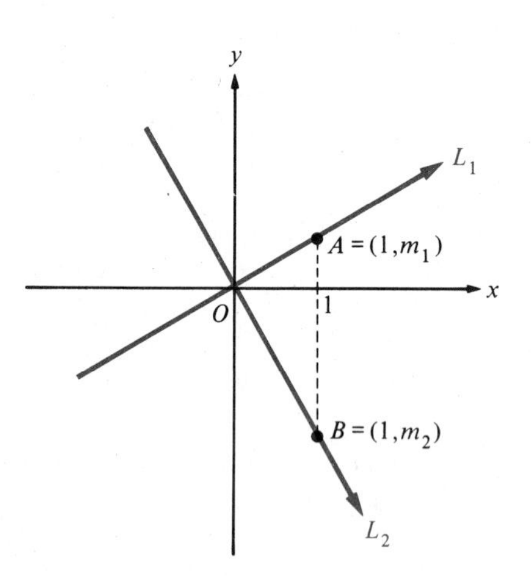

Using the distance formula (page 25), we find that

$$|\overline{AB}|^2 = (1 - 1)^2 + (m_1 - m_2)^2 = m_1^2 - 2m_1 m_2 + m_2^2$$
$$|\overline{OA}|^2 = (1 - 0)^2 + (m_1 - 0)^2 = 1 + m_1^2$$
$$|\overline{OB}|^2 = (1 - 0)^2 + (m_2 - 0)^2 = 1 + m_2^2.$$

Therefore, the condition $|\overline{AB}|^2 = |\overline{OA}|^2 + |\overline{OB}|^2$ holds if and only if

$$m_1^2 - 2m_1 m_2 + m_2^2 = 1 + m_1^2 + 1 + m_2^2.$$

The last equation simplifies to $m_1 m_2 = -1$, and therefore the angle formed by L_1 and L_2 at O is a right angle if and only if $m_1 m_2 = -1$.

Example 1 Find the slope m_1 of a line that is perpendicular to the line segment $\overline{AB}$, where $A = (-1, 2)$ and $B = (4, -5)$.

Solution The slope m_2 of the line containing the points A and B is given by

$$m_2 = \frac{y_2 - y_1}{x_2 - x_1} = \frac{-5 - 2}{4 - (-1)} = -\frac{7}{5}.$$

Therefore, by the perpendicularity condition (Theorem 3),

$$m_1 = -\frac{1}{m_2} = -\frac{1}{(-7/5)} = \frac{5}{7}.$$

Example 2 Sketch the line L that contains the point $P = (2, 1)$ and is perpendicular to the line segment $\overline{AB}$, where $A = (-\frac{5}{3}, 0)$ and $B = (0, 5)$.

Solution The slope of the line through A and B is given by

$$\frac{y_2 - y_1}{x_2 - x_1} = \frac{5 - 0}{0 - (-5/3)} = \frac{5}{5/3} = 3.$$

Therefore, the slope m of L is

$$m = -\tfrac{1}{3}.$$

Starting at the point $P = (2, 1)$ on L, we move 3 units to the right and 1 unit down to the point $Q = (2 + 3, 1 - 1) = (5, 0)$ on L. Using a straightedge, we draw the line L through P and Q (Figure 8).

Figure 8

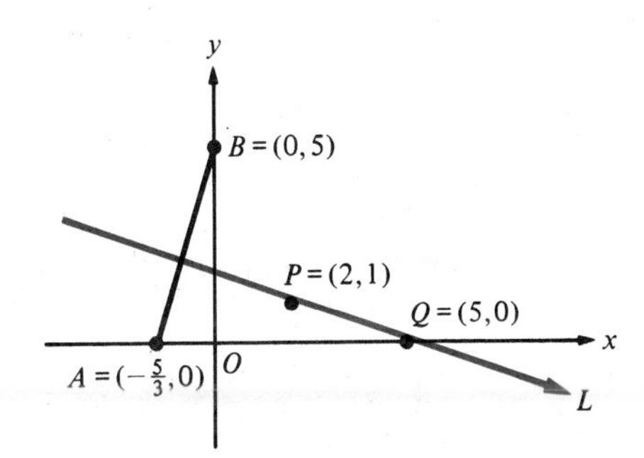

Now consider a nonvertical line L with slope m and containing the point $P_0 = (x_0, y_0)$ (Figure 9). If $P = (x, y)$ is any other point on L, then, by the slope formula (Theorem 1),

Figure 9

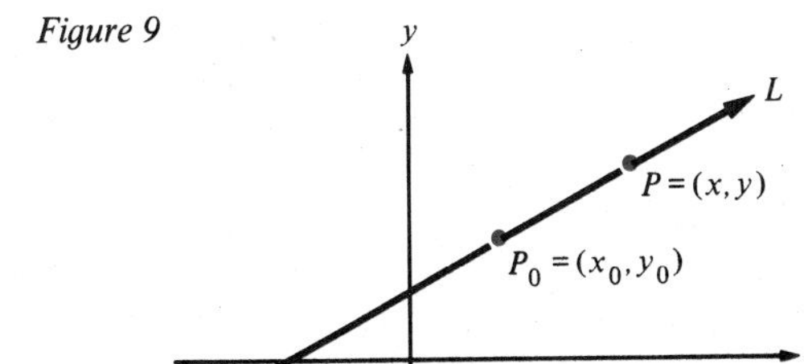

$$m = \frac{y - y_0}{x - x_0};$$

that is,

$$y - y_0 = m(x - x_0).$$

Notice that the last equation holds even if $P = P_0$, when it simply reduces to $0 = 0$.

Using the parallelism condition (Theorem 2) and some elementary geometry, you can show that, conversely, any point $P = (x, y)$ whose coordinates satisfy the equation $y - y_0 = m(x - x_0)$ lies on the line L (Problem 58). Therefore, we have the following theorem.

Theorem 4 **Point-Slope Equation of a Straight Line**

> The straight line L in the xy plane that contains the point $P_0 = (x_0, y_0)$ and has slope m is the graph of the equation
>
> $$y - y_0 = m(x - x_0).$$

You can write the equation $y - y_0 = m(x - x_0)$ as soon as you know the coordinates (x_0, y_0) of one point P_0 on the line L and the slope m of L. For this reason, the equation is said to be in **point-slope form.**

Examples Find an equation in point-slope form of the line L in the xy plane that satisfies the given conditions.

1 L contains the point $P_0 = (3,4)$ and has slope $m = 5$.

Solution Substituting $x_0 = 3$, $y_0 = 4$, and $m = 5$ in the point-slope equation

$$y - y_0 = m(x - x_0),$$

we have

$$y - 4 = 5(x - 3).$$

2 L contains the two points $A = (-1,3)$ and $B = (4, -2)$.

Solution By the slope formula, the slope m of L is given by

$$m = \frac{-2 - 3}{4 - (-1)} = \frac{-5}{5} = -1.$$

Because the point $A = (-1,3)$ belongs to L, we can take $(x_0, y_0) = (-1,3)$ in Theorem 4. Thus, an equation of L is

$$y - 3 = -1[x - (-1)] \quad \text{or} \quad y - 3 = -(x + 1).$$

A nonvertical line L must intersect the y axis at some point $(0, b)$ (Figure 10). The ordinate b of the intersection point is called the ***y* intercept** of L. Because $(0, b)$ belongs to L, we can write a point-slope equation

$$y - b = m(x - 0)$$

for L. This equation can be rewritten in the form

> $$y = mx + b,$$

Figure 10

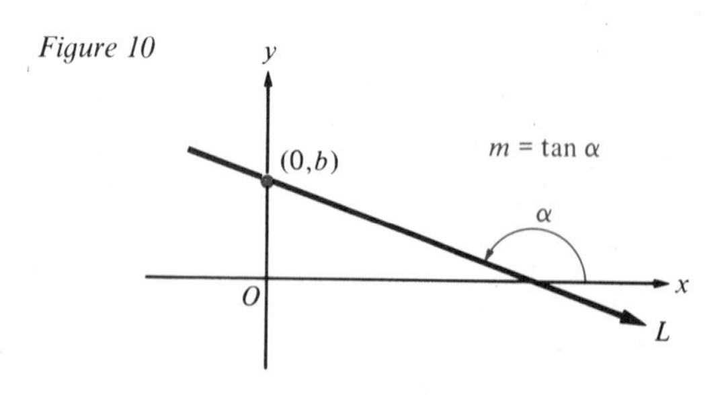

which is called the **slope-intercept form** of the equation for L. In the slope-intercept form, the coefficient of x is the slope b and the constant term is the y intercept.

Example Show that the graph of the equation

$$3x - 5y - 15 = 0$$

is a straight line by rewriting it in slope-intercept form. Find the slope m and the y intercept b, and sketch the graph.

Solution We begin by solving the equation for y in terms of x. Thus,

$$-5y = -3x + 15 \quad \text{or} \quad 5y = 3x - 15;$$

that is,

$$y = \tfrac{3}{5}x - \tfrac{15}{5} \quad \text{or} \quad y = \tfrac{3}{5}x - 3,$$

which is the equation in slope-intercept form of a line with slope $m = \frac{3}{5}$ and y intercept $b = -3$. The graph is obtained by drawing the line with slope $m = \frac{3}{5}$ through the point $(0, -3)$ (Figure 11).

Figure 11

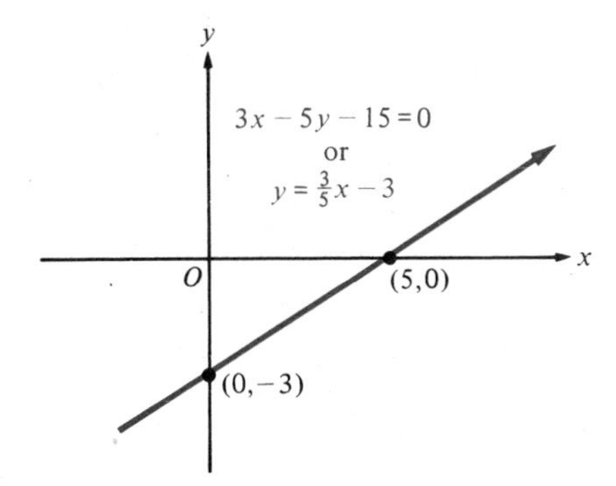

A horizontal line has slope $m = 0$; hence, in slope-intercept form, its equation is $y = 0(x) + b$, or simply $y = b$ (Figure 12). Because a vertical line has an undefined slope, its equation can't be written in slope-intercept form at all. However, as we saw in Chapter 1, the graph of $x = a$ is a vertical line consisting of all points with abscissa a (Figure 13).

Figure 12

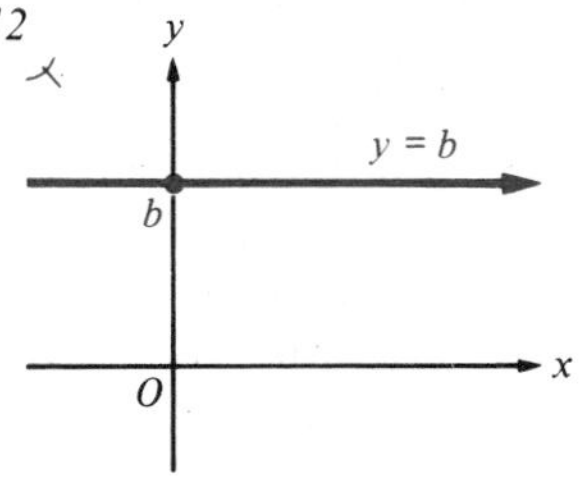

Figure 13

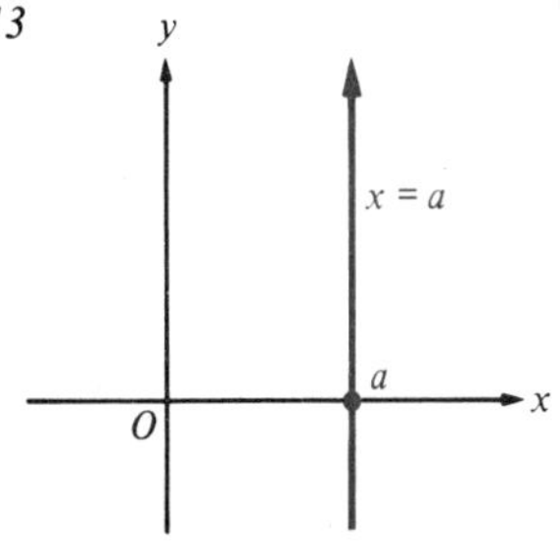

If A, B, and C are constants and if A and B are not both zero, an equation of the form

$$Ax + By + C = 0$$

represents a straight line. If $B \neq 0$, the equation can be rewritten as

$$y = \left(\frac{-A}{B}\right)x + \left(\frac{-C}{B}\right),$$

which is the equation of a line with slope $m = -A/B$ and y intercept $b = -C/B$. On the other hand, if $B = 0$, then $A \neq 0$ and the equation can be rewritten as

$$x = \frac{-C}{A},$$

which is the equation of a vertical line. The equation

$$\boxed{Ax + By + C = 0}$$

is called the **general form** of the equation of a line in the xy plane.

Example Let L be the line that contains the point $(-1, 2)$ and is perpendicular to the line L_1 whose equation is

$$3x - 2y + 5 = 0.$$

Find the equation of L in (a) point-slope form, (b) slope-intercept form, and (c) general form.

Solution We begin by solving the equation $3x - 2y + 5 = 0$ for y in terms of x. The result is

$$y = \tfrac{3}{2}x + \tfrac{5}{2},$$

which shows that the slope of L_1 is $m_1 = \frac{3}{2}$. By the perpendicularity condition (Theorem 3), the slope of L is given by

$$m = -\frac{1}{m_1} = -\frac{1}{(3/2)} = -\frac{2}{3}.$$

(a) Because L has slope $m = -\frac{2}{3}$ and contains the point $(-1, 2)$, its equation in point-slope form is

$$y - 2 = -\tfrac{2}{3}[x - (-1)]$$

or

$$y - 2 = -\tfrac{2}{3}(x + 1).$$

(b) Solving the equation in (a) for y in terms of x, we obtain the equation of L in slope-intercept form:

$$y = -\tfrac{2}{3}x + \tfrac{4}{3}.$$

(c) Multiplying both sides of the equation in (b) by 3 and rearranging terms, we obtain the equation of L in general form:

$$2x + 3y - 4 = 0.$$

PROBLEM SET 3

In problems 1–10, sketch the line L that contains the points A and B and find the slope of L using the slope formula.

1. $A = (-1, 8)$, $B = (4, 3)$
2. $A = (6, -1)$, $B = (-3, 4)$
3. $A = (-2, 2)$, $B = (2, -2)$
4. $A = (-3, 7)$, $B = (-1, 0)$
5. $A = (4, 0)$, $B = (0, -1)$
6. $A = (-1, 3)$, $B = (0, 0)$
7. $A = (-3, -\frac{1}{2})$, $B = (1, \frac{3}{2})$
8. $A = (-\frac{1}{3}, \frac{1}{4})$, $B = (\frac{1}{4}, \frac{3}{4})$
9. [C] $A = (2.6, -5.3)$, $B = (1.7, -1.1)$
10. [C] $A = (73.24, 31.53)$, $B = (1.71, 2.42)$

In problems 11–16, sketch the line L that contains the point P and has slope m.

11. $P = (1, 1)$, $m = 2$
12. $P = (-4, 1)$, $m = 0$
13. $P = (-3, 2)$, $m = -\frac{2}{5}$
14. $P = (\frac{1}{2}, 0)$, $m = \frac{1}{2}$
15. $P = (-\frac{2}{3}, -3)$, $m = -2$
16. $P = (-\frac{5}{3}, \frac{1}{2})$, $m = -\frac{3}{4}$

In problems 17 and 18, sketch the line L that contains the point P and is parallel to the line segment $\overline{AB}$.

17. $P = (4, -3)$, $A = (-2, 3)$, $B = (3, -7)$
18. $P = (\frac{2}{3}, \frac{5}{3})$, $A = (\frac{1}{5}, \frac{3}{5})$, $B = (-\frac{2}{5}, \frac{4}{5})$

In problems 19 and 20, find the slope of a line that is perpendicular to the line segment $\overline{AB}$.

19. $A = (-1, 4)$, $B = (-2, -1)$
20. $A = (2, \frac{7}{5})$, $B = (1, \frac{12}{5})$

In problems 21–24, sketch the line L that contains the point P and is perpendicular to the line segment $\overline{AB}$.

21. $P = (1, 2)$, $A = (-7, -3)$, $B = (-5, 0)$
22. $P = (5, 12)$, $A = (3, -2)$, $B = (\frac{4}{3}, -\frac{4}{3})$
23. $P = (\frac{3}{2}, \frac{5}{2})$, $A = (4, -\frac{1}{3})$, $B = (\frac{1}{2}, 6)$
24. $P = (0, 4)$, $A = (\frac{22}{7}, 2)$, $B = (\frac{1}{7}, 3)$

In problems 25–30, find the equation in point-slope form of the line L that satisfies the given conditions.

25. L contains the point $(3, 2)$ and has slope $m = \frac{3}{4}$.
26. L contains the point $(0, 2)$ and has slope $m = -\frac{2}{3}$.
27. L contains the points $(-3, 2)$ and $(4, 1)$.
28. L contains the points $(1, 3)$ and $(-1, 1)$.
29. L contains the point $(-3, 5)$ and is perpendicular to the line segment $\overline{AB}$, where $A = (3, 7)$ and $B = (-2, 2)$.
30. L contains the point $(7, 2)$ and is perpendicular to the line segment $\overline{AB}$, where $A = (\frac{1}{3}, 1)$ and $B = (-\frac{2}{3}, \frac{3}{5})$.

In problems 31–38, rewrite each equation in slope-intercept form, find the slope m and the y intercept b, and sketch the graph.

31. $3x - 2y = 6$
32. $5x - 2y - 10 = 0$
33. $y - 3x - 1 = 0$
34. $x = -3y + 9$
35. $2x + y + 3 = 0$
36. $y - 2x - 3 = 0$
37. $y + 1 = 0$
38. $x = -\frac{3}{5}y + \frac{7}{5}$

In problems 39–50, find the equation of the line L in (a) point-slope form, (b) slope-intercept form, and (c) general form.

39. L contains the point $(-5, 2)$ and has slope $m = 4$.
40. L contains the point $(3, -1)$ and has slope $m = 0$.
41. L has slope $m = -3$ and y intercept $b = 5$.
42. L has slope $m = \frac{4}{5}$ and intersects the x axis at the point $(-3, 0)$.
43. L intersects the x and y axes at the points $(3, 0)$ and $(0, 5)$.
44. L contains the points $(\frac{7}{2}, \frac{5}{3})$ and $(\frac{2}{5}, -6)$.

45. L contains the point $(4, -4)$ and is parallel to the line whose equation is $2x - 5y + 3 = 0$.
46. L contains the point $(-2, 5)$ and is perpendicular to the y axis.
47. L contains the point $(-3, \frac{2}{3})$ and is perpendicular to the line $5x + 3y - 1 = 0$.
48. L contains the point $(-6, -8)$ and has y intercept $b = 0$.
49. L contains the point $(\frac{2}{3}, \frac{5}{7})$ and is perpendicular to the line $7x + 3y - 12 = 0$.
50. L is the perpendicular bisector of the line segment $\overline{AB}$, where $A = (3, -2)$ and $B = (7, 6)$.

51. Find a real number B so that the graph of $3x + By - 5 = 0$ has y intercept $b = -4$.
52. If a line intersects the x axis at the point $(a, 0)$, we refer to a as its ***x* intercept**. If a line has nonzero x and y intercepts a and b, respectively, show that its equation can be written as

$$\frac{x}{a} + \frac{y}{b} = 1.$$

This is called the **intercept form** of the equation of the line.

53. A car rental company leases automobiles for a charge of \$22 per day plus \$0.20 per mile. Write an equation for the cost y dollars in terms of the distance x miles driven if the car is leased for N days. If $N = 3$, sketch a graph of this equation.
54. Prove Theorem 1 (page 231).
55. In 1980, the Solar Electric Company showed a profit of \$3.45 per share, and it expects this figure to increase annually by \$0.25 per share. Counting the years so that 1980 corresponds to $x = 0$ and successive years correspond to $x = 1, 2, 3$, and so on, find the equation $y = mx + b$ of the line that allows the company to predict its profit y dollars per share during the future years. Sketch a graph of the equation and find the predicted profit per share in 1988.
56. Show that two distinct lines with the same slope are necessarily parallel. [Hint: If they weren't parallel, they would meet at some point P_0.]
57. If a piece of property is *depreciated linearly* over a period of n years, then its value y dollars at the end of x years is given by the equation $y = c\left(1 - \frac{x}{n}\right)$, where c dollars is the original value of the property. An apartment building, built in 1975 and originally worth \$400,000, is being depreciated linearly over a period of $n = 40$ years. Sketch a graph showing the value y dollars of the apartment building x years after it was built and determine its value in the year 1995.
58. Let L be the line that contains the point $P_0 = (x_0, y_0)$ and has slope m. Suppose that (x, y) is a point such that $y - y_0 = m(x - x_0)$. Prove that (x, y) lies on the line L.
59. Use slopes to show that the triangle with vertices $(-4, -2)$, $(2, -8)$, and $(4, 6)$ is a right triangle.

4 The Conic Sections

If a cone with two nappes (Figure 1) is cut by a plane which does not pass through its vertex, the resulting curve of intersection will be a **circle,** an **ellipse,** a **parabola,** or a **hyperbola.** Therefore, these curves are called **conic sections** (Figure 1).

Figure 1

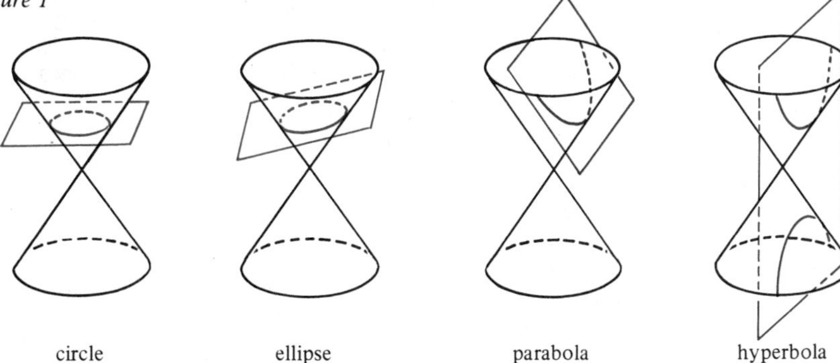

4.1 Circles and Ellipses

As we saw in Section 4.1 of Chapter 1, if $r > 0$, the graph of $x^2 + y^2 = r^2$ is a circle of radius r with center at the origin. This equation can be rewritten in the form

$$\frac{x^2}{r^2} + \frac{y^2}{r^2} = 1.$$

A slight modification of the latter equation, namely,

$$\frac{x^2}{a^2} + \frac{y^2}{b^2} = 1, \qquad \text{where } a > 0, \quad b > 0, \quad a \neq b,$$

produces the equation of an **ellipse** (Figure 2). The ellipse intersects the x axis at $(-a, 0)$ and $(a, 0)$, while it intersects the y axis at $(0, -b)$ and $(0, b)$. (Why?) These four points are called the **vertices** of the ellipse. The two line segments joining opposite pairs of vertices are called the **axes** of the ellipse. The longer axis is called the **major axis,** the shorter is called the **minor axis,** and the two axes intersect at the **center** of the ellipse. For the ellipse $(x^2/a^2) + (y^2/b^2) = 1$, the axes lie along the coordinate axes and the center is at the origin. If $a > b > 0$, the major axis is horizontal (Figure 2a), whereas, if $0 < a < b$, the major axis is vertical (Figure 2b).

Figure 2

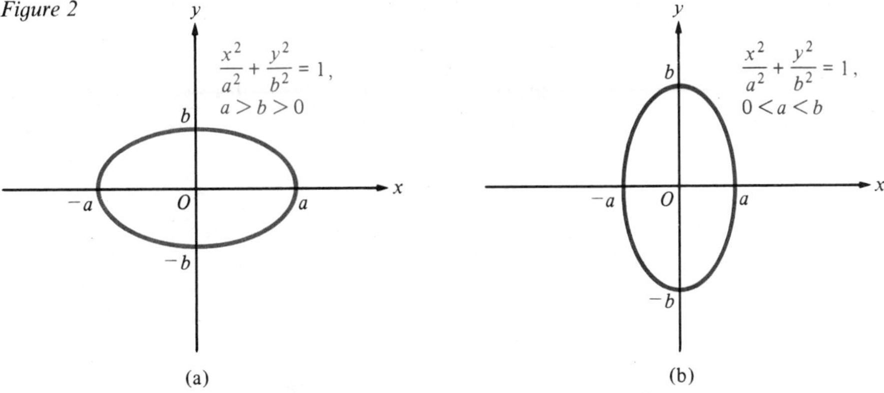

Example Find the vertices of the ellipse $4x^2 + 9y^2 = 36$ and sketch its graph.

Solution We divide both sides of the equation by 36 to obtain

$$\frac{x^2}{9} + \frac{y^2}{4} = 1.$$

This equation has the form

$$\frac{x^2}{a^2} + \frac{y^2}{b^2} = 1$$

with $a = 3$ and $b = 2$, so its graph is an ellipse (Figure 3) with its center at the origin and its vertices at $(-3, 0)$, $(3, 0)$, $(0, -2)$, and $(0, 2)$.

Figure 3

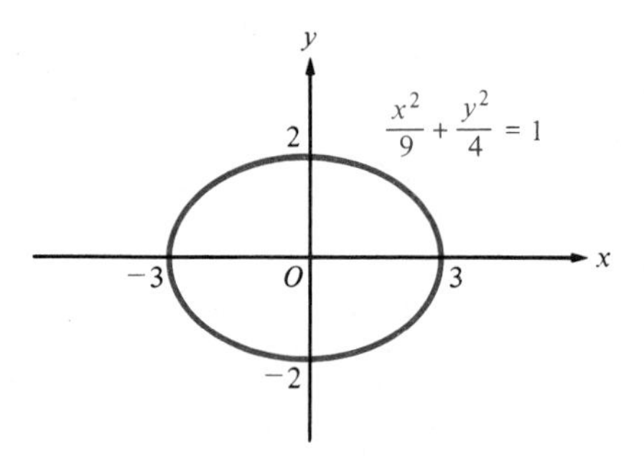

For $r > 0$, we saw in Section 4.1 of Chapter 1 that $(x - h)^2 + (y - k)^2 = r^2$ is an equation of a circle of radius r with center at the point (h, k). This equation can be rewritten as

$$\frac{(x - h)^2}{r^2} + \frac{(y - k)^2}{r^2} = 1.$$

Again, the slightly modified equation

$$\frac{(x - h)^2}{a^2} + \frac{(y - k)^2}{b^2} = 1, \qquad \text{where } a > 0, \quad b > 0, \quad a \neq b,$$

is the equation of an ellipse with its center at the point (h, k). This ellipse is obtained by shifting the ellipse $(x^2/a^2) + (y^2/b^2) = 1$ (Figure 2) so that its center moves from the origin to the point (h, k) (Figure 4). The vertices of the shifted ellipse are at $(h - a, k)$, $(h + a, k)$, $(h, k - b)$, and $(h, k + b)$, and its major and minor axes are parallel to the coordinate axes. We refer to

$$\boxed{\frac{(x - h)^2}{a^2} + \frac{(y - k)^2}{b^2} = 1, \qquad \text{where } a > 0, \quad b > 0, \quad a \neq b}$$

as the equation in **standard form** for an ellipse.

Figure 4

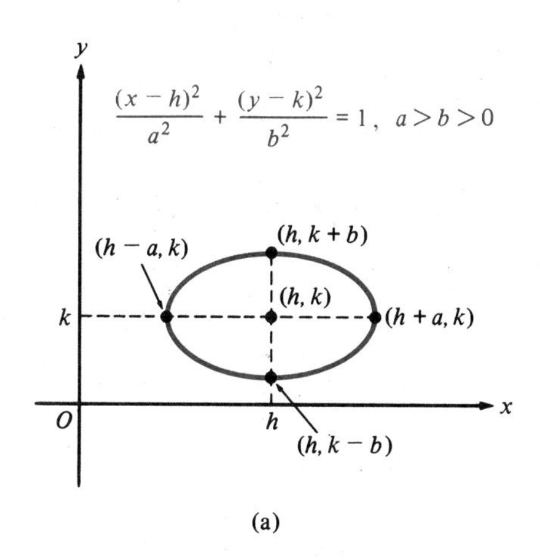

(a)

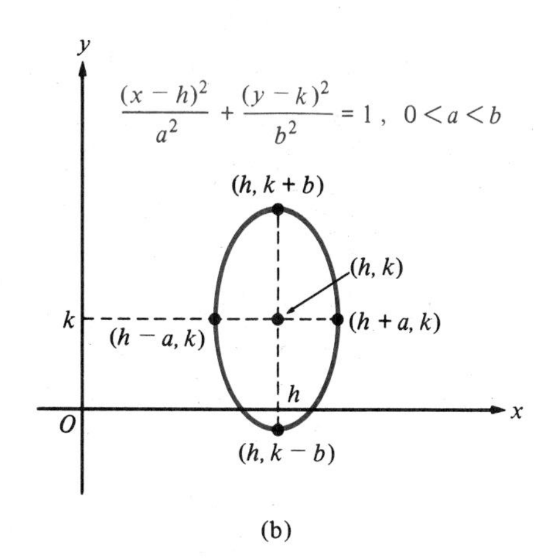

(b)

Example Find the equation in standard form for the ellipse with vertices at $(3, -2)$, $(9, -2)$, $(6, -6)$, and $(6, 2)$, and sketch the graph.

Solution We begin by plotting the four vertices. Evidently, $(3, -2)$ and $(9, -2)$ are the endpoints of the minor axis, while $(6, -6)$ and $(6, 2)$ are the endpoints of the major axis; therefore, the center is the point $(h, k) = (6, -2)$. In the standard form for the equation of an ellipse, a is the distance from the center to either vertex at the end of the horizontal axis and b is the distance from the center to either vertex at the end of the vertical axis. In this case, $a = 3$ and $b = 4$. Therefore, the equation in standard form for the ellipse is

$$\frac{(x-6)^2}{3^2}+\frac{[y-(-2)]^2}{4^2}=1 \quad \text{or} \quad \frac{(x-6)^2}{9}+\frac{(y+2)^2}{16}=1.$$

The graph is sketched in Figure 5.

Figure 5

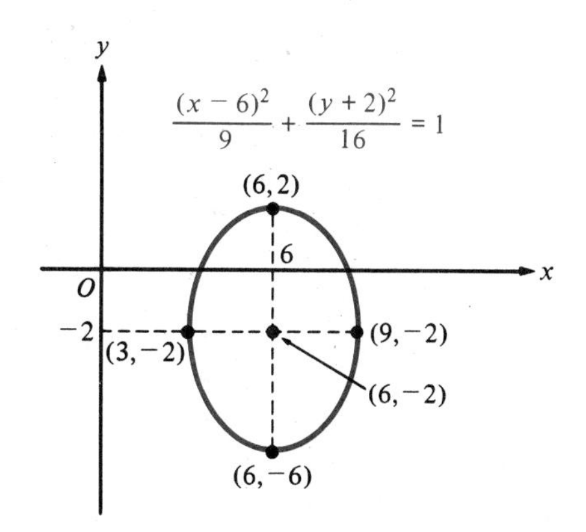

4.2 Parabolas

In Section 6.1 of Chapter 1, we mentioned that, if $a \neq 0$, the graph of the equation $y = a(x - h)^2 + k$ or $y - k = a(x - h)^2$ is a parabola with a vertical axis, vertex at the point (h, k), and opening upward or downward according to whether $a > 0$ or $a < 0$, respectively (Figure 6a). By interchanging the roles of x and y, we find that, if $a \neq 0$, $x - h = a(y - k)^2$ is an equation of a parabola with a horizontal axis, vertex at the point (h, k), and opening to the right or left according to whether $a > 0$ or $a < 0$, respectively (Figure 6b). We refer to

$$y - k = a(x - h)^2 \quad \text{and} \quad x - h = a(y - k)^2, \qquad a \neq 0,$$

as the equations in **standard form** for a parabola.

Figure 6

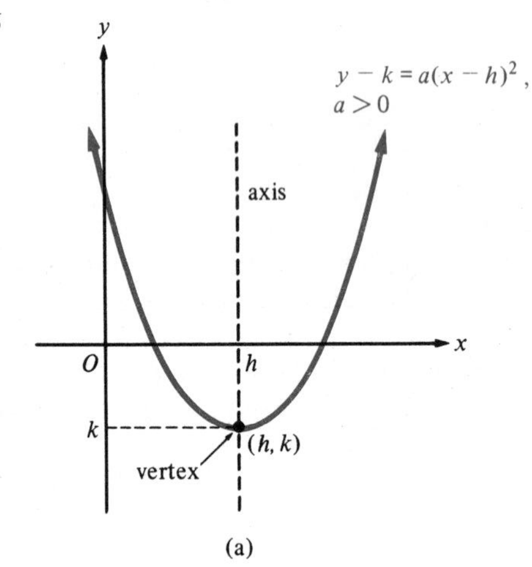

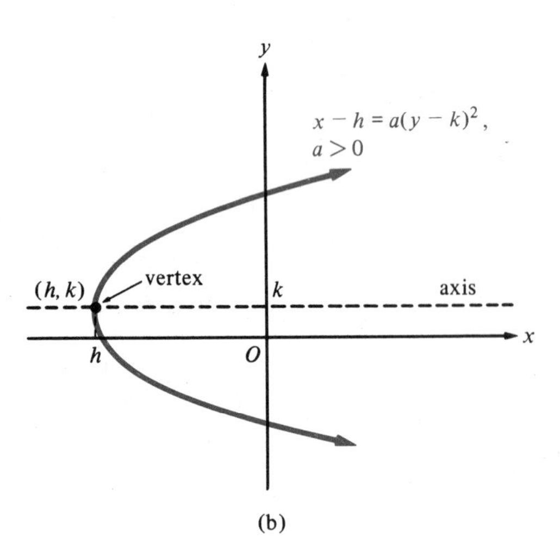

Example 1 A parabola with a horizontal axis has its vertex at the origin and contains the point (4, 3). Find its equation in standard form and sketch the graph.

Solution Since the parabola has a horizontal axis, the standard form of its equation is

$$x - h = a(y - k)^2, \quad a \neq 0.$$

Figure 7

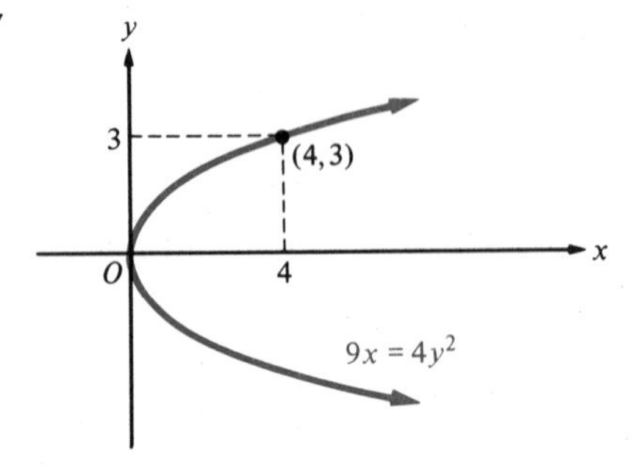

Because the vertex is at the origin, $h = 0$, $k = 0$, and the equation has the form $x = ay^2$. To determine a, we substitute $x = 4$ and $y = 3$ in the equation $x = ay^2$ to obtain $4 = a(9)$. Thus, $a = \frac{4}{9}$ and the parabola has the equation

$$x = \tfrac{4}{9}y^2 \quad \text{or} \quad 9x = 4y^2.$$

The graph is sketched in Figure 7.

Example 2 Sketch the graph of $(y+1)^2 = -12(x-2)$.

Solution The equation can be rewritten in the form

$$x - 2 = -\tfrac{1}{12}[y - (-1)]^2;$$

that is,

$$x - h = a(y - k)^2$$

with $h = 2$, $k = -1$, and $a = -\tfrac{1}{12}$. Therefore, the graph is a parabola (Figure 8) with a horizontal axis, its vertex at $(h, k) = (2, -1)$, and opening to the left.

Figure 8

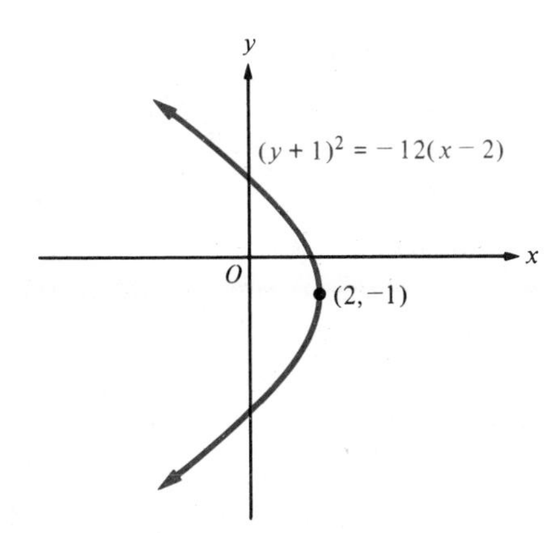

4.3 Hyperbolas

If, in the equation

$$\frac{x^2}{a^2} + \frac{y^2}{b^2} = 1$$

of an ellipse with center at the origin, we change the plus sign to a minus sign, we obtain the equation

$$\boxed{\frac{x^2}{a^2} - \frac{y^2}{b^2} = 1, \qquad a > 0, \quad b > 0,}$$

Figure 9

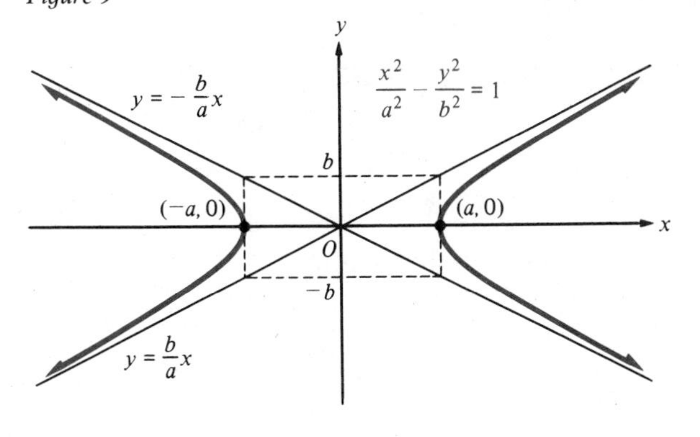

whose graph is a **hyperbola** (Figure 9). The hyperbola has two **branches,** one opening to the right and one opening to the left. The two points $(-a, 0)$ and $(a, 0)$, where the hyperbola intersects the x axis, are called its **vertices,** and the line segment between the two vertices is called the **transverse axis** of the hyperbola. The midpoint of the transverse axis, which is the origin in Figure 9, is called the **center** of the hyperbola.

The two straight lines

$$y = \frac{b}{a}x \quad \text{and} \quad y = -\frac{b}{a}x$$

are called the **asymptotes** of the hyperbola $\frac{x^2}{a^2} - \frac{y^2}{b^2} = 1$. As a point (x, y) on the hyperbola moves farther and farther from the center, it comes closer and closer to one of these asymptotes. Although the asymptotes are not part of the hyperbola itself, they are useful in sketching it. In fact, if we begin by drawing the so-called *fundamental rectangle* with height $2b$ and horizontal base $2a$ whose center is at the center of the hyperbola, then the straight lines through the diagonals of the rectangle are the asymptotes (Figure 9). If we use the fact that the vertices of the hyperbola are the midpoints of the left and right sides of the rectangle and that the hyperbola approaches the asymptotes as one moves out away from the vertices, it is easy to sketch the graph.

Example Find the asymptotes and vertices of the hyperbola $\frac{x^2}{4} - \frac{y^2}{1} = 1$; sketch its graph.

Solution The equation has the form $\frac{x^2}{a^2} - \frac{y^2}{b^2} = 1$, with $a = 2$ and $b = 1$. We begin by drawing the fundamental rectangle with height $2b = 2$, horizontal base $2a = 4$, and center at the origin. The diagonals of this rectangle determine the asymptotes, whose equations are $y = \frac{1}{2}x$ and $y = -\frac{1}{2}x$. The vertices are the midpoints, $(-2,0)$ and $(2,0)$, of the left and right sides of the rectangle (Figure 10).

Figure 10

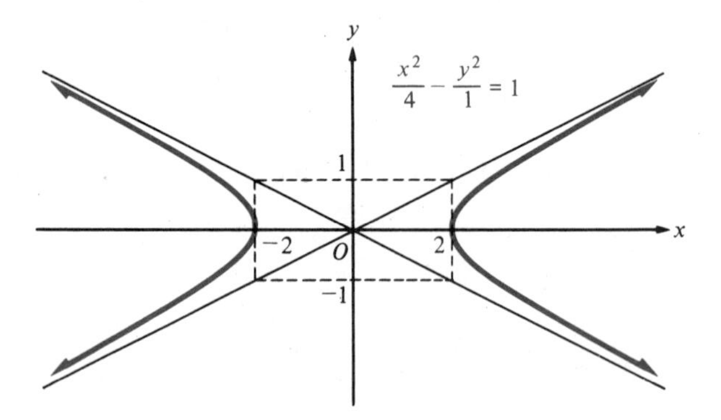

If we interchange the roles of x and y, we find that

$$\frac{y^2}{b^2} - \frac{x^2}{a^2} = 1, \qquad a > 0, \quad b > 0$$

is an equation of a hyperbola with a vertical transverse axis and center at the origin (Figure 11). The vertices are the points $(0, -b)$ and $(0, b)$ and the asymptotes are the straight lines $y = \frac{b}{a}x$ and $y = -\frac{b}{a}x$, just as before.

Figure 11

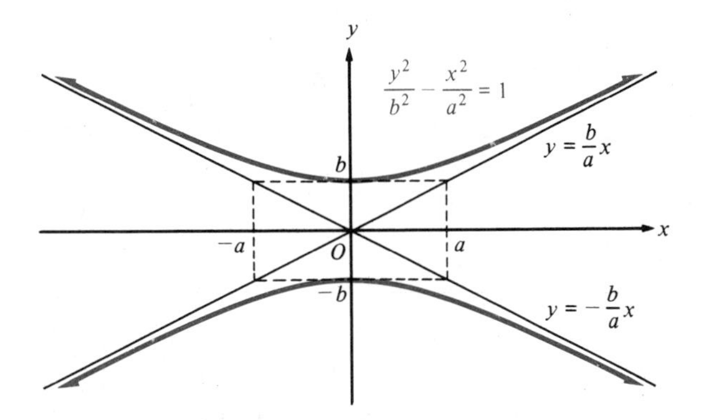

If we shift the hyperbola $\frac{x^2}{a^2} - \frac{y^2}{b^2} = 1$ (Figure 9) or $\frac{y^2}{b^2} - \frac{x^2}{a^2} = 1$ (Figure 11) so that the center is at the point (h, k), the resulting equation becomes either

$$\frac{(x-h)^2}{a^2} - \frac{(y-k)^2}{b^2} = 1 \quad \text{or} \quad \frac{(y-k)^2}{b^2} - \frac{(x-h)^2}{a^2} = 1,$$

depending on whether the transverse axis is horizontal or vertical. These are called equations in **standard form** for a hyperbola (Figure 12). For either equation, the asymptotes of the hyperbola are the straight lines

$$y - k = \frac{b}{a}(x - h) \quad \text{and} \quad y - k = -\frac{b}{a}(x - h).$$

Figure 12

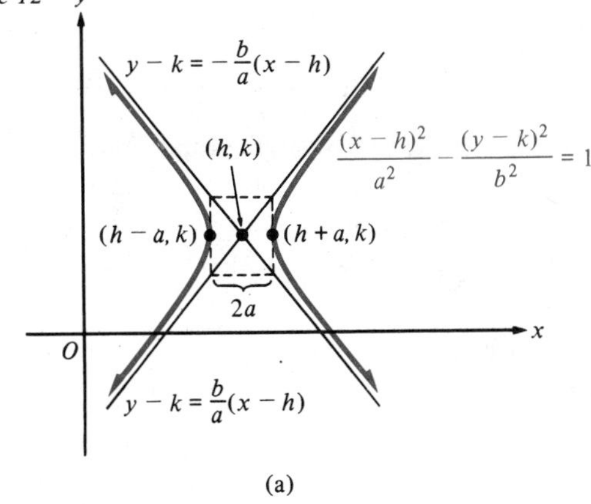

(a)

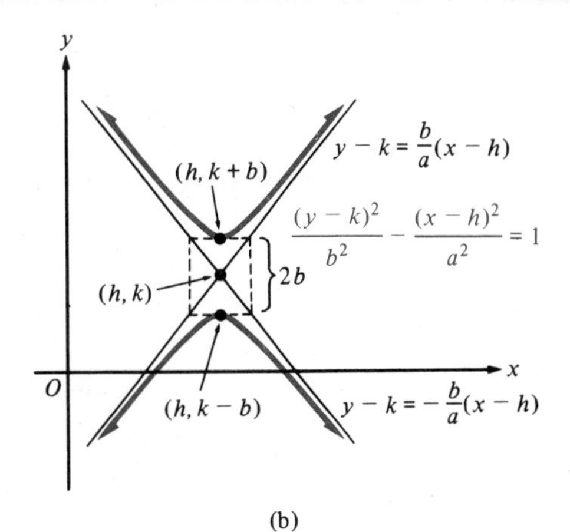

(b)

Example Find the center, the asymptotes, and the vertices of the hyperbola

$$\frac{(y-2)^2}{9} - \frac{(x+1)^2}{4} = 1$$

and sketch the graph.

Solution The equation has the form $\dfrac{(y-k)^2}{b^2} - \dfrac{(x-h)^2}{a^2} = 1$, with $h = -1$, $k = 2$, $a = 2$, and $b = 3$, so its graph is a hyperbola (Figure 13) with a vertical transverse axis, center at $(-1, 2)$, asymptotes $y - 2 = \frac{3}{2}(x + 1)$ and $y - 2 = -\frac{3}{2}(x + 1)$, and vertices at $(-1, -1)$ and $(-1, 5)$.

Figure 13

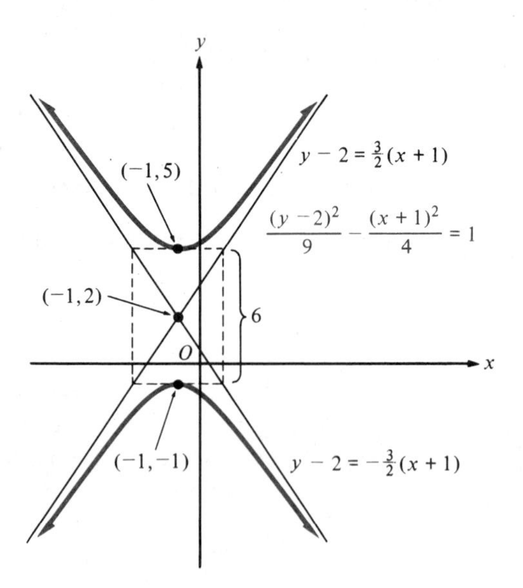

4.4 Restoring the Standard Form by Completing the Square

If the squares of the binomials in the standard form of an equation of a circle, an ellipse, a parabola, or a hyperbola are expanded, and if the resulting constant terms are combined, the equation will no longer be in standard form. By completing squares, as in the following examples, the standard form can be restored.

Examples By completing squares, rewrite each equation in standard form and sketch the graph of the resulting conic section.

1 $x^2 + 2x + y^2 - 4y = 4$

Solution By completing squares, we obtain

$$x^2 + 2x + 1 + y^2 - 4y + 4 = 4 + 1 + 4$$
$$(x + 1)^2 + (y - 2)^2 = 9.$$

The graph is a circle (Figure 14) of radius 3 with center at $(-1, 2)$.

Figure 14

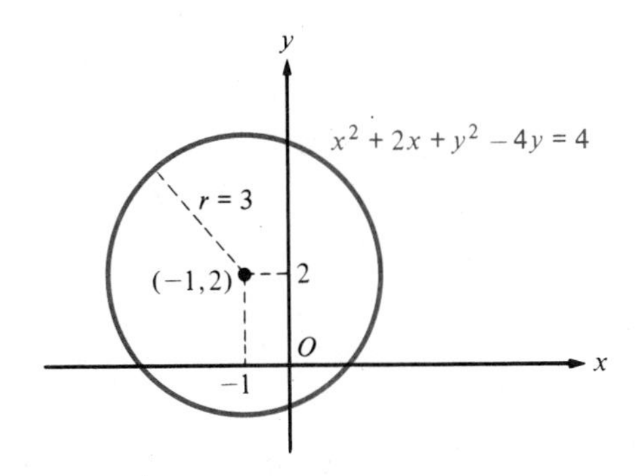

2 $25x^2 + 9y^2 - 100x - 54y - 44 = 0$

Solution

$$25(x^2 - 4x) + 9(y^2 - 6y) = 44$$
$$25(x^2 - 4x + 4) + 9(y^2 - 6y + 9) = 44 + 25(4) + 9(9)$$
$$25(x - 2)^2 + 9(y - 3)^2 = 225$$
$$\frac{(x - 2)^2}{225/25} + \frac{(y - 3)^2}{225/9} = 1$$
$$\frac{(x - 2)^2}{9} + \frac{(y - 3)^2}{25} = 1.$$

The graph is an ellipse (Figure 15) with center at $(2, 3)$ and vertices at $(-1, 3)$, $(5, 3)$, $(2, -2)$, and $(2, 8)$.

Figure 15

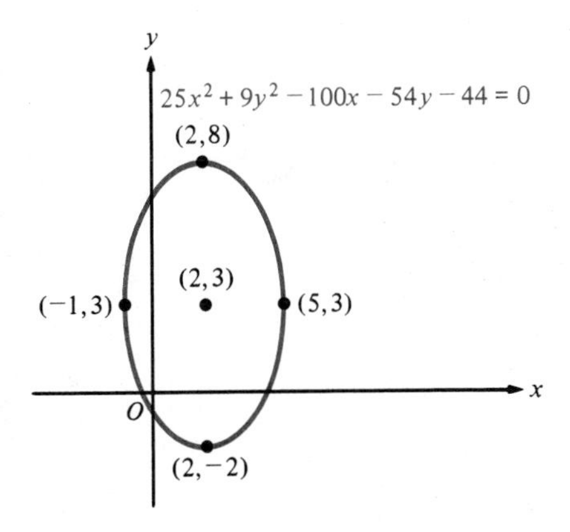

3 $x^2 + 4x - 10y + 34 = 0$

Solution

$$x^2 + 4x + 4 - 10y + 34 = 4$$
$$(x + 2)^2 = 10y - 30$$
$$10(y - 3) = (x + 2)^2$$
$$y - 3 = \tfrac{1}{10}(x + 2)^2$$

The graph is a parabola (Figure 16) with a vertical axis, its vertex at $(-2, 3)$, and opening upward.

Figure 16

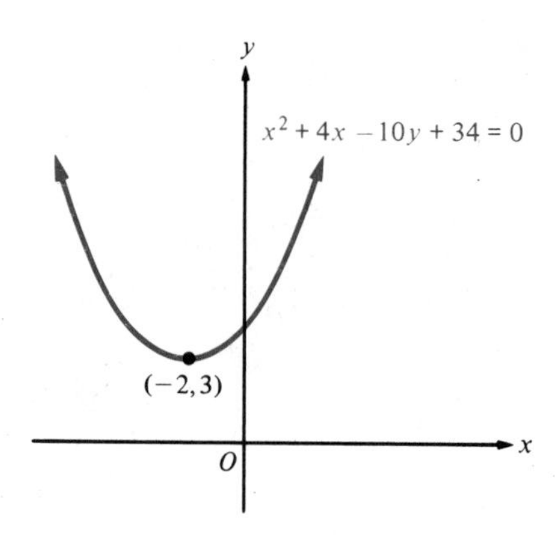

4 $4x^2 - 9y^2 - 24x - 90y - 225 = 0$

Solution

$$4(x^2 - 6x) - 9(y^2 + 10y) = 225$$
$$4(x^2 - 6x + 9) - 9(y^2 + 10y + 25) = 225 + 4(9) - 9(25)$$
$$4(x - 3)^2 - 9(y + 5)^2 = 36$$
$$\frac{(x - 3)^2}{36/4} - \frac{(y + 5)^2}{36/9} = 1$$
$$\frac{(x - 3)^2}{9} - \frac{(y + 5)^2}{4} = 1.$$

Figure 17

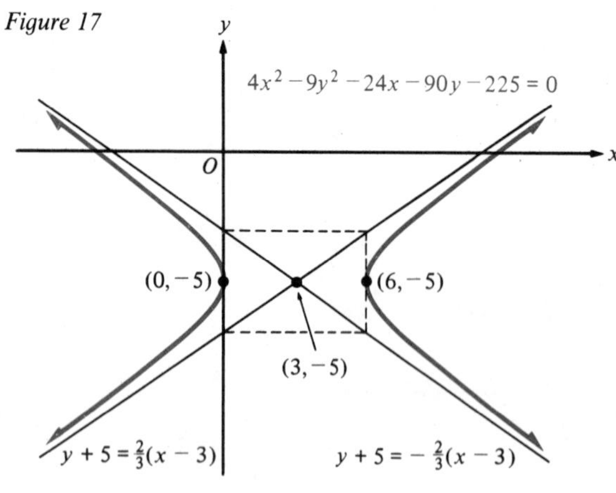

The graph is a hyperbola (Figure 17) with a horizontal transverse axis, center at $(3, -5)$, vertices at $(0, -5)$ and $(6, -5)$, and asymptotes

$$y + 5 = \tfrac{2}{3}(x - 3)$$

and

$$y + 5 = -\tfrac{2}{3}(x - 3).$$

PROBLEM SET 4

In problems 1–10, determine whether the graph of the equation is a circle or an ellipse with center at the origin. In the case of a circle, find the radius. In the case of an ellipse, find the vertices. Sketch the graph.

1. $\dfrac{x^2}{25} + \dfrac{y^2}{25} = 1$
2. $\dfrac{x^2}{16} + \dfrac{y^2}{4} = 1$
3. $\dfrac{x^2}{9} + \dfrac{y^2}{1} = 1$
4. $4x^2 + y^2 = 16$
5. $36x^2 + 9y^2 = 144$
6. $x^2 + 16y^2 = 16$
7. $16x^2 + 25y^2 = 400$
8. $x^2 + 4y^2 = 1$
9. $4x^2 + 4y^2 = 1$
10. $9x^2 + 36y^2 = 4$

In problems 11–16, find the equation in standard form for the ellipse that satisfies the indicated conditions.

11. Vertices at $(-8, 5)$, $(-2, 5)$, $(-5, -3)$ and $(-5, 7)$
12. Vertices at $(0, -8)$ and $(0, 8)$ and containing the point $(6, 0)$
13. Vertices at $(-2, -3)$, $(-2, 5)$, $(-7, 1)$, and $(3, 1)$
14. Center at the origin and containing the points $(4, 0)$ and $(3, 2)$
15. Center at $(1, -2)$, major axis parallel to the y axis and 6 units long, and minor axis 4 units long
16. Vertices at $(0, -3)$ and $(0, 3)$ and containing the point $(\frac{2}{3}, 2\sqrt{2})$

In problems 17–20, find the center and the vertices of each ellipse and sketch the graph.

17. $\dfrac{(x-1)^2}{9} + \dfrac{(y+2)^2}{4} = 1$
18. $\dfrac{(x+2)^2}{16} + \dfrac{(y-1)^2}{4} = 1$
19. $4(x+3)^2 + (y-1)^2 = 36$
20. $25(x+1)^2 + 16y^2 = 400$

In problems 21–30, find the vertex of each parabola, determine whether its axis is horizontal or vertical, and sketch the graph.

21. $y^2 = 4x$
22. $y^2 = -9x$
23. $x^2 - 2y = 0$
24. $3x^2 - 4y = 0$
25. $x^2 + 9y = 0$
26. $(y-2)^2 = 8(x+3)$
27. $(x-4)^2 = 12(y+7)$
28. $(y+1)^2 = -4(x-1)$
29. $(x+1)^2 = -8y$
30. $4x^2 + 3(y+1) = 0$

In problems 31–36, find the equation in standard form for the parabola that satisfies the indicated conditions.

31. Vertex at the origin, horizontal axis, and containing the point $(2, 4)$
32. Vertex at $(3, 2)$, vertical axis, and containing the point $(6, 1)$
33. Vertex at $(5, -1)$, horizontal axis, and containing the point $(-4, 2)$
34. Axis coincides with the y axis and containing the points $(2, 3)$ and $(-1, -2)$
35. Vertex at $(-1, -\frac{1}{2})$, vertical axis, and containing the point $(2, \frac{5}{8})$
36. Horizontal axis and containing the points $(-1, 0)$, $(0, -1)$, and $(0, 1)$
37. In each case, determine whether the transverse axis of the hyperbola is horizontal or vertical.

(a) $\dfrac{y^2}{16} - \dfrac{x^2}{9} = 1$
(b) $\dfrac{y^2}{9} - \dfrac{x^2}{16} = 1$
(c) $\dfrac{y^2}{9} - \dfrac{x^2}{9} = 1$
(d) $\dfrac{x^2}{16} - \dfrac{y^2}{9} = 1$
(e) $\dfrac{x^2}{9} - \dfrac{y^2}{16} = 1$
(f) $\dfrac{x^2}{9} - \dfrac{y^2}{9} = 1$

38. In each case, determine whether the transverse axis of the hyperbola is horizontal or vertical.

(a) $\dfrac{(x-3)^2}{25} - \dfrac{(y+2)^2}{16} = 1$ (b) $\dfrac{(x+2)^2}{16} - \dfrac{(y-3)^2}{25} = 1$

(c) $\dfrac{(y-3)^2}{9} - \dfrac{(x-3)^2}{16} = 1$ (d) $\dfrac{(y+3)^2}{9} - \dfrac{(x+3)^2}{9} = 1$

In problems 39–48, find the center, the asymptotes, and the vertices of each hyperbola and sketch the graph.

39. $\dfrac{x^2}{9} - \dfrac{y^2}{4} = 1$

40. $x^2 - \dfrac{y^2}{9} = 1$

41. $\dfrac{y^2}{16} - \dfrac{x^2}{4} = 1$

42. $49x^2 - 14y^2 = 196$

43. $4x^2 - 16y^2 = 64$

44. $36y^2 - 10x^2 = 360$

45. $\dfrac{(x-1)^2}{9} - \dfrac{(y+2)^2}{4} = 1$

46. $(x+3)^2 - \dfrac{(y-1)^2}{9} = 1$

47. $\dfrac{(y+1)^2}{16} - \dfrac{(x+2)^2}{25} = 1$

48. $\dfrac{(y+3)^2}{4} - (x-1)^2 = 1$

In problems 49–52, find the equation in standard form for the hyperbola that satisfies the indicated conditions.

49. Vertices at $(-4, 0)$ and $(4, 0)$ and asymptotes $y = -\frac{5}{4}x$ and $y = \frac{5}{4}x$

50. Asymptotes $y = -2x$ and $y = 2x$ and containing the point $(1, 1)$

51. Center at $(3, -2)$, vertices at $(3, -4)$ and $(3, 0)$, and asymptotes having slopes -4 and 4.

52. Asymptotes $y = x + 5$ and $y = -x + 3$ and containing the point $(2, 4)$

In problems 53–68, rewrite the given equation in standard form and identify the conic section.

53. $x^2 + y^2 + 2x + 4y + 4 = 0$

54. $4x^2 - y^2 - 8x + 2y + 7 = 0$

55. $y^2 - 8y - 6x - 2 = 0$

56. $2y + x = y^2$

57. $x^2 + 2y^2 + 6x + 7 = 0$

58. $4x^2 + y^2 - 8x + 4y - 8 = 0$

59. $2x^2 + 5y^2 + 20x - 30y + 75 = 0$

60. $4x^2 + 4y^2 = 4y - 8x - 1$

61. $x^2 - 4y^2 - 4x - 8y - 4 = 0$

62. $9x^2 - 25y^2 + 72x - 100y + 269 = 0$

63. $16x^2 + 180y = 9y^2 + 612$

64. $3x^2 + 3y^2 - 6x + 9y = 27$

65. $9x^2 + 4y^2 + 18x - 16y = 11$

66. $9x^2 - 16y^2 = 223 + 90x + 256y$

67. $2x^2 + 8x - 3y + 4 = 0$

68. $x = y^2 + 10y + 21$

69. The point $P = (x, y)$ moves so that the sum of its distances from the points $(3, 0)$ and $(-3, 0)$ is 8. Show that P moves on an ellipse.

70. A point P moves so that it is equidistant from the point $(2, 0)$ and a circle of radius 3 with center at $(-2, 0)$. Find an equation of the path of P.

71. The roadway of a suspension bridge is 400 meters long and is supported by a main cable in the shape of a parabola which is 100 meters above the roadway at the ends and 4 meters above the roadway at the center. Vertical supporting cables run at 50-meter intervals from the roadway to the main cable. Find the lengths of each of these vertical cables.

5 Translation and Rotation of the Coordinate Axes

An equation of a curve in the plane can often be simplified by changing to a new pair of coordinate axes. In practice, this is usually done by choosing one or both of the new coordinate axes to coincide with an axis of symmetry of the curve.

5.1 Translation of the Coordinate Axes

If two cartesian coordinate systems have corresponding axes that are parallel and have the same positive directions, then we say that these coordinate systems are obtained from one another by **translation.** Figure 1 shows a translation of an "old" xy coordinate system to a "new" $\bar{x}\bar{y}$ coordinate system whose origin $\bar{O}$ has the "old" coordinates (h, k). Consider the point P in Figure 1 having old coordinates (x, y), but having new coordinates $(\bar{x}, \bar{y})$. Evidently,

$$\begin{cases} x = h + \bar{x} \\ y = k + \bar{y} \end{cases} \quad \text{or} \quad \begin{cases} \bar{x} = x - h \\ \bar{y} = y - k. \end{cases}$$

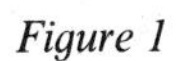

Figure 1

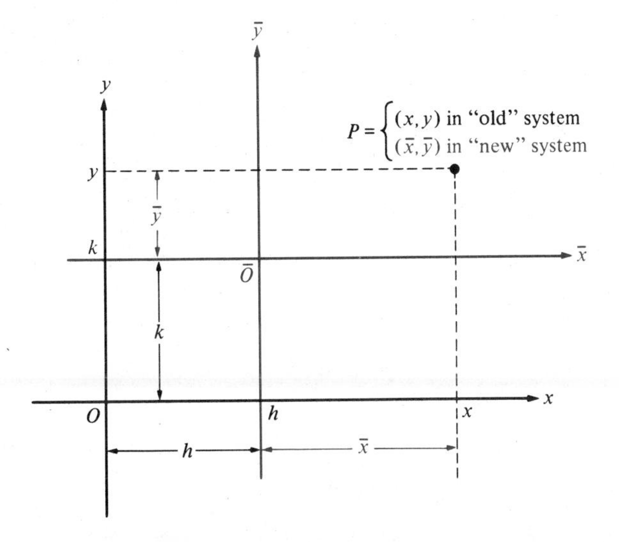

These equations, which allow us to change from the new to the old coordinates of a point P or vice versa, are called the **translation equations.** Notice that a translation of the coordinate axes does not change the position of a point P in the plane—it only changes the numerical "address" of the point.

Example Let the $\bar{x}\bar{y}$ axes be obtained from the xy axes by a translation so that the new origin $\bar{O}$ has coordinates $(h, k) = (-3, 4)$ in the old xy coordinate system. Let P be the point whose old coordinates are $(x, y) = (2, 1)$. Find the new coordinates $(\bar{x}, \bar{y})$ of P.

Solution By the translation equations,

$$\bar{x} = x - h = 2 - (-3) = 5$$

and

$$\bar{y} = y - k = 1 - 4 = -3,$$

so, in the new coordinate system, $P = (\bar{x}, \bar{y}) = (5, -3)$.

Figure 2

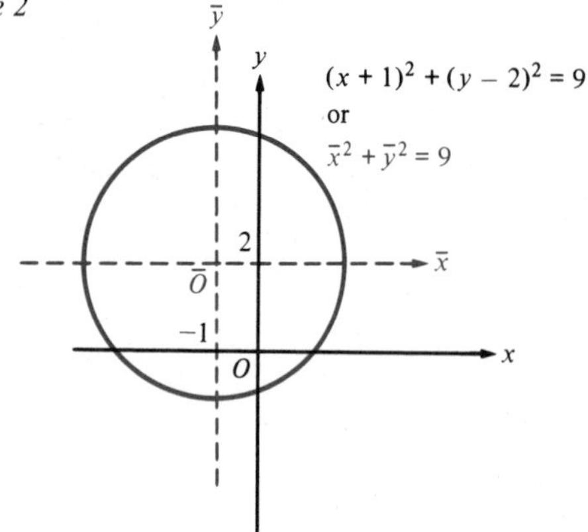

An equation of a curve in the plane depends not only on the set of points comprising it but also on our choice of the coordinate system. For instance, the circle in Figure 2 has the equation

$$(x + 1)^2 + (y - 2)^2 = 9$$

with respect to the old xy coordinate system. However, the very same circle has the simpler equation

$$\bar{x}^2 + \bar{y}^2 = 9$$

with respect to the new $\bar{x}\bar{y}$ coordinate system whose origin $\bar{O}$ is at the center of the circle. Notice that a translation of the coordinate axes does not change the position or the shape of a geometric curve in the plane—it only changes the *equation* of the curve.

Example Find a translation of axes that will reduce the equation

$$4x^2 + 9y^2 - 8x + 36y + 4 = 0$$

to an equation involving no first-degree terms in $\bar{x}$ or $\bar{y}$. Identify the graph.

Solution The simplest procedure is to complete the squares, so that the equation becomes

$$4(x^2 - 2x + 1) + 9(y^2 + 4y + 4) + 4 = 4(1) + 9(4)$$

or

$$4(x - 1)^2 + 9(y + 2)^2 = 36.$$

Then we let $\bar{x} = x - 1$ and $\bar{y} = y + 2$ to obtain

$$4\bar{x}^2 + 9\bar{y}^2 = 36 \quad \text{or} \quad \frac{\bar{x}^2}{9} + \frac{\bar{y}^2}{4} = 1,$$

the equation in standard form of an ellipse with center at the origin $\bar{O}$ of the new coordinate system. An alternative solution is obtained by letting

$$x = \bar{x} + h \quad \text{and} \quad y = \bar{y} + k,$$

where h and k are yet to be determined, and then substituting into the equation $4x^2 + 9y^2 - 8x + 36y + 4 = 0$. Routine algebraic simplification then yields

$$4\bar{x}^2 + 9\bar{y}^2 + 8(h - 1)\bar{x} + 18(k + 2)\bar{y} + 4h^2 + 9k^2 - 8h + 36k + 4 = 0.$$

The first-degree terms in $\bar{x}$ and $\bar{y}$ drop out if we let $h = 1$ and $k = -2$, and the equation becomes

$$4\bar{x}^2 + 9\bar{y}^2 - 36 = 0 \quad \text{or} \quad \frac{\bar{x}^2}{9} + \frac{\bar{y}^2}{4} = 1.$$

5.2 Rotation of the Coordinate Axes

In some situations we can simplify an equation of a curve by **rotating** the coordinate system rather than translating it. Of course, the polar coordinate system (Chapter 4, Section 5) is naturally adapted to rotation about the pole. Figure 3 shows an "old" polar axis and a "new" polar axis obtained by rotating the old polar axis counterclockwise about the pole $O = \bar{O}$ through the angle ϕ. Consider the point P in Figure 3 having old polar coordinates (r, θ), but having new polar coordinates $(\bar{r}, \bar{\theta})$.

Figure 3

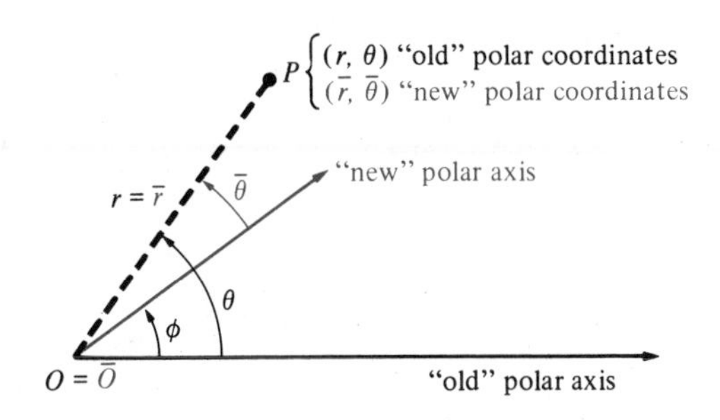

Evidently,

$$\begin{cases} r = \bar{r} \\ \theta = \phi + \bar{\theta} \end{cases} \quad \text{or} \quad \begin{cases} \bar{r} = r \\ \bar{\theta} = \theta - \phi. \end{cases}$$

These equations are called the **rotation equations for polar coordinates.**

The rotation equations for cartesian coordinates are

$$x = \bar{x}\cos\phi - \bar{y}\sin\phi$$
$$y = \bar{x}\sin\phi + \bar{y}\cos\phi,$$

where the "new" $\bar{x}\bar{y}$ coordinate system is obtained by rotating the "old" xy coordinate system counterclockwise about the origin through the angle ϕ (Figure 4). These cartesian rotation equations can be derived by converting from cartesian to polar coordinates, rotating the polar coordinate system through the angle ϕ by using the rotation equations in polar coordinates, and then converting back to cartesian coordinates. We leave the derivation of these cartesian rotation equations as an exercise (Problem 36).

Figure 4

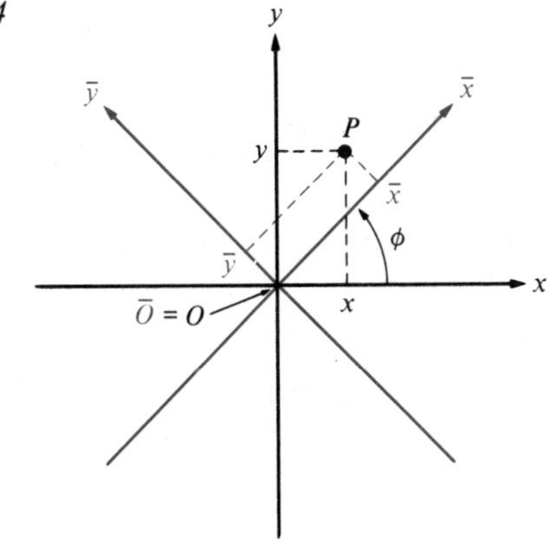

The cartesian rotation equations given above can be solved for $\bar{x}$ and $\bar{y}$ in terms of x and y to obtain

$$\begin{cases} \bar{x} = x\cos\phi + y\sin\phi \\ \bar{y} = -x\sin\phi + y\cos\phi. \end{cases}$$

(See Problem 37.)

Example The old xy coordinate system is rotated through $\pi/6$ radian to obtain a new $\bar{x}\bar{y}$ coordinate system (Figure 5). Find the new $\bar{x}\bar{y}$ coordinates of the point P whose old xy coordinates are $(1, 2)$.

Figure 5

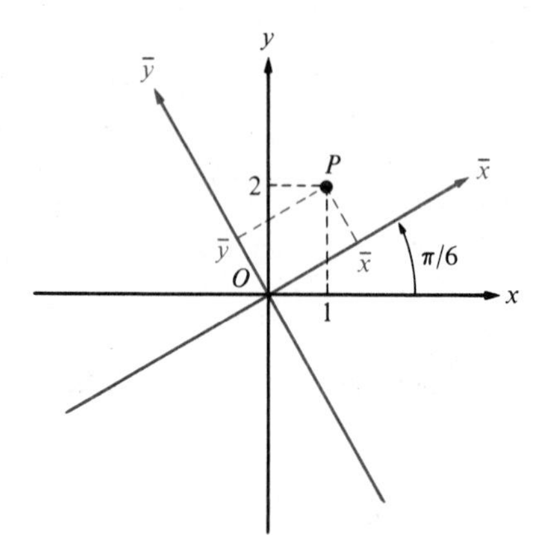

Solution We use the cartesian rotation equations, solved for $\bar{x}$ and $\bar{y}$ in terms of x and y, with $\phi = \pi/6$, $x = 1$, and $y = 2$ to obtain

$$\bar{x} = x\cos\phi + y\sin\phi = 1\cos\frac{\pi}{6} + 2\sin\frac{\pi}{6} = \frac{\sqrt{3}}{2} + 2\left(\frac{1}{2}\right) = \frac{\sqrt{3}}{2} + 1$$

$$\bar{y} = -x\sin\phi + y\cos\phi = -1\sin\frac{\pi}{6} + 2\cos\frac{\pi}{6} = -\frac{1}{2} + 2\left(\frac{\sqrt{3}}{2}\right) = -\frac{1}{2} + \sqrt{3}.$$

Hence, the new coordinates of P are $(\bar{x}, \bar{y}) = \left(1 + \frac{\sqrt{3}}{2}, \sqrt{3} - \frac{1}{2}\right)$.

By a suitable rotation of the coordinate axes, an equation of a curve can often be "simplified" or brought into recognizable form.

Example Suppose that the old xy coordinate system is rotated through $\phi = 45°$ to obtain a new $\bar{x}\bar{y}$ coordinate system (Figure 6). Find an equation of the curve $xy = 1$ in the new $\bar{x}\bar{y}$ coordinate system and identify the curve.

Figure 6

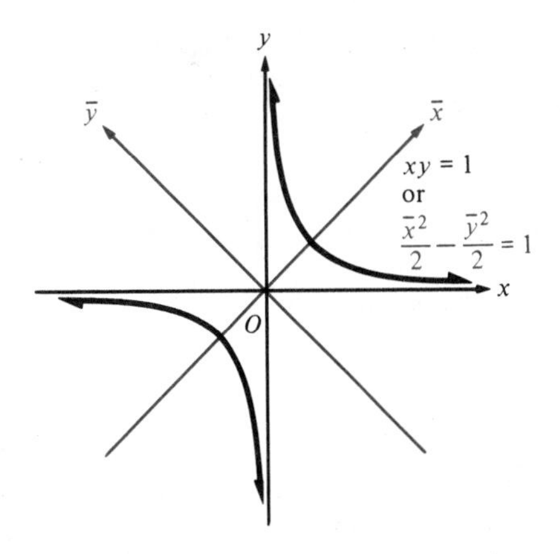

Solution The graph of $xy = 1$ looks suspiciously like a hyperbola whose transverse axis makes an angle of 45° with the x axis. Our calculations will confirm this. We substitute the cartesian rotation equations with $\phi = 45°$, that is,

$$x = \bar{x}\cos 45° - \bar{y}\sin 45° = \frac{\sqrt{2}}{2}(\bar{x} - \bar{y})$$

$$y = \bar{x}\sin 45° + \bar{y}\cos 45° = \frac{\sqrt{2}}{2}(\bar{x} + \bar{y}),$$

into the equation $xy = 1$ to obtain

$$\frac{\sqrt{2}}{2}(\bar{x} - \bar{y})\frac{\sqrt{2}}{2}(\bar{x} + \bar{y}) = 1 \quad \text{or} \quad \frac{\bar{x}^2}{2} - \frac{\bar{y}^2}{2} = 1.$$

Thus, just as we suspected, the curve is a hyperbola since its equation has the standard form in the new $\bar{x}\bar{y}$ coordinate system.

In the example above we were able to identify the curve $xy = 1$ as a hyperbola because rotation of the coordinate axis through the 45° angle brought the equation into a familiar form. This occurred because the new $\bar{x}\bar{y}$ coordinate axes were aligned with the axes of symmetry of the hyperbola. Essentially the same procedure can be applied to any equation of the form

$$\boxed{Ax^2 + Bxy + Cy^2 + Dx + Ey + F = 0}$$

containing a **mixed term Bxy.** The latter equation, in which the coefficients denote constant real numbers, is called the **general second-degree equation in x and y.**

The equation in standard form of a circle, an ellipse, a parabola, or a hyperbola can always be rewritten in the form of a general second-degree equation *containing no mixed term* (see Problem 38). For instance, the equation in standard form

$$\frac{(x - 3)^2}{4} - \frac{(y + 1)^2}{9} = 1$$

of a hyperbola can be rewritten as

$$\frac{x^2 - 6x + 9}{4} - \frac{y^2 + 2y + 1}{9} = 1$$

or

$$9(x^2 - 6x + 9) - 4(y^2 + 2y + 1) = 36;$$

that is,

$$9x^2 - 54x + 81 - 4y^2 - 8y - 4 - 36 = 0$$

or

$$9x^2 - 4y^2 - 54x - 8y + 41 = 0.$$

The latter equation has the general second-degree form, with $A = 9$, $B = 0$, $C = -4$, $D = -54$, $E = -8$, and $F = 41$.

Now consider a curve in the plane whose equation has the general second-degree form, but with a mixed term Bxy, with $B \neq 0$. The following theorem shows that the mixed term can always be removed by a suitable rotation of the coordinate axes (see Problems 40 and 42).

Theorem 1 **Removal of the Mixed Term by Rotation**

If the old xy coordinate system is rotated about the origin through the angle ϕ to obtain a new $\bar{x}\bar{y}$ coordinate system, then the curve whose old equation was

$$Ax^2 + Bxy + Cy^2 + Dx + Ey + F = 0$$

will have a new equation of the form

$$\bar{A}\bar{x}^2 + \bar{B}\bar{x}\bar{y} + \bar{C}\bar{y}^2 + \bar{D}\bar{x} + \bar{E}\bar{y} + \bar{F} = 0.$$

If $B \neq 0$ and if ϕ is chosen so that $0 < \phi < \frac{\pi}{2}$ and $\cot 2\phi = \frac{A - C}{B}$, then $\bar{B} = 0$, and the mixed term will not appear in the new equation.

Example Rotate the old xy coordinate axes to remove the mixed term from the equation $x^2 - 4xy + y^2 - 6 = 0$ and sketch the graph showing both the old xy coordinate system and the new $\bar{x}\bar{y}$ coordinate system.

Solution The equation

$$x^2 - 4xy + y^2 - 6 = 0$$

has the form

$$Ax^2 + Bxy + Cy^2 + Dx + Ey + F = 0,$$

with $A = 1, B = -4, C = 1, D = 0, E = 0$, and $F = -6$. According to Theorem 1, the mixed term can be removed by rotating the coordinate system through the angle ϕ, where $0 < \phi < \frac{\pi}{2}$ and $\cot 2\phi = \frac{A - C}{B} = \frac{1 - 1}{-4} = 0$. Therefore, we take

$2\phi = \pi/2$ or $\phi = \pi/4$. The rotation equations with $\phi = \pi/4$ are

$$x = \bar{x}\cos\phi - \bar{y}\sin\phi = \frac{\sqrt{2}}{2}(\bar{x} - \bar{y})$$

$$y = \bar{x}\sin\phi + \bar{y}\cos\phi = \frac{\sqrt{2}}{2}(\bar{x} + \bar{y}).$$

Substituting these into the equation $x^2 - 4xy + y^2 - 6 = 0$, we obtain

$$\left[\frac{\sqrt{2}}{2}(\bar{x} - \bar{y})\right]^2 - 4\left[\frac{\sqrt{2}}{2}(\bar{x} - \bar{y})\right]\left[\frac{\sqrt{2}}{2}(\bar{x} + \bar{y})\right] + \left[\frac{\sqrt{2}}{2}(\bar{x} + \bar{y})\right]^2 - 6 = 0,$$

which simplifies to

$$\tfrac{1}{2}(\bar{x} - \bar{y})^2 - 2(\bar{x} - \bar{y})(\bar{x} + \bar{y}) + \tfrac{1}{2}(\bar{x} + \bar{y})^2 = 6$$

or

$$\tfrac{1}{2}(\bar{x}^2 - 2\bar{x}\bar{y} + \bar{y}^2) - 2(\bar{x}^2 - \bar{y}^2) + \tfrac{1}{2}(\bar{x}^2 + 2\bar{x}\bar{y} + \bar{y}^2) = 6.$$

Collecting terms in the latter equation, we get

$$-\bar{x}^2 + 3\bar{y}^2 = 6 \quad \text{or} \quad \frac{\bar{y}^2}{2} - \frac{\bar{x}^2}{6} = 1,$$

an equation of a hyperbola (Figure 7).

Figure 7

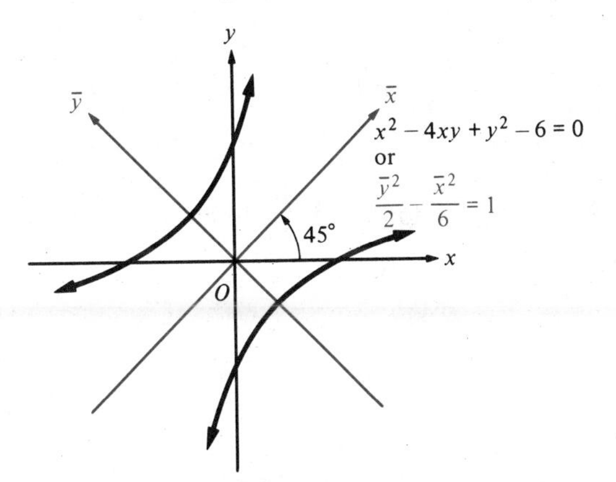

If $0 < \phi < \dfrac{\pi}{2}$, then the trigonometric identities

$\cos 2\phi = \dfrac{\cot 2\phi}{\sqrt{\cot^2 2\phi + 1}}$,	$\cos\phi = \sqrt{\dfrac{1 + \cos 2\phi}{2}}$,	$\sin\phi = \sqrt{\dfrac{1 - \cos 2\phi}{2}}$

permit us to find $\cos\phi$ and $\sin\phi$ algebraically in terms of the value of $\cot 2\phi$. This is useful in applying Theorem 1, as the following example shows.

Example Ⓒ Rotate the old xy coordinate axes to remove the mixed term from the equation

$$8x^2 - 4xy + 5y^2 = 144$$

and sketch the graph showing both the old and the new coordinate systems.

Solution Here $A = 8$, $B = -4$, $C = 5$, $D = 0$, $E = 0$, and $F = -144$. Hence,

$$\cot 2\phi = \frac{A - C}{B} = \frac{8 - 5}{-4} = -\frac{3}{4}.$$

Thus,

$$\cos 2\phi = \frac{\cot 2\phi}{\sqrt{\cot^2 2\phi + 1}} = \frac{-3/4}{\sqrt{(-3/4)^2 + 1}} = \frac{-3/4}{\sqrt{(9/16) + 1}} = \frac{-3/4}{5/4} = -\frac{3}{5},$$

so that

$$\cos\phi = \sqrt{\frac{1 + \cos 2\phi}{2}} = \sqrt{\frac{1 - (3/5)}{2}} = \sqrt{\frac{2}{10}} = \sqrt{\frac{1}{5}} = \frac{\sqrt{5}}{5},$$

and

$$\sin\phi = \sqrt{\frac{1 - \cos 2\phi}{2}} = \sqrt{\frac{1 + (3/5)}{2}} = \sqrt{\frac{8}{10}} = \sqrt{\frac{4}{5}} = \frac{2\sqrt{5}}{5}.$$

Now, substituting the rotation equations

$$x = \bar{x}\cos\phi - \bar{y}\sin\phi = \frac{\sqrt{5}}{5}\bar{x} - \frac{2\sqrt{5}}{5}\bar{y}$$

$$y = \bar{x}\sin\phi + \bar{y}\cos\phi = \frac{2\sqrt{5}}{5}\bar{x} + \frac{\sqrt{5}}{5}\bar{y}$$

into the given equation $8x^2 - 4xy + 5y^2 = 144$, we obtain

$$8\left(\frac{\sqrt{5}}{5}\bar{x} - \frac{2\sqrt{5}}{5}\bar{y}\right)^2 - 4\left(\frac{\sqrt{5}}{5}\bar{x} - \frac{2\sqrt{5}}{5}\bar{y}\right)\left(\frac{2\sqrt{5}}{5}\bar{x} + \frac{\sqrt{5}}{5}\bar{y}\right) + 5\left(\frac{2\sqrt{5}}{5}\bar{x} + \frac{\sqrt{5}}{5}\bar{y}\right)^2 = 144.$$

Figure 8

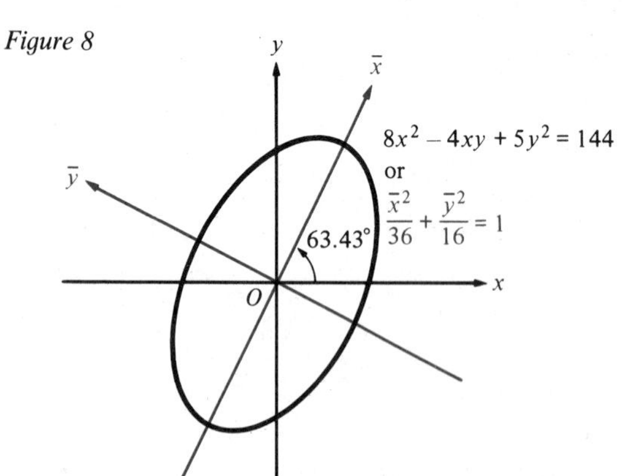

This equation simplifies to

$$4\bar{x}^2 + 9\bar{y}^2 = 144 \quad \text{or} \quad \frac{\bar{x}^2}{36} + \frac{\bar{y}^2}{16} = 1,$$

an equation of an ellipse. Since $\sin\phi = 2\sqrt{5}/5$, it follows (using a calculator or Table 1 of Section 3 of the Appendix) that

$$\phi = \sin^{-1}\frac{2\sqrt{5}}{5} \approx 63.43^\circ$$

(Figure 8).

We now have the means to sketch the graph of any second-degree equation in x and y. If the equation contains a mixed term, a suitable rotation of the coordinate axes removes it (Theorem 1). Then, by completing the squares as in Section 4.4, we *usually* obtain the equation in standard form for a circle, an ellipse, a parabola, or a hyperbola. The rotation of axes that removes the mixed term lines up the new coordinate system with the axes of symmetry of the conic section (see Figures 7 and 8).

There are some exceptional cases in which, even after completing the squares, an equation cannot be brought into any of the standard forms mentioned above. In these cases, we say that the graph is a **degenerate conic.** The only possible degenerate conics are (i) the whole plane, (ii) the empty set, (iii) a pair of straight lines, (iv) a straight line, or (v) one point. (See Problems 39, 41, 43, and 45.)

PROBLEM SET 5

1. A new $\bar{x}\bar{y}$ coordinate system is obtained by translating the old xy coordinate system so that the origin $\bar{O}$ of the new system has old xy coordinates $(-1, 2)$. Find the new $\bar{x}\bar{y}$ coordinates of the points whose old xy coordinates are:
(a) $(0,0)$ (b) $(-2,1)$ (c) $(3,-3)$ (d) $(-3,-2)$ (e) $(5,5)$ (f) $(6,0)$.
2. A new $\bar{x}\bar{y}$ coordinate system is obtained by translating the old xy coordinate system so that the origin $\bar{O}$ of the new system has old xy coordinates $(2, -3)$. Find the old xy coordinates of the points whose new $\bar{x}\bar{y}$ coordinates are:
(a) $(0,0)$ (b) $(3,2)$ (c) $(-3,4)$ (d) $(-2,3)$ (e) $(\pi, \sqrt{2})$ (f) $(0, \pi)$.

In problems 3–16, find a translation of axes $x = \bar{x} + h$ and $y = \bar{y} + k$ that will reduce each equation to an equation involving no first-degree terms in $\bar{x}$ or $\bar{y}$. Identify the graph.

3. $x^2 + y^2 + 4x - 2y + 1 = 0$
4. $3x^2 + 3y^2 + 7x - 5y + 3 = 0$
5. $x^2 + 4y^2 + 2x - 8y + 1 = 0$
6. $9x^2 + y^2 - 18x + 2y + 9 = 0$
7. $6x^2 + 9y^2 - 24x - 54y + 51 = 0$
8. $9x^2 + 4y^2 - 18x + 16y - 11 = 0$
9. $x^2 - 2x - y^2 + 6 = 0$
10. $4x^2 + 24x + 39 - 3y^2 = 0$
11. $x^2 - 10y - 4x + 21 = 0$
12. $3x^2 - y^2 + 12x + 8y = 7$
13. $4x^2 - 25y^2 + 24x + 50y + 22 = 0$
14. $5y^2 - 9x^2 + 10y + 54x = 112$
15. $x^2 - 4y^2 - 4x - 8y - 4 = 0$
16. $9x^2 - y^2 - 18x - 4y + 5 = 0$

In problems 17–24, the old xy axis has been rotated counterclockwise about the origin through the angle ϕ to form a new $\bar{x}\bar{y}$ coordinate system. The point P has coordinates (x, y) in the old system and coordinates $(\bar{x}, \bar{y})$ in the new system. Supply the missing information.

17. $(x, y) = (4, -7)$, $\phi = 90°$, $(\bar{x}, \bar{y}) = ?$
18. $(x, y) = (2, 0)$, $(\bar{x}, \bar{y}) = (1, \sqrt{3})$, $0 < \phi < 90°$, $\phi = ?$
19. $(\bar{x}, \bar{y}) = (-3, -3)$, $\phi = \pi/3$, $(x, y) = ?$
20. $(x, y) = (5\sqrt{2}, \sqrt{2})$, $\phi = 45°$, $(\bar{x}, \bar{y}) = ?$
21. $(\bar{x}, \bar{y}) = (-4, -2)$, $\phi = 30°$, $(x, y) = ?$
22. $(\bar{x}, \bar{y}) = (-3, \sqrt{2})$, $\phi = 3\pi/4$, $(x, y) = ?$
23. $(x, y) = (-4, 0)$, $\phi = \pi$, $(\bar{x}, \bar{y}) = ?$
24. $(x, y) = (1, -7)$, $\phi = 240°$, $(\bar{x}, \bar{y}) = ?$

25. The xy coordinate system is rotated 30° about the origin to form the $\bar{x}\bar{y}$ coordinate system. Rewrite each equation as an equation in the other coordinate system.
(a) $y^2 = 3x$ (b) $\bar{y} = 3\bar{x}$ (c) $\bar{x}^2 + \bar{y}^2 = 1$ (d) $5x - y = 4$ (e) $x^2 + y^2 = 1$ (f) $x^2 = 25$

26. Find an angle ϕ (if one exists) for which the cartesian rotation equations give each of the following.
(a) $\bar{x} = y$ and $\bar{y} = -x$
(b) $\bar{x} = -x$ and $\bar{y} = -y$
(c) $\bar{x} = y$ and $\bar{y} = x$
(d) $\bar{x} = -y$ and $\bar{y} = -x$

In problems 27–35, (a) use Theorem 1 to find an angle ϕ through which to rotate the coordinate system so as to remove the mixed term from the equation, (b) find x and y in terms of $\bar{x}$ and $\bar{y}$, (c) find the new equation in terms of $\bar{x}$ and $\bar{y}$, and (d) sketch the graph showing both the old xy axes and the new $\bar{x}\bar{y}$ axes.

[C] **27.** $x^2 + 4xy - 2y^2 = 12$
28. $x^2 + 2xy + y^2 + x + y = 0$
29. $x^2 + 2xy + y^2 = 1$
30. $x^2 + 2xy + y^2 - 4\sqrt{2}x + 4\sqrt{2}y = 0$
[C] **31.** $9x^2 - 24xy + 16y^2 = 144$
32. $2x^2 + 4\sqrt{3}xy - 2y^2 - 4 = 0$
[C] **33.** $6x^2 - 6xy + 14y^2 = 45$
[C] **34.** $17x^2 - 12xy + 8y^2 - 68x + 24y - 12 = 0$
[C] **35.** $2x^2 + 6xy - 6y^2 + 2\sqrt{10}x + 3\sqrt{10}y - 16 = 0$

36. Derive the cartesian rotation equations by converting from cartesian to polar coordinates, rotating the polar coordinate system through the angle ϕ, and then converting back to cartesian coordinates.

37. Solve the simultaneous equations

$$\begin{cases} \bar{x}\cos\phi - \bar{y}\sin\phi = x \\ \bar{x}\sin\phi + \bar{y}\cos\phi = y \end{cases}$$

for $\bar{x}$ and $\bar{y}$ in terms of x and y.

38. Show that an equation in standard form of a circle, an ellipse, a parabola, or a hyperbola can be rewritten in the form of a general second-degree equation *with no mixed term.*

39. The following second-degree equations in x and y have graphs that are degenerate conics. In each case, verify that the graph is as described.
(a) $0x^2 + 0yx + 0y^2 + 0x + 0y + 0 = 0$ (The whole xy plane)
(b) $x^2 + y^2 + 1 = 0$ (The empty set)
(c) $2x^2 - 4xy + 2y^2 = 0$ (A straight line)
(d) $4x^2 - y^2 + 16x + 2y + 15 = 0$ (Two intersecting straight lines)
(e) $x^2 - 2xy + y^2 - 18 = 0$ (Two parallel straight lines)
(f) $x^2 + y^2 - 6x + 4y + 13 = 0$ (A single point)

40. In Theorem 1, show by direct calculation that:
(a) $\bar{A} = A\cos^2\phi + B\cos\phi\sin\phi + C\sin^2\phi$
(b) $\bar{B} = 2(C - A)\cos\phi\sin\phi + B(\cos^2\phi - \sin^2\phi) = (C - A)\sin 2\phi + B\cos 2\phi$
(c) $\bar{C} = A\sin^2\phi - B\cos\phi\sin\phi + C\cos^2\phi$
(d) $\bar{D} = D\cos\phi + E\sin\phi$
(e) $\bar{E} = -D\sin\phi + E\cos\phi$
(f) $\bar{F} = F$.

41. If $AC > 0$, show that the graph of $Ax^2 + Cy^2 + Dx + Ey + F = 0$ is an ellipse, a circle, a single point, or the empty set.

42. Using the results of part (b) of problem 40, prove Theorem 1.

43. If $AC < 0$, show that the graph of $Ax^2 + Cy^2 + Dx + Ey + F = 0$ is a hyperbola or a pair of intersecting straight lines.

44. Using the results of problem 40, show that $A + C = \bar{A} + \bar{C}$.

45. If $AC = 0$, show that the graph of $Ax^2 + Cy^2 + Dx + Ey + F = 0$ is a parabola, a pair of parallel straight lines, a single straight line, the whole plane, or the empty set.

46. Using the results of problem 40, show that $B^2 - 4AC = \bar{B}^2 - 4\bar{A}\bar{C}$.

REVIEW PROBLEM SET

In problems 1–24, express each complex number in the form $a + bi$, where a and b are real numbers.

1. $(3 + 2i) + (7 + 3i)$
2. $(2 - 3i) + (1 + 2i)$
3. $(5 - 7i) - (4 + 2i)$
4. $(3 + 5i) - (5 - 3i)$
5. $(7 - 4i) - (-6 + 4i)$
6. $(-\frac{5}{2} + 6i) + (-\frac{7}{4} - 3i)$
7. $(5 - 11i)(5 + 2i)$
8. $(5 + 2i)(7 + 3i)$
9. $(2 + 5i)(-2 + 4i)$
10. $(7 - 2i)(2 + 3i)$
11. $\dfrac{2 + 5i}{3 + 2i}$
12. $\dfrac{1 + 4i}{\sqrt{3} + 2i}$
13. $\dfrac{4 + 2i}{(2 + 3i)(4 + i)}$
14. $\dfrac{5 + 15i}{(3 - i)(1 + i)}$
15. i^{403}
16. i^{-21}
17. $(1/3i)^3$
18. $1/(3 + 2i)^2$
19. $(2 - 3i)\overline{(3 - 2i)}$
20. $(3 + 5i)\overline{(3 + 5i)}$
21. $3(\cos 45^\circ + i \sin 45^\circ)$
22. $6[\cos(-30^\circ) + i \sin(-30^\circ)]$
23. $16\left(\cos \dfrac{5\pi}{3} + i \sin \dfrac{5\pi}{3}\right)$
24. $2\left(\cos \dfrac{71\pi}{4} + i \sin \dfrac{71\pi}{4}\right)$

In problems 25–30, find each absolute value (modulus).

25. $|-4 - 3i|$
26. $|6 - 8i|$
27. $|8i^7|$
28. $|i^{17}|$
29. $\left|\dfrac{3 + 2i}{3 - 4i}\right|$
30. $\left|\dfrac{-3 - 2i}{-6 + 8i}\right|$

In problems 31–34, express each complex number in polar form.

31. $-4 + 4i$
32. $3 + 0i$
33. $2 + \sqrt{3}i$
34. $7 - 7i$

In problems 35–42, find (a) zw and (b) $\dfrac{z}{w}$. Express your answers in polar form.

35. $z = 6(\cos 22^\circ + i \sin 22^\circ)$, $w = 4(\cos 8^\circ + i \sin 8^\circ)$
36. $z = 8(\cos 85^\circ + i \sin 85^\circ)$, $w = 2(\cos 95^\circ + i \sin 95^\circ)$
37. $z = 12(\cos 235^\circ + i \sin 235^\circ)$, $w = 4(\cos 125^\circ + i \sin 125^\circ)$
38. $z = 14(\cos 78^\circ + i \sin 78^\circ)$, $w = 7(\cos 18^\circ + i \sin 18^\circ)$
39. $z = 10\left(\cos \dfrac{13\pi}{36} + i \sin \dfrac{13\pi}{36}\right)$, $w = 5\left(\cos \dfrac{2\pi}{9} + i \sin \dfrac{2\pi}{9}\right)$
40. $z = 20\left(\cos \dfrac{143\pi}{180} + i \sin \dfrac{143\pi}{180}\right)$, $w = 4\left(\cos \dfrac{2\pi}{45} + i \sin \dfrac{2\pi}{45}\right)$
41. $z = 3 + \sqrt{3}i$, $w = \sqrt{3} + i$
42. $z = -3 - \sqrt{3}i$, $w = 2 + 2\sqrt{3}i$

In problems 43–48, use De Moivre's theorem to evaluate each expression. Write the answer in cartesian form.

43. $[2(\cos 300^\circ + i \sin 300^\circ)]^5$
44. $\left[2\left(\cos \dfrac{\pi}{3} + i \sin \dfrac{\pi}{3}\right)\right]^4$
45. $\left[2\left(\cos \dfrac{5\pi}{6} + i \sin \dfrac{5\pi}{6}\right)\right]^6$
46. $(1 - \sqrt{3}i)^9$
47. $(-1 + i)^6$
48. $(1 - i)^{12}$

In problems 49–52, find all of the indicated complex nth roots.

49. The cube roots of i
50. The fifth roots of $-\sqrt{3} - i$
51. The fourth roots of $\dfrac{1}{2} - \dfrac{\sqrt{3}}{2}i$
52. The seventh roots of 1

53. Find all complex solutions of the equation $z^4 - 1 = 0$.
54. Find all complex solutions of the equation $z^3 - i = 0$.

In problems 55–58, find the slope of the line L containing the two points A and B and find the equation in point-slope form of L.

55. $A = (3, -5), \quad B = (2, 2)$
56. $A = (0, 7), \quad B = (5, 0)$
57. $A = (1, 2), \quad B = (-3, -4)$
58. $A = \left(\frac{3}{2}, \frac{2}{3}\right), \quad B = \left(\frac{1}{6}, \frac{-5}{6}\right)$

In problems 59 and 60, sketch the line that contains the point P and has slope m.

59. $P = (5, 2), \quad m = -\frac{3}{5}$
60. $P = (-\frac{2}{3}, \frac{1}{2}), \quad m = \frac{3}{2}$

In problems 61 and 62, (a) find an equation of the line that contains the point P and is parallel to line segment $\overline{AB}$, and (b) find an equation of the line that contains the point P and is perpendicular to line segment $\overline{AB}$.

61. $P = (7, -5), \quad A = (1, 8), \quad B = (-3, 2)$
62. $P = (\frac{2}{5}, \frac{1}{3}), \quad A = (\frac{7}{3}, -\frac{3}{5}), \quad B = (1, \frac{2}{5})$

In problems 63 and 64, rewrite each equation in slope-intercept form, find the slope m and the y intercept b of the graph, and sketch it.

63. $4x - 3y + 2 = 0$
64. $\frac{2}{3}x - \frac{1}{5}y + 3 = 0$

In problems 65–68, find an equation of the line L in (a) point-slope form, (b) slope-intercept form, and (c) general form.

65. L contains the point $(-7, 1)$ and has slope $m = 3$.
66. L contains the points $(2, 5)$ and $(1, -3)$.
67. L contains the point $(1, -2)$ and is parallel to the line whose equation is $7x - 3y + 2 = 0$.
68. L contains the point $(3, -4)$ and is perpendicular to the line $2x - 5y + 4 = 0$.
69. If (a, b) is a point on the circle $x^2 + y^2 = r^2$, find an equation of the tangent line to the circle at (a, b). [Hint: The tangent line to a circle at a point is perpendicular to the radius drawn from the center to the point.]
70. Show that the straight line containing the points (a, b) and (c, d) has the equation $(b - d)x - (a - c)y + ad - bc = 0$.
71. If L is a straight line with inclination angle α and slope $m = \tan\alpha$, and if θ is the smallest *clockwise* angle from the positive x axis to L (Figure 1), show that $m = \tan\theta$.

Figure 1

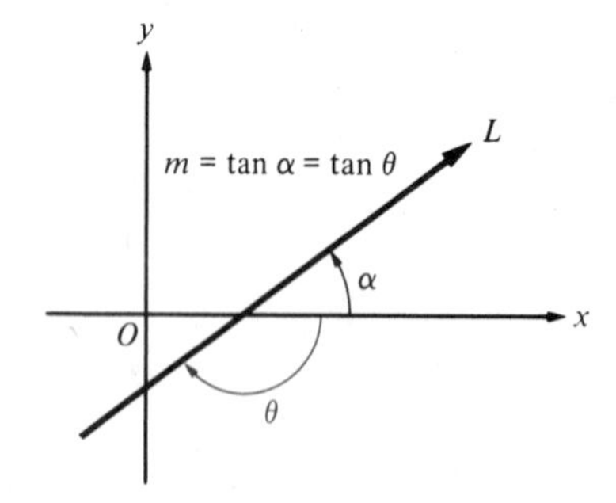

72. In 1980, tests showed that water in a lake was polluted with 7 milligrams of mercury compounds per 1000 liters of water. Efforts to clean up the lake were immediately implemented by environmentalists, who determined that the pollution level would drop at the rate of 0.75 milligram of mercury compounds per 1000 liters of water per year if all of their recommendations were followed. Counting the years so that 1980 corresponds to $x = 0$ and successive years correspond to $x = 1, 2, 3$, and so on, find the equation $y = mx + b$ of the line that allows the environmentalists to predict the pollution level y in future years if their recommendations are followed. Sketch the graph of the equation and determine just when the lake will be free of mercury pollution according to this graph.

In problems 73–76, determine whether the graph of the equation is a circle or an ellipse with center at the origin. In the case of a circle, find the radius. In the case of an ellipse, find the vertices. Sketch the graph.

73. $9x^2 + 9y^2 = 1$ **74.** $x^2 + \frac{y^2}{4} = 1$ **75.** $\frac{x^2}{16} + \frac{y^2}{9} = 1$ **76.** $3x^2 + 2y^2 = 1$

In problems 77–80, find the equation in standard form for the ellipse that satisfies the indicated conditions.

77. Center at $(0, 0)$, horizontal major axis of length 16, and minor axis of length 8
78. Vertices at $(0, 0)$, $(6, 0)$, $(3, -5)$, and $(3, 5)$
79. Vertices at $(-3, 1)$, $(5, 1)$, $(1, -4)$, and $(1, 6)$
80. Center at $(0, 0)$, major axis horizontal, and containing the points $(4, 3)$ and $(6, 2)$

In problems 81–86, find the center and the vertices of each ellipse and sketch the graph.

81. $\frac{x^2}{8} + \frac{y^2}{12} = 1$ **82.** $144x^2 + 169y^2 = 24{,}336$
83. $9x^2 + 25y^2 + 18x - 50y = 191$ **84.** $3x^2 + 4y^2 - 28x - 16y + 48 = 0$
85. $9x^2 + 4y^2 + 72x - 48y + 144 = 0$ **86.** $9x^2 + 4y^2 + 36x - 24y = 252$

In problems 87–92, find the vertex of each parabola, determine whether its axis is horizontal or vertical, and sketch the graph.

87. $y^2 = 4x$ **88.** $-x^2 = 5y$ **89.** $x^2 = -4(y - 1)$
90. $x^2 = 8(y + 1)$ **91.** $x + 8 - y^2 + 2y = 0$ **92.** $x + 4 - y^2 + 2y = 0$

In problems 93 and 94, find the equation in standard form for the parabola that satisfies the indicated conditions.

93. Vertex at $(0, -6)$, horizontal axis, and containing the point $(-9, -3)$.
94. Vertical axis and containing the points $(9, -1)$, $(3, -4)$, and $(-9, 8)$.

In problems 95–100, find the center, the asymptotes, and the vertices of each hyperbola and sketch the graph.

95. $x^2 - 9y^2 = 72$ **96.** $y^2 - 9x^2 = 54$
97. $x^2 - 4y^2 + 4x + 24y - 48 = 0$ **98.** $16x^2 - 9y^2 - 96x = 0$
99. $4y^2 - x^2 - 24y + 2x + 34 = 0$ **100.** $11y^2 - 66y - 25x^2 + 100x = 276$

In problems 101 and 102, find the equation in standard form for the hyperbola that satisfies the indicated conditions.

101. Asymptotes $y = -2x$ and $y = 2x$ and containing the point $(1, 1)$.
102. Vertices at $(2, -8)$ and $(2, 2)$ and asymptotes $25x - 9y = 77$ and $25x + 9y = 23$.

In problems 103–108, rewrite the given equation in standard form and identify the conic section.

103. $16x^2 + y^2 - 32x + 4y - 44 = 0$ **104.** $x^2 + y^2 - 4x + 2y - 4 = 0$
105. $4x^2 + 4y^2 + 4x - 4y + 1 = 0$ **106.** $4x^2 + 9y^2 + 16x - 18y = 11$
107. $4y^2 - x^2 - 8y + 2x + 7 = 0$ **108.** $9x^2 + 25y^2 + 18x - 50y - 191 = 0$

109. A point P moves so that the product of the slopes of the line segments $\overline{PQ}$ and $\overline{PR}$ is -6, where $Q = (3, -2)$ and $R = (-2, 1)$. Find an equation of the curve traced out by P, identify the curve, and sketch it.

110. The cable of a suspension bridge assumes the shape of a parabola if the weight of the suspended roadbed (together with that of the cable) is uniformly distributed horizontally. Suppose that the towers of such a bridge are 240 meters apart and 60 meters high and that the lowest point of the cable is 20 meters above the roadway. Find the vertical distance from roadway to the cable at intervals of 20 meters.

In problems 111 and 112, assume that a new $\bar{x}\bar{y}$ coordinate system is obtained by translating the old xy coordinate system so that the origin $\bar{O}$ of the new system has old coordinates $(3, -2)$.

111. Find the new $\bar{x}\bar{y}$ coordinates of the points whose old xy coordinates are:
(a) $(0, 0)$ (b) $(5, -7)$ (c) $(4, 4)$
(d) $(3, -2)$ (e) $(0, 1)$ (f) $(-1, 0)$.

112. Find the old xy coordinates of the points whose new $\bar{x}\bar{y}$ coordinates are:
(a) $(0, 0)$ (b) $(-3, 2)$ (c) $(3, 10)$
(d) $(-2, 3)$ (e) $(5, 0)$ (f) $(0, -7)$

In problems 113–116, find a translation of axes $x = \bar{x} + h$ and $y = \bar{y} + k$ that will reduce each equation to an equation involving no first-degree terms in $\bar{x}$ or $\bar{y}$. Identify the graph.

113. $4x^2 + 4y^2 + 8x - 4y + 1 = 0$

114. $x^2 - 4y^2 + 6x + 24y = 31$

115. $9x^2 - 4y^2 - 54x + 45 = 0$

116. $2x^2 + 8y^2 - 8x - 16y + 9 = 0$

In problems 117–122, the old xy axis has been rotated counterclockwise about the origin through the angle ϕ to form a new $\bar{x}\bar{y}$ coordinate system. The point P has coordinates (x, y) in the old system and $(\bar{x}, \bar{y})$ in the new system. Supply the missing information.

117. $(x, y) = (-4, 6)$, $\phi = 30°$, $(\bar{x}, \bar{y}) = ?$

118. $(x, y) = (-1, 4)$, $\phi = -\pi/2$, $(\bar{x}, \bar{y}) = ?$

119. $(\bar{x}, \bar{y}) = (2, -4)$, $\phi = \pi/3$, $(x, y) = ?$

120. $(\bar{x}, \bar{y}) = (-2, -6)$, $\phi = 240°$, $(x, y) = ?$

121. $(x, y) = (4, -4)$, $\phi = 300°$, $(\bar{x}, \bar{y}) = ?$

122. $(x, y) = (-3, -4)$, $(\bar{x}, \bar{y}) = (-4, 3)$, $0 < \phi < \pi$, $\phi = ?$

In problems 123–134, (a) find an angle ϕ through which to rotate the coordinate system so as to remove the mixed term from the equation, (b) find x and y in terms of $\bar{x}$ and $\bar{y}$, (c) find the new equation in terms of $\bar{x}$ and $\bar{y}$, and (d) sketch the graph.

123. $4x^2 + 4xy + 4y^2 = 24$

124. $3x^2 + 4\sqrt{3}xy - y^2 = 15$

125. $\sqrt{3}x^2 + xy = 11$

126. $y^2 + xy + 2x^2 = 7$

127. $13x^2 + 6\sqrt{3}xy + 7y^2 = 16$

128. $x^2 + 6xy + y^2 + 8 = 0$

129. $x^2 - 3xy + 5y^2 = 16$

130. $7x^2 + 2\sqrt{3}xy + 5y^2 - 1 = 0$

131. $41x^2 - 24xy + 34y^2 - 25 = 0$

132. $9x^2 + 24xy + 16y^2 + 80x - 60y = 0$

133. $x^2 + 4xy + 4y^2 = 9$

134. $x^2 + 6xy + 9y^2 - 30x + 10y = 0$

Appendix

Appendix

1 Exponents and Logarithms

In Section 1 of Chapter 1, we reviewed the definitions of b^n and $b^{1/n}$ for a positive base b and a positive integer n. The concept of the nth root of a number b enables us to extend the definition of b^n from integers to rational numbers.

Definition

Rational Exponents

Let b be a positive real number, and suppose that m and n are integers, that n is positive, and that the fraction m/n is reduced to lowest terms. Then, if $b^{1/n}$ exists,

$$b^{m/n} = (b^{1/n})^m.$$

It is also possible to define b^x for every real number x (rational or irrational) in such a way that the following properties of exponents remain true.

Properties of Exponents

Let x and y be real numbers and suppose that a and b are positive real numbers. Then

(i) $b^x \cdot b^y = b^{x+y}$ (ii) $\dfrac{b^x}{b^y} = b^{x-y}$ (iii) $(b^x)^y = b^{xy}$ (iv) $(ab)^x = a^x b^x$

(v) $\left(\dfrac{a}{b}\right)^x = \dfrac{a^x}{b^x}$ (vi) $\dfrac{b^{-x}}{b^{-y}} = \dfrac{b^y}{b^x}$ (vii) $b^{-x} = \dfrac{1}{b^x}$

Example

Simplify each expression and write the answer so that it contains only positive exponents. Assume that all variables are restricted to values for which all expressions are defined.

(a) $(x^4)^{-2}$ (b) $5^{-1/3} \cdot 5^{7/3}$ (c) $\left(\dfrac{2^{-1}}{3}\right)^{-1}$

(d) $(64x^{-3})^{-2/3}$ (e) $(x^{-2}y^{-3})^{-4}$ (f) $\dfrac{8^{-1/3}}{4^{-3/2}}$

Solution

(a) $(x^4)^{-2} = x^{4(-2)} = x^{-8} = \dfrac{1}{x^8}$

(b) $5^{-1/3} \cdot 5^{7/3} = 5^{(-1/3)+(7/3)} = 5^{6/3} = 5^2 = 25$

(c) $\left(\dfrac{2^{-1}}{3}\right)^{-1} = \dfrac{(2^{-1})^{-1}}{3^{-1}} = 2(3) = 6$

(d) $(64x^{-3})^{-2/3} = 64^{-2/3}x^{(-3)(-2/3)}$

$$= \frac{x^2}{64^{2/3}} = \frac{x^2}{(\sqrt[3]{64})^2} = \frac{x^2}{4^2} = \frac{x^2}{16}$$

(e) $(x^{-2}y^{-3})^{-4} = (x^{-2})^{-4}(y^{-3})^{-4} = x^8y^{12}$

(f) $\dfrac{8^{-1/3}}{4^{-3/2}} = \dfrac{4^{3/2}}{8^{1/3}} = \dfrac{\sqrt{4^3}}{\sqrt[3]{8}} = \dfrac{8}{2} = 4$

If $b > 0$, $b \neq 1$, and $c > 0$, the solution x of the exponential equation

$$b^x = c$$

is denoted by

$$x = \log_b c,$$

which is read "x equals the **logarithm of c to the base b**." In other words,

$\log_b c$ is the power to which you must raise b to obtain c.

Example Find

(a) $\log_2 16$ (b) $\log_3 27$

(c) $\log_{10} \frac{1}{10}$ (d) $\log_b b^5$

(e) $\log_b 1$ for $b > 0$ and $b \neq 1$.

Solution

(a) We ask ourselves, "To what power must we raise 2 to obtain 16?" Since $2^4 = 16$, the answer is 4. Therefore, $\log_2 16 = 4$.

(b) Since $3^3 = 27$, it follows that $\log_3 27 = 3$.

(c) Since $10^{-1} = \frac{1}{10}$, it follows that $\log_{10} \frac{1}{10} = -1$.

(d) We ask ourselves, "To what power must we raise b to obtain b^5?" The answer is 5, so $\log_b b^5 = 5$.

(e) Since $b^0 = 1$, it follows that $\log_b 1 = 0$.

Using the fact that for $b > 0$, $b \neq 1$, and $c > 0$,

$$x = \log_b c \quad \text{if and only if} \quad b^x = c,$$

you can convert equations from logarithmic form to exponential form and vice versa. For instance:

Logarithmic Form	Exponential Form
$2 = \log_2 4$	$2^2 = 4$
$\log_{10} 10{,}000 = 4$	$10{,}000 = 10^4$
$-\frac{1}{2} = \log_{64} \frac{1}{8}$	$64^{-1/2} = \frac{1}{8}$
$\log_b x = y$	$b^y = x \quad (b > 0, b \neq 1, x > 0)$
$k = \log_x d$	$x^k = d \quad (x > 0, x \neq 1, d > 0)$

Notice that whenever you write $\log_b c$, you must make sure that c is positive and that b is positive and not equal to 1.

Examples Solve each equation.

1 $\log_3 x^2 = 4$

Solution The equation $\log_3 x^2 = 4$ is equivalent to $3^4 = x^2$; that is, $x^2 = 81$. The solutions are $x = 9$ and $x = -9$.

2 $\log_x 25 = 2$

Solution The equation $\log_x 25 = 2$ is equivalent to $x^2 = 25$, with the restriction that $x > 0$ and $x \neq 1$; hence, $x = 5$ is the solution.

Using the connection between logarithms and exponents, we can translate properties of exponents into properties of logarithms. Some of these properties are as follows.

Properties of Logarithms

Let M, N, and b be positive numbers, $b \neq 1$, and let y be any real number. Then:

(i) $b^{\log_b N} = N$ (ii) $\log_b b^y = y$

(iii) $\log_b(MN) = \log_b M + \log_b N$ (iv) $\log_b \frac{M}{N} = \log_b M - \log_b N$

(v) $\log_b N^y = y \log_b N$ (vi) $\log_b \frac{1}{N} = -\log_b N$

Properties (i) and (ii) are direct consequences of the definition of logarithms. We verify Property (iii) here and leave it to you to check Properties (iv), (v), and (vi) (Problem 60). To prove Property (iii), let

$$x = \log_b M \quad \text{and} \quad y = \log_b N,$$

so that

$$b^x = M \quad \text{and} \quad b^y = N.$$

Then,

$$\log_b(MN) = \log_b(b^x b^y) = \log_b(b^{x+y})$$
$$= x + y = \log_b M + \log_b N.$$

Examples Use the Properties of Logarithms to work the following.

1 Rewrite each expression as a sum or difference of multiples of logarithms.

(a) $\log_b \dfrac{z}{uv}$ (b) $\log_2 \dfrac{(x^2 + 5)(2x + 5)^{3/2}}{\sqrt[4]{3x + 1}}$

Solution (a) Assuming that z, u, and v are positive, we have

$$\log_b \frac{z}{uv} = \log_b z - \log_b(uv) \qquad \text{[Property (iv)]}$$
$$= \log_b z - (\log_b u + \log_b v) \qquad \text{[Property (iii)]}$$
$$= \log_b z - \log_b u - \log_b v.$$

(b) Assuming that $2x + 5$ and $3x + 1$ are positive, we have

$$\log_2 \frac{(x^2 + 5)(2x + 5)^{3/2}}{\sqrt[4]{3x + 1}} = \log_2[(x^2 + 5)(2x + 5)^{3/2}] - \log_2 \sqrt[4]{3x + 1}$$
$$= \log_2(x^2 + 5) + \log_2(2x + 5)^{3/2} - \log_2(3x + 1)^{1/4}$$
$$= \log_2(x^2 + 5) + \tfrac{3}{2}\log_2(2x + 5) - \tfrac{1}{4}\log_2(3x + 1),$$

where we applied Property (v) in the last step.

2 Rewrite each expression as a single logarithm.

(a) $2\log_{10} x + 3\log_{10}(x + 1)$ (b) $\log_b\left(x + \dfrac{x}{y}\right) - \log_b\left(z + \dfrac{z}{y}\right)$

Solution Assuming that all quantities whose logarithms are taken are positive, we have the following:

(a) $2\log_{10} x + 3\log_{10}(x + 1) = \log_{10} x^2 + \log_{10}(x + 1)^3$
$$= \log_{10}[x^2(x + 1)^3]$$

(b) $$\log_b\left(x + \frac{x}{y}\right) - \log_b\left(z + \frac{z}{y}\right) = \log_b \frac{x + \dfrac{x}{y}}{z + \dfrac{z}{y}} = \log_b \frac{\left(1 + \dfrac{1}{y}\right)x}{\left(1 + \dfrac{1}{y}\right)z} = \log_b \frac{x}{z}.$$

3 Suppose that $\log_b 2 = 0.48$ and $\log_b 3 = 0.76$. Find
(a) $\log_b 6$ (b) $\log_b \frac{3}{2}$ (c) $\log_b \sqrt[4]{2}$.

Solution (a) By Property (iii),

$$\log_b 6 = \log_b(2 \cdot 3) = \log_b 2 + \log_b 3$$
$$= 0.48 + 0.76 = 1.24.$$

(b) By Property (iv),

$$\log_b \tfrac{3}{2} = \log_b 3 - \log_b 2$$
$$= 0.76 - 0.48 = 0.28.$$

(c) By Property (v),

$$\log_b \sqrt[4]{2} = \log_b 2^{1/4} = \tfrac{1}{4}\log_b 2 = \tfrac{1}{4}(0.48) = 0.12.$$

4 Solve each equation:
(a) $\log_{10} x + \log_{10}(x + 21) = 2$
(b) $\log_7(3t^2 - 5t - 2) - \log_7(t - 2) = 1$.

Solution (a) We begin by noticing that x must be positive for $\log_{10} x$ to be defined. If x is positive, so is $x + 21$, and $\log_{10}(x + 21)$ is also defined. Applying Property (iii), we rewrite

$$\log_{10} x + \log_{10}(x + 21) = 2$$

as

$$\log_{10}[x(x + 21)] = 2.$$

The last equation can be rewritten in exponential form as

$$x(x + 21) = 10^2;$$

that is,

$$x^2 + 21x - 100 = 0.$$

Factoring, we have

$$(x + 25)(x - 4) = 0,$$

so

$$x = -25 \quad \text{or} \quad x = 4.$$

Since x must be positive, we can eliminate $x = -25$ as an extraneous root. Therefore, the solution is $x = 4$.

(b) Applying Property (iv), we can rewrite the given equation

$$\log_7(3t^2 - 5t - 2) - \log_7(t - 2) = 1$$

as

$$\log_7 \frac{3t^2 - 5t - 2}{t - 2} = 1,$$

provided that both $3t^2 - 5t - 2$ and $t - 2$ are positive. The last equation can be simplified by reducing the fraction,

$$\frac{3t^2 - 5t - 2}{t - 2} = \frac{(3t + 1)\cancel{(t - 2)}}{\cancel{t - 2}}$$

$$= 3t + 1,$$

so

$$\log_7(3t + 1) = 1;$$

that is,

$$3t + 1 = 7^1 = 7.$$

The solution of this equation is

$$t = \frac{7 - 1}{3} = 2.$$

We must now check this answer against the original restrictions on the variables. We find that if $t = 2$, then $t - 2 = 0$, so $\log_7(t - 2)$ is undefined. Thus, $t = 2$ is an extraneous root, and the original equation has no solution.

Because the usual positional system for writing numerals is based on 10, arithmetic calculation is easiest when logarithms with base 10 are used. Logarithms with base 10 are called **common logarithms**, and the symbol "log x" (with no subscript) is often used as an abbreviation for $\log_{10} x$. In advanced mathematics, many formulas become simpler if logarithms with a special base $e \approx 2.718$ are used. Logarithms with base e are called **natural logarithms**, and the symbol "ln x" is often used as an abbreviation for $\log_e x$.

PROBLEM SET 1

In problems 1–14, simplify each expression and write the answer so that it contains only positive exponents. Assume that all variables are restricted to values for which all expressions are defined.

1. $\left(\frac{1}{3}\right)^{-3}$
2. $\frac{1}{7^{-2}}$
3. $(8x^0)^{-2}$
4. $(x^{-3})^6(x^0)^{-2}$
5. $\left(\frac{a^{-3}}{b^{-3}}\right)^{-n}$
6. $\left(\frac{5x^{-1}}{y}\right)^{-1}\left(\frac{y}{5x^{-1}}\right)$
7. $a^{1/2} \cdot a^{3/2}$
8. $\left(\frac{x^{-1/3}}{x^{3/2}}\right)^6$
9. $\left(-\frac{8}{27}x^3\right)^{2/3}$
10. $\left(\frac{-x^{10}}{32}\right)^{-3/5}$
11. $\frac{4^{-1/2}}{8^{-2/3}}$
12. $\frac{16^{-3/2}}{25^{3/2}}$
13. $\left(\frac{x^{1/2}}{y^2}\right)^4\left(\frac{y^{-1/3}}{x^{2/3}}\right)^3$
14. $[(a^4b^{-3})^{1/5}]^{1/10}$
15. Find:
    (a) $\log_2 4$
    (b) $\log_2 8$
    (c) $\log_3 81$
    (d) $\log_9 9^5$
    (e) $\log_3 \frac{1}{9}$
    (f) $\log_8 \frac{1}{64}$
    (g) $\log_{10} 100{,}000$

16. Find:
(a) $\log_2 \frac{1}{4}$ (b) $\log_3 \sqrt{3}$ (c) $\log_9 1$ (d) $\log_e e^{\pi}$
(e) $\log_2 4^3$ (f) $\log_3 9^{(-0.5)}$ (g) $\log_{10} \frac{1}{100,000}$

17. Rewrite each logarithmic equation as an equivalent exponential equation:
(a) $\log_2 32 = 5$ (b) $\log_{16} 2 = \frac{1}{4}$ (c) $\log_9 \frac{1}{3} = -\frac{1}{2}$ (d) $\log_e e = 1$
(e) $\log_{\sqrt{3}} 9 = 4$ (f) $\log_{10} 10^n = n$ (g) $\log_x x^5 = 5$

18. Rewrite each exponential equation as an equivalent logarithmic equation:
(a) $8^0 = 1$ (b) $10^{-4} = 0.0001$ (c) $4^4 = 256$ (d) $27^{(-1/3)} = \frac{1}{3}$
(e) $8^{2/3} = 4$ (f) $a^c = y$

In problems 19–42, solve each equation.

19. $x = \log_6 36$ **20.** $\log_5 x = 2$ **21.** $\log_x 125 = 3$
22. $x = \log_3 729$ **23.** $\log_7 x = 1$ **24.** $\log_3 x = 1$
25. $\log_x 16 = -\frac{4}{3}$ **26.** $\log_{\sqrt{3}} x = 6$ **27.** $\log_{\sqrt{2}} x = -6$
28. $\log_x \frac{27}{8} = -\frac{3}{2}$ **29.** $x = \log_{10} 10^{-7}$ **30.** $x = \log_e e^{-0.01}$
31. $x = \log_{27} \frac{1}{9}$ **32.** $x = \log_2(\log_4 256)$ **33.** $x = \log_5 \sqrt[4]{5}$
34. $x = \log_{3/4} \frac{4}{3}$ **35.** $\log_2(2x - 1) = 3$ **36.** $\log_5(2x - 3) = 2$
37. $\log_3(3x - 4) = 4$ **38.** $\log_7(2x - 7) = 0$ **39.** $\log_2(t^2 + 3t + 4) = 1$
40. $\log_5(y^2 - 4y) = 1$ **41.** $\log_4(9u^2 + 6u + 1) = 2$ **42.** $\log_3|3 - 2t| = 2$

In problems 43–50, rewrite each expression as a sum or difference of multiples of logarithms. (Make the necessary assumptions about the values of the variables.)

43. $\log_b[x(x + 1)]$ **44.** $\log_a(x^4 \sqrt{y})$ **45.** $\log_{10}[x^2(x + 1)]$

46. $\log_c \sqrt{\dfrac{x}{x + 7}}$ **47.** $\log_3 \dfrac{x^3 y^2}{z}$ **48.** $\log_e \dfrac{t(t + 1)}{(t + 2)^3}$

49. $\log_e \sqrt{x(x + 3)}$ **50.** $\log_b \sqrt[3]{(x + 1)^2 \sqrt{x + 7}}$

In problems 51–58, rewrite each expression as a single logarithm. (Make the necessary assumptions about the values of the variables.)

51. $2\log_3 x + 7\log_3 x$ **52.** $\log_{10} \dfrac{a^3}{b} + \log_{10} \dfrac{b^2}{5a}$ **53.** $\frac{1}{2}[\log_5 a - \log_5 3b]$

54. $\log_x \dfrac{y^5}{z^4} - \log_x \dfrac{y^3}{z^2}$ **55.** $\log_e \dfrac{x}{x - 1} + \log_e \dfrac{x^2 - 1}{x}$ **56.** $\log_b \dfrac{x + y}{z} - \log_b \dfrac{1}{x + y}$

57. $\log_3 \dfrac{x^2 + 14x - 15}{x^2 + 4x - 5} - \log_3 \dfrac{x^2 + 12x - 45}{x^2 + 6x - 27}$

58. $\log_e \dfrac{m^2 - 2m - 24}{m^2 - m - 30} + \log_e \dfrac{(m + 5)^2}{m^2 - 16}$

59. Suppose that $\log_b 2 = 0.53$, $\log_b 3 = 0.83$, $\log_b 5 = 1.22$, and $\log_b 7 = 1.48$. Find:
(a) $\log_b 21$ (b) $\log_b 35$ (c) $\log_b \frac{2}{7}$ (d) $\log_b \frac{35}{3}$
(e) $\log_b \sqrt{7}$ (f) $\log_b \sqrt[3]{42}$ (g) $\log_b 3\sqrt{8}$.
Round off all answers to two decimal places.

60. Verify Properties (iv), (v), and (vi) on page 271.

In problems 61–68, solve each equation.

61. $\log_4 x + \log_4(x + 6) = 2$ **62.** $\log_{10} x + \log_{10}(x + 3) = 1$
63. $\log_7 x + \log_7(18x + 61) = 1$ **64.** $\log_2 x + \log_2(x - 2) = \log_2(9 - 2x)$
65. $\log_3(x^2 + x) - \log_3(x^2 - x) = 1$ **66.** $\log_5(4x^2 - 1) = 2 + \log_5(2x + 1)$
67. $\log_8(x^2 - 9) - \log_8(x + 3) = 2$ **68.** $2\log_2 x - \log_2(x - 1) = 2$

2 Tables of Logarithms and Exponentials

In this section, we present a table of natural logarithms (Table 1), a table of common logarithms (Table 2), and a table of exponential functions (Table 3), as well as examples illustrating their use.

Table 1 gives values of $\ln x$, rounded off to four decimal places, corresponding to values of x between 1 and 9.99 in steps of 0.01.

Example Use Table 1 to find the value of ln 3.47 rounded off to four decimal places.

Solution We begin by locating the number 3.4 in the vertical column on the left side of the table. The numbers in the horizontal row to the right of 3.4 are the natural logarithms of the numbers from 3.40 to 3.49 in steps of 0.01. The entry in this horizontal row below the heading 0.07 is the natural logarithm of 3.47. Thus,

$$\ln 3.47 = 1.2442.$$

For values of x lying between two numbers whose natural logarithms are shown in Table 1, it is possible to find the approximate value of $\ln x$ (again rounded off to four decimal places) by using **linear interpolation**. The basic idea of linear interpolation is that small changes in the independent variable cause approximately proportional changes in the values of a function.

Example Use linear interpolation and Table 1 to find the approximate value of ln 2.724.

Solution We arrange the work as follows:

$$0.01\left[\begin{array}{ll} 0.004\left[\begin{array}{l} 2.72 \\ 2.724 \end{array}\right. & \\ 2.73 & \end{array}\right. \qquad \begin{array}{l} \ln x \\ 1.0006 \\ \ln 2.724 \\ 1.0043 \end{array} \left.\begin{array}{l} \left.\right] d \\ \\ \end{array}\right] 0.0037.$$

We have used Table 1 to determine that $\ln 2.72 = 1.0006$ and $\ln 2.73 = 1.0043$. The notation

$$0.004\left[\begin{array}{l} 2.72 \\ 2.724 \end{array}\right.$$

indicates that the difference between 2.72 and 2.724 is

$$2.724 - 2.72 = 0.004.$$

Other such differences are indicated similarly. For linear interpolation, we assume that corresponding differences are (approximately) proportional, so that

$$\frac{0.004}{0.01} = \frac{d}{0.0037}$$

or

$$d = 0.0037\left(\frac{0.004}{0.01}\right) = 0.00148.$$

Thus, rounded off to four decimal places, we have

$$\begin{aligned} \ln 2.724 &= 1.0006 + d \\ &= 1.0006 + 0.00148 \\ &= 1.0021. \end{aligned}$$

Table 2 gives values of $\log x$, rounded off to four decimal places, corresponding to values of x between 1 and 9.99 in steps of 0.01.

Example 1 Using Table 2, find the value of log 7.36 rounded off to four decimal places.

Solution $\log 7.36 = 0.8669$

Example 2 Use linear interpolation and Table 2 to find the approximate value of log 5.068.

Solution We have

$$0.01\left[\begin{array}{l} 0.008\left[\begin{array}{l} 5.06 \\ 5.068 \end{array}\right. \\ \ 5.07 \end{array}\right. \quad \begin{array}{ll} x & \log x \\ 5.06 & 0.7042 \\ 5.068 & \log 5.068 \\ 5.07 & 0.7050 \end{array} \quad \left.\begin{array}{l} \left.\begin{array}{l} \\ \end{array}\right] d \\ \end{array}\right] 0.0008.$$

Thus,

$$\frac{0.008}{0.01} = \frac{d}{0.0008}$$

or

$$d = 0.0008\left(\frac{0.008}{0.01}\right) = 0.00064.$$

Hence, rounded off to four decimal places,

$$\begin{aligned} \log 5.068 &= 0.7042 + d \\ &= 0.7042 + 0.00064 \\ &= 0.7048. \end{aligned}$$

Now, suppose that x is a positive number less than 1 or greater than 9.99. To find $\log x$, begin by writing x in scientific notation

$$x = p \times 10^n,$$

where $1 \le p < 10$ and n is an integer. Thus,

$$\log x = \log p + \log 10^n = \log p + n.$$

The quantity $\log p$ is called the **mantissa** of $\log x$, whereas the integer n is called the **characteristic** of $\log x$. Thus,

$$\log x = \text{mantissa} + \text{characteristic}.$$

You can find the mantissa by using Table 2.

Examples Using Table 2, find the approximate value of each logarithm.

1 $\log 371.4$

Solution In scientific notation,

$$371.4 = 3.714 \times 10^2.$$

Here, the mantissa

$$\log 3.714 = 0.5698$$

is obtained by using linear interpolation and Table 2; the characteristic is $n = 2$, and so, rounded off to four decimal places,

$$\log 371.4 = 0.5698 + 2$$
$$= 2.5698.$$

2 $\log 0.05422$

Solution In scientific notation

$$0.05422 = 5.422 \times 10^{-2}.$$

Here, the mantissa

$$\log 5.422 = 0.7342$$

is obtained by using linear interpolation and Table 2; the characteristic is $n = -2$, and so, rounded off to four decimal places,

$$\log 0.05422 = 0.7342 + (-2)$$
$$= -1.2658.$$

Although a scientific calculator gives $\log 0.05422 = -1.2658$, rounded off to four decimal places, it is customary to leave the answer in the form

$$\log 0.05422 = 0.7342 + (-2)$$

when using a table of logarithms.

By reading Table 2 "backwards," you can find 10^x, which is sometimes called the **antilogarithm** of x.

Examples Using Table 2, find each antilogarithm.

1 $10^{0.9159}$

Solution In the body of Table 2, we find 0.9159 and see (by reading the table "backwards") that it is the common logarithm of 8.24; hence,

$$\log 8.24 = 0.9159$$

or

$$10^{0.9159} = 8.24.$$

2 $10^{0.03}$

Solution The number $0.03 = 0.0300$ does not appear in the body of Table 2. The numbers closest to it that do appear are 0.0294 and 0.0334. Thus, we must use linear interpolation. We have

		x	10^x		
0.0040	0.0006	0.0294	1.07	d	0.01.
		0.0300	$10^{0.03}$		
		0.0334	1.08		

Thus,

$$\frac{0.0006}{0.0040} = \frac{d}{0.01}$$

or

$$d = 0.01\left(\frac{0.0006}{0.0040}\right) = 0.0015;$$

so that, rounded off to four significant digits,

$$\begin{aligned} 10^{0.03} &= 1.07 + d \\ &= 1.07 + 0.0015 \\ &= 1.072. \end{aligned}$$

3 $10^{4.3312}$

Solution The given number can be rewritten as

$$10^{4.3312} = 10^{4+0.3312} = 10^4 \cdot 10^{0.3312}.$$

Using Table 2 "backwards" with linear interpolation, we find that, rounded off to four significant digits,

$$10^{0.3312} = 2.144.$$

Therefore,

$$\begin{aligned} 10^{4.3312} &= 10^4 \cdot 10^{0.3312} \\ &= 10{,}000(2.144) \\ &= 21{,}440. \end{aligned}$$

4 $10^{-5.7074}$

Solution We begin by writing -5.7074 in the form

$$-5.7074 = \text{mantissa} + \text{characteristic},$$

where the characteristic is negative and the mantissa is between 0 and 1. Evidently, the characteristic is -6; hence,

$$\text{mantissa} = -5.7074 - (-6) = 0.2926.$$

Using Table 2 "backwards" with linear interpolation, we find that, rounded off to four significant digits,

$$10^{0.2926} = 1.961.$$

Therefore, rounded off to four significant digits,

$$\begin{aligned} 10^{-5.7074} &= 10^{0.2926+(-6)} \\ &= 10^{0.2926} \cdot 10^{-6} \\ &= 1.961(10^{-6}) \\ &= 0.000{,}001{,}961. \end{aligned}$$

By reading Table 1 "backwards," you can find values of e^x; however, for convenience, we present a separate table of such values (Table 3).

Example Use linear interpolation and Table 3 to find the approximate value of $e^{1.96}$.

Solution We have

$$\begin{array}{cccc} & x & e^x & \\ 0.1\left[\begin{array}{l} 0.06\left[\begin{array}{l}1.9 \\ 1.96\end{array}\right. \\ \\ \end{array}\right. & \begin{array}{c} \\ \\ 2.0 \end{array} & \begin{array}{c} 6.6859 \\ e^{1.96} \\ 7.3891 \end{array} & \left.\begin{array}{l} \left.\right]d \\ \\ \end{array}\right]0.7032. \end{array}$$

Thus,

$$\frac{0.06}{0.1} = \frac{d}{0.7032}$$

or

$$d = 0.7032\left(\frac{0.06}{0.1}\right) = 0.4219,$$

so

$$\begin{aligned} e^{1.96} &= 6.6859 + 0.4219 \\ &= 7.1078. \end{aligned}$$

Because Table 3 is just a "short table" of values of e^x, linear interpolation may not be particularly accurate here. (A calculator gives $e^{1.96} = 7.0993$, rounded off to five significant digits.)

PROBLEM SET 2

In problems 1–6, use Table 1 to find the value of each natural logarithm rounded off to four decimal places.

1. $\ln 5.78$
2. $\ln 3.95$
3. $\ln 8.62$
4. $\ln 7.41$
5. $\ln 9.53$
6. $\ln 6.45$

In problems 7–12, use linear interpolation and Table 1 to find the approximate value of each natural logarithm.

7. $\ln 3.456$
8. $\ln 6.891$
9. $\ln 4.643$
10. $\ln 8.562$
11. $\ln 5.436$
12. $\ln 7.325$

In problems 13–24, use Table 2 to find the value of each logarithm rounded off to four decimal places.

13. $\log 317$
14. $\log 17.1$
15. $\log 6.83$
16. $\log 3910$
17. $\log 50$
18. $\log 1.18$
19. $\log 0.315$
20. $\log 0.613$
21. $\log 0.0612$
22. $\log 0.000812$
23. $\log 0.000143$
24. $\log 0.000052$

In problems 25–30, use linear interpolation and Table 2 to approximate the value of each logarithm.

25. $\log 1543$
26. $\log 444.4$
27. $\log 0.6132$
28. $\log 0.05347$
29. $\log 1.681$
30. $\log 0.0006481$

In problems 31–42, use Table 2 and linear interpolation (if necessary) to find each antilogarithm.

31. $10^{0.4133}$
32. $10^{0.4871}$
33. $10^{1.7825}$
34. $10^{1.2945}$
35. $10^{2.9795}$
36. $10^{3.8993}$
37. $10^{3.5514}$
38. $10^{3.7348}$
39. $10^{0.1453}$
40. $10^{1.5375}$
41. $10^{1.5425}$
42. $10^{2.4167}$

TABLE 1 **NATURAL LOGARITHMS**

x	0.00	0.01	0.02	0.03	0.04	0.05	0.06	0.07	0.08	0.09
1.0	0.0000	0.0100	0.0198	0.0296	0.0392	0.0488	0.0583	0.0677	0.0770	0.0862
1.1	0.0953	0.1044	0.1133	0.1222	0.1310	0.1398	0.1484	0.1570	0.1655	0.1740
1.2	0.1823	0.1906	0.1989	0.2070	0.2151	0.2231	0.2311	0.2390	0.2469	0.2546
1.3	0.2624	0.2700	0.2776	0.2852	0.2927	0.3001	0.3075	0.3148	0.3221	0.3293
1.4	0.3365	0.3436	0.3507	0.3577	0.3646	0.3716	0.3784	0.3853	0.3920	0.3988
1.5	0.4055	0.4121	0.4187	0.4253	0.4318	0.4383	0.4447	0.4511	0.4574	0.4637
1.6	0.4700	0.4762	0.4824	0.4886	0.4947	0.5008	0.5068	0.5128	0.5188	0.5247
1.7	0.5306	0.5365	0.5423	0.5481	0.5539	0.5596	0.5653	0.5710	0.5766	0.5822
1.8	0.5878	0.5933	0.5988	0.6043	0.6098	0.6152	0.6206	0.6259	0.6313	0.6366
1.9	0.6419	0.6471	0.6523	0.6575	0.6627	0.6678	0.6729	0.6780	0.6831	0.6881
2.0	0.6931	0.6981	0.7031	0.7080	0.7130	0.7178	0.7227	0.7275	0.7324	0.7372
2.1	0.7419	0.7467	0.7514	0.7561	0.7608	0.7655	0.7701	0.7747	0.7793	0.7839
2.2	0.7885	0.7930	0.7975	0.8020	0.8065	0.8109	0.8154	0.8198	0.8242	0.8286
2.3	0.8329	0.8372	0.8416	0.8459	0.8502	0.8544	0.8587	0.8629	0.8671	0.8713
2.4	0.8755	0.8796	0.8838	0.8879	0.8920	0.8961	0.9002	0.9042	0.9083	0.9123
2.5	0.9163	0.9203	0.9243	0.9282	0.9322	0.9361	0.9400	0.9439	0.9478	0.9517
2.6	0.9555	0.9594	0.9632	0.9670	0.9708	0.9746	0.9783	0.9821	0.9858	0.9895
2.7	0.9933	0.9969	1.0006	1.0043	1.0080	1.0116	1.0152	0.0188	1.0225	1.0260
2.8	1.0296	1.0332	1.0367	1.0403	1.0438	1.0473	1.0508	1.0543	1.0578	1.0613
2.9	1.0647	1.0682	1.0716	1.0750	1.0784	1.0818	1.0852	1.0886	1.0919	1.0953
3.0	1.0986	1.1019	1.1053	1.1086	1.1119	1.1151	1.1184	1.1217	1.1249	1.1282
3.1	1.1314	1.1346	1.1378	1.1410	1.1442	1.1474	1.1506	1.1537	1.1569	1.1600
3.2	1.1632	1.1663	1.1694	1.1725	1.1756	1.1787	1.1817	1.1848	1.1878	1.1909
3.3	1.1939	1.1970	1.2000	1.2030	1.2060	1.2090	1.2119	1.2149	1.2179	1.2208
3.4	1.2238	1.2267	1.2296	1.2326	1.2355	1.2384	1.2413	1.2442	1.2470	1.2499
3.5	1.2528	1.2556	1.2585	1.2613	1.2641	1.2669	1.2698	1.2726	1.2754	1.2782
3.6	1.2809	1.2837	1.2865	1.2892	1.2920	1.2947	1.2975	1.3002	1.3029	1.3056
3.7	1.3083	1.3110	1.3137	1.3164	1.3191	1.3218	1.3244	1.3271	1.3297	1.3324
3.8	1.3350	1.3376	1.3403	1.3429	1.3455	1.3481	1.3507	1.3533	1.3558	1.3584
3.9	1.3610	1.3635	1.3661	1.3686	1.3712	1.3737	1.3762	1.3788	1.3813	1.3838
4.0	1.3863	1.3888	1.3913	1.3938	1.3962	1.3987	1.4012	1.4036	1.4061	1.4085
4.1	1.4110	1.4134	1.4159	1.4183	1.4207	1.4231	1.4255	1.4279	1.4303	1.4327
4.2	1.4351	1.4375	1.4398	1.4422	1.4446	1.4469	1.4493	1.4516	1.4540	1.4563
4.3	1.4586	1.4609	1.4633	1.4656	1.4679	1.4702	1.4725	1.4748	1.4770	1.4793
4.4	1.4816	1.4839	1.4861	1.4884	1.4907	1.4929	1.4952	1.4974	1.4996	1.5019
4.5	1.5041	1.5063	1.5085	1.5107	1.5129	1.5151	1.5173	1.5195	1.5217	1.5239
4.6	1.5261	1.5282	1.5304	1.5326	1.5347	1.5369	1.5390	1.5412	1.5433	1.5454
4.7	1.5476	1.5497	1.5518	1.5539	1.5560	1.5581	1.5602	1.5623	1.5644	1.5665
4.8	1.5686	1.5707	1.5728	1.5748	1.5769	1.5790	1.5810	1.5831	1.5851	1.5872
4.9	1.5892	1.5913	1.5933	1.5953	1.5974	1.5994	1.6014	1.6034	1.6054	1.6074
5.0	1.6094	1.6114	1.6134	1.6154	1.6174	1.6194	1.6214	1.6233	1.6253	1.6273
5.1	1.6292	1.6312	1.6332	1.6351	1.6371	1.6390	1.6409	1.6429	1.6448	1.6467
5.2	1.6487	1.6506	1.6525	1.6544	1.6563	1.6582	1.6601	1.6620	1.6639	1.6658
5.3	1.6677	1.6696	1.6715	1.6734	1.6752	1.6771	1.6790	1.6808	1.6827	1.6845
5.4	1.6864	1.6882	1.6901	1.6919	1.6938	1.6956	1.6974	1.6993	1.7011	1.7029
5.5	1.7047	1.7066	1.7084	1.7102	1.7120	1.7138	1.7156	1.7174	1.7192	1.7210
5.6	1.7228	1.7246	1.7263	1.7281	1.7299	1.7317	1.7334	1.7352	1.7370	1.7387
5.7	1.7405	1.7422	1.7440	1.7457	1.7475	1.7492	1.7509	1.7527	1.7544	1.7561
5.8	1.7579	1.7596	1.7613	1.7630	1.7647	1.7664	1.7682	1.7699	1.7716	1.7733
5.9	1.7750	1.7766	1.7783	1.7800	1.7817	1.7834	1.7851	1.7867	1.7884	1.7901

x	0.00	0.01	0.02	0.03	0.04	0.05	0.06	0.07	0.08	0.09
6.0	1.7918	1.7934	1.7951	1.7967	1.7984	1.8001	1.8017	1.8034	1.8050	1.8066
6.1	1.8083	1.8099	1.8116	1.8132	1.8148	1.8165	1.8181	1.8197	1.8213	1.8229
6.2	1.8245	1.8262	1.8278	1.8294	1.8310	1.8326	1.8342	1.8358	1.8374	1.8390
6.3	1.8406	1.8421	1.8437	1.8453	1.8469	1.8485	1.8500	1.8516	1.8532	1.8547
6.4	1.8563	1.8579	1.8594	1.8610	1.8625	1.8641	1.8656	1.8672	1.8687	1.8703
6.5	1.8718	1.8733	1.8749	1.8764	1.8779	1.8795	1.8810	1.8825	1.8840	1.8856
6.6	1.8871	1.8886	1.8901	1.8916	1.8931	1.8946	1.8961	1.8976	1.8991	1.9006
6.7	1.9021	1.9036	1.9051	1.9066	1.9081	1.9095	1.9110	1.9125	1.9140	1.9155
6.8	1.9169	1.9184	1.9199	1.9213	1.9228	1.9242	1.9257	1.9272	1.9286	1.9301
6.9	1.9315	1.9330	1.9344	1.9359	1.9373	1.9387	1.9402	1.9416	1.9430	1.9445
7.0	1.9459	1.9473	1.9488	1.9502	1.9516	1.9530	1.9544	1.9559	1.9573	1.9587
7.1	1.9601	1.9615	1.9629	1.9643	1.9657	1.9671	1.9685	1.9699	1.9713	1.9727
7.2	1.9741	1.9755	1.9769	1.9782	1.9796	1.9810	1.9824	1.9838	1.9851	1.9865
7.3	1.9879	1.9892	1.9906	1.9920	1.9933	1.9947	1.9961	1.9974	1.9988	2.0001
7.4	2.0015	2.0028	2.0042	2.0055	2.0069	2.0082	2.0096	2.0109	2.0122	2.0136
7.5	2.0149	2.0162	2.0176	2.0189	2.0202	2.0215	2.0229	2.0242	2.0255	2.0268
7.6	2.0282	2.0295	2.0308	2.0321	2.0334	2.0347	2.0360	2.0373	2.0386	2.0399
7.7	2.0412	2.0425	2.0438	2.0451	2.0464	2.0477	2.0490	2.0503	2.0516	2.0528
7.8	2.0541	2.0554	2.0567	2.0580	2.0592	2.0605	2.0618	2.0631	2.0643	2.0665
7.9	2.0669	2.0681	2.0694	2.0707	2.0719	2.0732	2.0744	2.0757	2.0769	2.0782
8.0	2.0794	2.0807	2.0819	2.0832	2.0844	2.0857	2.0869	2.0882	2.0894	2.0906
8.1	2.0919	2.0931	2.0943	2.0956	2.0968	2.0980	2.0992	2.1005	2.1017	2.1029
8.2	2.1041	2.1054	2.1066	2.1078	2.1090	2.1102	2.1114	2.1126	2.1138	2.1150
8.3	2.1163	2.1175	2.1187	2.1199	2.1211	2.1223	2.1235	2.1247	2.1258	2.1270
8.4	2.1282	2.1294	2.1306	2.1318	2.1330	2.1342	2.1353	2.1365	2.1377	2.1389
8.5	2.1401	2.1412	2.1424	2.1436	2.1448	2.1459	2.1471	2.1483	2.1494	2.1506
8.6	2.1518	2.1529	2.1541	2.1552	2.1564	2.1576	2.1587	2.1599	2.1610	2.1622
8.7	2.1633	2.1645	2.1656	2.1668	2.1679	2.1691	2.1702	2.1713	2.1725	2.1736
8.8	2.1748	2.1759	2.1770	2.1782	2.1793	2.1804	2.1815	2.1827	2.1838	2.1849
8.9	2.1861	2.1872	2.1883	2.1894	2.1905	2.1917	2.1928	2.1939	2.1950	2.1961
9.0	2.1972	2.1983	2.1994	2.2006	2.2017	2.2028	2.2039	2.2050	2.2061	2.2072
9.1	2.2083	2.2094	2.2105	2.2116	2.2127	2.2138	2.2148	2.2159	2.2170	2.2181
9.2	2.2192	2.2203	2.2214	2.2225	2.2235	2.2246	2.2257	2.2268	2.2279	2.2289
9.3	2.2300	2.2311	2.2322	2.2332	2.2343	2.2354	2.2364	2.2375	2.2386	2.2396
9.4	2.2407	2.2418	2.2428	2.2439	2.2450	2.2460	2.2471	2.2481	2.2492	2.2502
9.5	2.2513	2.2523	2.2534	2.2544	2.2555	2.2565	2.2576	2.2586	2.2597	2.2607
9.6	2.2618	2.2628	2.2638	2.2649	2.2659	2.2670	2.2680	2.2690	2.2701	2.2711
9.7	2.2721	2.2732	2.2742	2.2752	2.2762	2.2773	2.2783	2.2793	2.2803	2.2814
9.8	2.2824	2.2834	2.2844	2.2854	2.2865	2.2875	2.2885	2.2895	2.2905	2.2915
9.9	2.2925	2.2935	2.2946	2.2956	2.2966	2.2976	2.2986	2.2996	2.3006	2.3016

TABLE 2 COMMON LOGARITHMS

x	0.00	0.01	0.02	0.03	0.04	0.05	0.06	0.07	0.08	0.09
1.0	0.0000	0.0043	0.0086	0.0128	0.0170	0.0212	0.0253	0.0294	0.0334	0.0374
1.1	0.0414	0.0453	0.0492	0.0531	0.0569	0.0607	0.0645	0.0682	0.0719	0.0755
1.2	0.0792	0.0828	0.0864	0.0899	0.0934	0.0969	0.1004	0.1038	0.1072	0.1106
1.3	0.1139	0.1173	0.1206	0.1239	0.1271	0.1303	0.1335	0.1367	0.1399	0.1430
1.4	0.1461	0.1492	0.1523	0.1553	0.1584	0.1614	0.1644	0.1673	0.1703	0.1732
1.5	0.1761	0.1790	0.1818	0.1847	0.1875	0.1903	0.1931	0.1959	0.1987	0.2014
1.6	0.2041	0.2068	0.2095	0.2122	0.2148	0.2175	0.2201	0.2227	0.2253	0.2279
1.7	0.2304	0.2330	0.2355	0.2380	0.2405	0.2430	0.2455	0.2480	0.2504	0.2529
1.8	0.2553	0.2577	0.2601	0.2625	0.2648	0.2672	0.2695	0.2718	0.2742	0.2765
1.9	0.2788	0.2810	0.2833	0.2856	0.2878	0.2900	0.2923	0.2945	0.2967	0.2989
2.0	0.3010	0.3032	0.3054	0.3075	0.3096	0.3118	0.3139	0.3160	0.3181	0.3201
2.1	0.3222	0.3243	0.3263	0.3284	0.3304	0.3324	0.3345	0.3365	0.3385	0.3404
2.2	0.3424	0.3444	0.3464	0.3483	0.3502	0.3522	0.3541	0.3560	0.3579	0.3598
2.3	0.3617	0.3636	0.3655	0.3674	0.3692	0.3711	0.3729	0.3747	0.3766	0.3784
2.4	0.3802	0.3820	0.3838	0.3856	0.3874	0.3892	0.3909	0.3927	0.3945	0.3962
2.5	0.3979	0.3997	0.4014	0.4031	0.4048	0.4065	0.4082	0.4099	0.4116	0.4133
2.6	0.4150	0.4166	0.4183	0.4200	0.4216	0.4232	0.4249	0.4265	0.4281	0.4298
2.7	0.4314	0.4330	0.4346	0.4362	0.4378	0.4393	0.4409	0.4425	0.4440	0.4456
2.8	0.4472	0.4487	0.4502	0.4518	0.4533	0.4548	0.4564	0.4579	0.4594	0.4609
2.9	0.4624	0.4639	0.4654	0.4669	0.4683	0.4698	0.4713	0.4728	0.4742	0.4757
3.0	0.4771	0.4786	0.4800	0.4814	0.4829	0.4843	0.4857	0.4871	0.4886	0.4900
3.1	0.4914	0.4928	0.4942	0.4955	0.4969	0.4983	0.4997	0.5011	0.5024	0.5038
3.2	0.5051	0.5065	0.5079	0.5092	0.5105	0.5119	0.5132	0.5145	0.5159	0.5172
3.3	0.5185	0.5198	0.5211	0.5224	0.5237	0.5250	0.5263	0.5276	0.5289	0.5302
3.4	0.5315	0.5328	0.5340	0.5353	0.5366	0.5378	0.5391	0.5403	0.5416	0.5428
3.5	0.5441	0.5453	0.5465	0.5478	0.5490	0.5502	0.5514	0.5527	0.5539	0.5551
3.6	0.5563	0.5575	0.5587	0.5599	0.5611	0.5623	0.5635	0.5647	0.5658	0.5670
3.7	0.5682	0.5694	0.5705	0.5717	0.5729	0.5740	0.5752	0.5763	0.5775	0.5786
3.8	0.5798	0.5809	0.5821	0.5832	0.5843	0.5855	0.5866	0.5877	0.5888	0.5899
3.9	0.5911	0.5922	0.5933	0.5944	0.5955	0.5966	0.5977	0.5988	0.5999	0.6010
4.0	0.6021	0.6031	0.6042	0.6053	0.6064	0.6075	0.6085	0.6096	0.6107	0.6117
4.1	0.6128	0.6138	0.6149	0.6160	0.6170	0.6180	0.6191	0.6201	0.6212	0.6222
4.2	0.6232	0.6243	0.6253	0.6263	0.6274	0.6284	0.6294	0.6304	0.6314	0.6325
4.3	0.6335	0.6345	0.6355	0.6365	0.6375	0.6385	0.6395	0.6405	0.6415	0.6425
4.4	0.6435	0.6444	0.6454	0.6464	0.6474	0.6484	0.6493	0.6503	0.6513	0.6522
4.5	0.6532	0.6542	0.6551	0.6561	0.6571	0.6580	0.6590	0.6599	0.6609	0.6618
4.6	0.6628	0.6637	0.6646	0.6656	0.6665	0.6675	0.6684	0.6693	0.6702	0.6712
4.7	0.6721	0.6730	0.6739	0.6749	0.6758	0.6767	0.6776	0.6785	0.6794	0.6803
4.8	0.6812	0.6821	0.6830	0.6839	0.6848	0.6857	0.6866	0.6875	0.6884	0.6893
4.9	0.6902	0.6911	0.6920	0.6928	0.6937	0.6946	0.6955	0.6964	0.6972	0.6981
5.0	0.6990	0.6998	0.7007	0.7016	0.7024	0.7033	0.7042	0.7050	0.7059	0.7067
5.1	0.7076	0.7084	0.7093	0.7101	0.7110	0.7118	0.7126	0.7135	0.7143	0.7152
5.2	0.7160	0.7168	0.7177	0.7185	0.7193	0.7202	0.7210	0.7218	0.7226	0.7235
5.3	0.7243	0.7251	0.7259	0.7267	0.7275	0.7284	0.7292	0.7300	0.7308	0.7316
5.4	0.7324	0.7332	0.7340	0.7348	0.7356	0.7364	0.7372	0.7380	0.7388	0.7396
5.5	0.7404	0.7412	0.7419	0.7427	0.7435	0.7443	0.7451	0.7459	0.7466	0.7474
5.6	0.7482	0.7490	0.7497	0.7505	0.7513	0.7520	0.7528	0.7536	0.7543	0.7551
5.7	0.7559	0.7566	0.7574	0.7582	0.7589	0.7597	0.7604	0.7612	0.7619	0.7627
5.8	0.7634	0.7642	0.7649	0.7657	0.7664	0.7672	0.7679	0.7686	0.7694	0.7701
5.9	0.7709	0.7716	0.7723	0.7731	0.7738	0.7745	0.7752	0.7760	0.7767	0.7774

x	0.00	0.01	0.02	0.03	0.04	0.05	0.06	0.07	0.08	0.09
6.0	0.7782	0.7789	0.7796	0.7803	0.7810	0.7818	0.7825	0.7832	0.7839	0.7846
6.1	0.7853	0.7860	0.7868	0.7875	0.7882	0.7889	0.7896	0.7903	0.7910	0.7917
6.2	0.7924	0.7931	0.7938	0.7945	0.7952	0.7959	0.7966	0.7973	0.7980	0.7987
6.3	0.7993	0.8000	0.8007	0.8014	0.8021	0.8028	0.8035	0.8041	0.8048	0.8055
6.4	0.8062	0.8069	0.8075	0.8082	0.8089	0.8096	0.8102	0.8109	0.8116	0.8122
6.5	0.8129	0.8136	0.8142	0.8149	0.8156	0.8162	0.8169	0.8176	0.8182	0.8189
6.6	0.8195	0.8202	0.8209	0.8215	0.8222	0.8228	0.8235	0.8241	0.8248	0.8254
6.7	0.8261	0.8267	0.8274	0.8280	0.8287	0.8293	0.8299	0.8306	0.8312	0.8319
6.8	0.8325	0.8331	0.8338	0.8344	0.8351	0.8357	0.8363	0.8370	0.8376	0.8382
6.9	0.8388	0.8395	0.8401	0.8407	0.8414	0.8420	0.8426	0.8432	0.8439	0.8445
7.0	0.8451	0.8457	0.8463	0.8470	0.8476	0.8482	0.8488	0.8494	0.8500	0.8506
7.1	0.8513	0.8519	0.8525	0.8531	0.8537	0.8543	0.8549	0.8555	0.8561	0.8567
7.2	0.8573	0.8579	0.8585	0.8591	0.8597	0.8603	0.8609	0.8615	0.8621	0.8627
7.3	0.8633	0.8639	0.8645	0.8651	0.8657	0.8663	0.8669	0.8675	0.8681	0.8686
7.4	0.8692	0.8698	0.8704	0.8710	0.8716	0.8722	0.8727	0.8733	0.8739	0.8745
7.5	0.8751	0.8756	0.8762	0.8768	0.8774	0.8779	0.8785	0.8791	0.8797	0.8802
7.6	0.8808	0.8814	0.8820	0.8825	0.8831	0.8837	0.8842	0.8848	0.8854	0.8859
7.7	0.8865	0.8871	0.8876	0.8882	0.8887	0.8893	0.8899	0.8904	0.8910	0.8915
7.8	0.8921	0.8927	0.8932	0.8938	0.8943	0.8949	0.8954	0.8960	0.8965	0.8971
7.9	0.8976	0.8982	0.8987	0.8993	0.8998	0.9004	0.9009	0.9015	0.9020	0.9025
8.0	0.9031	0.9036	0.9042	0.9047	0.9053	0.9058	0.9063	0.9069	0.9074	0.9079
8.1	0.9085	0.9090	0.9096	0.9101	0.9106	0.9112	0.9117	0.9122	0.9128	0.9133
8.2	0.9138	0.9143	0.9149	0.9154	0.9159	0.9165	0.9170	0.9175	0.9180	0.9186
8.3	0.9191	0.9196	0.9201	0.9206	0.9212	0.9217	0.9222	0.9227	0.9232	0.9238
8.4	0.9243	0.9248	0.9253	0.9258	0.9263	0.9269	0.9274	0.9279	0.9284	0.9289
8.5	0.9294	0.9299	0.9304	0.9309	0.9315	0.9320	0.9325	0.9330	0.9335	0.9340
8.6	0.9345	0.9350	0.9355	0.9360	0.9365	0.9370	0.9375	0.9380	0.9385	0.9390
8.7	0.9395	0.9400	0.9405	0.9410	0.9415	0.9420	0.9425	0.9430	0.9435	0.9440
8.8	0.9445	0.9450	0.9455	0.9460	0.9465	0.9469	0.9474	0.9479	0.9484	0.9489
8.9	0.9494	0.9499	0.9504	0.9509	0.9513	0.9518	0.9523	0.9528	0.9533	0.9538
9.0	0.9542	0.9547	0.9552	0.9557	0.9562	0.9566	0.9571	0.9576	0.9581	0.9586
9.1	0.9590	0.9595	0.9600	0.9605	0.9609	0.9614	0.9619	0.9624	0.9628	0.9633
9.2	0.9638	0.9643	0.9647	0.9652	0.9657	0.9661	0.9666	0.9671	0.9675	0.9680
9.3	0.9685	0.9689	0.9694	0.9699	0.9703	0.9708	0.9713	0.9717	0.9722	0.9727
9.4	0.9731	0.9736	0.9741	0.9745	0.9750	0.9754	0.9759	0.9763	0.9768	0.9773
9.5	0.9777	0.9782	0.9786	0.9791	0.9795	0.9800	0.9805	0.9809	0.9814	0.9818
9.6	0.9823	0.9827	0.9832	0.9836	0.9841	0.9845	0.9850	0.9854	0.9859	0.9863
9.7	0.9868	0.9872	0.9877	0.9881	0.9886	0.9890	0.9894	0.9899	0.9903	0.9908
9.8	0.9912	0.9917	0.9921	0.9926	0.9930	0.9934	0.9939	0.9943	0.9948	0.9952
9.9	0.9956	0.9961	0.9965	0.9969	0.9974	0.9978	0.9983	0.9987	0.9991	0.9996

TABLE 3 EXPONENTIAL FUNCTIONS

x	e^x	e^{-x}	x	e^x	e^{-x}
0.00	1.0000	1.0000	3.0	20.086	0.0498
0.05	1.0513	0.9512	3.1	22.198	0.0450
0.10	1.1052	0.9048	3.2	24.533	0.0408
0.15	1.1618	0.8607	3.3	27.113	0.0369
0.20	1.2214	0.8187	3.4	29.964	0.0334
0.25	1.2840	0.7788	3.5	33.115	0.0302
0.30	1.3499	0.7408	3.6	36.598	0.0273
0.35	1.4191	0.7047	3.7	40.447	0.0247
0.40	1.4918	0.6703	3.8	44.701	0.0224
0.45	1.5683	0.6376	3.9	49.402	0.0202
0.50	1.6487	0.6065	4.0	54.598	0.0183
0.55	1.7333	0.5769	4.1	60.340	0.0166
0.60	1.8221	0.5488	4.2	66.686	0.0150
0.65	1.9155	0.5220	4.3	73.700	0.0136
0.70	2.0138	0.4966	4.4	81.451	0.0123
0.75	2.1170	0.4724	4.5	90.017	0.0111
0.80	2.2255	0.4493	4.6	99.484	0.0101
0.85	2.3396	0.4274	4.7	109.95	0.0091
0.90	2.4596	0.4066	4.8	121.51	0.0082
0.95	2.5857	0.3867	4.9	134.29	0.0074
1.0	2.7183	0.3679	5.0	148.41	0.0067
1.1	3.0042	0.3329	5.1	164.02	0.0061
1.2	3.3201	0.3012	5.2	181.27	0.0055
1.3	3.6693	0.2725	5.3	200.34	0.0050
1.4	4.0552	0.2466	5.4	221.41	0.0045
1.5	4.4817	0.2231	5.5	244.69	0.0041
1.6	4.9530	0.2019	5.6	270.43	0.0037
1.7	5.4739	0.1827	5.7	298.87	0.0033
1.8	6.0496	0.1653	5.8	330.30	0.0030
1.9	6.6859	0.1496	5.9	365.04	0.0027
2.0	7.3891	0.1353	6.0	403.43	0.0025
2.1	8.1662	0.1225	6.5	665.14	0.0015
2.2	9.0250	0.1108	7.0	1096.6	0.0009
2.3	9.9742	0.1003	7.5	1808.0	0.0006
2.4	11.023	0.0907	8.0	2981.0	0.0003
2.5	12.182	0.0821	8.5	4914.8	0.0002
2.6	13.464	0.0743	9.0	8103.1	0.0001
2.7	14.880	0.0672	9.5	13,360	0.00007
2.8	16.445	0.0608	10.0	22,026	0.00004
2.9	18.174	0.0550			

3 Tables of Trigonometric Functions

In this section, we present tables of trigonometric functions of angles measured in degrees (Table 1) and radians (Table 2), and we give examples illustrating their use.

Table 1 gives values of sine, tangent, cotangent, and cosine, rounded off to four or five significant digits, corresponding to angles between 0° and 90° in steps of 0.1°. Angles between 0° and 45° are found in the vertical column on the left side of the table, and the headings at the top of the table apply to these angles. Angles between 45° and 90° are found in the vertical column on the right side of the table, and the headings at the bottom of the table apply to these angles.

Examples Use Table 1 to find the approximate value of each trigonometric function.

1 $\cos 29.3°$

Solution Since $29.3° < 45°$, we begin by locating 29.3 in the vertical column on the *left* side of Table 1. As can be seen from the headings *above* the table, the numbers to the *right* of 29.3 are the sine, tangent, cotangent, and cosine of 29.3°. Looking in the vertical column with the heading "cosine," we find that $\cos 29.3° = 0.8721$.

2 $\tan 128.8°$

Solution The angle 128.8° lies in quadrant II, so its tangent is negative. The reference angle corresponding to 128.8° is

$$180° - 128.8° = 51.2°.$$

Thus,

$$\tan 128.8° = -\tan 51.2°.$$

Since $51.2° > 45°$, we first locate 51.2 in the vertical column on the *right* side of Table 1. As can be seen from the headings *below* the table, the numbers to the *left* of 51.2 are the sine, tangent, cotangent, and cosine of 51.2°. Thus,

$$\tan 51.2° = 1.2437,$$

so that,

$$\tan 128.8° = -1.2437.$$

3 $\csc 72.3°$

Solution Since values of the cosecant do not appear in Table 1, we use the fact that

$$\csc 72.3° = \frac{1}{\sin 72.3°}.$$

From Table 1, $\sin 72.3° = 0.9527$, so

$$\csc 72.3° = \frac{1}{0.9527} = 1.050,$$

rounded off to four significant digits.

Linear interpolation can be used to find the values of trigonometric functions of angles lying between two angles whose values are given in Table 1. The basic idea of linear interpolation is that small changes in the angle cause approximately proportional changes in the values of the function.

Example Use linear interpolation and Table 1 to find the approximate value of $\sin 19.74^\circ$.

Solution We arrange the work as follows:

		x	$\sin x$		
0.1°	0.04°	19.7°	0.3371	d	$0.0016.$
		19.74°	$\sin 19.74^\circ$		
		19.8°	0.3387		

Here, we have used Table 1 to determine that

$$\sin 19.7^\circ = 0.3371 \quad \text{and} \quad \sin 19.8^\circ = 0.3387.$$

The notation

$$0.04^\circ \left[\begin{matrix} 19.7^\circ \\ 19.74^\circ \end{matrix} \right.$$

indicates that the difference between 19.7° and 19.74° is

$$19.74^\circ - 19.7^\circ = 0.04^\circ.$$

Other differences are indicated similarly. For linear interpolation, we assume that corresponding differences are approximately proportional, so that

$$\frac{0.04^\circ}{0.1^\circ} = \frac{d}{0.0016}$$

or

$$d = 0.0016\left(\frac{0.04}{0.1}\right) = 0.00064.$$

Thus, rounded off to four significant digits,

$$\begin{aligned} \sin 19.74^\circ &= 0.3371 + d \\ &= 0.3371 + 0.00064 = 0.3377. \end{aligned}$$

By reading Table 1 "backwards," you can find values in degrees of the inverse trigonometric functions.

Examples Using Table 1, find the value of each inverse trigonometric function rounded off to the nearest 0.1°.

1 $\cos^{-1} 0.3714$

Solution Looking in the body of the table, we find 0.3714 in the vertical column with the *lower* heading "cos." By reading the table "backwards," we see that

$$\cos 68.2^\circ = 0.3714 \quad \text{or} \quad \cos^{-1} 0.3714 = 68.2^\circ.$$

2 $\sin^{-1} 0.7320$

Solution The number 0.7320 does not appear in the vertical sine column in Table 2. The numbers closest to it are

$$0.7314 = \sin 47.0^\circ \quad \text{and} \quad 0.7325 = \sin 47.1^\circ.$$

Using linear interpolation, we have

$$0.0011\left[\begin{array}{ll} 0.0006\left[\begin{array}{l} 0.7314 \\ 0.7320 \end{array}\right. & \begin{array}{l} 47.0^\circ \\ \sin^{-1} 0.7320 \end{array}\Big]d \\ \quad 0.7325 & \quad 47.1^\circ \end{array}\right]0.1^\circ.$$

(columns: x, $\sin^{-1} x$)

Thus,

$$\frac{0.0006}{0.0011} = \frac{d}{0.1^\circ}$$

or

$$d = 0.1^\circ\left(\frac{0.0006}{0.0011}\right) = 0.05^\circ,$$

rounded off to the nearest 0.01°, so that

$$\begin{aligned} \sin^{-1} 0.7320 &= 47.0^\circ + d \\ &= 47.0^\circ + 0.05^\circ \\ &= 47.05^\circ \end{aligned}$$

Table 2 gives values of the six trigonometric functions, rounded off to four or five significant digits, corresponding to angles between 0 and $\pi/2$ radians in steps of 0.01 radian.

Examples Use Table 2 to find the approximate value of each trigonometric function.

1 csc 1.02

Solution From Table 2

$$\csc 1.02 = 1.174.$$

2 sin 3.538

Solution In radians, $\pi < 3.538 < (3\pi/2)$, so 3.538 is a quadrant III angle and its sine is negative. The reference angle corresponding to 3.538 is

$$3.538 - \pi = 0.396,$$

rounded off to three decimal places, so

$$\sin 3.538 = -\sin 0.396.$$

Using Table 2 and linear interpolation, we have

$$
0.01\left[\begin{array}{c} 0.006\left[\begin{array}{c} 0.39 \\ 0.396 \end{array}\right. \\ \;\;0.40 \end{array}\right.
\qquad
\left.\begin{array}{c} \left.\begin{array}{c} 0.3802 \\ \sin 0.396 \end{array}\right] d \\ 0.3894 \end{array}\right] 0.0092.
$$

(columns headed x and $\sin x$)

Thus,

$$\frac{0.006}{0.01} = \frac{d}{0.0092}$$

or

$$d = 0.0092\left(\frac{0.006}{0.01}\right) = 0.0055;$$

so that, rounded off to four decimal places,

$$\begin{aligned} \sin 0.396 &= 0.3802 + d \\ &= 0.3802 + 0.0055 \\ &= 0.3857. \end{aligned}$$

Therefore,

$$\sin 3.538 = -0.3857.$$

(Because of all the rounding off and the linear interpolation, this value isn't very accurate—a calculator gives

$$\sin 3.538 = -0.3861.$$

However, it's the best we can do with our table.)

By reading Table 2 "backwards," you can find values in radians of the inverse trigonometric functions.

Examples Using Table 2, find the value of each inverse trigonometric function rounded off to the nearest 0.01 radian.

1 $\arctan 2.427$

Solution Looking in the tangent column of Table 2, we find 2.427 in the position corresponding to 1.18 radians. Hence,

$$\arctan 2.427 = 1.18.$$

2 $\cos^{-1} 0.8932$

Solution Since 0.8932 does not appear in the column for cosines, we must use linear interpolation. Thus,

$$\begin{array}{ll|ll|l} & & x & \cos^{-1} x & \\ -0.0045 & -0.0029 & 0.8961 & 0.46 & d \quad 0.01. \\ & & 0.8932 & \cos^{-1} 0.8932 & \\ & & 0.8916 & 0.47 & \end{array}$$

It follows that

$$\frac{-0.0029}{-0.0045} = \frac{d}{0.01}$$

or

$$d = 0.01\left(\frac{-0.0029}{-0.0045}\right) = 0.006,$$

rounded off to three decimal places, and so

$$\cos^{-1} 0.8932 = 0.46 + d = 0.46 + 0.006 = 0.466.$$

PROBLEM SET 3

In problems 1–12, use Table 1 to find the approximate value of each trigonometric function.

1. $\sin 25.3°$ **2.** $\cos 67.2°$ **3.** $\tan 38.7°$ **4.** $\cot 81.6°$
5. $\sec 65.4°$ **6.** $\csc 57.9°$ **7.** $\cos 113.2°$ **8.** $\cot 169.7°$
9. $\csc 118.3°$ **10** $\sin 175.5°$ **11.** $\cot 129.7°$ **12.** $\sec 145.6°$

In problems 13–18, use linear interpolation and Table 1 to find the approximate value of each trigonometric function.

13. $\tan 41.32°$ **14.** $\cos 49.21°$ **15.** $\cot 38.72°$ **16.** $\csc 114.57°$
17. $\sin 129.56°$ **18.** $\sec 153.42°$

In problems 19–30, use Table 2 and linear interpolation (if necessary) to find the approximate value of each trigonometric function.

19. $\sin 0.89$ **20.** $\cos 0.93$ **21.** $\cot 1.44$ **22.** $\cos 1.12$
23. $\sec 1.54$ **24.** $\csc 1.39$ **25.** $\tan 1.34$ **26.** $\sin 3.592$
27. $\sec 4.821$ **28.** $\cot 3.245$ **29.** $\cos 2.713$ **30.** $\csc 3.572$

In problems 31–42, use Table 1 and linear interpolation (if necessary) to find the value of each inverse trigonometric function rounded off to the nearest 0.1°.

31. $\sin^{-1} 0.1736$ **32.** $\tan^{-1} 1.0355$ **33.** $\cos^{-1} 0.6481$ **34.** $\cot^{-1} 0.8156$
35. $\tan^{-1} 1.4335$ **36.** $\sin^{-1} 0.7869$ **37.** $\cos^{-1} 0.4924$ **38.** $\cot^{-1} 2.1450$
39. $\sin^{-1} 0.7325$ **40.** $\tan^{-1} 2.765$ **41.** $\cos^{-1} 0.4921$ **42.** $\sin^{-1} 0.8807$

In problems 43–54, use Table 2 and linear interpolation (if necessary) to find the approximate value of each inverse trigonometric function.

43. $\sin^{-1} 0.9713$ **44.** $\arccos 0.9902$ **45.** $\text{arccsc}\, 2.348$ **46.** $\cos^{-1} 0.6294$
47. $\cot^{-1} 4.086$ **48.** $\arctan 3.467$ **49.** $\arcsin 0.7891$ **50.** $\sec^{-1} 3.603$
51. $\csc^{-1} 1.027$ **52.** $\text{arccot}\, 0.6425$ **53.** $\arcsin 0.9711$ **54.** $\arccos 0.5891$

TABLE 1 TRIGONOMETRIC FUNCTIONS—DEGREE MEASURE

Deg.	sin	tan	cot	cos	
0.0	0.00000	0.00000	∞	1.0000	90.0
.1	.00175	.00175	573.0	1.0000	89.9
.2	.00349	.00349	286.5	1.0000	.8
.3	.00524	.00524	191.0	1.0000	.7
.4	.00698	.00698	143.24	1.0000	.6
.5	.00873	.00873	114.59	1.0000	.5
.6	.01047	.01047	95.49	0.9999	.4
.7	.01222	.01222	81.85	.9999	.3
.8	.01396	.01396	71.62	.9999	.2
.9	.01571	.01571	63.66	.9999	89.1
1.0	0.01745	0.01746	57.29	0.9998	89.0
.1	.01920	.01920	52.08	.9998	88.9
.2	.02094	.02095	47.74	.9998	.8
.3	.02269	.02269	44.07	.9997	.7
.4	.02443	.02444	40.92	.9997	.6
.5	.02618	.02619	38.19	.9997	.5
.6	.02792	.02793	35.80	.9996	.4
.7	.02967	.02968	33.69	.9996	.3
.8	.03141	.03143	31.82	.9995	.2
.9	.03316	.03317	30.14	.9995	88.1
2.0	0.03490	0.03492	28.64	0.9994	88.0
.1	.03664	.03667	27.27	.9993	87.9
.2	.03839	.03842	26.03	.9993	.8
.3	.04013	.04016	24.90	.9992	.7
.4	.04188	.04191	23.86	.9991	.6
.5	.04362	.04366	22.90	.9990	.5
.6	.04536	.04541	22.02	.9990	.4
.7	.04711	.04716	21.20	.9989	.3
.8	.04885	.04891	20.45	.9988	.2
.9	.05059	.05066	19.74	.9987	87.1
3.0	0.05234	0.05241	19.081	0.9986	87.0
.1	.05408	.05416	18.464	.9985	86.9
.2	.05582	.05591	17.886	.9984	.8
.3	.05756	.05766	17.343	.9983	.7
.4	.05931	.05941	16.832	.9982	.6
.5	.06105	.06116	16.350	.9981	.5
.6	.06279	.06291	15.895	.9980	.4
.7	.06453	.06467	15.464	.9979	.3
.8	.06627	.06642	15.056	.9978	.2
.9	.06802	.06817	14.669	.9977	86.1
4.0	0.06976	0.06993	14.301	0.9976	86.0
.1	.07150	.07168	13.951	.9974	85.9
.2	.07324	.07344	13.617	.9973	.8
.3	.07498	.07519	13.300	.9972	.7
.4	.07672	.07695	12.996	.9971	.6
.5	.07846	.07870	12.706	.9969	.5
.6	.08020	.08046	12.429	.9968	.4
.7	.08194	.08221	12.163	.9966	.3
.8	.08368	.08397	11.909	.9965	.2
.9	.08542	.08573	11.664	.9963	85.1
5.0	0.08716	0.08749	11.430	0.9962	85.0
	cos	cot	tan	sin	Deg.

Deg.	sin	tan	cot	cos	
5.0	0.08716	0.08749	11.430	0.9962	85.0
.1	.08889	.08925	11.205	.9960	84.9
.2	.09063	.09101	10.988	.9959	.8
.3	.09237	.09277	10.780	.9957	.7
.4	.09411	.09453	10.579	.9956	.6
.5	.09585	.09629	10.385	.9954	.5
.6	.09758	.09805	10.199	.9952	.4
.7	.09932	.09981	10.019	.9951	.3
.8	.10106	.10158	9.845	.9949	.2
.9	.10279	.10334	9.677	.9947	84.1
6.0	0.10453	0.10510	9.514	0.9945	84.0
.1	.10626	.10687	9.357	.9943	83.9
.2	.10800	.10863	9.205	.9942	.8
.3	.10973	.11040	9.058	.9940	.7
.4	.11147	.11217	8.915	.9938	.6
.5	.11320	.11394	8.777	.9936	.5
.6	.11494	.11570	8.643	.9934	.4
.7	.11667	.11747	8.513	.9932	.3
.8	.11840	.11924	8.386	.9930	.2
.9	.12014	.12101	8.264	.9928	83.1
7.0	0.12187	0.12278	8.144	0.9925	83.0
.1	.12360	.12456	8.028	.9923	82.9
.2	.12533	.12633	7.916	.9921	.8
.3	.12706	.12810	7.806	.9919	.7
.4	.12880	.12988	7.700	.9917	.6
.5	.13053	.13165	7.596	.9914	.5
.6	.13226	.13343	7.495	.9912	.4
.7	.13399	.13521	7.396	.9910	.3
.8	.13572	.13698	7.300	.9907	.2
.9	.13744	.13876	7.207	.9905	82.1
8.0	0.13917	0.14054	7.115	0.9903	82.0
.1	.14090	.14232	7.026	.9900	81.9
.2	.14263	.14410	6.940	.9898	.8
.3	.14436	.14588	6.855	.9895	.7
.4	.14608	.14767	6.772	.9893	.6
.5	.14781	.14945	6.691	.9890	.5
.6	.14954	.15124	6.612	.9888	.4
.7	.15126	.15302	6.535	.9885	.3
.8	.15299	.15481	6.460	.9882	.2
.9	.15471	.15660	6.386	.9880	81.1
9.0	0.15643	0.15838	6.314	0.9877	81.0
.1	.15816	.16017	6.243	.9874	80.9
.2	.15988	.16196	6.174	.9871	.8
.3	.16160	.16376	6.107	.9869	.7
.4	.16333	.16555	6.041	.9866	.6
.5	.16505	.16734	5.976	.9863	.5
.6	.16677	.16914	5.912	.9860	.4
.7	.16849	.17093	5.850	.9857	.3
.8	.17021	.17273	5.789	.9854	.2
.9	.17193	.17453	5.730	.9851	80.1
10.0	0.1736	0.1763	5.671	0.9848	80.0
	cos	cot	tan	sin	Deg.

Deg.	sin	tan	cot	cos	
10.0	0.1736	0.1763	5.671	0.9848	80.0
.1	.1754	.1781	5.614	.9845	79.9
.2	.1771	.1799	5.558	.9842	.8
.3	.1788	.1817	5.503	.9839	.7
.4	.1805	.1835	5.449	.9836	.6
.5	.1822	.1853	5.396	.9833	.5
.6	.1840	.1871	5.343	.9829	.4
.7	.1857	.1890	5.292	.9826	.3
.8	.1874	.1908	5.242	.9823	.2
.9	.1891	.1926	5.193	.9820	79.1
11.0	0.1908	0.1944	5.145	0.9816	79.0
.1	.1925	.1962	5.079	.9813	78.9
.2	.1942	.1980	5.050	.9810	.8
.3	.1959	.1998	5.005	.9806	.7
.4	.1977	.2016	4.959	.9803	.6
.5	.1994	.2035	4.915	.9799	.5
.6	.2011	.2053	4.872	.9796	.4
.7	.2028	.2071	4.829	.9792	.3
.8	.2045	.2089	4.787	.9789	.2
.9	.2062	.2107	4.745	.9785	78.1
12.0	0.2079	0.2126	4.705	0.9781	78.0
.1	.2096	.2144	4.665	.9778	77.9
.2	.2113	.2162	4.625	.9774	.8
.3	.2130	.2180	4.586	.9770	.7
.4	.2147	.2199	4.548	.9767	.6
.5	.2164	.2217	4.511	.9763	.5
.6	.2181	.2235	4.474	.9759	.4
.7	.2198	.2254	4.437	.9755	.3
.8	.2215	.2272	4.402	.9751	.2
.9	.2233	.2290	4.366	.9748	77.1
13.0	0.2250	0.2309	4.331	0.9744	77.0
.1	.2267	.2327	4.297	.9740	76.9
.2	.2284	.2345	4.264	.9736	.8
.3	.2300	.2364	4.230	.9732	.7
.4	.2317	.2382	4.198	.9728	.6
.5	.2334	.2401	4.165	.9724	.5
.6	.2351	.2419	4.134	.9720	.4
.7	.2368	.2438	4.102	.9715	.3
.8	.2385	.2456	4.071	.9711	.2
.9	.2402	.2475	4.041	.9707	76.1
14.0	0.2419	0.2493	4.011	0.9703	76.0
.1	.2436	.2512	3.981	.9699	75.9
.2	.2453	.2530	3.952	.9694	.8
.3	.2470	.2549	3.923	.9690	.7
.4	.2487	.2568	3.895	.9686	.6
.5	.2504	.2586	3.867	.9681	.5
.6	.2521	.2605	3.839	.9677	.4
.7	.2538	.2623	3.812	.9673	.3
.8	.2554	.2642	3.785	.9668	.2
.9	.2571	.2661	3.758	.9664	75.1
15.0	0.2588	0.2679	3.732	0.9659	75.0
	cos	cot	tan	sin	Deg.

Deg.	sin	tan	cot	cos	
15.0	0.2588	0.2679	3.732	0.9659	75.0
.1	.2605	.2698	3.706	.9655	74.9
.2	.2622	.2717	3.681	.9650	.8
.3	.2639	.2736	3.655	.9646	.7
.4	.2656	.2754	3.630	.9641	.6
.5	.2672	.2773	3.606	.9636	.5
.6	.2689	.2792	3.582	.9632	.4
.7	.2706	.2811	3.558	.9627	.3
.8	.2723	.2830	3.534	.9622	.2
.9	.2740	.2849	3.511	.9617	74.1
16.0	0.2756	0.2867	3.487	0.9613	74.0
.1	.2773	.2886	3.465	.9608	73.9
.2	.2790	.2905	3.442	.9603	.8
.3	.2807	.2924	3.420	.9598	.7
.4	.2823	.2943	3.398	.9593	.6
.5	.2840	.2962	3.376	.9588	.5
.6	.2857	.2981	3.354	.9583	.4
.7	.2874	.3000	3.333	.9578	.3
.8	.2890	.3019	3.312	.9573	.2
.9	.2907	.3038	3.291	.9568	73.1
17.0	0.2924	0.3057	3.271	0.9563	73.0
.1	.2940	.3076	3.251	.9558	72.9
.2	.2957	.3096	3.230	.9553	.8
.3	.2974	.3115	3.211	.9548	.7
.4	.2990	.3134	3.191	.9542	.6
.5	.3007	.3153	3.172	.9537	.5
.6	.3024	.3172	3.152	.9532	.4
.7	.3040	.3191	3.133	.9527	.3
.8	.3057	.3211	3.115	.9521	.2
.9	.3074	.3230	3.096	.9516	72.1
18.0	0.3090	0.3249	3.078	0.9511	72.0
.1	.3107	.3269	3.060	.9505	71.9
.2	.3123	.3288	3.042	.9500	.8
.3	.3140	.3307	3.024	.9494	.7
.4	.3156	.3327	3.006	.9489	.6
.5	.3173	.3346	2.989	.9483	.5
.6	.3190	.3365	2.971	.9478	.4
.7	.3206	.3385	2.954	.9472	.3
.8	.3223	.3404	2.937	.9466	.2
.9	.3239	.3424	2.921	.9461	71.1
19.0	0.3256	0.3443	2.904	0.9455	71.0
.1	.3272	.3463	2.888	.9449	70.9
.2	.3289	.3482	2.872	.9444	.8
.3	.3305	.3502	2.856	.9438	.7
.4	.3322	.3522	2.840	.9432	.6
.5	.3338	.3541	2.824	.9426	.5
.6	.3355	.3561	2.808	.9421	.4
.7	.3371	.3581	2.793	.9415	.3
.8	.3387	.3600	2.778	.9409	.2
.9	.3404	.3620	2.762	.9403	70.1
20.0	0.3420	0.3640	2.747	0.9397	70.0
	cos	cot	tan	sin	Deg.

TABLE 1 TRIGONOMETRIC FUNCTIONS—DEGREE MEASURE

Deg.	sin	tan	cot	cos	
20.0	0.3420	0.3640	2.747	0.9397	70.0
.1	.3437	.3659	2.733	.9391	69.9
.2	.3453	.3679	2.718	.9385	.8
.3	.3469	.3699	2.703	.9379	.7
.4	.3486	.3719	2.689	.9373	.6
.5	.3502	.3739	2.675	.9367	.5
.6	.3518	.3759	2.660	.9361	.4
.7	.3535	.3779	2.646	.9354	.3
.8	.3551	.3799	2.633	.9348	.2
.9	.3567	.3819	2.619	.9342	69.1
21.0	0.3584	0.3839	2.605	0.9336	69.0
.1	.3600	.3859	2.592	.9330	68.9
.2	.3616	.3879	2.578	.9323	.8
.3	.3633	.3899	2.565	.9317	.7
.4	.3649	.3919	2.552	.9311	.6
.5	.3665	.3939	2.539	.9304	.5
.6	.3681	.3959	2.526	.9298	.4
.7	.3697	.3979	2.513	.9291	.3
.8	.3714	.4000	2.500	.9285	.2
.9	.3730	.4020	2.488	.9278	68.1
22.0	0.3746	0.4040	2.475	0.9272	68.0
.1	.3762	.4061	2.463	.9265	67.9
.2	.3778	.4081	2.450	.9259	.8
.3	.3795	.4101	2.438	.9252	.7
.4	.3811	.4122	2.426	.9245	.6
.5	.3827	.4142	2.414	.9239	.5
.6	.3843	.4163	2.402	.9232	.4
.7	.3859	.4183	2.391	.9225	.3
.8	.3875	.4204	2.379	.9219	.2
.9	.3891	.4224	2.367	.9212	67.1
23.0	0.3907	0.4245	2.356	0.9205	67.0
.1	.3923	.4265	2.344	.9198	66.9
.2	.3939	.4286	2.333	.9191	.8
.3	.3955	.4307	2.322	.9184	.7
.4	.3971	.4327	2.311	.9178	.6
.5	.3987	.4348	2.300	.9171	.5
.6	.4003	.4369	2.289	.9164	.4
.7	.4019	.4390	2.278	.9157	.3
.8	.4035	.4411	2.267	.9150	.2
.9	.4051	.4431	2.257	.9143	66.1
24.0	0.4067	0.4452	2.246	0.9135	66.0
.1	.4083	.4473	2.236	.9128	65.9
.2	.4099	.4494	2.225	.9121	.8
.3	.4115	.4515	2.215	.9114	.7
.4	.4131	.4536	2.204	.9107	.6
.5	.4147	.4557	2.194	.9100	.5
.6	.4163	.4578	2.184	.9092	.4
.7	.4179	.4599	2.174	.9085	.3
.8	.4195	.4621	2.164	.9078	.2
.9	.4210	.4642	2.154	.9070	65.1
25.0	0.4226	0.4663	2.145	0.9063	65.0
	cos	cot	tan	sin	Deg.

Deg.	sin	tan	cot	cos	
25.0	0.4226	0.4663	2.145	0.9063	65.0
.1	.4242	.4684	2.135	.9056	64.9
.2	.4258	.4706	2.125	.9048	.8
.3	.4274	.4727	2.116	.9041	.7
.4	.4289	.4748	2.106	.9033	.6
.5	.4305	.4770	2.097	.9026	.5
.6	.4321	.4791	2.087	.9018	.4
.7	.4337	.4813	2.078	.9011	.3
.8	.4352	.4834	2.069	.9003	.2
.9	.4368	.4856	2.059	.8996	64.1
26.0	0.4384	0.4887	2.050	0.8988	64.0
.1	.4399	.4899	2.041	.8980	63.9
.2	.4415	.4921	2.032	.8973	.8
.3	.4431	.4942	2.023	.8965	.7
.4	.4446	.4964	2.014	.8957	.6
.5	.4462	.4986	2.006	.8949	.5
.6	.4478	.5008	1.997	.8942	.4
.7	.4493	.5029	1.988	.8934	.3
.8	.4509	.5051	1.980	.8926	.2
.9	.4524	.5073	1.971	.8918	63.1
27.0	0.4540	0.5095	1.963	0.8910	63.0
.1	.4555	.5117	1.954	.8902	62.9
.2	.4571	.5139	1.946	.8894	.8
.3	.4586	.5161	1.937	.8886	.7
.4	.4602	.5184	1.929	.8878	.6
.5	.4617	.5206	1.921	.8870	.5
.6	.4633	.5228	1.913	.8862	.4
.7	.4648	.5250	1.905	.8854	.3
.8	.4664	.5272	1.897	.8846	.2
.9	.4679	.5295	1.889	.8838	62.1
28.0	0.4695	0.5317	1.881	0.8829	62.0
.1	.4710	.5340	1.873	.8821	61.9
.2	.4726	.5362	1.865	.8813	.8
.3	.4741	.5384	1.857	.8805	.7
.4	.4756	.5407	1.849	.8796	.6
.5	.4772	.5430	1.842	.8788	.5
.6	.4787	.5452	1.834	.8780	.4
.7	.4802	.5475	1.827	.8771	.3
.8	.4818	.5498	1.819	.8763	.2
.9	.4833	.5520	1.811	.8755	61.1
29.0	0.4848	0.5543	1.804	0.8746	61.0
.1	.4863	.5566	1.797	.8738	60.9
.2	.4879	.5589	1.789	.8729	.8
.3	.4894	.5612	1.782	.8721	.7
.4	.4909	.5635	1.775	.8712	.6
.5	.4924	.5658	1.767	.8704	.5
.6	.4939	.5681	1.760	.8695	.4
.7	.4955	.5704	1.753	.8686	.3
.8	.4970	.5727	1.746	.8678	.2
.9	.4985	.5750	1.739	.8669	60.1
30.0	0.5000	0.5774	1.732	0.8660	60.0
	cos	cot	tan	sin	Deg.

Deg.	sin	tan	cot	cos	
30.0	0.5000	0.5774	1.7321	0.8660	60.0
.1	.5015	.5797	1.7251	.8652	59.9
.2	.5030	.5820	1.7182	.8643	.8
.3	.5045	.5844	1.7113	.8634	.7
.4	.5060	.5867	1.7045	.8625	.6
.5	.5075	.5890	1.6977	.8616	.5
.6	.5090	.5914	1.6909	.8607	.4
.7	.5105	.5938	1.6842	.8599	.3
.8	.5120	.5961	1.6775	.8590	.2
.9	.5135	.5985	1.6709	.8581	59.1
31.0	0.5150	0.6009	1.6643	0.8572	59.0
.1	.5165	.6032	1.6577	.8563	58.9
.2	.5180	.6056	1.6512	.8554	.8
.3	.5195	.6080	1.6447	.8545	.7
.4	.5210	.6104	1.6383	.8536	.6
.5	.5225	.6128	1.6319	.8526	.5
.6	.5240	.6152	1.6255	.8517	.4
.7	.5255	.6176	1.6191	.8508	.3
.8	.5270	.6200	1.6128	.8499	.2
.9	.5284	.6224	1.6066	.8490	58.1
32.0	0.5299	0.6249	1.6003	0.8480	58.0
.1	.5314	.6273	1.5941	.8471	57.9
.2	.5329	.6297	1.5880	.8462	.8
.3	.5344	.6322	1.5818	.8453	.7
.4	.5358	.6346	1.5757	.8443	.6
.5	.5373	.6371	1.5697	.8434	.5
.6	.5388	.6395	1.5637	.8425	.4
.7	.5402	.6420	1.5577	.8415	.3
.8	.5417	.6445	1.5517	.8406	.2
.9	.5432	.6469	1.5458	.8396	57.1
33.0	0.5446	0.6494	1.5399	0.8387	57.0
.1	.5461	.6519	1.5340	.8377	56.9
.2	.5476	.6544	1.5282	.8368	.8
.3	.5490	.6569	1.5224	.8358	.7
.4	.5505	.6594	1.5166	.8348	.6
.5	.5519	.6619	1.5108	.8339	.5
.6	.5534	.6644	1.5051	.8329	.4
.7	.5548	.6669	1.4994	.8320	.3
.8	.5563	.6694	1.4938	.8310	.2
.9	.5577	.6720	1.4882	.8300	56.1
34.0	0.5592	0.6745	1.4826	0.8290	56.0
.1	.5606	.6771	1.4770	.8281	55.9
.2	.5621	.6796	1.4715	.8271	.8
.3	.5635	.6822	1.4659	.8261	.7
.4	.5650	.6847	1.4605	.8251	.6
.5	.5664	.6873	1.4550	.8241	.5
.6	.5678	.6899	1.4496	.8231	.4
.7	.5693	.6924	1.4442	.8221	.3
.8	.5707	.6950	1.4388	.8211	.2
.9	.5721	.6976	1.4335	.8202	55.1
35.0	0.5736	0.7002	1.4281	0.8192	55.0
	cos	cot	tan	sin	Deg.

Deg.	sin	tan	cot	cos	
35.0	0.5736	0.7002	1.4281	0.8192	55.0
.1	.5750	.7028	1.4229	.8181	54.9
.2	.5764	.7054	1.4176	.8171	.8
.3	.5779	.7080	1.4124	.8161	.7
.4	.5793	.7107	1.4071	.8151	.6
.5	.5807	.7133	1.4019	.8141	.5
.6	.5821	.7159	1.3968	.8131	.4
.7	.5835	.7186	1.3916	.8121	.3
.8	.5850	.7212	1.3865	.8111	.2
.9	.5864	.7239	1.3814	.8100	54.1
36.0	0.5878	0.7265	1.3764	0.8090	54.0
.1	.5892	.7292	1.3713	.8080	53.9
.2	.5906	.7319	1.3663	.8070	.8
.3	.5920	.7346	1.3613	.8059	.7
.4	.5934	.7373	1.3564	.8049	.6
.5	.5948	.7400	1.3514	.8039	.5
.6	.5962	.7427	1.3465	.8028	.4
.7	.5976	.7454	1.3416	.8018	.3
.8	.5990	.7481	1.3367	.8007	.2
.9	.6004	.7508	1.3319	.7997	53.1
37.0	0.6018	0.7536	1.3270	0.7986	53.0
.1	.6032	.7563	1.3222	.7976	52.9
.2	.6046	.7590	1.3175	.7965	.8
.3	.6060	.7618	1.3127	.7955	.7
.4	.6074	.7646	1.3079	.7944	.6
.5	.6088	.7673	1.3032	.7934	.5
.6	.6101	.7701	1.2985	.7923	.4
.7	.6115	.7729	1.2938	.7912	.3
.8	.6129	.7757	1.2892	.7902	.2
.9	.6143	.7785	1.2846	.7891	52.1
38.0	0.6157	0.7813	1.2799	0.7880	52.0
.1	.6170	.7841	1.2753	.7869	51.9
.2	.6184	.7869	1.2708	.7859	.8
.3	.6198	.7898	1.2662	.7848	.7
.4	.6211	.7926	1.2617	.7837	.6
.5	.6225	.7954	1.2572	.7826	.5
.6	.6239	.7983	1.2527	.7815	.4
.7	.6252	.8012	1.2482	.7804	.3
.8	.6266	.8040	1.2437	.7793	.2
.9	.6280	.8069	1.2393	.7782	51.1
39.0	0.6293	0.8098	1.2349	0.7771	51.0
.1	.6307	.8127	1.2305	.7760	50.9
.2	.6320	.8156	1.2261	.7749	.8
.3	.6334	.8185	1.2218	.7738	.7
.4	.6347	.8214	1.2174	.7727	.6
.5	.6361	.8243	1.2131	.7716	.5
.6	.6374	.8273	1.2088	.7705	.4
.7	.6388	.8302	1.2045	.7694	.3
.8	.6401	.8332	1.2002	.7683	.2
.9	.6414	.8361	1.1960	.7672	50.1
40.0	0.6428	0.8391	1.1918	0.7660	50.0
	cos	cot	tan	sin	Deg.

TABLE 1 TRIGONOMETRIC FUNCTIONS—DEGREE MEASURE

Deg.	sin	tan	cot	cos	
40.0	0.6428	0.8391	1.1918	0.7660	50.0
.1	.6441	.8421	1.1875	.7649	49.9
.2	.6455	.8451	1.1833	.7638	.8
.3	.6468	.8481	1.1792	.7627	.7
.4	.6481	.8511	1.1750	.7615	.6
.5	.6494	.8541	1.1708	.7604	.5
.6	.6508	.8571	1.1667	.7593	.4
.7	.6521	.8601	1.1626	.7581	.3
.8	.6534	.8632	1.1585	.7570	.2
.9	.6547	.8662	1.1544	.7559	49.1
41.0	0.6561	0.8693	1.1504	0.7547	49.0
.1	.6574	.8724	1.1463	.7536	48.9
.2	.6587	.8754	1.1423	.7524	.8
.3	.6600	.8785	1.1383	.7513	.7
.4	.6613	.8816	1.1343	.7501	.6
.5	.6626	.8847	1.1303	.7490	.5
.6	.6639	.8878	1.1263	.7478	.4
.7	.6652	.8910	1.1224	.7466	.3
.8	.6665	.8941	1.1184	.7455	.2
.9	.6678	.8972	1.1145	.7443	48.1
42.0	0.6691	0.9004	1.1106	0.7431	48.0
.1	.6704	.9036	1.1067	.7420	47.9
.2	.6717	.9067	1.1028	.7408	.8
.3	.6730	.9099	1.0990	.7396	.7
.4	.6743	.9131	1.0951	.7385	.6
42.5	0.6756	0.9163	1.0913	0.7373	0.5
	cos	cot	tan	sin	Deg.

Deg.	sin	tan	cot	cos	
42.5	0.6756	0.9163	1.0913	0.7373	0.5
.6	.6769	.9195	1.0875	.7361	.4
.7	.6782	.9228	1.0837	.7349	.3
.8	.6794	.9260	1.0799	.7337	.2
.9	.6807	.9293	1.0761	.7325	47.1
43.0	0.6820	0.9325	1.0724	0.7314	47.0
.1	.6833	.9358	1.0686	.7302	46.9
.2	.6845	.9391	1.0649	.7290	.8
.3	.6858	.9424	1.0612	.7278	.7
.4	.6871	.9457	1.0575	.7266	.6
.5	.6884	.9490	1.0538	.7254	.5
.6	.6896	.9523	1.0501	.7242	.4
.7	.6909	.9556	1.0464	.7230	.3
.8	.6921	.9590	1.0428	.7218	.2
.9	.6934	.9623	1.0392	.7206	46.1
44.0	0.6947	0.9657	1.0355	0.7193	46.0
.1	.6959	.9691	1.0319	.7181	45.9
.2	.6972	.9725	1.0283	.7169	.8
.3	.6984	.9759	1.0247	.7157	.7
.4	.6997	.9793	1.0212	.7145	.6
.5	.7009	.9827	1.0176	.7133	.5
.6	.7022	.9861	1.0141	.7120	.4
.7	.7034	.9896	1.0105	.7108	.3
.8	.7046	.9930	1.0070	.7096	.2
.9	.7059	.9965	1.0035	.7083	45.1
45.0	0.7071	1.0000	1.0000	0.7071	45.0
	cos	cot	tan	sin	Deg.

TABLE 2

TRIGONOMETRIC FUNCTIONS—RADIAN MEASURE

x	$\sin x$	$\cos x$	$\tan x$	$\cot x$	$\sec x$	$\csc x$
0.00	0.0000	1.0000	0.0000	—	1.000	—
0.01	0.0100	1.0000	0.0100	99.997	1.000	100.00
0.02	0.0200	0.9998	0.0200	49.993	1.000	50.00
0.03	0.0300	0.9996	0.0300	33.323	1.000	33.34
0.04	0.0400	0.9992	0.0400	24.987	1.001	25.01
0.05	0.0500	0.9988	0.0500	19.983	1.001	20.01
0.06	0.0600	0.9982	0.0601	16.647	1.002	16.68
0.07	0.0699	0.9976	0.0701	14.262	1.002	14.30
0.08	0.0799	0.9968	0.0802	12.473	1.003	12.51
0.09	0.0899	0.9960	0.0902	11.081	1.004	11.13
0.10	0.0998	0.9950	0.1003	9.967	1.005	10.02
0.11	0.1098	0.9940	0.1104	9.054	1.006	9.109
0.12	0.1197	0.9928	0.1206	8.293	1.007	8.353
0.13	0.1296	0.9916	0.1307	7.649	1.009	7.714
0.14	0.1395	0.9902	0.1409	7.096	1.010	7.166
0.15	0.1494	0.9888	0.1511	6.617	1.011	6.692
0.16	0.1593	0.9872	0.1614	6.197	1.013	6.277
0.17	0.1692	0.9856	0.1717	5.826	1.015	5.911
0.18	0.1790	0.9838	0.1820	5.495	1.016	5.586
0.19	0.1889	0.9820	0.1923	5.200	1.018	5.295
0.20	0.1987	0.9801	0.2027	4.933	1.020	5.033
0.21	0.2085	0.9780	0.2131	4.692	1.022	4.797
0.22	0.2182	0.9759	0.2236	4.472	1.025	4.582
0.23	0.2280	0.9737	0.2341	4.271	1.027	4.386
0.24	0.2377	0.9713	0.2447	4.086	1.030	4.207
0.25	0.2474	0.9689	0.2553	3.916	1.032	4.042
0.26	0.2571	0.9664	0.2660	3.759	1.035	3.890
0.27	0.2667	0.9638	0.2768	3.613	1.038	3.749
0.28	0.2764	0.9611	0.2876	3.478	1.041	3.619
0.29	0.2860	0.9582	0.2984	3.351	1.044	3.497
0.30	0.2955	0.9553	0.3093	3.233	1.047	3.384
0.31	0.3051	0.9523	0.3203	3.122	1.050	3.278
0.32	0.3146	0.9492	0.3314	3.018	1.053	3.179
0.33	0.3240	0.9460	0.3425	2.920	1.057	3.086
0.34	0.3335	0.9428	0.3537	2.827	1.061	2.999
0.35	0.3429	0.9394	0.3650	2.740	1.065	2.916
0.36	0.3523	0.9359	0.3764	2.657	1.068	2.839
0.37	0.3616	0.9323	0.3879	2.578	1.073	2.765
0.38	0.3709	0.9287	0.3994	2.504	1.077	2.696
0.39	0.3802	0.9249	0.4111	2.433	1.081	2.630
0.40	0.3894	0.9211	0.4228	2.365	1.086	2.568
0.41	0.3986	0.9171	0.4346	2.301	1.090	2.509
0.42	0.4078	0.9131	0.4466	2.239	1.095	2.452
0.43	0.4169	0.9090	0.4586	2.180	1.100	2.399
0.44	0.4259	0.9048	0.4708	2.124	1.105	2.348

TABLE 2

TRIGONOMETRIC FUNCTIONS—RADIAN MEASURE

x	$\sin x$	$\cos x$	$\tan x$	$\cot x$	$\sec x$	$\csc x$
0.45	0.4350	0.9004	0.4831	2.070	1.111	2.299
0.46	0.4439	0.8961	0.4954	2.018	1.116	2.253
0.47	0.4529	0.8916	0.5080	1.969	1.122	2.208
0.48	0.4618	0.8870	0.5206	1.921	1.127	2.166
0.49	0.4706	0.8823	0.5334	1.875	1.133	2.125
0.50	0.4794	0.8776	0.5463	1.830	1.139	2.086
0.51	0.4882	0.8727	0.5594	1.788	1.146	2.048
0.52	0.4969	0.8678	0.5726	1.747	1.152	2.013
$\pi/6$	0.5000	0.8660	0.5774	1.732	1.155	2.000
0.53	0.5055	0.8628	0.5859	1.707	1.159	1.978
0.54	0.5141	0.8577	0.5994	1.668	1.166	1.945
0.55	0.5227	0.8525	0.6131	1.631	1.173	1.913
0.56	0.5312	0.8473	0.6269	1.595	1.180	1.883
0.57	0.5396	0.8419	0.6410	1.560	1.188	1.853
0.58	0.5480	0.8365	0.6552	1.526	1.196	1.825
0.59	0.5564	0.8309	0.6696	1.494	1.203	1.797
0.60	0.5646	0.8253	0.6841	1.462	1.212	1.771
0.61	0.5729	0.8196	0.6989	1.431	1.220	1.746
0.62	0.5810	0.8139	0.7139	1.401	1.229	1.721
0.63	0.5891	0.8080	0.7291	1.372	1.238	1.697
0.64	0.5972	0.8021	0.7445	1.343	1.247	1.674
0.65	0.6052	0.7961	0.7602	1.315	1.256	1.652
0.66	0.6131	0.7900	0.7761	1.288	1.266	1.631
0.67	0.6210	0.7838	0.7923	1.262	1.276	1.610
0.68	0.6288	0.7776	0.8087	1.237	1.286	1.590
0.69	0.6365	0.7712	0.8253	1.212	1.297	1.571
0.70	0.6442	0.7648	0.8423	1.187	1.307	1.552
0.71	0.6518	0.7584	0.8595	1.163	1.319	1.534
0.72	0.6594	0.7518	0.8771	1.140	1.330	1.517
0.73	0.6669	0.7452	0.8949	1.117	1.342	1.500
0.74	0.6743	0.7385	0.9131	1.095	1.354	1.483
0.75	0.6816	0.7317	0.9316	1.073	1.367	1.467
0.76	0.6889	0.7248	0.9505	1.052	1.380	1.452
0.77	0.6961	0.7179	0.9697	1.031	1.393	1.437
0.78	0.7033	0.7109	0.9893	1.011	1.407	1.422
$\pi/4$	0.7071	0.7071	1.000	1.000	1.414	1.414
0.79	0.7104	0.7038	1.009	0.9908	1.421	1.408
0.80	0.7174	0.6967	1.030	0.9712	1.435	1.394
0.81	0.7243	0.6895	1.050	0.9520	1.450	1.381
0.82	0.7311	0.6822	1.072	0.9331	1.466	1.368
0.83	0.7379	0.6749	1.093	0.9146	1.482	1.355
0.84	0.7446	0.6675	1.116	0.8964	1.498	1.343
0.85	0.7513	0.6600	1.138	0.8785	1.515	1.331
0.86	0.7578	0.6524	1.162	0.8609	1.533	1.320
0.87	0.7643	0.6448	1.185	0.8437	1.551	1.308
0.88	0.7707	0.6372	1.210	0.8267	1.569	1.297
0.89	0.7771	0.6294	1.235	0.8100	1.589	1.287

x	$\sin x$	$\cos x$	$\tan x$	$\cot x$	$\sec x$	$\csc x$
0.90	0.7833	0.6216	1.260	0.7936	1.609	1.277
0.91	0.7895	0.6137	1.286	0.7774	1.629	1.267
0.92	0.7956	0.6058	1.313	0.7615	1.651	1.257
0.93	0.8016	0.5978	1.341	0.7458	1.673	1.247
0.94	0.8076	0.5898	1.369	0.7303	1.696	1.238
0.95	0.8134	0.5817	1.398	0.7151	1.719	1.229
0.96	0.8192	0.5735	1.428	0.7001	1.744	1.221
0.97	0.8249	0.5653	1.459	0.6853	1.769	1.212
0.98	0.8305	0.5570	1.491	0.6707	1.795	1.204
0.99	0.8360	0.5487	1.524	0.6563	1.823	1.196
1.00	0.8415	0.5403	1.557	0.6421	1.851	1.188
1.01	0.8468	0.5319	1.592	0.6281	1.880	1.181
1.02	0.8521	0.5234	1.628	0.6142	1.911	1.174
1.03	0.8573	0.5148	1.665	0.6005	1.942	1.166
1.04	0.8624	0.5062	1.704	0.5870	1.975	1.160
$\pi/3$	0.8660	0.5000	1.732	0.5774	2.000	1.155
1.05	0.8674	0.4976	1.743	0.5736	2.010	1.153
1.06	0.8724	0.4889	1.784	0.5604	2.046	1.146
1.07	0.8772	0.4801	1.827	0.5473	2.083	1.140
1.08	0.8820	0.4713	1.871	0.5344	2.122	1.134
1.09	0.8866	0.4625	1.917	0.5216	2.162	1.128
1.10	0.8912	0.4536	1.965	0.5090	2.205	1.122
1.11	0.8957	0.4447	2.014	0.4964	2.249	1.116
1.12	0.9001	0.4357	2.066	0.4840	2.295	1.111
1.13	0.9044	0.4267	2.120	0.4718	2.344	1.106
1.14	0.9086	0.4176	2.176	0.4596	2.395	1.101
1.15	0.9128	0.4085	2.234	0.4475	2.448	1.096
1.16	0.9168	0.3993	2.296	0.4356	2.504	1.091
1.17	0.9208	0.3902	2.360	0.4237	2.563	1.086
1.18	0.9246	0.3809	2.427	0.4120	2.625	1.082
1.19	0.9284	0.3717	2.498	0.4003	2.691	1.077
1.20	0.9320	0.3624	2.572	0.3888	2.760	1.073
1.21	0.9356	0.3530	2.650	0.3773	2.833	1.069
1.22	0.9391	0.3436	2.733	0.3659	2.910	1.065
1.23	0.9425	0.3342	2.820	0.3546	2.992	1.061
1.24	0.9458	0.3248	2.912	0.3434	3.079	1.057
1.25	0.9490	0.3153	3.010	0.3323	3.171	1.054
1.26	0.9521	0.3058	3.113	0.3212	3.270	1.050
1.27	0.9551	0.2963	3.224	0.3102	3.375	1.047
1.28	0.9580	0.2867	3.341	0.2993	3.488	1.044
1.29	0.9608	0.2771	3.467	0.2884	3.609	1.041
1.30	0.9636	0.2675	3.602	0.2776	3.738	1.038
1.31	0.9662	0.2579	3.747	0.2669	3.878	1.035
1.32	0.9687	0.2482	3.903	0.2562	4.029	1.032
1.33	0.9711	0.2385	4.072	0.2456	4.193	1.030
1.34	0.9735	0.2288	4.256	0.2350	4.372	1.027

TABLE 2

TRIGONOMETRIC FUNCTIONS—RADIAN MEASURE

x	$\sin x$	$\cos x$	$\tan x$	$\cot x$	$\sec x$	$\csc x$
1.35	0.9757	0.2190	4.455	0.2245	4.566	1.025
1.36	0.9779	0.2092	4.673	0.2140	4.779	1.023
1.37	0.9799	0.1994	4.913	0.2035	5.014	1.021
1.38	0.9819	0.1896	5.177	0.1931	5.273	1.018
1.39	0.9837	0.1798	5.471	0.1828	5.561	1.017
1.40	0.9854	0.1700	5.798	0.1725	5.883	1.015
1.41	0.9871	0.1601	6.165	0.1622	6.246	1.013
1.42	0.9887	0.1502	6.581	0.1519	6.657	1.011
1.43	0.9901	0.1403	7.055	0.1417	7.126	1.010
1.44	0.9915	0.1304	7.602	0.1315	7.667	1.009
1.45	0.9927	0.1205	8.238	0.1214	8.299	1.007
1.46	0.9939	0.1106	8.989	0.1113	9.044	1.006
1.47	0.9949	0.1006	9.887	0.1011	9.938	1.005
1.48	0.9959	0.0907	10.983	0.0910	11.029	1.004
1.49	0.9967	0.0807	12.350	0.0810	12.390	1.003
1.50	0.9975	0.0707	14.101	0.0709	14.137	1.003
1.51	0.9982	0.0608	16.428	0.0609	16.458	1.002
1.52	0.9987	0.0508	19.670	0.0508	19.695	1.001
1.53	0.9992	0.0408	24.498	0.0408	24.519	1.001
1.54	0.9995	0.0308	32.461	0.0308	32.476	1.000
1.55	0.9998	0.0208	48.078	0.0208	48.089	1.000
1.56	0.9999	0.0108	92.620	0.0108	92.626	1.000
1.57	1.0000	0.0008	1255.8	0.0008	1255.8	1.000
$\pi/2$	1.0000	0.0000	—	0.0000	—	1.000

Answers to Selected Problems

Answers to Selected Problems

Chapter 1

Problem Set 1, page 8

1. False **3.** True **5.** True **7.** False **9.** True **11.** False **13.** False **15.** False **17.** True
19. False **21.** $x = -3, x = 1$ **23.** $x = 0, x = 7$ **25.** $x = -3, x = 7$ **27.** $(x + 3)^2$ **29.** $(x - \frac{5}{2})^2$
31. $(x + \frac{3}{8})^2$ **33.** $x = 2, x = -\frac{3}{5}$ **35.** $x = \dfrac{-5 \pm \sqrt{13}}{6}$ **37.** $x = \dfrac{1 \pm \sqrt{17}}{4}$ **39.** $y = \dfrac{3 \pm \sqrt{57}}{8}$
41. $x = 1$ **43.** $x = 2, x = 3$ **45.** No solution

Problem Set 2, page 16

1.

3.

5.

7.

9.
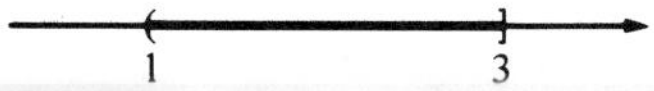

11.
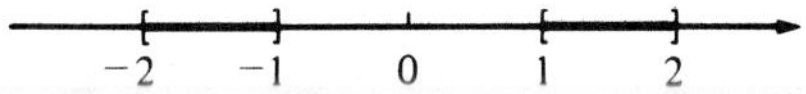

13.

15. Addition property **17.** Multiplication property **19.** Division property **21.** Order-reversing property
23. (a) $x + 4 > 1$ (b) $x - 4 > -7$ (c) $5x > -15$ (d) $-5x < 15$ **25.** $x < 5$ or $(-\infty, 5)$ **27.** $x \geq -5$ or $[-5, \infty)$
29. $1 \leq x < 2$ or $[1, 2)$ **31.** $53 \leq x < 93$ **33.** 2 **35.** 2 **37.** 1 **39.** 8 **41.** 32 **43.** 8 **45.** 0
49. $x = 5$ and $y = -2$

Problem Set 3, page 20

1. 1.55×10^4 **3.** 5.87×10^{-4} **5.** 1.86×10^{11} **7.** 9.01×10^{-7} **9.** 33,300 **11.** 0.00004102
13. 10,010,000 **15.** 6.2×10^{-2} **17.** 9.29×10^7; 1.92×10^{13} **19.** 2.51×10^{-10} **21.** 4.8×10^8 **23.** 7×10^{31}
25. 6.56×10^{-4} **27.** 2 **29.** 5 **31.** 5300 **33.** 0.015 **35.** 110,000 **37.** 2.14×10^{-13}
39. (a) 1.327×10^{12} (b) 9.6×10^3

Problem Set 4, page 29

7. 5 **9.** 0.75 **11.** 31.98 **13.** Yes **15.** No **17.** 10 **19.** 4 **21.** $\sqrt{145}$ **23.** 5 **25.** 6.224
27. (b) 4, 6, $\sqrt{52}$ (d) 12 **29.** (b) $\sqrt{18}, \sqrt{26}, \sqrt{8}$ (d) 6 **31.** Yes **33.** 13
35. (a) $x^2 + y^2 = 16$ (b) $(x + 1)^2 + (y - 3)^2 = 4$

37.

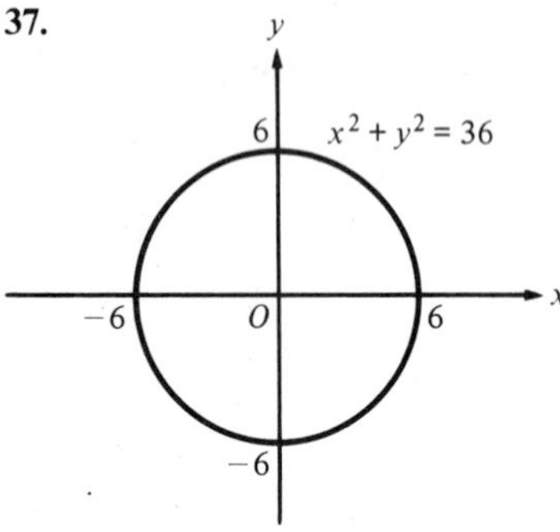

39.

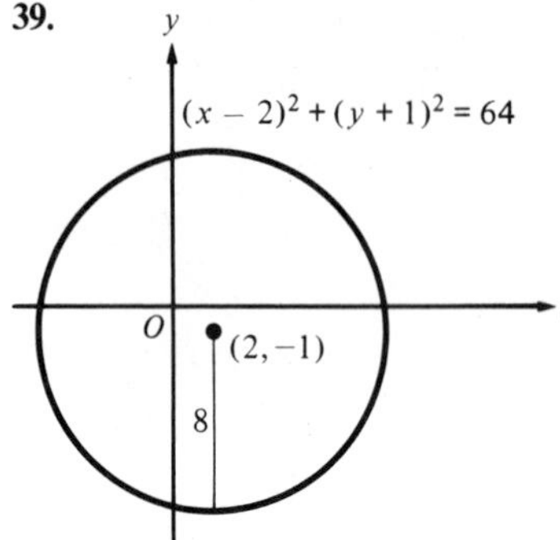

41.

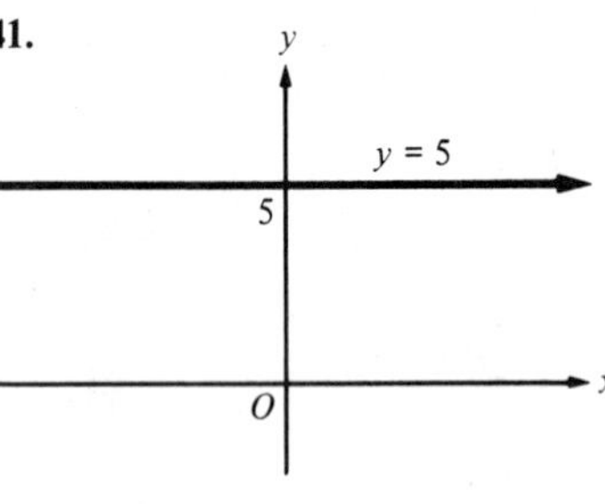

43.

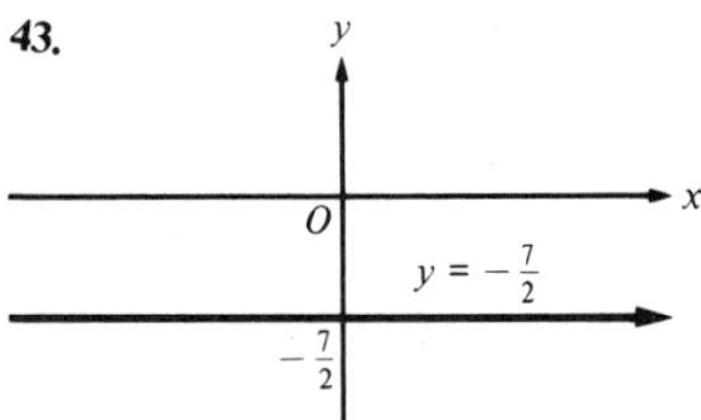

45.

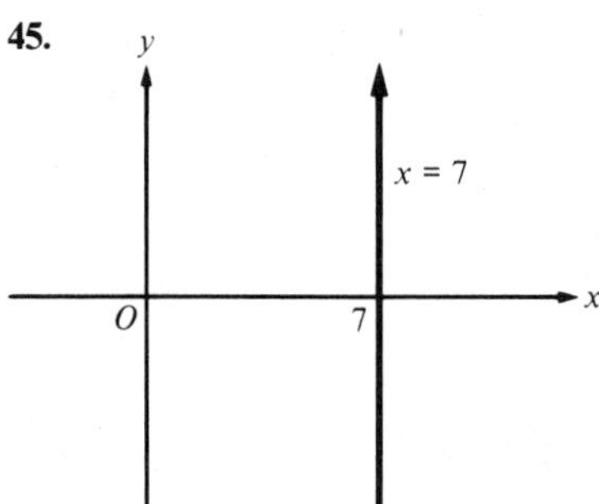

47.

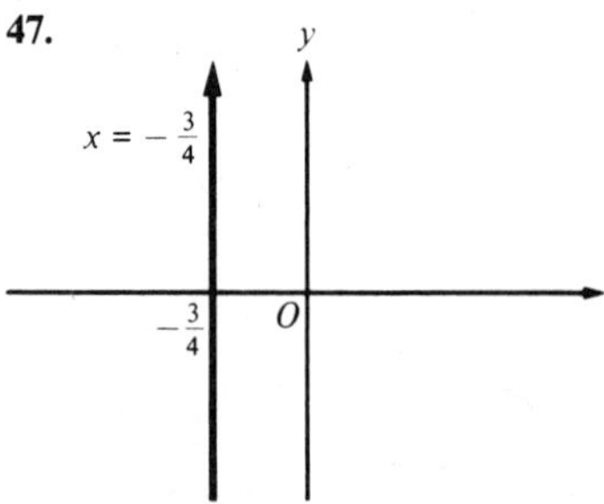

49.

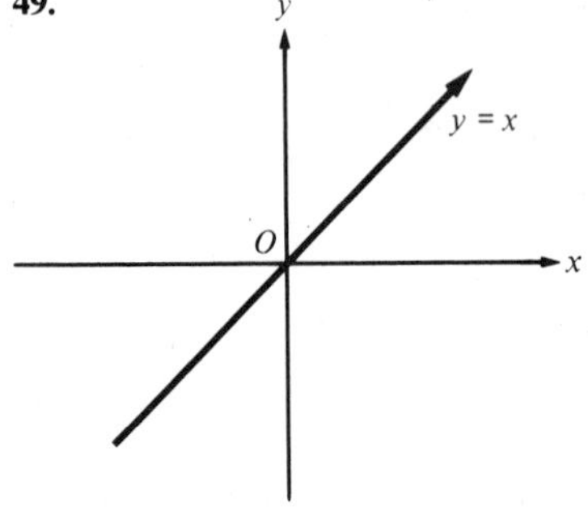

Problem Set 5, page 38

1. -5 **3.** 2 **5.** $\frac{7}{24}$ **7.** 0 **9.** 1 **11.** $x^2 + 3x - 4$ **13.** a **15.** $\dfrac{x + 2}{3x + 7}$ **17.** $\sqrt{3x^4 + 5}$ **19.** $x + 1$
21. x **23.** $(a - b)(a + b - 3)$ **25.** $\sqrt{3x + 5} - \sqrt{5}$ **27.** 2 **29.** $\mathbb{R}$ **31.** $[0, \infty)$ **33.** $[-3, 3]$
35. $(0, \infty)$ **37.** All real numbers except 0 **39.** (a) $\mathbb{R}$ (b) Odd (d) Always increasing (e) $\mathbb{R}$
41. (a) $\mathbb{R}$ (b) Odd (d) Always increasing (e) $\mathbb{R}$ **43.** (a) $\mathbb{R}$ (b) Even (d) Constant (e) $\{2\}$
45. (a) $[0, \infty)$ (b) Neither even nor odd (d) Increasing on $[0, \infty)$ (e) $[0, \infty)$

47. (a) Domain $= \mathbb{R}$; range $= \mathbb{R}$; increasing on $(-\infty, -2]$ and $[0, 2]$; decreasing on $[-2, 0]$ and $[2, \infty)$; even
(b) Domain $= [-5, 5]$; range $= [-3, 3]$; increasing on $[-1, 1]$ and $[3, 5]$; decreasing on $[-5, -1]$ and $[1, 3]$; neither
(c) Domain $= \left[-\frac{3\pi}{2}, \frac{3\pi}{2}\right]$; range $= [-1, 1]$; increasing on $\left[-\frac{3\pi}{2}, -\frac{\pi}{2}\right]$ and $\left[\frac{\pi}{2}, \frac{3\pi}{2}\right]$; decreasing on $\left[-\frac{\pi}{2}, \frac{\pi}{2}\right]$; odd
(d) Domain $= \mathbb{R}$; range $= [-2, 1]$; increasing on $[0, \pi]$; constant on $(-\infty, 0]$ and $[\pi, \infty)$; neither odd nor even
49. Even, symmetric about the y axis **51.** Neither even nor odd **53.** Odd, symmetric about the origin
55. (a) Yes (b) No (c) Yes (d) No

Problem Set 6, page 49

1.

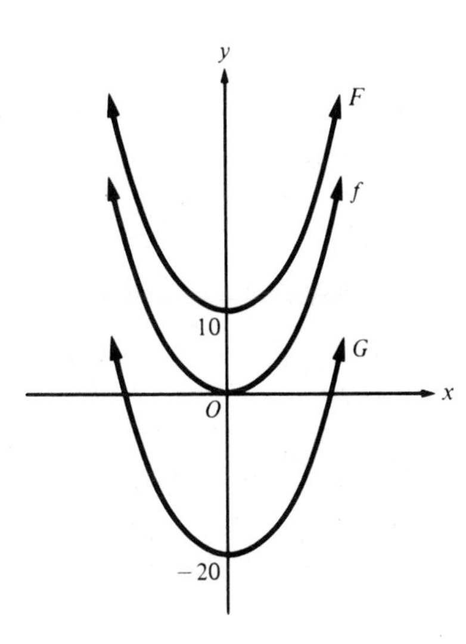

3.

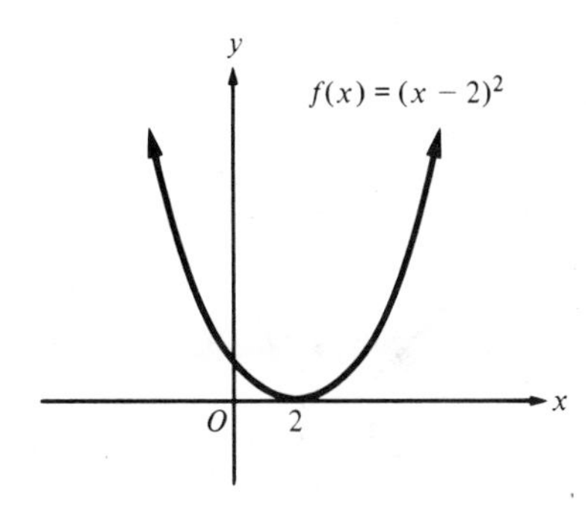

5.

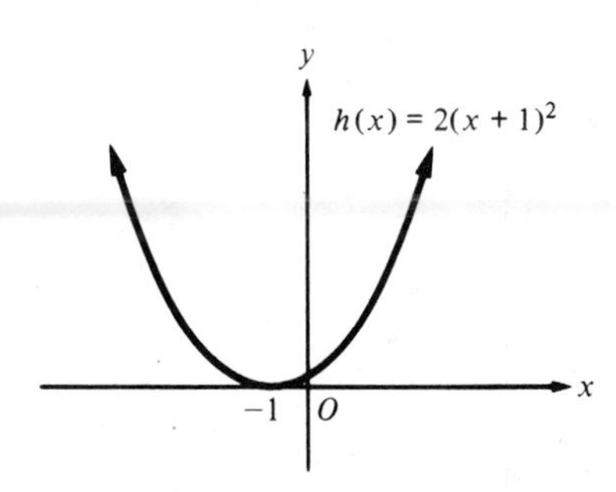

7.

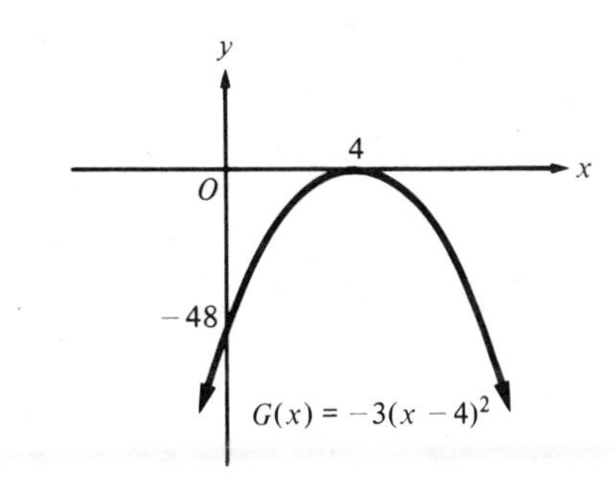

9.

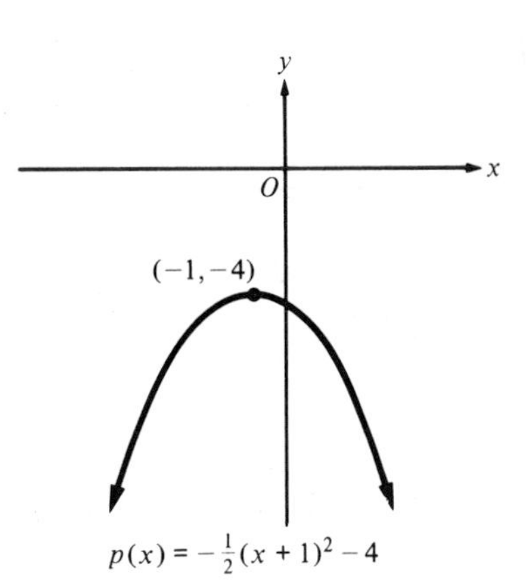

11.

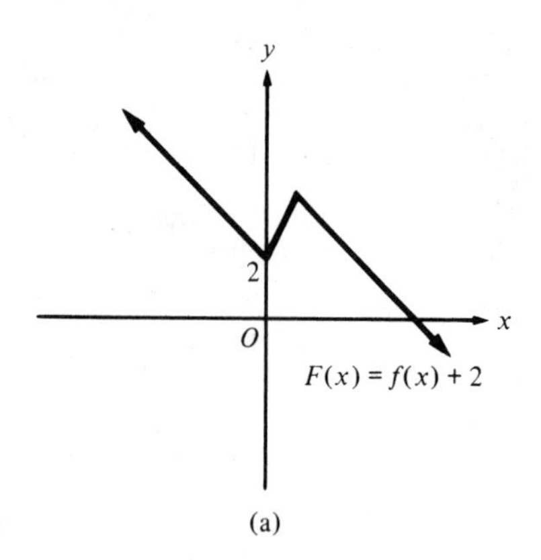

y
2
O
x
F(x) = f(x) + 2

(a)

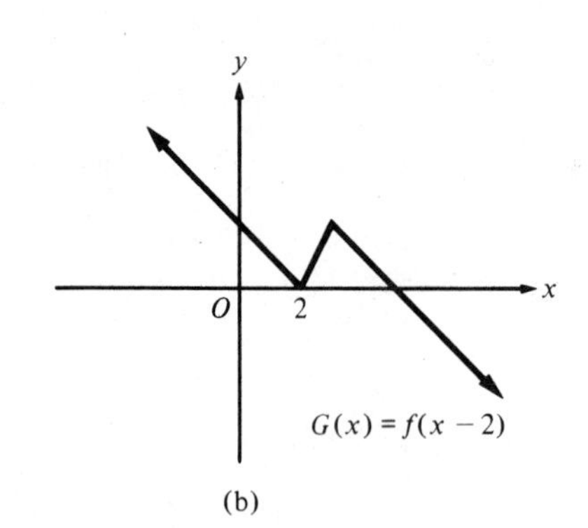

y
O
2
x
G(x) = f(x − 2)

(b)

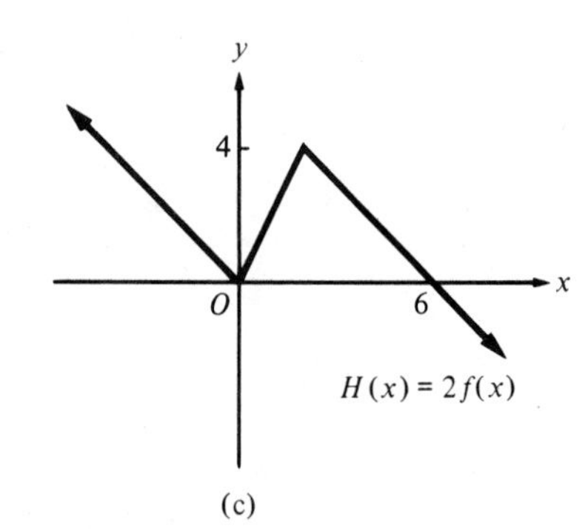

y
4
O
6
x
H(x) = 2f(x)

(c)

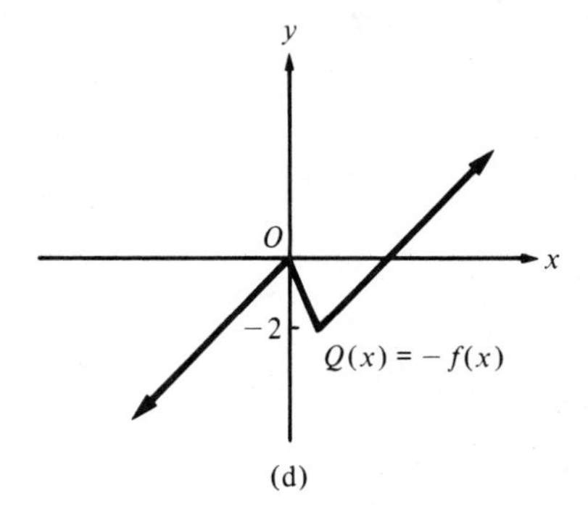

y
O
−2
x
Q(x) = − f(x)

(d)

13.

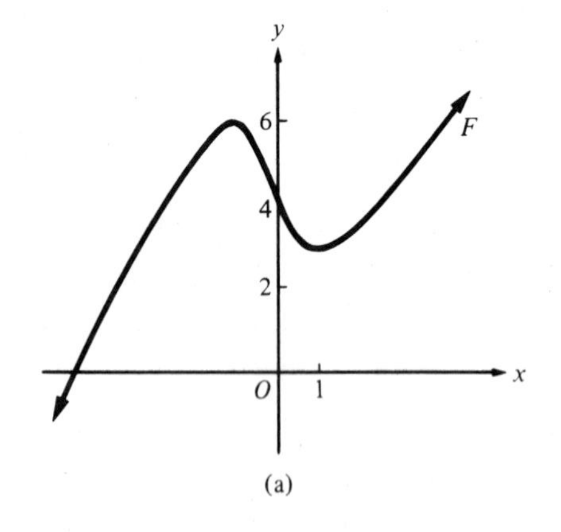

y
6
4
2
O
1
x
F

(a)

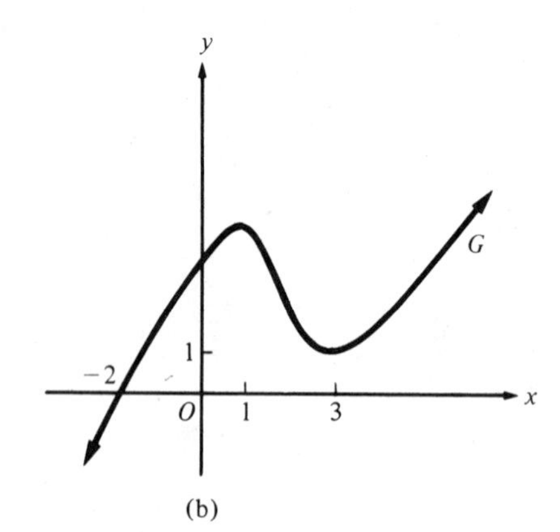

y
1
−2
O
1
3
x
G

(b)

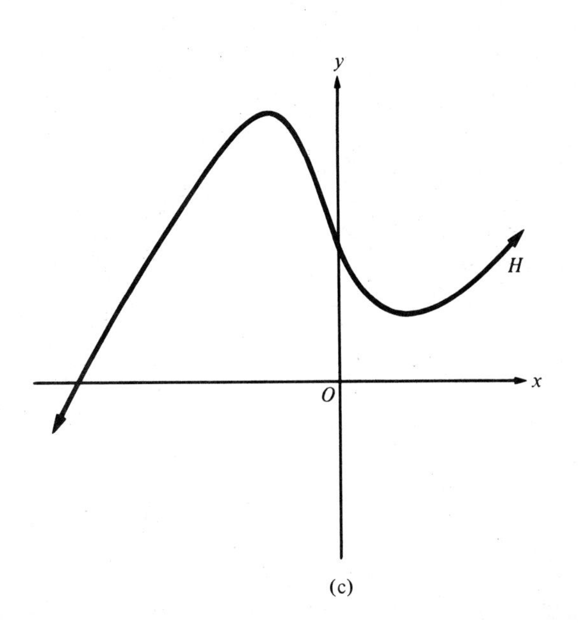

(c)

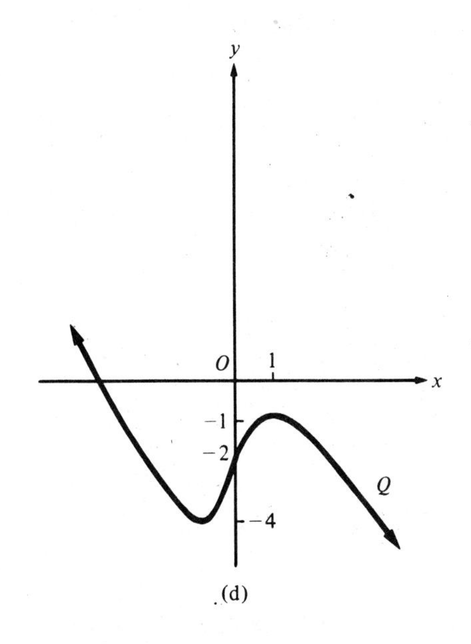

(d)

15.

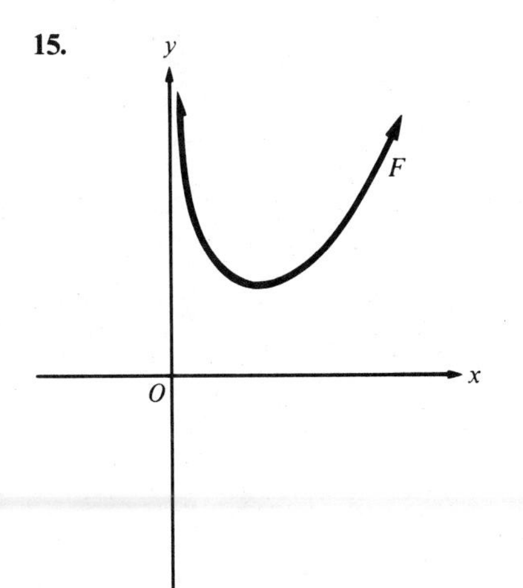

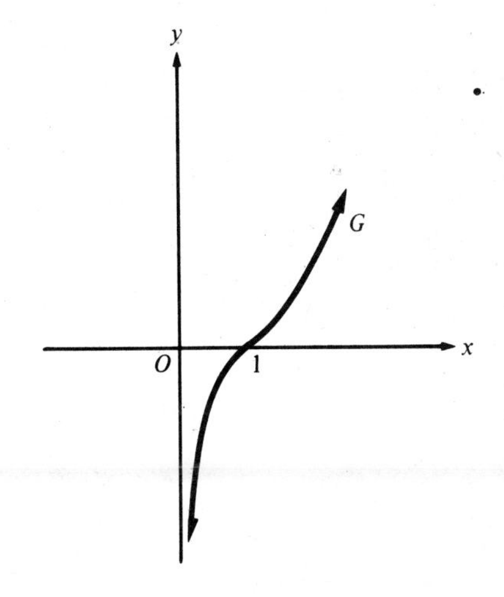

17.

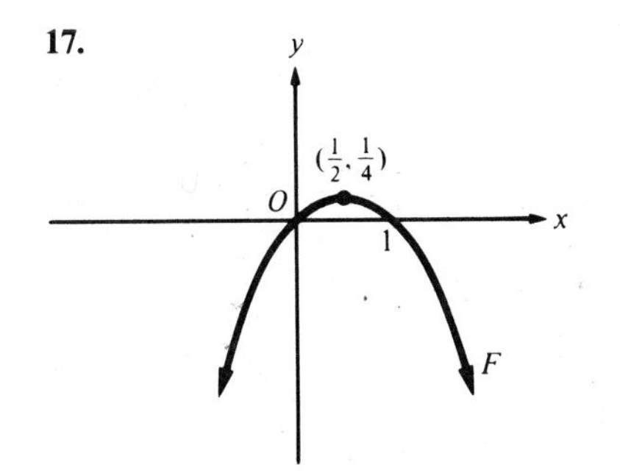

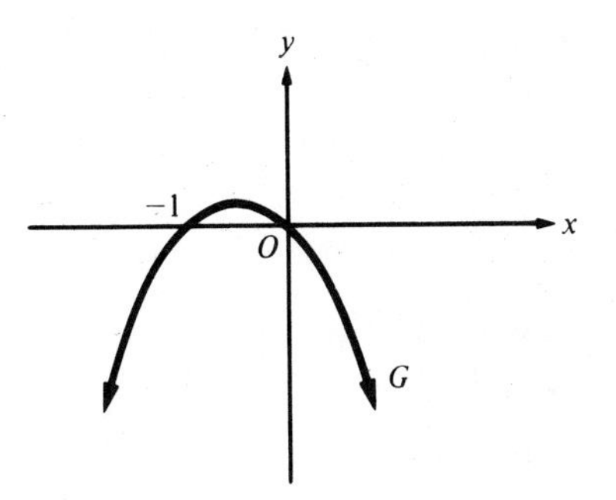

19.

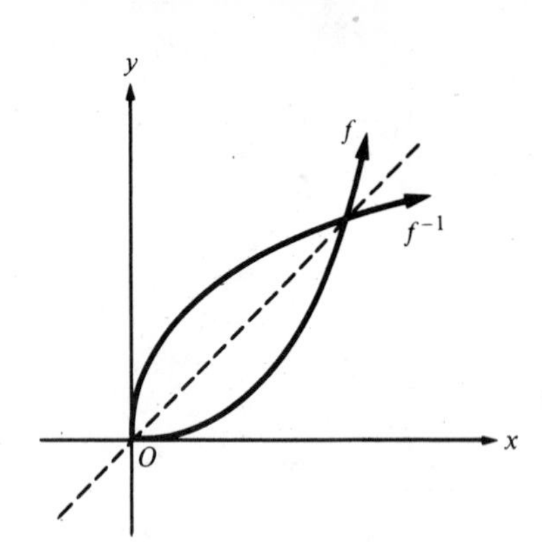

21.

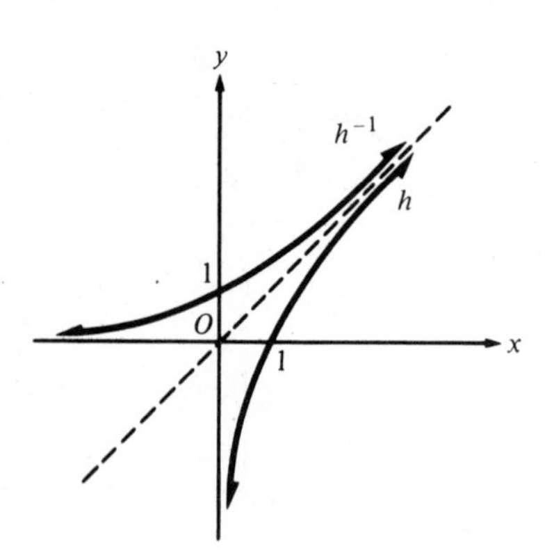

23. Not invertible

25. Invertible

Review Problem Set, page 51

1. False **3.** False **5.** False **7.** True **9.** False **11.** False **13.** True **15.** $x = 4, x = 3$

17. $x = -5, x = 21$ **19.** $y = \frac{7}{2}, y = -\frac{5}{2}$ **21.** $(x + \frac{9}{2})^2$ **23.** $(x - \frac{1}{4})^2$ **25.** $x = \frac{7}{2}, x = -1$ **27.** $x = \dfrac{7 \pm \sqrt{89}}{10}$

29. $r = \dfrac{1 \pm \sqrt{17}}{4}$ **31.** $x = 6$ **33.** $x = \frac{1}{3}$

35. (a) -1 to 3 (b) -2 to 5

(c) -7 (d) ••• 4

(e) $-\frac{1}{3}$ to 5 (f) $-\frac{5}{2}$ to $\frac{3}{2}$

(g) $-\frac{2}{3}$ to $\frac{1}{3}$ (h) ••• $\sqrt{2}$

37.

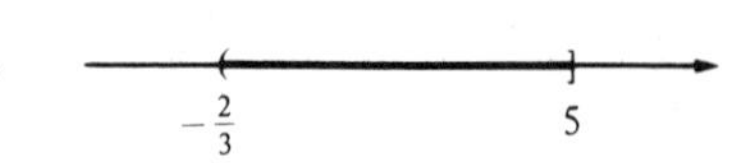

39. Transitivity **41.** Trichotomy **43.** 7 **45.** $\frac{11}{6}$ **47.** True for all values of x
49. True only for non-negative numbers **51.** 5.712×10^{10} **53.** 7.14×10^{-7} **55.** 17,320,000 **57.** 0.0000000312
59. 2 **61.** 7 **63.** 17,000 **65.** 7.23×10^5 **69.** 3.78 **71.** 5 **73.** 13 **75.** $\sqrt{41} + \sqrt{80} + 15$
77. $(x + 5)^2 + (y - 2)^2 = 36$

79.

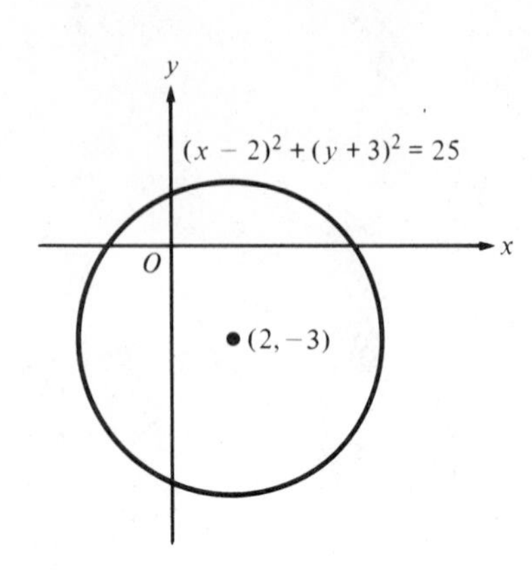

81.

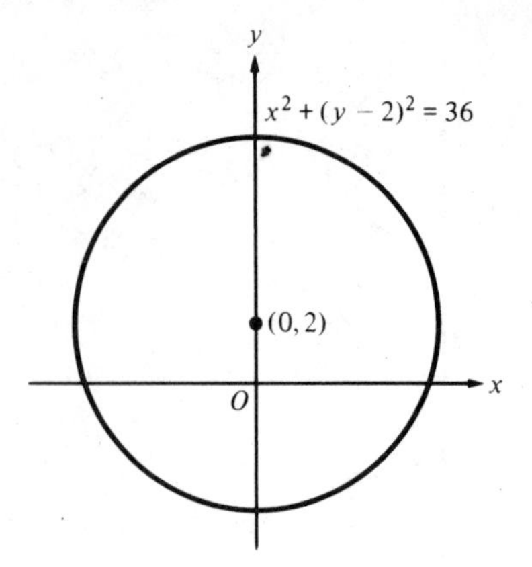

83.

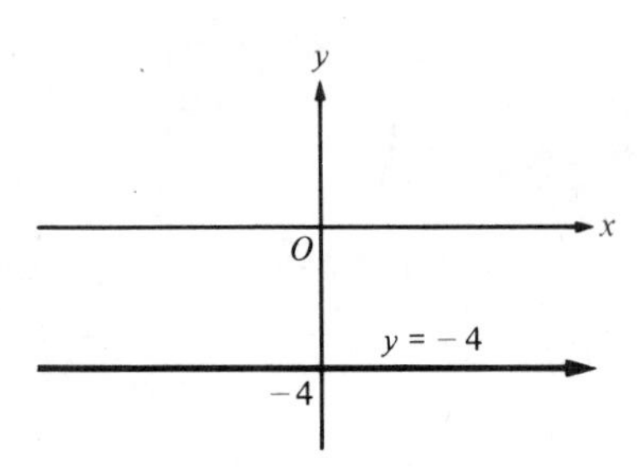

85. 23 **87.** 0 **89.** -3 **91.** $\dfrac{19+6k}{4+k}$ **93.** All real numbers except 1 **95.** $[-1, \infty)$

97 (a) Domain $= \mathbb{R}$ (b) Even (d) Increasing on $[0, \infty)$ and decreasing on $(-\infty, 0]$ (e) Range $= [0, \infty)$
99. (a) Domain = all real numbers except 0 (b) Odd (d) Decreasing on the domain (e) Range = all real numbers except 0
101. (a) Domain $= \mathbb{R}$ (b) Even (d) Constant (e) Range $= \{-16\}$
103. (a) Domain $= \mathbb{R}$; range $= [-1, 1]$; the graph is constant on $(-\infty, -2]$ and $[2, \infty)$, decreasing on $[-2, 0]$, and increasing on $(0, 2)$; even **105.** (a) Odd; symmetric about the origin (b) Even; symmetric about the y axis (c) Neither even nor odd (d) Odd; symmetric about the origin (e) Even; symmetric about the y axis (f) Neither even nor odd; no symmetry
107. (a) Yes (b) No (c) Yes (d) No

109.

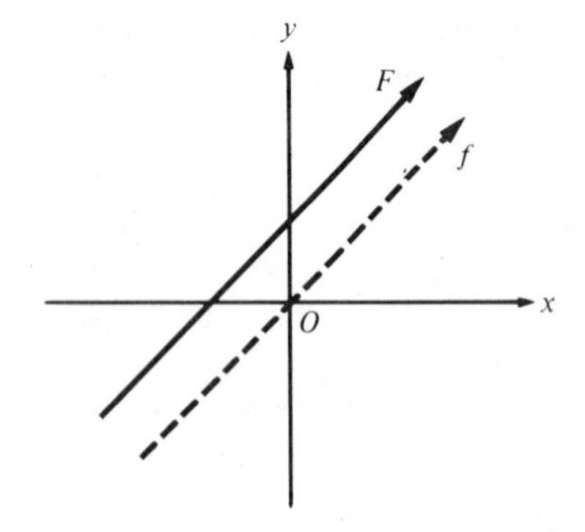

111.

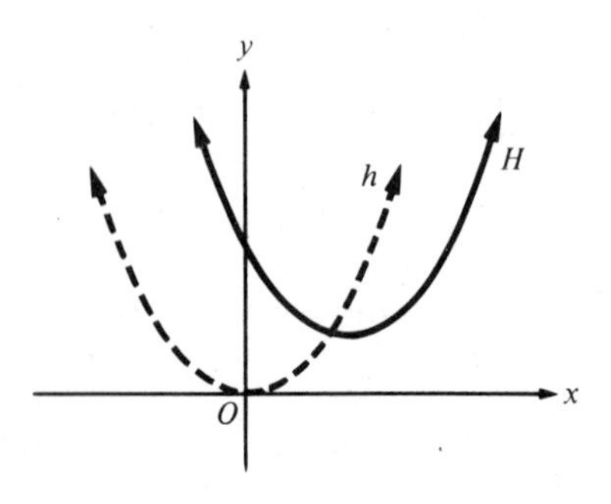

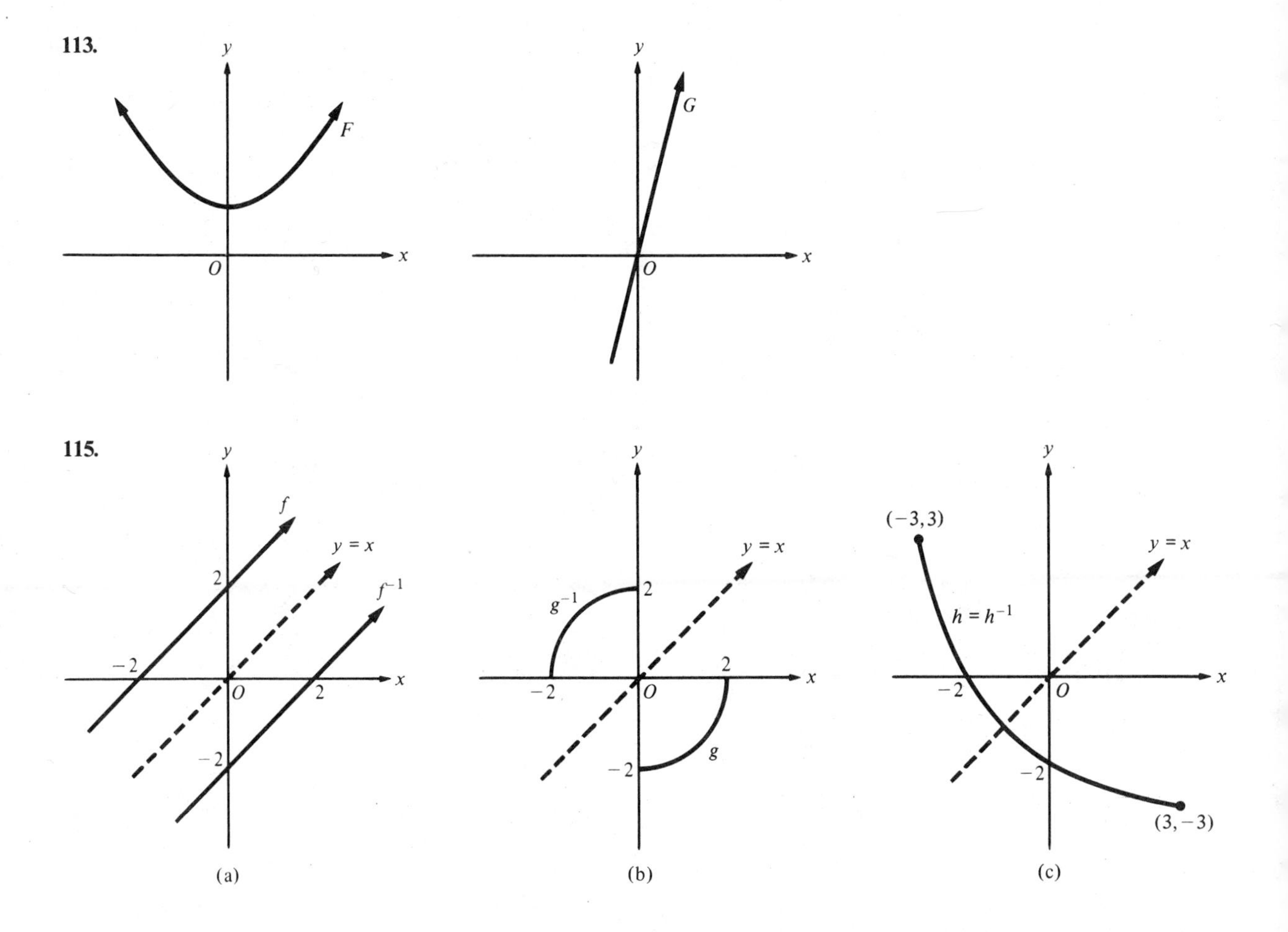

Chapter 2

Problem Set 1, page 63

1. 60° **3.** -792° **5.** $\frac{1}{8}$ counterclockwise **7.** $\frac{1}{3}$ counterclockwise **9.** $\frac{11}{12}$ clockwise **11.** $\frac{1}{30}$ clockwise **13.** $0.0\overline{3}^\circ$ **15.** $0.0008\overline{3}^\circ$ **17.** $100.50\overline{5}^\circ$ **19.** $62^\circ 15'$ **21.** $-0^\circ 7' 30''$ **23.** $21^\circ 9' 36''$ **25.** 3.3 meters **27.** $\frac{4}{3}$ radians **29.** $10\pi/3$ inches **31.** (a) $\pi/6$ (b) $\pi/4$ (c) $\pi/2$ (d) $2\pi/3$ (e) $-5\pi/6$ (f) $26\pi/9$ (g) $\frac{2}{5}\pi$ (h) $3\pi/8$ (i) $-11\pi/6$ (j) $5\pi/2$ (k) $7\pi/60$ (l) -2π **33.** (a) 90° (b) 60° (c) 45° (d) 30° (e) 120° (f) -180° (g) 108° (h) -450° (i) 405° (j) -67.5° (k) 1260° (l) $-(\frac{90}{7})^\circ$ **35.** (a) $-3\pi/4$, -135° (b) $25\pi/3$, 1500° (c) $2\pi/3$, 120° **37.** $10\pi/3$ meters **39.** $11{,}057^\circ$
41. (a) $4\pi/3$ radians per hour (b) $10{,}240\pi \approx 32{,}169.91$ kilometers per hour **43.** $760{,}320/13 \approx 58{,}486.15$ radians per hour
45. $146\pi/9 \approx 50.96$ centimeters per second **47.** $1{,}550{,}000\pi/73 \approx 66{,}705$ miles per hour
49. (a) $21\pi/4$ square centimeters (b) $117\pi/2$ square inches

Problem Set 2, page 71

1. $\sin\theta = \frac{3}{5}$; $\cos\theta = \frac{4}{5}$; $\tan\theta = \frac{3}{4}$; $\cot\theta = \frac{4}{3}$; $\sec\theta = \frac{5}{4}$; $\csc\theta = \frac{5}{3}$
3. $\sin\theta = \frac{3}{4}$; $\cos\theta = \sqrt{7}/4$; $\tan\theta = 3\sqrt{7}/7$, $\cot\theta = \sqrt{7}/3$; $\sec\theta = 4\sqrt{7}/7$; $\csc\theta = \frac{4}{3}$
5. $\sin\theta = \frac{1}{2}$; $\cos\theta = \sqrt{3}/2$; $\tan\theta = \sqrt{3}/3$; $\cot\theta = \sqrt{3}$; $\sec\theta = 2\sqrt{3}/3$; $\csc\theta = 2$
7. $\sin\theta = \sqrt{2}/2$; $\cos\theta = \sqrt{2}/2$; $\tan\theta = 1$; $\cot\theta = 1$; $\sec\theta = \sqrt{2}$; $\csc\theta = \sqrt{2}$ **9.** $55°$ **11.** $3\pi/10$ **13.** $\pi/12$
15. $67.66°$ **17.** $\cos 59°$ **19.** $\cot(7\pi/18)$ **21.** $\tan(\pi/10)$ **23.** $\sec 12.97°$
25. $\sin 48° \approx 0.7431448255$ $\csc 48° \approx 1.345632730$
$\cos 48° \approx 0.6691306064$ $\sec 48° \approx 1.494476550$
$\tan 48° \approx 1.110612515$ $\cot 48° \approx 0.900404044$
27. $\sin 23°12'33'' \approx 0.3940889557$ $\csc 23°12'33'' \approx 2.537498160$
$\cos 23°12'33'' \approx 0.9190723017$ $\sec 23°12'33'' \approx 1.088053680$
$\tan 23°12'33'' \approx 0.4287899387$ $\cot 23°12'33'' \approx 2.332144273$
29. $\sin 16.19° \approx 0.2788234989$ $\csc 16.19° \approx 3.586498283$
$\cos 16.19° \approx 0.9603423642$ $\sec 16.19° \approx 1.041295310$
$\tan 16.19° \approx 0.2903376018$ $\cot 16.19° \approx 3.444266240$
31. $\sin(2\pi/7) \approx 0.7818314825$ $\csc(2\pi/7) \approx 1.279048008$
$\cos(2\pi/7) \approx 0.6234898018$ $\sec(2\pi/7) \approx 1.603875472$
$\tan(2\pi/7) \approx 1.253960338$ $\cot(2\pi/7) \approx 0.7974733887$
33. $\sin 0.7764 \approx 0.7007155788$ $\csc 0.7764 \approx 1.427112555$
$\cos 0.7764 \approx 0.7134407317$ $\sec 0.7764 \approx 1.401658127$
$\tan 0.7764 \approx 0.9821636859$ $\cot 0.7764 \approx 0.1018160226$
39. Evaluate sin 30 on the calculator; if result is 0.50, then 30 must be 30 degrees; and if result is not 0.50, then calculator must be in radian mode.
43. $\cot\theta = \dfrac{\text{adj}}{\text{opp}} = \dfrac{1}{\text{opp/adj}} = \dfrac{1}{\tan\theta}$ **45.** $\text{adj}^2 + \text{opp}^2 = \text{hyp}^2$, so $\left(\dfrac{\text{adj}}{\text{hyp}}\right)^2 + \left(\dfrac{\text{opp}}{\text{hyp}}\right)^2 = 1$; thus, $(\cos\theta)^2 + (\sin\theta)^2 = 1$

Problem Set 3, page 80

1. Quadrant I **3.** Quadrantal **5.** Quadrant IV **7.** Quadrant II **9.** Quadrant III **11.** $420°$, $-300°$, $780°$
13. $7\pi/4$, $-9\pi/4$, $-17\pi/4$ **15.** $-252°$, $-972°$, $108°$ **17.** $17\pi/6$, $-7\pi/6$, $-19\pi/6$
19. $\sin\theta = \frac{3}{5}$; $\cos\theta = \frac{4}{5}$; $\tan\theta = \frac{3}{4}$; $\cot\theta = \frac{4}{3}$; $\sec\theta = \frac{5}{4}$; $\csc\theta = \frac{5}{3}$
21. $\sin\theta = \frac{12}{13}$; $\cos\theta = -\frac{5}{13}$; $\tan\theta = -\frac{12}{5}$; $\cot\theta = -\frac{5}{12}$; $\sec\theta = -\frac{13}{5}$; $\csc\theta = \frac{13}{12}$
23. $\sin\theta = -\frac{4}{5}$; $\cos\theta = -\frac{3}{5}$; $\tan\theta = \frac{4}{3}$; $\cot\theta = \frac{3}{4}$; $\sec\theta = -\frac{5}{3}$; $\csc\theta = -\frac{5}{4}$
25. $\sin\theta = 3\sqrt{58}/58$; $\cos\theta = 7\sqrt{58}/58$; $\tan\theta = \frac{3}{7}$; $\cot\theta = \frac{7}{3}$; $\sec\theta = \sqrt{58}/7$; $\csc\theta = \sqrt{58}/3$
27. $\sin\theta = -\sqrt{10}/5$; $\cos\theta = -\sqrt{15}/5$; $\tan\theta = \sqrt{6}/3$; $\cot\theta = \sqrt{6}/2$; $\sec\theta = -\sqrt{15}/3$; $\csc\theta = -\sqrt{10}/2$
29. No, since $\sin^2\theta + \cos^2\theta = 1$ implies that $\sin^2\theta \le 1$; that is, $-1 \le \sin\theta \le 1$. **31.** (a) I (b) II (c) II (d) IV (e) III (f) III (g) IV (h) I **33.** (a) Negative (b) Negative (c) Positive (d) Negative (e) Positive (f) Negative (g) Negative
35. $\tan\theta = \dfrac{y}{x} = \dfrac{y/r}{x/r} = \dfrac{\sin\theta}{\cos\theta}$ **37.** (a) $\sec\theta = \dfrac{13}{12}$ (b) $\csc\theta = -\dfrac{13}{5}$ (c) $\tan\theta = -\dfrac{5}{12}$ (d) $\cot\theta = -\dfrac{12}{5}$
39. $\cos\theta = \frac{3}{5}$; $\tan\theta = \frac{4}{3}$; $\cot\theta = \frac{3}{4}$; $\sec\theta = \frac{5}{3}$; $\csc\theta = \frac{5}{4}$
41. $\cos\theta = -\sqrt{7}/4$; $\tan\theta = 3\sqrt{7}/7$; $\cot\theta = \sqrt{7}/3$; $\sec\theta = -4\sqrt{7}/7$; $\csc\theta = -\frac{4}{3}$
43. $\sin\theta = -\sqrt{33}/7$; $\tan\theta = -\sqrt{33}/4$; $\cot\theta = -4\sqrt{33}/33$; $\sec\theta = \frac{7}{4}$; $\csc\theta = -7\sqrt{33}/33$
45. $\sin\theta = \frac{2}{3}$; $\cos\theta = \sqrt{5}/3$; $\tan\theta = 2\sqrt{5}/5$; $\cot\theta = \sqrt{5}/2$; $\sec\theta = 3\sqrt{5}/5$
47. $\sin\theta = \frac{4}{5}$; $\cos\theta = \frac{3}{5}$; $\cot\theta = \frac{3}{4}$; $\sec\theta = \frac{5}{3}$; $\csc\theta = \frac{5}{4}$
49. $\sin\theta = \frac{5}{13}$; $\cos\theta = -\frac{12}{13}$; $\tan\theta = -\frac{5}{12}$; $\sec\theta = -\frac{13}{12}$; $\csc\theta = \frac{13}{5}$
51. $\sin\theta = 2\sqrt{2}/3$; $\cos\theta = -\frac{1}{3}$; $\tan\theta = -2\sqrt{2}$; $\cot\theta = -\sqrt{2}/4$; $\csc\theta = 3\sqrt{2}/4$

53. $\cos\theta = \sqrt{1-\sin^2\theta} = 0.8829$; $\tan\theta = 0.4695/0.8829 = 0.5318$; $\cot\theta = 1.8805$; $\sec\theta = 1.1326$; $\csc\theta = 2.1299$
55. $\sin\theta = 1/\csc\theta = 0.3827$; $\cos\theta = -\sqrt{1-\sin^2\theta} = -0.9239$; $\tan\theta = -0.4142$; $\cot\theta = -2.4142$; $\sec\theta = -1.0824$
57. Yes, tan and cot.

Problem Set 4, page 87

3. $0°$; see Table 1 **5.** $180°$; see Table 1
7. $30°$; $\sin 30° = \frac{1}{2}$; $\cos 30° = \sqrt{3}/2$; $\tan 30° = \sqrt{3}/3$; $\cot 30° = \sqrt{3}$; $\sec 30° = 2\sqrt{3}/3$; $\csc 30° = 2$
9. $45°$; $\sin 45° = \sqrt{2}/2$; $\cos 45° = \sqrt{2}/2$; $\tan 45° = 1$; $\cot 45° = 1$; $\sec 45° = \sqrt{2}$; $\csc 45° = \sqrt{2}$ **11.** π; see Table 1
13. $\pi/3$; $\sin\pi/3 = \sqrt{3}/2$; $\cos\pi/3 = \frac{1}{2}$; $\tan\pi/3 = \sqrt{3}$; $\cot\pi/3 = \sqrt{3}/3$; $\sec\pi/3 = 2$; $\csc\pi/3 = 2\sqrt{3}/3$
17. $\theta_R = 30°$; $\sin\theta = \frac{1}{2}$; $\cos\theta = -\sqrt{3}/2$; $\tan\theta = -\sqrt{3}/3$; $\cot\theta = -\sqrt{3}$; $\sec\theta = -2\sqrt{3}/3$; $\csc\theta = 2$
19. $\theta_R = 60°$; $\sin\theta = -\sqrt{3}/2$; $\cos\theta = -\frac{1}{2}$; $\tan\theta = \sqrt{3}$; $\cot\theta = \sqrt{3}/3$; $\sec\theta = -2$; $\csc\theta = -2\sqrt{3}/3$
21. $\theta_R = 45°$; $\sin\theta = -\sqrt{2}/2$; $\cos\theta = \sqrt{2}/2$; $\tan\theta = -1$; $\cot\theta = -1$; $\sec\theta = \sqrt{2}$; $\csc\theta = -\sqrt{2}$
23. $\theta_R = 30°$; $\sin\theta = -\frac{1}{2}$; $\cos\theta = -\sqrt{3}/2$; $\tan\theta = \sqrt{3}/3$; $\cot\theta = \sqrt{3}$; $\sec\theta = -2\sqrt{3}/3$; $\csc\theta = -2$
25. $\theta_R = 60°$; $\sin\theta = -\sqrt{3}/2$; $\cos\theta = \frac{1}{2}$; $\tan\theta = -\sqrt{3}$; $\cot\theta = -\sqrt{3}/3$; $\sec\theta = 2$; $\csc\theta = -2\sqrt{3}/3$
27. $\theta_R = \pi/4$; $\sin\theta = -\sqrt{2}/2$; $\cos\theta = \sqrt{2}/2$; $\tan\theta = -1$; $\cot\theta = -1$; $\sec\theta = \sqrt{2}$; $\csc\theta = -\sqrt{2}$
29. $\theta_R = \pi/3$; see problem 19 **31.** $\theta_R = \pi/4$; see problem 21 **33.** $\theta_R = \pi/3$; see problem 25
35. $\theta_R = 60°$; see problem 25

37.

θ	$\sin\theta$	$\cos\theta$	$\tan\theta$	$\cot\theta$	$\sec\theta$	$\csc\theta$
210°	$-\frac{1}{2}$	$-\sqrt{3}/2$	$\sqrt{3}/3$	$\sqrt{3}$	$-2\sqrt{3}/3$	-2
225°	$-\sqrt{2}/2$	$-\sqrt{2}/2$	1	1	$-\sqrt{2}$	$-\sqrt{2}$
240°	$-\sqrt{3}/2$	$-\frac{1}{2}$	$\sqrt{3}$	$\sqrt{3}/3$	-2	$-2\sqrt{3}/3$
300°	$-\sqrt{3}/2$	$\frac{1}{2}$	$-\sqrt{3}$	$-\sqrt{3}/3$	2	$-2\sqrt{3}/3$
315°	$-\sqrt{2}/2$	$\sqrt{2}/2$	-1	-1	$\sqrt{2}$	$-\sqrt{2}$
330°	$-\frac{1}{2}$	$\sqrt{3}/2$	$-\sqrt{3}/3$	$-\sqrt{3}$	$2\sqrt{3}/3$	-2

39. $\sin 46° \approx 0.7193398003$ $\cos 46° \approx 0.6946583705$ $\tan 46° \approx 1.035530314$
$\cot 46° \approx 0.9656887746$ $\sec 46° \approx 1.439556540$ $\csc 46° \approx 1.390163591$
41. $\sin 143° \approx 0.6018150232$; $\cos 143° \approx -0.7986355100$; $\tan 143° \approx -0.7535540502$; $\cot 143° \approx -1.327044821$; $\sec 143° \approx -1.252135658$; $\csc 143° \approx 1.661640141$
43. $\sin(-61.37°) \approx -0.8777322126$; $\cos(-61.37°) \approx 0.4791515031$; $\tan(-61.37°) \approx -1.831846936$; $\cot(-61.37°) \approx -0.5458971382$; $\sec(-61.37°) \approx 2.087022567$; $\csc(-61.37°) \approx -1.139299647$
45. $\sin 61°35' \approx 0.8795101821$; $\cos 61°35' \approx 0.4758800684$; $\tan 61°35' \approx 1.848176128$; $\cot 61°35' \approx 0.5410739728$; $\sec 61°35' \approx 2.101369791$; $\csc 61°35' \approx 1.136996501$
47. $\sin(-97°9'8'') \approx -0.9922188692$; $\cos(-97°9'8'') \approx -0.1245058859$; $\tan(-97°9'8'') \approx 7.969252714$; $\cot(-97°9'8'') \approx 0.1254822799$; $\sec(-97°9'8'') \approx -8.031748803$; $\csc(-97°9'8'') \approx -1.007842152$
49. $\sin(-2\pi/7) \approx -0.7818314825$; $\cos(-2\pi/7) \approx 0.6234898018$; $\tan(-2\pi/7) \approx -1.253960338$; $\cot(-2\pi/7) \approx -0.7974733887$; $\sec(-2\pi/7) \approx 1.603875472$; $\csc(-2\pi/7) \approx -1.279048008$
51. $\sin(1.67) \approx 0.9950833498$; $\cos(1.67) \approx -0.0990410366$; $\tan(1.67) \approx -10.04718230$; $\cot(1.67) \approx -0.0995303927$; $\sec(1.67) \approx -10.09682485$; $\csc(1.67) \approx 1.004940943$
57. $\sin(9.673) \approx -0.2456808808$; $\cos(9.673) \approx -0.9693507646$; $\tan(9.673) \approx 0.2534488957$; $\cot(9.673) \approx 3.945568582$; $\sec(9.673) \approx -1.031618313$; $\csc(9.673) \approx -4.070320803$
61. Check that $[\sin(3\pi/5)]^2 + [\cos(3\pi/5)]^2 = 1$.

Problem Set 5, page 96

7. $-3\pi/2, \pi/2, 1$ **9.** $-3\pi/2, -\pi/2, \pi/2, 3\pi/2$ **11.** $-\pi/2, 3\pi/2, -1$
13. Consider the *number* to be a number of radians. **15.** Amplitude 2, period 2π **17.** Amplitude $\frac{1}{2}$, period 2π
19. Amplitude $\frac{2}{3}$, period 2π; one cycle starts at $x = \pi$ and ends with $x = 3\pi$ **21.** Amplitude 1, period π
23. Amplitude 1, period $\pi/3$ **25.** Amplitude 1, period 4π
27. Amplitude π, period 2; one cycle starts at $x = 1$ and ends with $x = 3$
29. Amplitude 1, period 2π; one cycle starts at $x = \pi/6$ and ends with $x = 13\pi/6$
31. Amplitude 2, period 2π; one cycle starts at $x = \pi/3$ and ends with $x = 7\pi/3$
33. Amplitude 1, period 2π; one cycle starts at $x = -\pi/2$ and ends with $x = 3\pi/2$
35. For $0 \le bx - c \le 2\pi$, it follows that $c \le bx \le 2\pi + c$. Since $b > 0$, then $c/b \le x \le (2\pi + c)/b$. Now $(2\pi + c)/b - c/b = (2\pi + c - c)/b = 2\pi/b$; therefore, the period is $2\pi/b$. Since $-1 \le \sin(bx - c) \le 1$ and $a > 0$, it follows that $-a \le a\sin(bx - c) \le a$, and so the amplitude is a.

Problem Set 6, page 102

1. (a)

x	$\frac{\pi}{12}$	$\frac{\pi}{6}$	$\frac{\pi}{4}$	$\frac{\pi}{3}$	$\frac{5\pi}{12}$	$\frac{\pi}{2}$	$\frac{7\pi}{12}$	$\frac{2\pi}{3}$	$\frac{3\pi}{4}$	$\frac{5\pi}{6}$	$\frac{11\pi}{12}$
$\cot x$	3.73	1.73	1	0.58	0.27	0	−0.27	−0.58	−1	−1.73	−3.73

11. $-2\pi, -\pi, 0, \pi, 2\pi$ **13.** $-2\pi, 0, 2\pi; 1$ **15.** π **17.** $\sec(-x) = 1/\cos(-x) = 1/\cos x = \sec x$
19. Multiply the ordinates of points on the graph of $y = \tan x$ by 2.
21. Multiply the ordinates of points on the graph of $y = \cot x$ by $\frac{1}{2}$, reflect the resulting graph about the x axis, and then shift this graph 1 unit upward.
23. Multiply the ordinates of points on the graph of $y = \sec x$ by $\frac{1}{2}$ and shift the resulting graph 1 unit upward.
25. Shift the graph of $y = \tan x$ exactly $\pi/4$ units to the right.
27. Multiply the ordinates of points on the graph of $y = \cot x$ by $\frac{2}{3}$ and shift the resulting graph $\pi/2$ units to the right.
29. Shift the graph of $y = \sec x$ exactly $\pi/6$ units to the left.
31. Multiply the ordinates of points on the graph of $y = \csc x$ by 2 and then shift the resulting graph $\pi/3$ units to the right.

Problem Set 7, page 107

1. (a) 2 (b) 1 (c) 2π (d) $\pi/3$ (e) $\pi/3$ (f) 1 (g) One cycle starts at $x = \pi/3$ and ends at $x = 7\pi/3$, with trough at $x = 4\pi/3$.
3. (a) 4 (b) 1 (c) 2π (d) $-\pi/4$ (e) $-\pi/4$ (f) 2 (g) One cycle starts at $x = -\pi/4$, ends at $x = 7\pi/4$, with trough at $x = 3\pi/4$.
5. (a) 3 (b) 3 (c) $2\pi/3$ (d) $-5\pi/2$ (e) $-5\pi/6$ (f) 0 (g) One cycle starts at $x = -5\pi/6$, ends at $x = -\pi/6$, with trough at $x = -\pi/2$.
7. (a) $\frac{3}{4}$ (b) 4 (c) $\pi/2$ (d) -12 (e) -3 (f) $-\frac{3}{4}$ (g) One cycle starts at $x = -3$, ends at $x = (\pi - 6)/2$, with trough at $x = (\pi - 12)/4$.
9. (a) 110 (b) 120π (c) $\frac{1}{60}$ (d) $-3\pi/2$ (e) $-\frac{1}{80}$ (f) 0 (g) One cycle starts at $x = -\frac{1}{80}$, ends at $x = \frac{1}{240}$, with trough at $x = -\frac{1}{240}$.
11. (1) $1/2\pi$ (3) $1/2\pi$ (5) $3/2\pi$ (7) $2/\pi$ (9) 60 **13.** (a) 2 (b) 2π (c) $2\pi/3$ (d) 1 (e) $2\pi/3$ (f) 3 (g) $y = 2\cos[t - (2\pi/3)] + 3$
15. (a) $\frac{3}{4}$ (b) 4 (c) $\frac{3}{4}$ (d) $\pi/2$ (e) $3\pi/8$ (f) 0 (g) $y = \frac{3}{4}\cos[(\pi/2)t - (3\pi/8)]$
17. (a) 11 (b) 4 (c) 8 (d) $\pi/2$ (e) 4π (f) 11 (g) $y = 11\cos[(\pi/2)t - 4\pi] + 11$ **19.** (13) $1/2\pi$ (15) $\frac{1}{4}$ (17) $\frac{1}{4}$
21. (a) The axis is the line $y = (d + c)/2$, and the distance between the crest and the axis is $d - [(d + c)/2] = (d - c)/2$, so $a = (d - c)/2$ (b) $\omega = 2\pi/\text{period} = 2\pi/(\beta - \alpha)$ (c) $\phi = S\omega$, where $S = \alpha$; thus, $\phi = 2\pi\alpha/(\beta - \alpha)$
(d) $k = d - a = d - [(d - c)/2] = (2d - d + c)/2 = (d + c)/2$

23. (c) $\phi = S\omega = \left[\alpha + \dfrac{(\beta - \alpha)}{4}\right]\omega = \left(\dfrac{3\alpha + \beta}{4}\right)\left(\dfrac{2\pi}{\beta - \alpha}\right) = \dfrac{(3\alpha + \beta)\pi}{2(\beta - \alpha)}$

25.

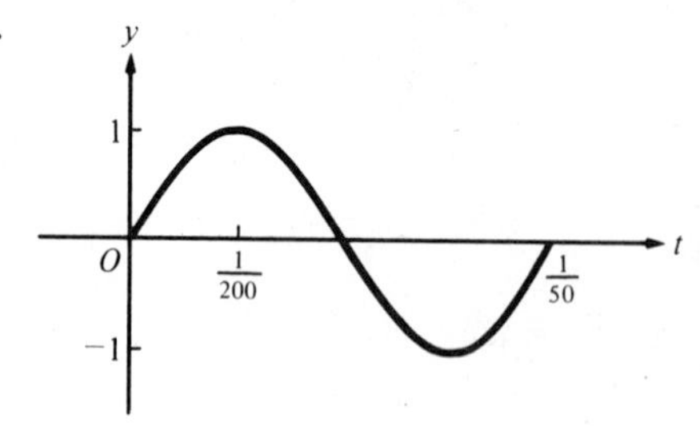

27.

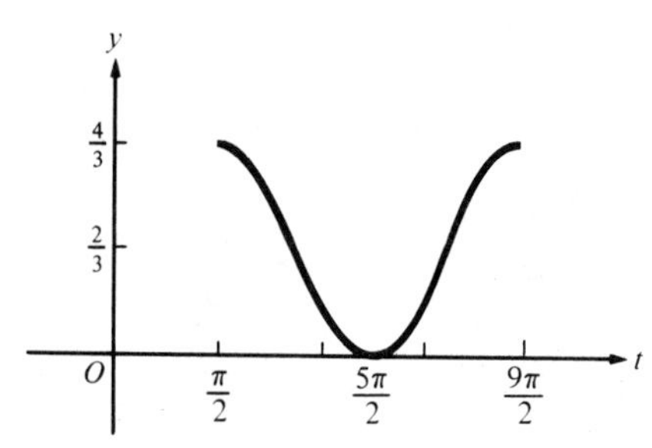

29.

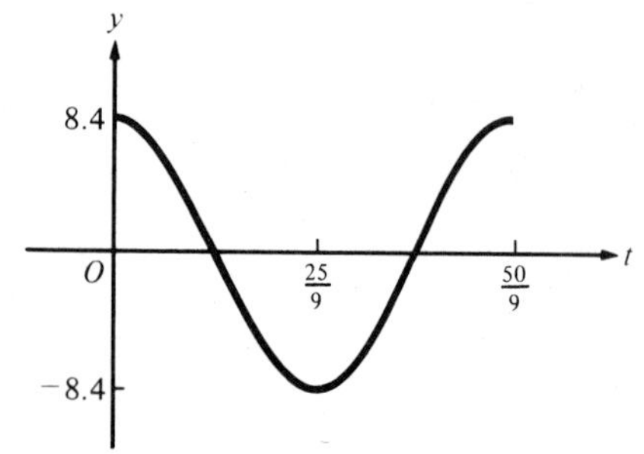

31. $T \approx \pi/4$ second, $\nu \approx 4/\pi$ Hz, $\omega = 8$, $y = 0.1 \cos 8t$
33. Construct a pendulum of length $8/\pi^2$ feet $\approx$ 9.7 inches (or 0.25 meter).
35. Sketch two cycles of the graph $y = 0.04 \cos 5t + k$. Choose $k = 1$, say. The first cycle starts at $x = 0$; the second cycle starts at $x = 2\pi/5$ and ends at $x = 4\pi/5$. The troughs are at $x = \pi/5$ and at $x = 3\pi/5$.
37. $E = 100 \cos(10\sqrt{2}t)$, $T = (\sqrt{2}/10)\pi$ second, $\nu = 5\sqrt{2}/\pi$ Hz

Review Problem Set, page 110

1. 36° **3.** −288° **5.** $\frac{1}{6}$ counterclockwise **7.** 4 clockwise **9.** $\frac{31}{12}$ counterclockwise **11.** 2.0500° **13.** 55.7597 **15.** 87°21′ **17.** −24°31′48″ **19.** 2.85 meters **21.** 3 feet **23.** $\pi/2$ radians **25.** $4\pi/9$

27. $-71\pi/36$ **29.** $-31\pi/18$ **31.** $72°$ **33.** $-\frac{315°}{2}$ **35.** $2295°$ **37.** (a) 0.0873 (b) 0.4844 (c) -0.2997 (d) 0.6158
39. 6.28 feet **41.** $45°$ **43.** $625\pi/12 \approx 163.6$ square centimeters **45.** $0.08\pi \approx 0.25$ meter
47. $120\pi \approx 377$ feet per minute **49.** 1207 miles above the earth **51.** 2094 centimeters per minute
53. $\sin\theta = \sqrt{91}/10$; $\cos\theta = \frac{3}{10}$; $\tan\theta = \sqrt{91}/3$; $\cot\theta = 3\sqrt{91}/91$; $\sec\theta = \frac{10}{3}$; $\csc\theta = 10\sqrt{91}/91$ **55.** (a) $\cos 40°$ (b) $\sin(5\pi/14)$ (c) $\csc 1°$ (d) $\tan(11\pi/28)$ **57.** (a) Quadrant I (b) Quadrant II (c) Quadrant III (d) Quadrantal (e) Quadrant IV (f) Quadrantal **59.** (a) $345°$; $705°$; $-375°$ (b) $100°$; $820°$; $-260°$ (c) $530°$; $-190°$; $-550°$ (d) $-620°$; $-260°$; $100°$ (e) $11\pi/3$; $-(\pi/3)$; $-(7\pi/3)$ **61.** (a) $\sin\theta = 5\sqrt{34}/34$; $\cos\theta = -3\sqrt{34}/34$; $\tan\theta = -\frac{5}{3}$; $\cot\theta = -\frac{3}{5}$; $\sec\theta = -\sqrt{34}/3$; $\csc\theta = \sqrt{34}/5$ (b) $\sin\theta = -3\sqrt{13}/13$; $\cos\theta = 2\sqrt{13}/13$; $\tan\theta = -\frac{3}{2}$; $\cot\theta = -\frac{2}{3}$; $\sec\theta = \sqrt{13}/2$; $\csc\theta = -\sqrt{13}/3$ (c) $\sin\theta = -\frac{4}{5}$; $\cos\theta = -\frac{3}{5}$; $\tan\theta = \frac{4}{3}$; $\cot\theta = \frac{3}{4}$; $\sec\theta = -\frac{5}{3}$; $\csc\theta = -\frac{5}{4}$ (d) $\sin\theta = -\frac{1}{2}$; $\cos\theta = \sqrt{3}/2$; $\tan\theta = -\sqrt{3}/3$; $\cot\theta = -\sqrt{3}$; $\sec\theta = 2\sqrt{3}/3$; $\csc\theta = -2$ **63.** (a) II (b) III (c) IV (d) IV
65. $\tan\theta = -\frac{4}{3}$; $\cot\theta = -\frac{3}{4}$; $\sec\theta = \frac{5}{3}$; $\csc\theta = -\frac{5}{4}$
67. $\cos\theta = \frac{12}{13}$; $\tan\theta = -\frac{5}{12}$; $\cot\theta = -\frac{12}{5}$; $\sec\theta = \frac{13}{12}$; $\csc\theta = -\frac{13}{5}$
69. $\sin\theta = \frac{12}{13}$; $\cos\theta = -\frac{5}{13}$; $\tan\theta = -\frac{12}{5}$; $\cot\theta = -\frac{5}{12}$; $\sec\theta = -\frac{13}{5}$
71. $\cos\theta = -\frac{4}{5}$; $\tan\theta = -\frac{3}{4}$; $\cot\theta = -\frac{4}{3}$; $\sec\theta = -\frac{5}{4}$; $\csc\theta = \frac{5}{3}$
73. $\tan\theta = 1.6003$; $\sec\theta = -1.8870$; $\cos\theta = -0.5299$; $\sin\theta = -0.8480$; $\csc\theta = -1.1792$
75. (a) $\sin(-180°) = 0$; $\cos(-180°) = -1$; $\tan(-180°) = 0$; $\sec(-180°) = -1$
(b) same as for part (a)
(c) $\sin 990° = -1$; $\cos 990° = 0$; $\cot 990° = 0$; $\csc 990° = -1$
(d) $\sin(-360°) = 0$; $\cos(-360°) = 1$; $\tan(-360°) = 0$; $\sec(-360°) = 1$
(e) same as for part (c)
(f) same as for part (a)
(g) same as for part (a)
(h) same as for part (d)
77. (a) $\sin(-150°) = -\frac{1}{2}$; $\cos(-150°) = -\sqrt{3}/2$; $\tan(-150°) = \sqrt{3}/3$; $\cot(-150°) = \sqrt{3}$; $\sec(-150°) = -2\sqrt{3}/3$; $\csc(-150°) = -2$ (b) $\sin(-315°) = \sqrt{2}/2$; $\cos(-315°) = \sqrt{2}/2$; $\tan(-315°) = 1$; $\cot(-315°) = 1$; $\sec(-315°) = \sqrt{2}$; $\csc(-315°) = \sqrt{2}$ (c) $\sin 780° = \sqrt{3}/2$; $\cos 780° = \frac{1}{2}$; $\tan 780° = \sqrt{3}$; $\cot 780° = \sqrt{3}/3$; $\sec 780° = 2$; $\csc 780° = 2\sqrt{3}/3$ (d) $\sin(13\pi/3) = \sqrt{3}/2$; $\cos(13\pi/3) = 1/2$; $\tan(13\pi/3) = \sqrt{3}$; $\cot(13\pi/3) = \sqrt{3}/3$; $\sec(13\pi/3) = 2$; $\csc(13\pi/3) = 2\sqrt{3}/3$ (e) $\sin(-15\pi/4) = \sqrt{2}/2$; $\cos(-15\pi/4) = \sqrt{2}/2$; $\tan(-15\pi/4) = 1$; $\cot(-15\pi/4) = 1$; $\sec(-15\pi/4) = \sqrt{2}$; $\csc(-15\pi/4) = \sqrt{2}$
79. 0.4591664533 **81.** 0.5952436037 **83.** 1.940926426 **85.** 1.042572391 **87.** 1.701301619 **89.** -26.02388181
91. 0.9346780153 **93.** -0.9545616245 **95.** 0.4440158399
97. Multiply the ordinates of points on the graph of $y = \sin x$ by $\frac{1}{3}$ and shift the resulting graph 1 unit upward.
99. Multiply the ordinates of points on the graph of $y = \cos x$ by $\frac{1}{2}$, shift the resulting graph $\pi/2$ units to the right, and then shift this resulting graph 2 units upward.
101. Shift the graph of $y = \tan x$ exactly $\pi/6$ unit to the right and then reflect the resulting graph about the x axis.
103. Shift the graph of $y = \sec x$ exactly π units to the right.
105. Multiply ordinates of points on the graph of $y = \cot x$ by $\frac{2}{3}$ and shift the resulting graph $\pi/2$ units to the right.
107. Maximum value 1 is reached at $x = -2\pi$, at $x = 0$, and at $x = 2\pi$; minimum value -1 is reached at $x = -\pi$ and at $x = \pi$.
109.

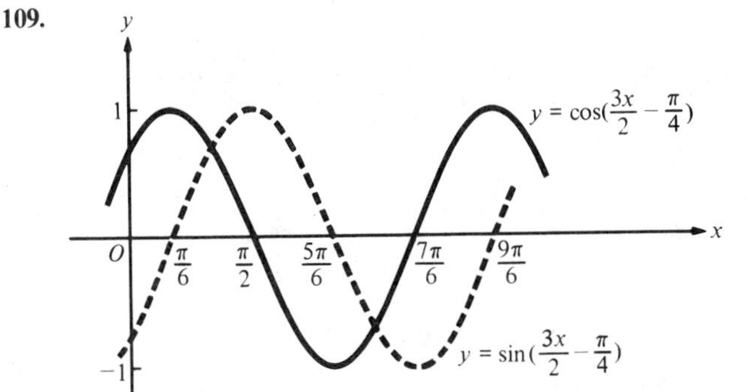

111. (a) 2 (b) π (c) 2 (d) $-\pi$ (e) -1 (f) 1 (g) $\frac{1}{2}$ (h) One cycle starts at $x = -1$ and ends at $x = 1$, with trough at $x = 0$.

113. (a) 3 (b) $\frac{1}{2}$ (c) 4π (d) $\pi/2$ (e) π (f) -3 (g) $1/4\pi$ (h) One cycle starts at $x = \pi$ and ends at $x = 5\pi$, with trough at $x = 3\pi$.
115. (a) 0.2 (b) 0.25 (c) 8π (d) π (e) 4π (f) 0 (g) $1/8\pi$ (h) One cycle starts at $x = 4\pi$ and ends at $x = 12\pi$, with trough at $x = 8\pi$.
117. $T \approx 4.6$ seconds, $\nu \approx 0.22$ Hz, $\omega \approx 1.4$, $y = 0.2\cos(1.4t)$; one cycle starts at $t = 0$ and ends at $t = T$, and the next cycle starts at $t = T$ and ends at $t = 2T$; the troughs are at $x = T/2$ and $x = 3T/2$.
119. $y = 0.03\cos(880\pi t - \pi/2) + 10^5$ **121.** 10,626 days $\approx$ 29.1 years

Chapter 3

Problem Set 1, page 119

1. $-\sin\theta\cos\theta$ **3.** 0 **5.** $\dfrac{1 - \csc\alpha}{1 + \cot\beta}$ **7.** $\tan\theta$ **9.** $\csc v$ **11.** $\cot\beta$ **13.** -1 **15.** $\cot^2 u$ **17.** 1
19. 1 **21.** 1 **23.** $\sin x\cos x$ **25.** $\tan t$ **27.** $2\csc t$ **29.** $\sin x$ **31.** -1 **33.** $\sin^2\alpha$ **35.** $\cos\theta + \sin\theta$
37. $\dfrac{\sin^2 u}{\cos u}$ **39.** $\pm\dfrac{\sqrt{1 - \cos^2\theta}}{\cos^3\theta}$ **41.** $\dfrac{1}{\cos^2 x}$ **43.** $-\sin\alpha$ **45.** $a\sec\theta$ **47.** $8\sin^3 t$

Problem Set 2, page 123

1. $\sin\theta\sec\theta = \sin\theta(1/\cos\theta) = \tan\theta$ **3.** $\tan x\cos x = (\sin x/\cos x)\cos x = \sin x$
5. $\csc(-t)\tan(-t) = (-1/\sin t)(-\sin t/\cos t) = 1/\cos t = \sec t$
7. $\tan\alpha\sin\alpha + \cos\alpha = \dfrac{\sin\alpha}{\cos\alpha}\sin\alpha + \cos\alpha = \dfrac{\sin^2\alpha + \cos^2\alpha}{\cos\alpha} = \dfrac{1}{\cos\alpha} = \sec\alpha$ **53.** Not true for $\theta = \pi/4$
55. Not true for $\theta = \pi/3$ **57.** Not true for $\theta = \pi/2$ **59.** Not true for $t = 2$

Problem Set 3, page 130

1. $(\sqrt{2}/4)(\sqrt{3} + 1)$ **3.** $(-\sqrt{2}/4)(\sqrt{3} - 1)$ **5.** $(-\sqrt{2}/4)(\sqrt{3} - 1)$ **7.** $(\sqrt{2}/4)(\sqrt{3} - 1)$ **9.** $2 - \sqrt{3}$
11. $-(2 + \sqrt{3})$ **13.** $\sin\theta$ **15.** $-\cos\alpha$ **17.** $-\sin s$ **19.** $-\tan t$ **21.** $-\sec\beta$ **23.** $\frac{1}{2}$ **25.** $\frac{1}{2}$ **27.** $\cos x$
29. $\cos\alpha$ **31.** $\sqrt{3}$ **33.** $\tan x$ **35.** (a) $\frac{3}{5}$ (b) $-\frac{5}{13}$ (c) $\frac{16}{65}$ (d) $-\frac{63}{65}$ (e) $-\frac{56}{65}$ (f) $-\frac{16}{63}$ (g) Quadrant II **37.** $\frac{3}{5}$
51. $\cos(\alpha - \beta) = 0.3090169938$; $\cos\alpha\cos\beta + \sin\alpha\sin\beta = 0.3090169947$

Problem Set 4, page 138

1. $2\sin 38^\circ\cos 38^\circ$ **3.** $\cos^2 72^\circ - \sin^2 72^\circ$ **5.** $2\sin\dfrac{\pi}{9}\cos\dfrac{\pi}{9}$ **7.** $\sin 58^\circ$ **9.** $\cos 5\theta$ **11.** $\cos\dfrac{2\pi}{17}$
13. (a) $\frac{24}{25}$ (b) $-\frac{7}{25}$ (c) $-\frac{24}{7}$ **15.** (a) $\frac{120}{169}$ (b) $-\frac{119}{169}$ (c) $-\frac{120}{119}$ **17.** (a) $-\frac{120}{169}$ (b) $-\frac{119}{169}$ (c) $\frac{120}{119}$ **19.** $\cos x$ **21.** $1 + 2\sin 2t$
23. $\dfrac{\sin 2x}{2}$ **25.** $\dfrac{1 - \cos 2\theta}{2}$ or $\sin^2\theta$ **27.** (a) $\sqrt{\dfrac{1 + \cos 30^\circ}{2}}$ (b) $\dfrac{\sqrt{2 + \sqrt{3}}}{2}$
29. (a) $-\sqrt{\dfrac{1 + \cos(5\pi/4)}{2}}$ (b) $-\dfrac{\sqrt{2 - \sqrt{2}}}{2}$ **31.** (a) $\dfrac{1 - \cos 315^\circ}{\sin 315^\circ}$ (b) $1 - \sqrt{2}$ **33.** (a) $-\sqrt{\dfrac{1 - \cos 405^\circ}{2}}$ (b) $-\dfrac{\sqrt{2 - \sqrt{2}}}{2}$
35. (a) $\dfrac{3}{5}$ (b) $\dfrac{4}{5}$ (c) $\dfrac{3}{4}$ **37.** (a) $\dfrac{3\sqrt{13}}{13}$ (b) $-\dfrac{2\sqrt{13}}{13}$ (c) $-\dfrac{3}{2}$ **39.** $\cos 125^\circ$ **41.** $\tan 3x$ **43.** $\cos\dfrac{\pi}{5}$ **57.** $\dfrac{1 + z^2}{1 - z^2}$
59. $\dfrac{1 - z^2}{2z}$

Problem Set 5, page 146

1. $\frac{1}{2}\sin 145^\circ + \frac{1}{2}\sin 65^\circ$ **3.** $\cos(3\pi/4) + \cos(\pi/2)$ or $-\sqrt{2}/2$ **5.** $\frac{1}{2}\sin 8\theta - \frac{1}{2}\sin 2\theta$
7. $\frac{1}{2}\cos(7s - 5t) - \frac{1}{2}\cos(7s + 5t)$ **9.** $\frac{1}{2}\sin 4x - \frac{1}{2}\sin 2x$ **11.** $2\sin 50^\circ \cos 30^\circ$ **13.** $2\cos(\pi/4)\cos(\pi/8)$
15. $2\sin 3\theta \cos\theta$ **17.** $2\sin 3t \sin t$ **19.** $2\cos\dfrac{2\alpha + \beta}{2}\sin\dfrac{\beta}{2}$ **21.** (a) $-\dfrac{2\sin 45^\circ \sin 30^\circ}{2\sin 45^\circ \cos 30^\circ} = -\dfrac{\sqrt{3}}{3}$ (b) $\dfrac{2\sin 45^\circ \cos 35^\circ}{2\cos 45^\circ \cos 35^\circ} = 1$
23. $\sin 50^\circ + \sin 10^\circ = 2\sin 30^\circ \cos 20^\circ = \cos 20^\circ = \sin 70^\circ$

41.

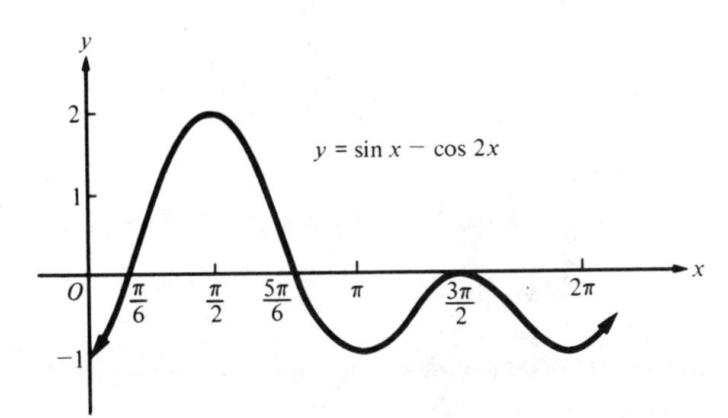

43.

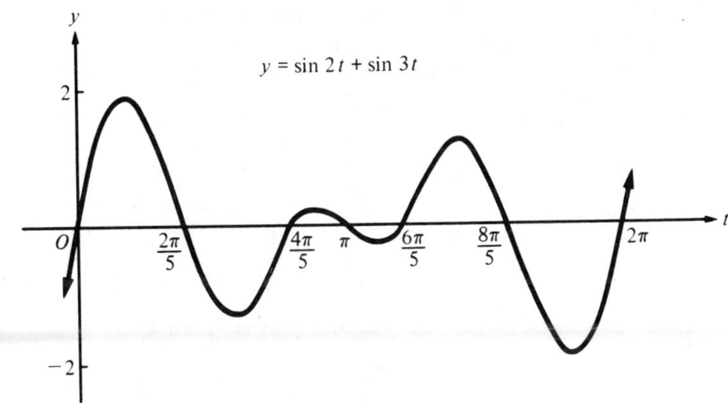

45. $y = \sqrt{2}\cos[2t - (\pi/4)]$ **47.** $y = 2\cos[2\pi t - (5\pi/3)]$ **49.** $y = \frac{1}{2}\sin 6x - \frac{1}{2}\sin 2x$ **51.** $y = \cos 2t - \cos 4t$
53. $y = -\frac{1}{2}\cos 3\pi x - \frac{1}{2}\cos \pi x$ **55.** 4 **57.** $2\sqrt{2\pi}$
59. The graph actually falls *below* the x axis, between 0 and about $\pi/10$.
61. No, maximum occurs a bit *to the right* of $2\pi/3$.

Problem Set 6, page 153

1. $\pi/2$ **3.** $-\pi/4$ **5.** 0 **7.** $\pi/4$ **9.** $\pi/3$ **11.** $-\pi/4$ **13.** 0.6999768749 **15.** 0.4391814802
17. 1.107148718 **19.** −1.270032196 **21.** −0.5829630403 **23.** 0.5931997761 **25.** $\frac{3}{4}$ **27.** $\pi/6$ **29.** $\frac{4}{5}$
31. $3\sqrt{10}/10$ **33.** $\frac{4}{3}$ **35.** $\frac{10}{7}$ **45.** (a) $-1 \le x \le 1$ (b) $0 \le x \le \pi$ **47.** (a) All real values of x (b) $-\pi/2 < x < \pi/2$

Problem Set 7, page 161

1. 90°, 270° **3.** 45°, 225° **5.** $\pi/6, 5\pi/6$ **7.** $2\pi/3, 5\pi/3$ **9.** $3\pi/2$ **11.** $k \cdot 180°, k = 0, \pm 1, \pm 2, \ldots$ **13.** $(\pi/4) + 2\pi k; (7\pi/4) + 2\pi k, k = 0, \pm 1, \pm 2, \ldots$ **15.** $(\pi/3) + k\pi, k = 0, \pm 1, \pm 2, \ldots$ **17.** 48.5904°, 131.4096° **19.** 0.8411, 5.4421 **21.** $1.2900 + k\pi, k = 0, \pm 1, \pm 2, \ldots$ **23.** 15°, 75°, 195°, 255° **25.** 20°, 80°, 140°, 200°, 260°, 320° **27.** $2\pi/3, 4\pi/3$ **29.** $3\pi/2$ **31.** 210°, 330° **33.** $0, \pi/6, 5\pi/6, \pi$ **35.** 30°, 150°, 210°, 330° **37.** $\pi/4, 3\pi/4, 5\pi/4, 7\pi/4$ **39.** $0, \pi/3, \pi, 4\pi/3$ **41.** 0°, 45°, 180°, 225° **43.** $\pi/4, 3\pi/4, 5\pi/4, 7\pi/4$ **45.** 90°, 150°, 210° **47.** 0.4429, 1.5708, 2.6987 **49.** 54.7356°, 90°, 270°, 305.2644° **51.** 1.288×10^{-3} second **53.** 11.20° **55.** (a) $t = (2n + 1) \cdot 2\pi$ and $t = (4\pi/3) + (2n + 1) \cdot 2\pi, n = 0, \pm 1, \pm 2, \ldots$ (b) 30 or 45 wild pigs

Review Problem Set, page 162

1. $-\tan\theta$ **3.** $\sin x$ **5.** $\tan^2 t$ **7.** $1 + \tan u$ **9.** $\tan\theta$ **11.** $125\cos^3\theta$ **27.** $\cos\theta$ **29.** $-\cos\alpha$ **31.** $\sin t$ **33.** $\sqrt{3}/2$ **35.** $\sin(x + y)$ **37.** (a) $\dfrac{\sqrt{2}(1 + \sqrt{3})}{4}$ (b) $\dfrac{\sqrt{2}(1 - \sqrt{3})}{4}$ (c) $-(2 + \sqrt{3})$ **39.** $-\frac{36}{85}$ **41.** $-\frac{21}{221}$ **43.** $-\frac{304}{425}$ **45.** $\frac{297}{304}$ **47.** $-\frac{36}{325}$ **57.** (a) $\frac{336}{625}$ (b) $-\frac{527}{625}$ (c) $\frac{4}{5}$ (d) $\frac{3}{4}$ **59.** $\cos 4x$ **61.** $\sin t$ **63.** 1 **65.** $\frac{1}{2}\tan 2\omega t$ **67.** 1 **85.** $\frac{1}{2}\sin 2x + \frac{1}{2}\sin x$ **87.** $\frac{1}{2}\cos 30° - \frac{1}{2}\cos 45°$ or $(\sqrt{3}/4) - (\sqrt{2}/4)$ **89.** $2\sin 30°\cos 25°$ or $\cos 25°$ **91.** $2\cos(5\beta/2)\sin(3\beta/2)$ **105.** $y = 2\sin(7t/3) - 2\sin(t/3)$ **107.** $y = 2\sqrt{2}\cos[(\pi/3)t - (7\pi/4)]$ **109.** $y = 2\cos[2t - 3 + (\pi/6)]$ **111.** 30 **113.** $y = \cos 4\pi t + \cos 2\pi t; T = 1$ **115.** $-\pi/6$ **117.** $\pi/3$ **119.** 0.3843967745 **121.** 0.3836622700 **123.** 0.9588938924 **125.** −0.8480620790 **127.** 2.711892987 **129.** $\frac{3}{5}$ **131.** $-\frac{12}{13}$ **133.** $-5\pi/14$ **141.** 30°, 210° **143.** $\pi/6, 5\pi/6, 7\pi/6, 11\pi/6$ **145.** 0°, 180°, 225°, 315° **147.** $7\pi/6, 3\pi/2, 11\pi/6$ **149.** 15°, 60°, 105°, 150°, 195°, 240°, 285°, 330° **151.** 0°, 90°, 180°, 270° **153.** 26.5651°, 153.4349°, 206.5651°, 333.4349° **155.** 0.5236, 2.6180, 3.8713, 5.5535 **157.** 8.135

Chapter 4

Problem Set 1, page 173

1. $c = 6.403; \alpha = 51.34°; \beta = 38.66°$ **3.** $b = 11.18; \alpha = 41.81°; \beta = 48.19°$ **5.** $a = 1031; c = 1988; \beta = 58.77°$ **7.** $b = 5.826; \alpha = 50.76°; \beta = 39.24°$ **9.** $c = 4331; \alpha = 0.04°; \beta = 89.96°$ **11.** $a = 9299; c = 9647; \alpha = 74.55°$ **13.** $a = 1.387 \times 10^{-2}$ meter; $c = 1.498 \times 10^{-2}$ meter; $\beta = 22°15'$ **15.** $a = 70.79$ miles; $\alpha = 66.53°; \beta = 23.47°$ **17.** $c = 6.414 \times 10^{-6}$ meter; $\alpha = 50.26°; \beta = 39.74°$ **19.** 22 feet **21.** 34 meters **23.** 44 meters **25.** 35.4° **27.** 98.3 meters **29.** 100 meters **31.** 56° **33.** 162 feet **35.** 196 feet **37.** 24.5 miles **39.** S20.56°W

Problem Set 2, page 181

1. $\alpha = 85°; b = 26.94; c = 19.78$ **3.** $\alpha = 121°; a = 33.41; c = 13.97$ **5.** $\beta = 73.87°; a = 36.77; c = 41.67$ **7.** $\gamma = 23.55°; b = 52.91; c = 51.08$ **9.** $\beta = 25.1°; \gamma = 109.9°; c = 66.49$ **11.** $\beta = 67.21°; \gamma = 52.79°; c = 28.51$ or $\beta = 112.79°; \gamma = 7.21°; c = 4.49$ **13.** No triangle **15.** $\beta = 52.02°; \gamma = 16.48°; c = 19.97$ **17.** $\gamma = 75.65°; a = 40.28$ feet; $c = 57.38$ feet **19.** $\beta = 51.44°; \gamma = 69.23°; c = 119.58$ kilometers

21. $\beta = 89.53°; \gamma = 68.47°; c = 5190$ meters or $\beta = 90.47°; \gamma = 67.53°; c = 5156$ meters **23.** No triangle
25. $|\overline{AC}| = 316.6$ meters; $|\overline{CB}| = 345.5$ meters **27.** 91.43 meters **29.** 72.86 meters **31.** 0.9906 kilometer
33. 2.11 kilometers **35.** 88 minutes, 16 seconds **41.** $\gamma = 38°; c = 5.935; b = 9.219$
43. $\beta = 32°; b = 8.048; c = 13.98$

Problem Set 3, page 186

1. 8.888 **3.** 44.99 **5.** 3.336 **7.** 104.48° **9.** 44.89° **11.** 33.72° **13.** $\alpha = 63.95°; \beta = 43.60°; c = 347.0$
15. $\beta = 60.75°; \gamma = 85.50°; c = 559.9$ or $\beta = 119.25°; \gamma = 27.00°; c = 255.9$ **17.** $\alpha = 26.75°; \beta = 39.75°; \gamma = 113.50°$
19. 69.14 meters **21.** 815.9 meters **23.** 744.5 kilometers **25.** 57.49 miles **27.** 17.32 cm^2 **29.** 98.06 ft^2
31. 74.22 km^2 **33.** 85,733 m^2 **41.** $a + b > c; a + c > b; b + c > a$

Problem Set 4, page 195

1. $4\sqrt{2}$; 135° **3.** 5; 36.86989765° **5.** 2; 300° **7.** $\frac{5}{8}$; 53.13010235° **9.** 4; 123.9878436° **11.** 5; 53.13010235°
13. $\langle 5\sqrt{2}/2, 5\sqrt{2}/2\rangle$ **15.** $\langle -4, 4\rangle$ **17.** $\langle -\sqrt{3}, -1\rangle$ **19.** $\langle 1.789851696, 6.798332951\rangle$ **21.** (a) $\langle 2, 8\rangle$ (b) $\langle 4, 0\rangle$
(c) $\langle 5, 28\rangle$ (d) $\langle 9, -4\rangle$ **23.** (a) $\langle -\frac{1}{2}, \frac{13}{3}\rangle$ (b) $\langle -\frac{11}{2}, -\frac{1}{3}\rangle$ (c) $\langle 1, \frac{46}{3}\rangle$ (d) $\langle -\frac{27}{2}, -3\rangle$ **25.** (a) $\langle 11, -21\rangle$ (b) $\langle -7, 9\rangle$
(c) $\langle 42, -78\rangle$ (d) $\langle -23, 33\rangle$ **27.** (a) $-\bar{i} + 12\bar{j}$ (b) $5\bar{i} - 2\bar{j}$ (c) $-6\bar{i} + 43\bar{j}$ (d) $13\bar{i} - 11\bar{j}$ **29.** (a) $10\bar{i} - 11\bar{j}$ (b) $2\bar{i} + \bar{j}$
(c) $34\bar{i} - 39\bar{j}$ (d) $8\bar{j}$ **43.** (a) 13 (b) 18 (c) -7 (d) 0 (e) 1

Problem Set 5, page 201

21. (a) $(-5, 225°)$ (b) $(5, -315°)$ (c) $(-5, -135°)$ **23.** $(2\sqrt{2}, 2\sqrt{2})$ **25.** $(-5\sqrt{2}/2, -5\sqrt{2}/2)$ **27.** $(\sqrt{2}/2, -\sqrt{2}/2)$
29. $(2\sqrt{3}, -2)$ **31.** $(-1.495, -2.326)$ **33.** $(14.17, -10.71)$ **35.** $(2, 45°)$ **37.** $(2, 60°)$ **39.** $(3, 180°)$
41. $(13, 112.62°)$ **43.** $(20.29, -50.12°)$ **45.** $(4, -\pi/6)$ **47.** $(2, -\pi/2)$ **49.** $(116.3, -2.239)$

Problem Set 6, page 208

1. (a) 60.33 pounds (b) 27.96° **3.** (a) 29.10 newtons (b) 79.11° **5.** (a) 16.44 pounds (b) 12.29° **7.** (a) 816.5 km/hr
(b) 2.97° (c) N92.03°E **9.** (a) 3.6 m/sec (b) 56.3° **11.** 13.73 kg; N32.87°W **13.** A circle of radius 1, center at O
15. A straight line through O making the angle $\pi/3$ with the polar axis
17. A straight line through O perpendicular to the polar axis **19.** $r = 5$ **21.** $3r\cos\theta - 2r\sin\theta = 6$ **23.** $r = 2\cos\theta$
25. $x^2 + y^2 = 1$ **27.** $y - 2x = 1$ **29.** $x^2 = y$

31.

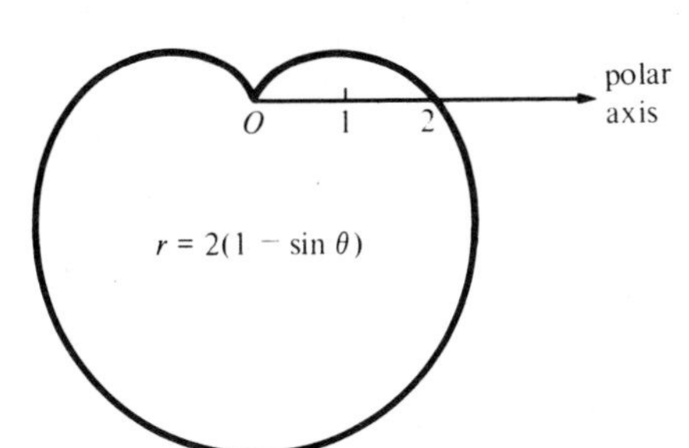

33.

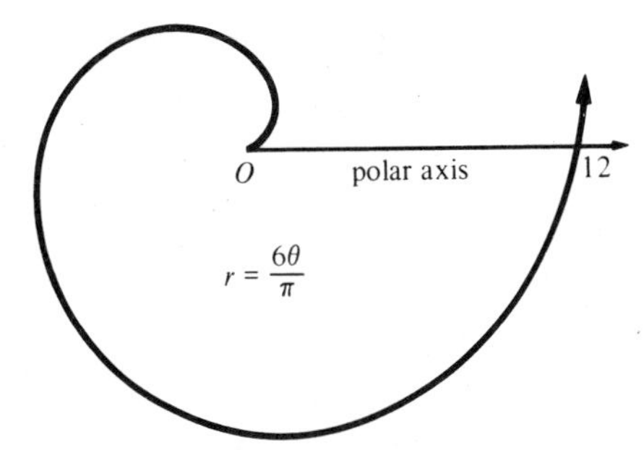

Review Problem Set, page 210

1. $\beta = 58.75°$; $a = 103.8$ meters; $b = 171.0$ meters **3.** $\alpha = 33.85°$; $a = 31.74$ microns; $c = 56.99$ microns **5.** 25.49 meters **7.** 658.2 meters **9.** 264 meters/minute **11.** 318.2 meters **13.** 85.48 nautical miles; S13°10′W **15.** 36.73 ft **17.** 17.94 miles **19.** 21.05° **21.** $\alpha = 90°$; $a = 100\sqrt{3} \approx 173.2$ ft; $\gamma = 60°$ or $\alpha = 30°$; $a = 50\sqrt{3} \approx 86.6$ ft; $\gamma = 120°$ **23.** $\alpha = 76.28°$; $a = 66.99$ cm; $\beta = 46.47°$ **25.** No triangle **27.** 60° **29.** 9.006 km **31.** 16.96 inches **33.** $\alpha = 23.94°$; $c = 80.44$ km; $\beta = 115.01°$ **35.** $\beta = 55.58°$; $b = 0.9869$ micron; $\alpha = 69.42°$ or $\beta = 14.42°$; $b = 0.2979$ micron; $\alpha = 110.58°$ **37.** $\gamma = 105.40°$; $a = 1.788$ nautical miles; $c = 2.555$ nautical miles **39.** 448.0 m^2 **41.** 392,100 square miles **43.** 2046 meters **45.** 1:15 P.M. **47.** 1275.3 meters **49.** 83.51° **51.** $\langle 10, -5\rangle$ **53.** $\langle 5, 2\rangle$ **55.** $\langle 13, 7\rangle$ **57.** $\langle -13, 5\rangle$ **59.** $4\sqrt{5}$ **61.** $2\sqrt{26}$ **63.** (a) $\langle 25\sqrt{3}, 25\rangle$ (b) $\langle 33.94, 61.22\rangle$ (c) $\langle 16.07, -19.15\rangle$ (d) $\langle 125\sqrt{3}, -125\rangle$ **65.** (a) 2, 60° (b) 19.21, 51.34° (c) 50, 126.87° (d) 0.3125, 53.13° (e) 4, 41.41° **67.** (a) 30.96 pounds, 33.35° (b) 221.2 newtons, 44.42° (c) 7.598 tonnes, 76° (d) 540.8 dynes, 303.69° **69.** (a) 486.6 knots (b) 7.99° (c) N37.01°W **71.** S29°26′E **73.** (a) $(\sqrt{2}, \sqrt{2})$ (b) $(-\sqrt{2}, -\sqrt{2})$ (c) $(-0.1736, 0.9848)$ (d) $(1.5, 0)$ (e) $(2\sqrt{3}, -2)$ (f) $(3\sqrt{2}, -3\sqrt{2})$ **75.** (a) $(1, \pi)$ (b) $(7\sqrt{2}, -3\pi/4)$ (c) $(13, 1.97)$ (d) $(5, -0.93)$ (e) $(17, 2.06)$ (f) $(34, -2.65)$ **77.** $r \sin\theta = 3r\cos\theta + 1$ **79.** $r = 4\sin\theta$ **81.** $r\sin^2\theta = 6\cos\theta$ **83.** $r^2(3\cos^2\theta + 1) = 4$ **85.** $(x - \frac{1}{2})^2 + (y - \frac{1}{2})^2 = \frac{1}{2}$ **87.** $x^2 - y^2 = 2$

Chapter 5

Problem Set 1, page 221

1. $9 + i$ **3.** $10 - i$ **5.** $12 - 6i$ **7.** $-2 - 2i$ **9.** $-5 + 2i$ **11.** $-11 - 4i$ **13.** $-1 - 14i$ **15.** $-3 + 11i$ **17.** $-3 + 54i$ **19.** $-7 + 9i$ **21.** $15 - 23i$ **23.** $64 - 40i$ **25.** -20 **27.** $\frac{13}{36}$ **29.** i **31.** i **33.** $12 + 16i$ **35.** $-\frac{1}{2} + (\sqrt{3}/2)i$ **37.** $\frac{2}{13} - \frac{3}{13}i$ **39.** $\frac{21}{53} - \frac{6}{53}i$ **41.** $\frac{1}{5} - \frac{11}{10}i$ **43.** $\frac{11}{34} + \frac{41}{34}i$ **45.** $\frac{23}{74} + \frac{27}{74}i$ **47.** $-\frac{13}{10} - \frac{19}{10}i$ **49.** $\frac{63}{290} - \frac{201}{290}i$ **51.** $\frac{6}{145} - \frac{43}{145}i$ **53.** (a) $2 - i$ (b) 4 (c) $2i$ (d) 5 (e) $\sqrt{5}$ **55.** (a) i (b) 0 (c) $-2i$ (d) 1 (e) 1 **57.** (a) $-i$ (b) 0 (c) $2i$ (d) 1 (e) 1 **59.** (a) $-12 - 5i$ (b) -24 (c) $10i$ (d) 169 (e) 13 **61.** (a) $-\frac{1}{5} + \frac{11}{10}i$ (b) $-\frac{2}{5}$ (c) $-\frac{11}{5}i$ (d) $\frac{5}{4}$ (e) $\sqrt{5}/2$ **63.** (a) 5 (b) 10 (c) 0 (d) 25 (e) 5

Problem Set 2, page 229

1. $2\sqrt{3} + 2i$ **3.** $-\frac{7}{2}\sqrt{2} + \frac{7}{2}\sqrt{2}i$ **5.** $3 - 3\sqrt{3}i$ **7.** $\sqrt{3} + i$ **9.** $\sqrt{2}\left(\cos\frac{3\pi}{4} + i\sin\frac{3\pi}{4}\right)$ **11.** $3(\cos 90° + i\sin 90°)$ **13.** $2(\cos 60° + i\sin 60°)$ **15.** $2(\cos 210° + i\sin 210°)$ **17.** $\cos 270°$ **19.** (a) $8(\cos 110° + i\sin 110°)$ (b) $2(\cos 30° + i\sin 30°)$ **21.** (a) $98\left(\cos\frac{11\pi}{4} + i\sin\frac{11\pi}{4}\right)$ (b) $2\left(\cos\frac{\pi}{4} + i\sin\frac{\pi}{4}\right)$ **23.** (a) $18(\cos 135° + i\sin 135°)$ (b) $2(\cos 45° + i\sin 45°)$ **25.** (a) $2\sqrt{2}(\cos 105° + i\sin 105°)$ (b) $\frac{\sqrt{2}}{2}[\cos(-15°) + i\sin(-15°)]$ **27.** $16\sqrt{2} + 16\sqrt{2}i$ **29.** -243 **31.** $-256\sqrt{2} + 256\sqrt{2}i$ **33.** $-16\sqrt{3} - 16i$ **35.** -256 **37.** $\frac{\sqrt{2}}{2} - \frac{\sqrt{2}}{2}i$ **39.** $-\frac{3\sqrt{3}}{2} + \frac{3}{2}i, \frac{3\sqrt{3}}{2} - \frac{3}{2}i$ **41.** $\sqrt[6]{2}(\cos 45° + i\sin 45°)$, $\sqrt[6]{2}(\cos 165° + i\sin 165°)$, $\sqrt[6]{2}(\cos 285° + i\sin 285°)$ **43.** $2(\cos 75° + i\sin 75°)$, $2(\cos 165° + i\sin 165°)$, $2(\cos 255° + i\sin 255°)$, $2(\cos 345° + i\sin 345°)$

45. $\frac{\sqrt{2}}{2}(1+i), \frac{\sqrt{2}}{2}(-1+i), \frac{\sqrt{2}}{2}(-1-i), \frac{\sqrt{2}}{2}(1-i)$

47. $2(\cos 45^\circ + i\sin 45^\circ), 2(\cos 117^\circ + i\sin 117^\circ), 2(\cos 189^\circ + i\sin 189^\circ), 2(\cos 261^\circ + i\sin 261^\circ), 2(\cos 333^\circ + i\sin 333^\circ)$

49. $-\frac{1}{12} \pm i\frac{\sqrt{71}}{12}$ **51.** $\frac{2}{5} \pm i\frac{\sqrt{6}}{5}$ **53.** $-\frac{1}{2} \pm i\frac{\sqrt{3}}{2}$ **55.** $\frac{1}{4} \pm i\frac{\sqrt{3}}{4}$ **57.** $-2+2i, 2-2i$ **59.** $-\frac{i}{2} \pm \frac{\sqrt{3}}{2}$

61. $1+i, 2$ **65.** (a) $1, -1$ (b) $1, -\frac{1}{2}+\frac{\sqrt{3}}{2}i, -\frac{1}{2}-\frac{\sqrt{3}}{2}i$ (c) $1, i, -1, -i$ **67.** (a) Reflect z about the real axis (b) Reflect z through the origin. (c) Reflect z about the imaginary axis. (d) Clockwise rotation of z through $\pi/2$. (e) Distance between z and w.

Problem Set 3, page 239

1. -1 **3.** -1 **5.** $\frac{1}{4}$ **7.** $\frac{1}{2}$ **9.** $-4.\overline{6}$

11.

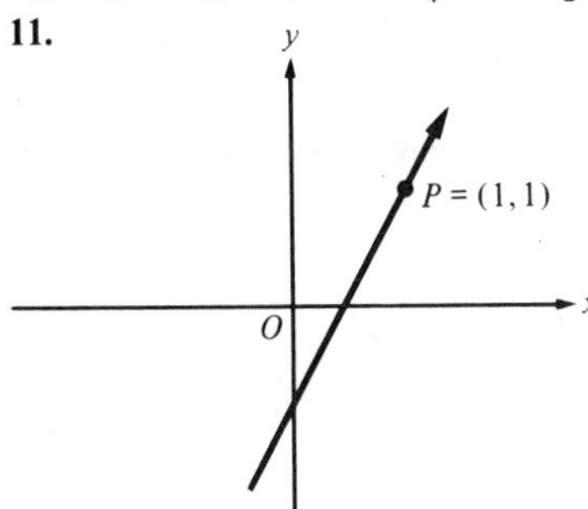

13.

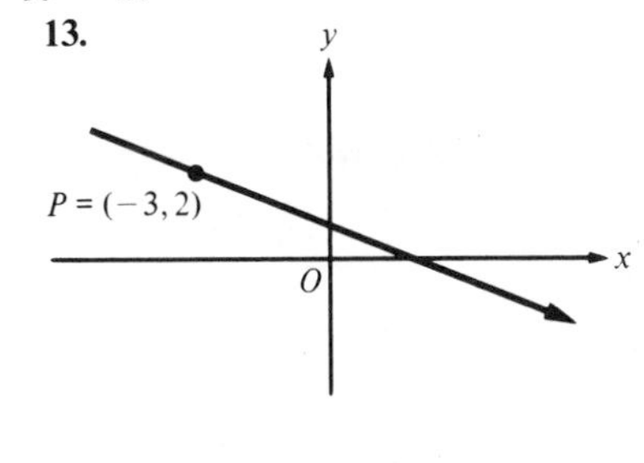

15.

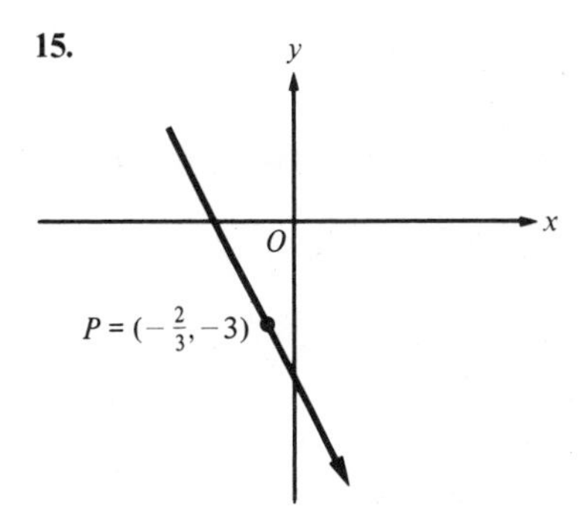

17. $m=-2$ **19.** $m=-\frac{1}{5}$ **21.** $m=-\frac{2}{3}$ **23.** $m=\frac{21}{38}$ **25.** $y-2=\frac{3}{4}(x-3)$ **27.** $y-1=-\frac{1}{7}(x-4)$
29. $y-5=x+3$ **31.** $m=\frac{3}{2}, b=-3$ **33.** $m=3, b=1$ **35.** $m=-2, b=-3$ **37.** $m=0, b=-1$
39. (a) $y-2=4(x+5)$ (b) $y=4x+22$ (c) $4x-y+22=0$
41. (a) $y-5=-3(x-0)$ (b) $y=-3x+5$ (c) $-3x-y+5=0$
43. (a) $y-0=-\frac{5}{3}(x-3)$ (b) $y=-\frac{5}{3}x+5$ (c) $-5x-3y+15=0$
45. (a) $y+4=\frac{2}{5}(x-4)$ (b) $y=\frac{2}{5}x-\frac{28}{5}$ (c) $2x-5y-28=0$
47. (a) $y-\frac{2}{3}=\frac{3}{5}(x+3)$ (b) $y=\frac{3}{5}x+\frac{37}{15}$ (c) $9x-15y+37=0$
49. (a) $y-\frac{5}{7}=-\frac{7}{3}(x-\frac{2}{3})$ (b) $y=-\frac{7}{3}x+\frac{143}{63}$ (c) $147x+63y-143=0$
51. $-\frac{5}{4}$ **53.** $y=22N+0.2x$ **55.** $y=0.25x+3.45$, \$5.45 **57.** \$200,000

Problem Set 4, page 251

1. Circle; $r=5$ **3.** Ellipse; vertices at $(3,0), (-3,0), (0,1)$, and $(0,-1)$ **5.** Ellipse; vertices at $(\pm 2, 0)$ and $(0, \pm 4)$

7. Ellipse; vertices at $(\pm 5, 0)$ and $(0, \pm 4)$ **9.** Circle; $r=\frac{1}{2}$ **11.** $\frac{(x+5)^2}{9}+\frac{(y-5)^2}{4}=1$

13. $\frac{(x+2)^2}{25}+\frac{(y-1)^2}{16}=1$ **15.** $\frac{(x-1)^2}{4}+\frac{(y+2)^2}{9}=1$

17. Center at $(1,2)$; vertices at $(4,-2), (-2,-2), (1,0)$, and $(1,-4)$

19. Center at $(-3, 1)$; vertices at $(-6, 1)$, $(0, 1)$, $(-3, -5)$, and $(-3, 7)$ **21.** Horizontal axis; $V = (0,0)$
23. Vertical axis; $V = (0,0)$ **25.** Vertical axis; $V = (0,0)$ **27.** Vertical axis; $V = (4, -7)$
29. Vertical axis; $V = (-1,0)$ **31.** $x = \frac{1}{8}y^2$ **33.** $x - 5 = -(y + 1)^2$ **35.** $y + \frac{1}{2} = \frac{1}{8}(x + 1)^2$
37. (a) Vertical (b) Vertical (c) Vertical (d) Horizontal (e) Horizontal (f) Horizontal
39. $C = (0,0)$; $V = (\pm 3,0)$; asymptotes are $y = \pm\frac{2}{3}x$ **41.** $C = (0,0)$; $V = (0, \pm 4)$; asymptotes are $y = \pm 2x$
43. $C = (0,0)$; $V = (\pm 4,0)$; asymptotes are $y = \pm\frac{1}{2}x$
45. $C = (1, -2)$; $V = (-2, -2)$ and $(4, -2)$; asymptotes are $y + 2 = \pm\frac{2}{3}(x - 1)$
47. $C = (-2, -1)$; $V = (-2, 3)$ and $(-2, -5)$; asymptotes are $y + 1 = \pm\frac{4}{5}(x + 2)$
49. $\frac{x^2}{16} - \frac{y^2}{25} = 1$ **51.** $\frac{(y + 2)^2}{4} - \frac{(x - 3)^2}{1/4} = 1$ **53.** $(x + 1)^2 + (y + 2)^2 = 1$; circle
55. $(y - 4)^2 = 6(x + 3)$; parabola **57.** $\frac{(x + 3)^2}{2} + \frac{y^2}{4} = 1$; ellipse **59.** $\frac{(x + 5)^2}{10} + \frac{(y - 3)^2}{4} = 1$; ellipse
61. $\frac{(x - 2)^2}{4} - \frac{(y + 1)^2}{1} = 1$; hyperbola **63.** $\frac{(y - 10)^2}{32} - \frac{x^2}{18} = 1$; hyperbola **65.** $\frac{(x + 1)^2}{24} + \frac{(y - 2)^2}{9} = 1$; ellipse
67. $2(x + 2)^2 = 3(y + \frac{4}{3})$; parabola **69.** $\frac{x^2}{16} + \frac{y^2}{7} = 1$ **71.** 100, 58, 28, 10, 4, 10, 28, 58, 100 meters

Problem Set 5, page 261

1. (a) $(1, -2)$ (b) $(-1, -1)$ (c) $(4, -5)$ (d) $(-2, -4)$ (e) $(6, 3)$ (f) $(7, -2)$ **3.** $\bar{x} = x + 2$; $\bar{y} = y - 1$; circle
5. $\bar{x} = x + 1$; $\bar{y} = y - 1$; ellipse **7.** $\bar{x} = x - 2$; $\bar{y} = y - 3$; ellipse **9.** $\bar{x} = x - 1$; $\bar{y} = y$; hyperbola
11. $\bar{x} = x + 1$; $\bar{y} = y - 5$; parabola **13.** $\bar{x} = x + 3$; $\bar{y} = y - 1$; hyperbola **15.** $\bar{x} = x - 2$; $\bar{y} = y + 1$; hyperbola
17. $(-7, -4)$ **19.** $\left(-\frac{3}{2} + \frac{3\sqrt{3}}{2}, -\frac{3\sqrt{3}}{2} - \frac{3}{2}\right)$ **21.** $(-2\sqrt{3} + 1, -2 - \sqrt{3})$ **23.** $(4, 0)$
25. (a) $(\bar{x} + \sqrt{3}\bar{y})^2 = 6(\sqrt{3}\bar{x} - \bar{y})$ (b) $\left(\frac{3\sqrt{3}}{2} + \frac{1}{2}\right)x + \left(\frac{3}{2} - \frac{\sqrt{3}}{2}\right)y = 0$ (c) $x^2 + y^2 = 1$ (d) $\left(\frac{5\sqrt{3}}{2} - \frac{1}{2}\right)\bar{x} - \left(\frac{5}{2} + \frac{\sqrt{3}}{2}\right)\bar{y} = 4$
(e) $\bar{x}^2 + \bar{y}^2 = 1$ (f) $(\sqrt{3}\bar{x} - \bar{y})^2 = 100$
27. (a) $\phi = \frac{1}{2}\cot^{-1}\frac{3}{4} \approx 26.6°$ (b) $x = \frac{\sqrt{5}}{5}(2\bar{x} - \bar{y})$, $y = \frac{\sqrt{5}}{5}(\bar{x} + 2\bar{y})$ (c) $\frac{\bar{x}^2}{6} - \frac{\bar{y}^2}{4} = 1$

(d)

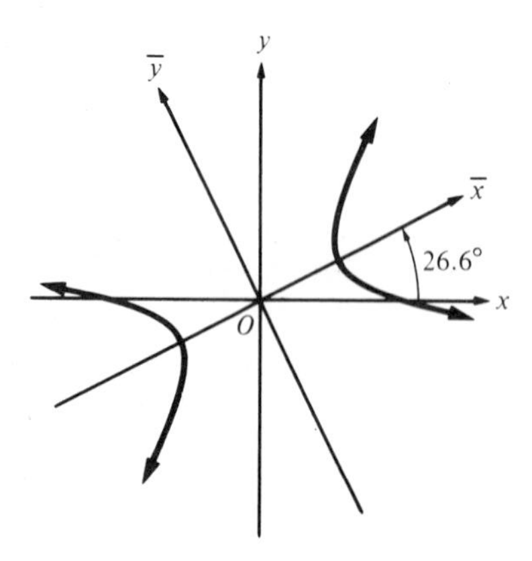

29. (a) $\phi = 45°$ (b) $x = \frac{\sqrt{2}}{2}(\bar{x} - \bar{y}),\ y = \frac{\sqrt{2}}{2}(\bar{x} + \bar{y})$ (c) $\bar{x} = \pm\frac{\sqrt{2}}{2}$

(d)

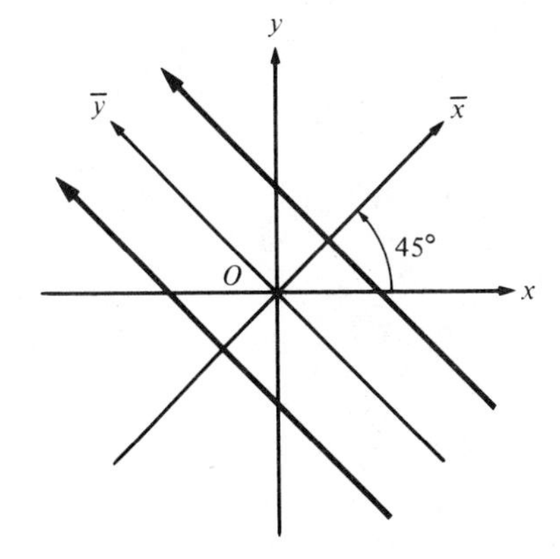

31. (a) $\phi = \frac{1}{2}\cot^{-1}\frac{7}{24} \approx 36.9°$ (b) $x = \frac{4}{5}\bar{x} - \frac{3}{5}\bar{y},\ y = \frac{3}{5}\bar{x} + \frac{4}{5}\bar{y}$ (c) $\bar{y} = \pm\frac{12}{5}$

(d)

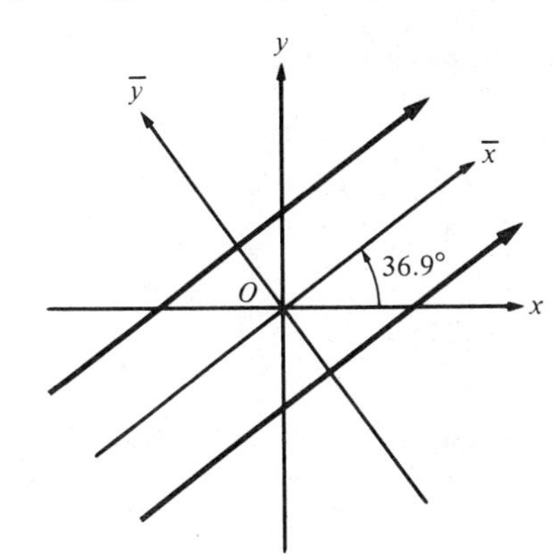

33. (a) $\phi \approx 18.4°$ (b) $x = \frac{3\sqrt{10}}{10}\bar{x} - \frac{\sqrt{10}}{10}\bar{y},\ y = \frac{\sqrt{10}}{10}\bar{x} + \frac{3\sqrt{10}}{10}\bar{y}$ (c) $\frac{\bar{x}^2}{9} + \frac{\bar{y}^2}{3} = 1$

(d)

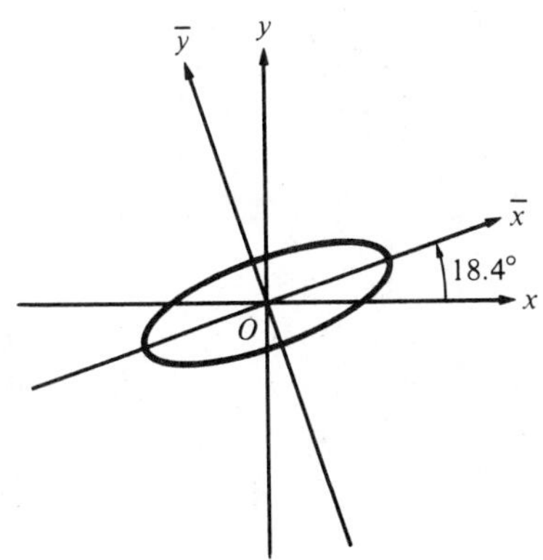

35. (a) $\phi \approx 18.4°$ (b) $x = \frac{3\sqrt{10}}{10}\bar{x} - \frac{\sqrt{10}}{10}\bar{y}$, $y = \frac{\sqrt{10}}{10}\bar{x} + \frac{3\sqrt{10}}{10}\bar{y}$ (c) $\frac{(\bar{x} + \frac{3}{20})^2}{\frac{321}{60}} - \frac{(\bar{y} - \frac{1}{20})^2}{\frac{321}{140}} = 1$

(d)

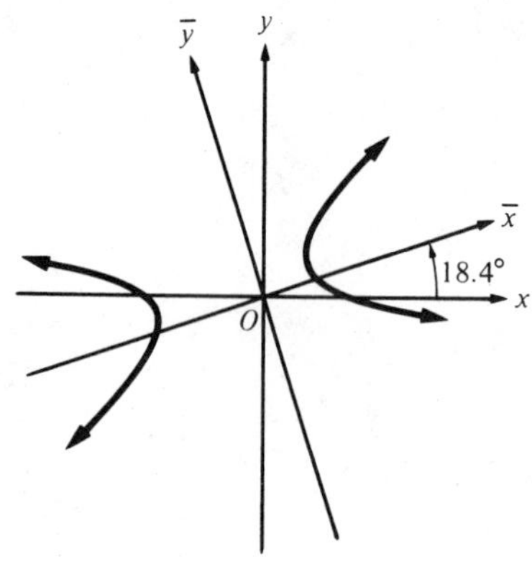

37. $\bar{x} = x\cos\phi + y\sin\phi$, $\bar{y} = -x\sin\phi + y\cos\phi$
39. (b) $x^2 + y^2 = -1$ (c) $(x - y)^2 = 0$ (d) $4(x + 2)^2 = (y - 1)^2$ (e) $(x - y)^2 = 18$ (f) $(x - 3)^2 + (y + 4)^2 = 0$

Review Problem Set, page 263

1. $10 + 5i$ **3.** $1 - 9i$ **5.** $13 - 8i$ **7.** $47 - 45i$ **9.** $-24 - 2i$ **11.** $\frac{16}{13} + \frac{11}{13}i$ **13.** $\frac{48}{221} - \frac{46}{221}i$ **15.** $-i$
17. $\frac{1}{27}i$ **19.** $12 - 5i$ **21.** $\frac{3\sqrt{2}}{2} + \frac{3\sqrt{2}}{2}i$ **23.** $8 - 8\sqrt{3}i$ **25.** 5 **27.** 8 **29.** $\frac{\sqrt{13}}{5}$
31. $4\sqrt{2}\left(\cos\frac{3\pi}{4} + i\sin\frac{3\pi}{4}\right)$ **33.** $\sqrt{7}(\cos 40.89° + i\sin 40.89°)$ **35.** (a) $24(\cos 30° + i\sin 30°)$ (b) $\frac{3}{2}(\cos 14° + i\sin 14°)$
37. (a) $48(\cos 360° + i\sin 360°)$ (b) $3(\cos 110° + i\sin 110°)$ **39.** (a) $50\left(\cos\frac{7\pi}{12} + i\sin\frac{7\pi}{12}\right)$ (b) $2\left(\cos\frac{5\pi}{6} + i\sin\frac{5\pi}{6}\right)$
41. (a) $4\sqrt{3}(\cos 60° + i\sin 60°)$ (b) $\sqrt{3}(\cos 0° + i\sin 0°)$ **43.** $16 + 16\sqrt{3}i$ **45.** -64 **47.** $8i$
49. $\frac{\sqrt{3}}{2} + \frac{1}{2}i, -\frac{\sqrt{3}}{2} + \frac{1}{2}i, -i$
51. $1(\cos 75° + i\sin 75°)$, $1(\cos 165° + i\sin 165°)$, $1(\cos 255° + i\sin 255°)$, $1(\cos 345° + i\sin 345°)$ **53.** $\pm i, \pm 1$
55. $y - 2 = -7(x - 2)$ **57.** $y - 2 = \frac{3}{2}(x - 1)$ **61.** (a) $3x - 2y - 31 = 0$ (b) $3y + 2x + 1 = 0$
63. $y = \frac{4}{3}x + \frac{2}{3}, m = \frac{4}{3}, b = \frac{2}{3}$ **65.** (a) $y - 1 = 3(x + 7)$ (b) $y = 3x + 22$ (c) $3x - y + 22 = 0$
67. (a) $y + 2 = \frac{7}{3}(x - 1)$ (b) $y = \frac{7}{3}x - \frac{13}{3}$ (c) $7x - 3y - 13 = 0$ **69.** $y - b = -\frac{a}{b}(x - a)$ **73.** Circle; $r = \frac{1}{3}$
75. Ellipse; vertices at $(\pm 4, 0)$ and $(0, \pm 3)$ **77.** $\frac{x^2}{64} + \frac{y^2}{16} = 1$ **79.** $\frac{(x - 1)^2}{16} + \frac{(y - 1)^2}{25} = 1$
81. $C = (0,0)$; vertices at $(0, \pm 2\sqrt{3})$ and $(\pm 2\sqrt{2}, 0)$ **83.** $C = (-1, 1)$; vertices at $(-1, -2)$, $(-1, 4)$, $(4, 1)$, and $(-6, 1)$
85. $C = (-4, 6)$; vertices at $(-4, 12)$, $(-4, 0)$, $(0, 6)$, and $(-8, 6)$ **87.** Horizontal axis; $V = (0, 0)$
89. Vertical axis; $V = (0, 1)$ **91.** Horizontal axis; $V = (-9, 1)$ **93.** $(y + 6)^2 = -x$
95. $C = (0,0)$; vertices at $(\pm 6\sqrt{2}, 0)$; asymptotes are $y = \pm\frac{1}{3}x$

97. $C = (-2, 3)$; vertices at $(2, 3)$ and $(-6, 3)$; asymptotes are $y - 3 = \pm\frac{1}{2}(x + 2)$

99. $C = (1, 3)$; vertices at $(1, \frac{7}{2})$ and $(1, \frac{5}{2})$; asymptotes are $y - 3 = \pm\frac{1}{2}(x - 1)$

101. $\dfrac{x^2}{3/4} - \dfrac{y^2}{3} = 1$

103. $\dfrac{(x-1)^2}{4} + \dfrac{(y+2)^2}{64} = 1$; ellipse

105. $(x + \frac{1}{2})^2 + (y - \frac{1}{2})^2 = \frac{1}{4}$; circle

107. $\dfrac{(x-1)^2}{4} - \dfrac{(y-1)^2}{1} = 1$; hyperbola

109. $\dfrac{(y + \frac{1}{2})^2}{\frac{159}{4}} + \dfrac{(x - \frac{1}{2})^2}{\frac{53}{8}} = 1$; ellipse

111. (a) $(-3, 2)$ (b) $(2, -5)$ (c) $(1, 6)$ (d) $(0, 0)$ (e) $(-3, 3)$ (f) $(-4, 2)$

113. $\bar{x} = x + 1, \bar{y} = y - \frac{1}{2}$; circle

115. $\bar{x} = x - 3, \bar{y} = y$; hyperbola

117. $(3 - 2\sqrt{3}, 2 + 3\sqrt{3})$

119. $(1 + 2\sqrt{3}, \sqrt{3} - 2)$

121. $(2 + 2\sqrt{3}, 2\sqrt{3} - 2)$

123. (a) $\phi = \dfrac{\pi}{4}$ (b) $x = \dfrac{\sqrt{2}}{2}(\bar{x} - \bar{y}), y = \dfrac{\sqrt{2}}{2}(\bar{x} + \bar{y})$ (c) $\dfrac{\bar{x}^2}{4} + \dfrac{\bar{y}^2}{12} = 1$

(d)

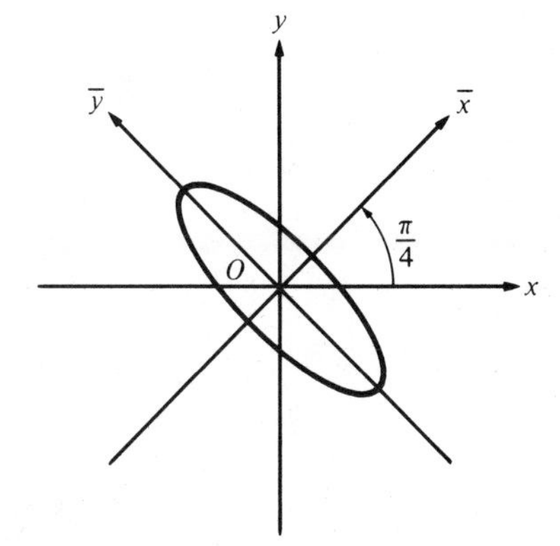

125. (a) $\phi = 15°$ (b) $x = \dfrac{\sqrt{2+\sqrt{3}}}{2}\bar{x} - \dfrac{\sqrt{2-\sqrt{3}}}{2}\bar{y}, y = \dfrac{\sqrt{2-\sqrt{3}}}{2}\bar{x} + \dfrac{\sqrt{2+\sqrt{3}}}{2}\bar{y}$ (c) $\dfrac{\sqrt{3}+2}{2}\bar{x}^2 - \dfrac{2-\sqrt{3}}{2}\bar{y}^2 = 11$

(d)

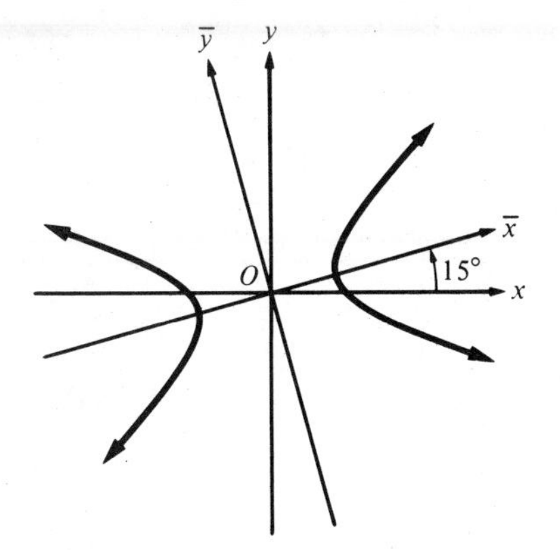

127. (a) $\phi = 30°$ (b) $x = \frac{1}{2}(\sqrt{3}\bar{x} - \bar{y})$, $y = \frac{1}{2}(\bar{x} + \sqrt{3}\bar{y})$ (c) $\bar{x}^2 + \frac{\bar{y}^2}{4} = 1$

(d)

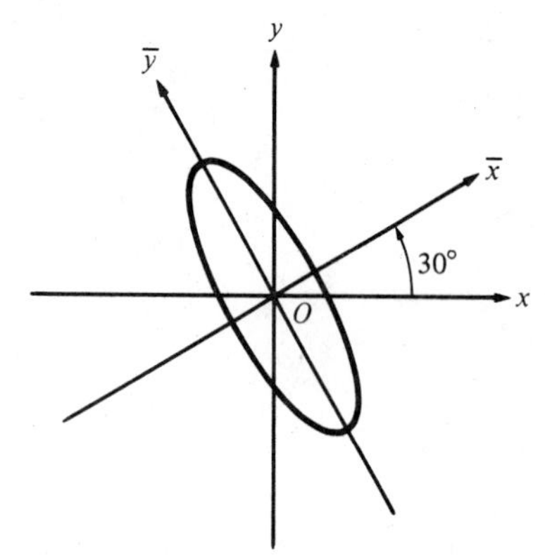

129. (a) $\phi \approx 18.4°$ (b) $x = \frac{\sqrt{10}}{10}(3\bar{x} - \bar{y})$, $y = \frac{\sqrt{10}}{10}(\bar{x} + 3\bar{y})$ (c) $\frac{\bar{x}^2}{32} + \frac{\bar{y}^2}{\frac{32}{11}} = 1$

(d)

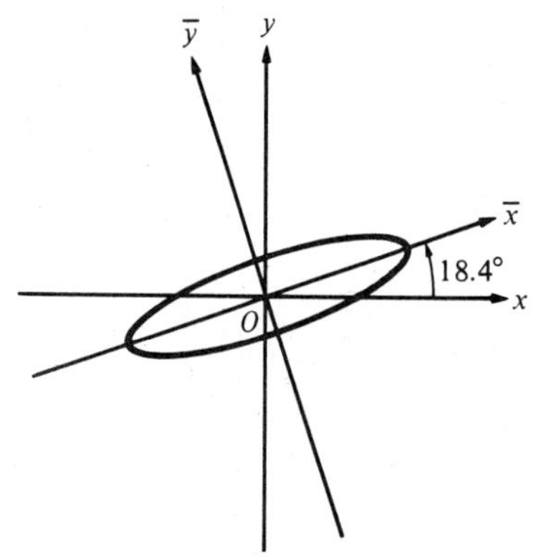

131. (a) $\phi \approx -36.87°$ (b) $x = \frac{3}{5}\bar{x} - \frac{4}{5}\bar{y}$, $y = \frac{4}{5}\bar{x} + \frac{3}{5}\bar{y}$ (c) $\bar{x}^2 + 2\bar{y}^2 = 1$

(d)

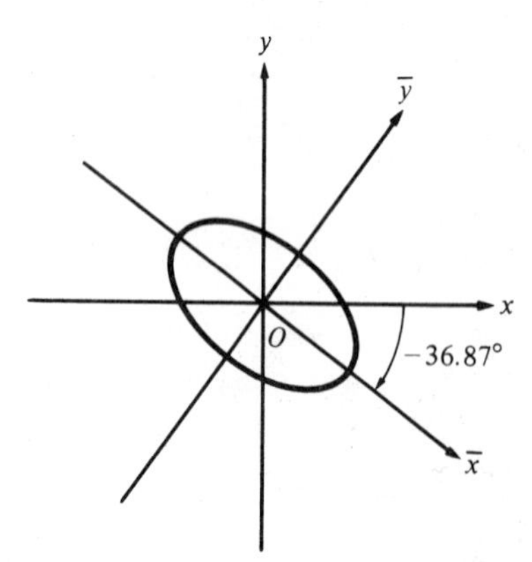

133. (a) $\phi \approx -26.57^\circ$ (b) $x = \frac{\sqrt{5}}{5}(\bar{x} - 2\bar{y})$, $y = \frac{\sqrt{5}}{5}(2\bar{x} + \bar{y})$ (c) $\bar{x} = \pm\frac{3}{5}\sqrt{5}$

(d)

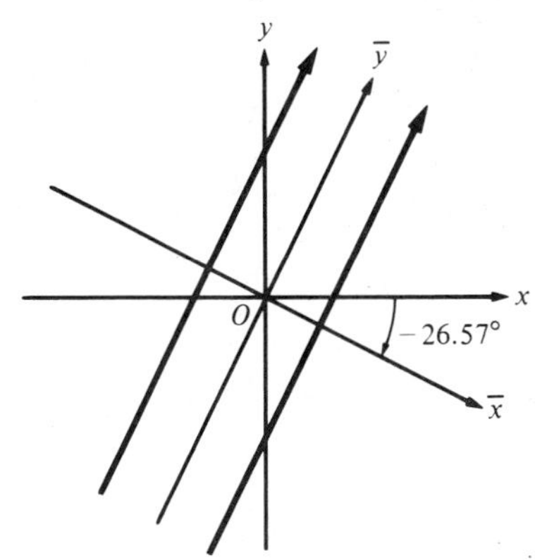

Appendix

Problem Set 1, page 274

1. 3^3 **3.** $\frac{1}{8^2}$ **5.** $\left(\frac{a}{b}\right)^{3n}$ **7.** a^2 **9.** $\frac{4}{9}x^2$ **11.** 2 **13.** $\frac{1}{y^9}$ **15.** (a) 2 (b) 3 (c) 4 (d) 5 (e) −2 (f) −2 (g) 5
17. (a) $2^5 = 32$ (b) $16^{1/4} = 2$ (c) $9^{-1/2} = \frac{1}{3}$ (d) $e^1 = e$ (e) $(\sqrt{3})^4 = 9$ (f) $10^n = 10^n$ (g) $x^5 = x^5$ **19.** $x = 2$ **21.** 5
23. 7 **25.** $\frac{1}{8}$ **27.** $\frac{1}{8}$ **29.** −7 **31.** −2/3 **33.** $\frac{1}{4}$ **35.** $\frac{9}{2}$ **37.** $\frac{85}{3}$ **39.** −1, −2 **41.** $-\frac{5}{3}$, 1
43. $\log_b x + \log_b(x + 1)$ **45.** $2\log_{10} x + \log_{10}(x + 1)$ **47.** $3\log_3 x + 2\log_3 y - \log_3 z$ **49.** $\frac{1}{2}[\log_e x + \log_e(x + 3)]$
51. $\log_3 x^9$ **53.** $\log_5 \sqrt{\frac{a}{3b}}$ **55.** $\log_e(x + 1)$ **57.** $\log_3 \frac{x + 9}{x + 5}$ **59.** (a) 2.31 (b) 2.70 (c) −0.95 (d) 1.87 (e) 0.74
(f) 0.95 (g) 1.63 **61.** 2 **63.** $\frac{1}{9}$ **65.** 2 **67.** 67

Problem Set 2, page 281

1. 1.7544 **3.** 2.1448 **5.** 2.2544 **7.** 1.2401 **9.** 1.5354 **11.** 1.6930 **13.** 2.5011 **15.** 0.8344
17. 1.6990 **19.** 0.4983 + (−1) **21.** 0.7868 + (−2) **23.** 0.1553 + (−4) **25.** 3.1883 **27.** 0.7876 + (−1)
29. 0.2256 **31.** 2.59 **33.** 60.6 **35.** 953 **37.** 3560 **39.** 1.40 **41.** 34.9

Problem Set 3, page 291

1. 0.4274 **3.** 0.8012 **5.** 2.4021 **7.** −0.3939 **9.** 1.1357 **11.** −0.8302 **13.** 0.8791 **15.** 1.2473
17. 0.7709 **19.** 0.7771 **21.** 0.1315 **23.** 32.476 **25.** 4.256 **27.** 9.223 **29.** −0.9094 **31.** 10°
33. 49.6° **35.** 55.1° **37.** 60.5° **39.** 47.1° **41.** 60.52° **43.** 1.331 **45.** 0.44 **47.** 0.24 **49.** 0.9094
51. 1.34 **53.** 1.33

Index of Applications

Index of Applications

(*Problem numbers are enclosed in parentheses.*)

Military

Navigation

Physics

Sports, Games, and Entertainment

Index

Trigonometric Functions

Acute Angles

$\sin\theta = \dfrac{\text{opp}}{\text{hyp}}$ $\qquad \csc\theta = \dfrac{\text{hyp}}{\text{opp}}$

$\cos\theta = \dfrac{\text{adj}}{\text{hyp}}$ $\qquad \sec\theta = \dfrac{\text{hyp}}{\text{adj}}$

$\tan\theta = \dfrac{\text{opp}}{\text{adj}}$ $\qquad \cot\theta = \dfrac{\text{adj}}{\text{opp}}$

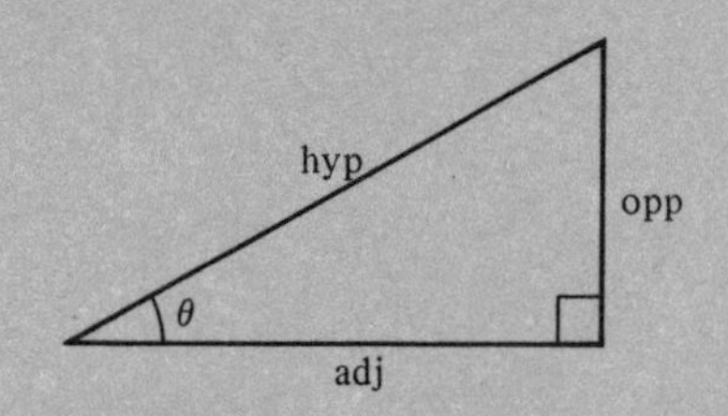

General Angles

$\sin\theta = \dfrac{y}{r}$ $\qquad \csc\theta = \dfrac{r}{y}$

$\cos\theta = \dfrac{x}{r}$ $\qquad \sec\theta = \dfrac{r}{x}$

$\tan\theta = \dfrac{y}{x}$ $\qquad \cot\theta = \dfrac{x}{y}$

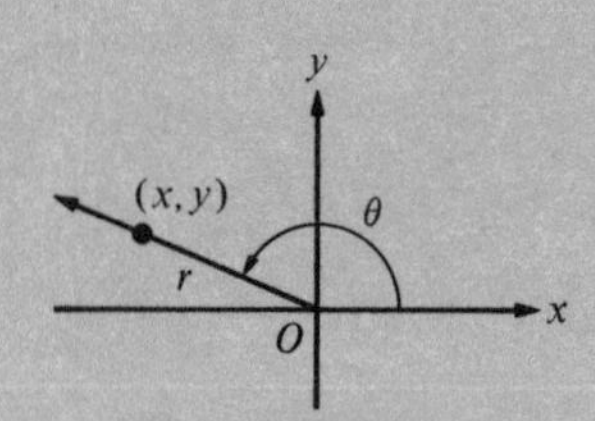

Trigonometric Identities

Fundamental Identities

1 $\csc\theta = \dfrac{1}{\sin\theta}$

2 $\sec\theta = \dfrac{1}{\cos\theta}$

3 $\cot\theta = \dfrac{1}{\tan\theta}$

4 $\tan\theta = \dfrac{\sin\theta}{\cos\theta}$

5 $\cot\theta = \dfrac{\cos\theta}{\sin\theta}$

6 $\cos^2\theta + \sin^2\theta = 1$

7 $1 + \tan^2\theta = \sec^2\theta$

8 $1 + \cot^2\theta = \csc^2\theta$

Even–Odd Identities

1 $\sin(-\theta) = -\sin\theta$

2 $\cos(-\theta) = \cos\theta$

3 $\tan(-\theta) = -\tan\theta$

4 $\cot(-\theta) = -\cot\theta$

5 $\sec(-\theta) = \sec\theta$

6 $\csc(-\theta) = -\csc\theta$

Addition Formulas

1 $\sin(\alpha + \beta) = \sin\alpha\cos\beta + \sin\beta\cos\alpha$

2 $\cos(\alpha + \beta) = \cos\alpha\cos\beta - \sin\alpha\sin\beta$

3 $\tan(\alpha + \beta) = \dfrac{\tan\alpha + \tan\beta}{1 - \tan\alpha\tan\beta}$

Subtraction Formulas

1 $\sin(\alpha - \beta) = \sin\alpha\cos\beta - \sin\beta\cos\alpha$

2 $\cos(\alpha - \beta) = \cos\alpha\cos\beta + \sin\alpha\sin\beta$

3 $\tan(\alpha - \beta) = \dfrac{\tan\alpha - \tan\beta}{1 + \tan\alpha\tan\beta}$